Ulam Type Stability

Janusz Brzdęk • Dorian Popa
Themistocles M. Rassias

Editors

Ulam Type Stability

 Springer

Editors
Janusz Brzdęk
Department of Applied Mathematics
AGH University of Science and Technology
Krakow, Poland

Dorian Popa
Department of Mathematics
Technical University of Cluj-Napoca
Cluj-Napoca, Romania

Themistocles M. Rassias
Department of Mathematics
National Technical University of Athens
Athens, Greece

ISBN 978-3-030-28974-4 ISBN 978-3-030-28972-0 (eBook)
https://doi.org/10.1007/978-3-030-28972-0

Mathematics Subject Classification (2010): 06B99, 20M99, 30D05, 30E25, 34D10, 37C25, 39A30, 39B12, 39B22, 39B32, 39B52, 39B62, 39B82, 43A22, 45M10, 46B20, 47H10, 47H14, 47J20, 47L05, 54H25, 60K25, 65J15, 65Q20, 91B32, 91B38

This Springer imprint is published by the registered company Springer Nature Switzerland AG.
The registered company address is: Gewerbestrasse 11, 6330 Cham, Switzerland

Preface

This volume presents papers written by experts, who are actively working in various areas of mathematics and its applications on issues in one way or another connected to Ulam-type stability problems, motivated by the famous question concerning approximate homomorphisms. These papers provide an insightful perspective on a large number of investigations in mathematical analysis.

The present book is the outcome of two Conferences on Ulam Type Stability (CUTS) organized in 2016 (July 4–9, Cluj-Napoca, Romania) and in 2018 (October 8–13, Timișoara, Romania).

The aim of the volume is not only to give an account of the present state of research on Ulam-type stability but also to stimulate further research in the area. Thus, alongside research papers containing new results, it includes surveys on various themes pointing to the potential for further future study and identifying several open problems and/or questions.

Let us recall that S.M. Ulam (April 13, 1909–May 13, 1984) was a prominent mathematician and physicist. In 1940, he posed his famous question concerning approximate homomorphisms, which gave rise to a long-lasting study of a field which we now call Ulam (or Hyers-Ulam) stability.

The book contains 21 articles written by 29 authors from 12 countries, all of whom have been intensively involved in active research in this area. Special emphasis has been placed on the topics which apply methods and techniques involving, or originating from, functional equations and inequalities (FEI).

We hope that this publication will serve as a kind of guidebook for both graduate students and researchers in various fields, including not only mathematics but also physics, engineering, and interdisciplinary research.

Subjects treated in this book are (in order of appearance in this volume) as follows:

- Stability and solutions of the Cauchy functional equation in lattice environments
- Fixed-point approach to the Hyers-Ulam stability and hyperstability of a general functional equation
- Reversing property of the Birkhoff-James orthogonality and its stability

- Optimal forward contract design for inventory (a value-of-waiting analysis via sensitivity analysis of a functional equation)
- Hyers-Ulam stability of functional equations in quasi-β-Banach spaces
- Stability of the functional equation of p-Wright affine functions in 2-Banach spaces
- Solutions and stability of a functional equation arising from a queueing system
- Approximately cubic mappings
- Solutions and stability of some functional equations on semigroups
- Bi-additive s-functional inequalities and quasi-$*$-multipliers on Banach $*$-algebras
- Ulam stability of a generalization of the Fréchet functional equation on a restricted domain
- Various remarks concerning the notion of stability of functional equations
- Subdominant eigenvalue location of a bordered diagonal matrix
- A fixed-point theorem in uniformizable spaces
- Symmetry of Birkhoff-James orthogonality of bounded linear operators
- Ulam stability of zero-point equations
- Cauchy difference operator in some Orlicz spaces
- Semi-inner products and parapreseminorms on groups and a generalization of a theorem of Maksa and Volkmann on additive functions
- Invariant means in Ulam-type stability theory
- Geometry of Banach function modules
- Exact and approximate orthogonalities based on norm derivatives

It is our pleasure to express warmest thanks to all the mathematicians, who participated in this publication. We would also wish to acknowledge the support of our referees.

Last but not least, we wish to acknowledge the superb assistance that the staff of Springer provided for the publication of this volume.

Krakow, Poland   Janusz Brzdęk
Cluj-Napoca, Romania   Dorian Popa
Athens, Greece   Themistocles M. Rassias
March 2019

Contents

Contributors

Nutefe Kwami Agbeko Institute of Mathematics, University of Miskolc, Miskolc, Hungary

Keltouma Belfakih Department of Mathematics, Faculty of Sciences, University Ibn Zohr, Agadir, Morocco

Jacek Chmieliński Department of Mathematics, Pedagogical University of Cracow, Kraków, Poland

Roy O. Davies School of Mathematics and Actuarial Science, University of Leicester, Leicester, UK

Mahdi Dehghani Department of Pure Mathematics, Faculty of Mathematical Sciences, University of Kashan, Kashan, Iran

Nguyen Van Dung Faculty of Mathematics Teacher Education, Dong Thap University, Cao Lanh, Dong Thap, Vietnam

El-Sayed El-hady Mathematics Department, College of Science, Jouf University, Sakaka, Kingdom of Saudi Arabia
Basic Science Department, Faculty of Computers and Informatics, Suez Canal University, Ismailia, Egypt

Elhoucien Elqorachi Department of Mathematics, Faculty of Sciences, University Ibn Zohr, Agadir, Morocco

Paşc Găvruţa Department of Mathematics, Politehnica University of Timişoara, Timişoara, Romania

Puja Ghosh Department of Mathematics, Sabang Sajanikanta Mahavidyalaya, Paschim Medinipur, West Bengal, India

Jung Rye Lee Daejin University, Pocheon-si, South Korea

Renata Malejki Institute of Mathematics, Pedagogical University of Cracow, Kraków, Poland

Laura Manolescu Department of Mathematics, Politehnica University of Timişoara, Timişoara, Romania

Zenon Moszner Institute of Mathematics, Pedagogical University of Cracow, Kraków, Poland

Adam J. Ostaszewski Department of Mathematics, London School of Economics, London, UK

Lahbib Oubbi Department of Mathematics, Team GrAAF, Laboratory LMSA, Center CeReMar, Ecole Normale Supérieure, Mohammed V University in Rabat, Takaddoum, Rabat, Morocco

Choonkil Park Hanyang University, Seoul, South Korea

Kallol Paul Department of Mathematics, Jadavpur University, Kolkata, India

Adrian Petruşel Babeş-Bolyai University, Cluj-Napoca, Romania
Academy of Romanian Scientists, Bucharest, Romania

Themistocles M. Rassias Department of Mathematics, National Technical University of Athens, Athens, Greece

Ioan A. Rus Babeş-Bolyai University, Cluj-Napoca, Romania

Debmalya Sain Department of Mathematics, Indian Institute of Science, Bangalore, India

Wutiphol Sintunavarat Department of Mathematics and Statistics, Faculty of Science and Technology, Thammasat University Rangsit Center, Pathumthani, Thailand

Stanisław Siudut Institute of Mathematics, Pedagogical University, Kraków, Poland

Árpád Száz Department of Mathematics, University of Debrecen, Debrecen, Hungary

László Székelyhidi Institute of Mathematics, University of Debrecen, Debrecen, Hungary

Paweł Wójcik Department of Mathematics, Pedagogical University of Cracow, Kraków, Poland

Ali Zamani Department of Mathematics, Farhangian University, Tehran, Iran

Chaimaa BenZarouala Department of Mathematics, Team GrAAF, Laboratory LMSA, Center CeReMar, Faculty of Sciences, Mohammed V University in Rabat, Rabat, Morocco

Chapter 1
Survey on Cauchy Functional Equation in Lattice Environments

Nutefe Kwami Agbeko

Abstract By replacing in Cauchy functional equation the addition with the lattice operations we are able to formulate the Ulam's stability problem in lattice environments. Various types of solution are formulated and proved similarly as their counterparts in addition environments. This survey contains a part of the habilitation thesis presented to the Department of Mathematics, University of Debrecen (cf. Agbeko, Studies on some addition-free environments. Habilitation Thesis submitted to the University of Debrecen. http://www.uni-miskolc.hu/~matagbek/Habilitation %20Thesis.pdf) and the material in Agbeko and Szokol (Extracta Math 33:1–10, 2018).

Keywords Functional equation · Functional inequality · Banach lattice · Ulam's stability · Lattice semigroup

Mathematics Subject Classification (2010) Primary 39B82, 06B99, 20M99; Secondary 39B42, 39B52, 46A40

1.1 Introduction

In the early 90s we substituted with the lattice join operation, the addition in the definition of measure as well as in the Lebesgue integral to obtain lattice-dependent operators which behave similarly as their counterparts in Measure Theory do, in the sense that existing major theorems in Measure Theory are also proved with the addition replaced by the join (or supremum). We refer the reader to [1–4] for the earliest results to [9] for other considerations. Later on we have studied the linear functional equation in lattice environments (by replacing in the Cauchy functional

N. K. Agbeko (✉)
Institute of Mathematics, University of Miskolc, Miskolc, Hungary
e-mail: matagbek@uni-miskolc.hu

© Springer Nature Switzerland AG 2019
J. Brzdęk et al. (eds.), *Ulam Type Stability*,
https://doi.org/10.1007/978-3-030-28972-0_1

equation the addition with lattice operations) and namely considered Ulam's type stability problem and separation theorem. Recalling it here as historical background the Cauchy (or linear) functional equation reads:

$$f(x + y) = f(x) + f(y), \tag{1.1}$$

where f is a real function.

The so-called Ulam's stability problem involving Eq. (1.1) was first posed by M. Ulam (see [33]) in the terms: "Give conditions in order for a linear mapping near an approximately linear mapping to exist." In a more precise formulation the problem reads:

Given two Banach algebras E and E', a transformation $f : E \to E'$ is called δ-linear if

$$\|f(x + y) - f(x) - f(y)\| < \delta, \tag{1.2}$$

for all $x, y \in E$. Does there exist for each $\varepsilon \in (0, 1)$ some $\delta > 0$ such that to each δ-linear transformation $f : E \to E'$ there corresponds a linear transformation $l : E \to E'$ satisfying the inequality

$$\|f(x) - l(x)\| < \varepsilon$$

for all $x \in E$? This question was answered in the affirmative for the first time by Hyers [20]. Ever since various problems of stability on various spaces have come to light. We shall just list few of them: [14, 19, 24, 25, 27, 32]. The lattice version of the Ulam's stability problem will be formulated in a more general form later in Sects. 1.3 and 1.5.

1.2 Stability of Maximum Preserving Functional Equations on Banach Lattices

We would like to stress the similitude between the present section and the result in [11].

If $\mathscr{B}$ is a Banach lattice, then $\mathscr{B}^+$ will stand for its positive cone, i.e.

$$\mathscr{B}^+ = \{x \in \mathscr{B} : x \geq 0\} = \{|x| : x \in \mathscr{B}\}.$$

Given two Banach lattices $\mathscr{X}$ and $\mathscr{Y}$ we say that a functional $F : \mathscr{X} \to \mathscr{Y}$ is *cone-related* if

$$F\left(\mathscr{X}^+\right) = \{F(|x|) : x \in \mathscr{X}\} \subset \mathscr{Y}^+,$$

also known in the literature as positive function.

Some Properties 1 *Let be given two Banach lattices $\mathscr{X}$ and $\mathscr{Y}$ and, a cone-related functional $F : \mathscr{X} \to \mathscr{Y}$.*

P1. **Maximum Preserving Functional Equation:**

$$F(|x| \vee |y|) = F(|x|) \vee F(|y|)$$

for all members $x, y \in \mathscr{X}$.

P2. **Semi-homogeneity:**

$$F(\tau |x|) = \tau F(|x|)$$

for all $x \in \mathscr{X}$ and every number $\tau \in [0, \infty)$.

P3. **Continuity From Below on the Positive Cone:** *The identity*

$$\lim_{n \to \infty} F(x_n) = F\left(\lim_{n \to \infty} x_n \right)$$

holds for every increasing sequence $(x_n)_{n \in \mathbb{N}} \subset \mathscr{X}^+$.

P4. *For any increasing sequence $(x_k) \subset \mathscr{X}^+$ the inequality hereafter holds*

$$\lim_{n \to \infty} \lim_{k \to \infty} \frac{F(2^n x_k)}{2^n} \leq \lim_{k \to \infty} \lim_{n \to \infty} \frac{F(2^n x_k)}{2^n}, \tag{1.3}$$

provided that the limits exist.

We should note that every functional, which solves the maximum preserving functional equation, is known as a *join homomorphism* in Lattice Theory.

Remark 1.1 Given two Banach lattices $\mathscr{X}$ and $\mathscr{Y}$ let a cone-related functional $F : \mathscr{X} \to \mathscr{Y}$ satisfy property P1. Then the following statements are valid.

1. $F(|x \vee y|) \leq F(|x|) \vee F(|y|)$ for all members $x, y \in \mathscr{X}$.
2. The semi-homogeneity implies that $F(0) = 0$.
3. F is an increasing operator, in the sense that if $x, y \in \mathscr{X}$ are such that $|x| \leq |y|$, then $F(|x|) \leq F(|y|)$.

Theorem 1.1 *Let be given a continuous function $p : [0, \infty) \to (0, \infty)$ and two Banach lattices $\mathscr{X}$ and $\mathscr{Y}$. Consider a cone-related functional $F : \mathscr{X} \to \mathscr{Y}$ for which there are numbers $\vartheta > 0$ and $\alpha \in [0, 1)$ such that*

$$\frac{\left\| F(\tau |x| \vee \eta |y|) - \frac{\tau p(\tau) F(|x|) \vee \eta p(\eta) F(|y|)}{p(\tau) \vee p(\eta)} \right\|}{\|x\|^\alpha + \|y\|^\alpha} \leq \vartheta \tag{1.4}$$

for all $x, y \in \mathscr{X}$ and $\tau, \eta \in \mathbb{R}^+$. Then there is a unique cone-related mapping $T : \mathscr{X} \to \mathscr{Y}$ which satisfies properties P1, P2 and inequality

$$\frac{\|T(|x|) - F(|x|)\|}{\|x\|^{\alpha}} \leq \frac{2\vartheta}{2 - 2^{\alpha}} \tag{1.5}$$

for every $x \in \mathcal{X}$.

Moreover, if F is continuous from below, then in order that T be continuous from below it is necessary and sufficient that F enjoy property P4.

Each of the following theorem is a variation of the above result.

Theorem 1.2 Let be given a continuous function $p : [0, \infty) \to (0, \infty)$ with $p(0) = 0$ and, two Banach lattices $\mathcal{X}$ and $\mathcal{Y}$. Consider a cone-related functional $F : \mathcal{X} \to \mathcal{Y}$ for which there are numbers $\vartheta > 0$ and $\alpha \in [0, 1)$ such that

$$\frac{\left\| F(\tau |x| \vee \eta |y|) - \frac{\tau p(\tau) F(|x|) \vee \eta p(\eta) F(|y|)}{p(\tau) + p(\eta)} \right\|}{\|x\|^{\alpha} + \|y\|^{\alpha}} \leq \vartheta \tag{1.6}$$

for all $x, y \in \mathcal{X}$ and $\tau, \eta \in \mathbb{R}^{+}$. Then there is a unique cone-related mapping $T : \mathcal{X} \to \mathcal{Y}$ which satisfies properties P1, P2 and inequality (1.5) is valid for every $x \in \mathcal{X}$. Moreover, if F is continuous from below, then in order that T be continuous from below it is necessary and sufficient that F enjoy property P4.

Theorem 1.3 Let be given a continuous function $p : [0, \infty) \to (0, \infty)$ and, two Banach lattices $\mathcal{X}$ and $\mathcal{Y}$. Consider a cone-related functional $F : \mathcal{X} \to \mathcal{Y}$ for which there are numbers $\vartheta > 0$ and $\alpha \in [0, 1)$ such that

$$\frac{\left\| F\left(\frac{\tau p(\tau)|x| \vee \eta p(\eta)|y|}{p(\tau) \vee p(\eta)}\right) - \frac{\tau p(\tau) F(|x|) \vee \eta p(\eta) F(|y|)}{p(\tau) \vee p(\eta)} \right\|}{\|x\|^{\alpha} + \|y\|^{\alpha}} \leq \vartheta \tag{1.7}$$

for all $x, y \in \mathcal{X}$ and $\tau, \eta \in \mathbb{R}^{+}$. Then there is a unique cone-related mapping $T : \mathcal{X} \to \mathcal{Y}$ which satisfies properties P1, P2 and inequality (1.5) is valid for every $x \in \mathcal{X}$. Moreover, if F is continuous from below, then in order that T be continuous from below it is necessary and sufficient that F enjoy property P4.

Theorem 1.4 Let be given a continuous function $p : [0, \infty) \to (0, \infty)$ with $p(0) = 0$ and, two Banach lattices $\mathcal{X}$ and $\mathcal{Y}$. Consider a cone-related functional $F : \mathcal{X} \to \mathcal{Y}$ for which there are numbers $\vartheta > 0$ and $\alpha \in [0, 1)$ such that

$$\frac{\left\| F\left(\frac{\tau p(\tau)|x| \vee \eta p(\eta)|y|}{p(\tau) + p(\eta)}\right) - \frac{\tau p(\tau) F(|x|) \vee \eta p(\eta) F(|y|)}{p(\tau) + p(\eta)} \right\|}{\|x\|^{\alpha} + \|y\|^{\alpha}} \leq \vartheta \tag{1.8}$$

for all $x, y \in \mathcal{X}$ and $\tau, \eta \in \mathbb{R}^{+}$. Then there is a unique cone-related mapping $T : \mathcal{X} \to \mathcal{Y}$ which satisfies properties P1, P2 and inequality (1.5) is valid for every $x \in \mathcal{X}$. Moreover, if F is continuous from below, then in order that T be continuous from below it is necessary and sufficient that F enjoy property P4.

We point out that the proof of Theorem 1.1 can be suitably adapted to show the validity of Theorems 1.2–1.4.

It is worth to ask the question: *Under what conditions inequalities* (1.4) *and* (1.6)–(1.8) *hold true?* The answer is formulated in the following results without proof, because of their easiness (in fact, only the triangle inequality of the norm is needed to check their validity).

Lemma 1.1 *Let be given a continuous function* $p : [0, \infty) \to (0, \infty)$ *and, two Banach lattices* $\mathscr{X}$ *and* $\mathscr{Y}$. *Consider a cone-related functional* $F : \mathscr{X} \to \mathscr{Y}$ *and define the functional* $F_c : \mathscr{X} \to \mathscr{Y}$ *by* $F_c(x) = F(x) \wedge c$, *where* $c \in \mathscr{Y}^+$. *Let* $\alpha \in [0, 1)$ *be some number,* $\mathscr{Y}_0 \subset \mathscr{Y}^+$ *be some non-empty subset and consider the following four quantities:*

$$
\sup_{c \in \mathscr{Y}_0} \sup_{\tau, \eta \in [0, \infty)} \sup_{x, y \in \mathscr{X}} \frac{\| F_c(\tau |x| \vee \eta |y|) - F(\tau |x| \vee \eta |y|) \|}{\|x\|^\alpha + \|y\|^\alpha},
$$

$$
\sup_{c \in \mathscr{Y}_0} \sup_{\tau, \eta \in [0, \infty)} \sup_{x, y \in \mathscr{X}} \frac{\left\| \frac{\tau p(\tau) F_c(|x|) \vee \eta p(\eta) F_c(|y|)}{p(\tau) \vee p(\eta)} - \frac{\tau p(\tau) F(|x|) \vee \eta p(\eta) F(|y|)}{p(\tau) \vee p(\eta)} \right\|}{\|x\|^\alpha + \|y\|^\alpha},
$$

$$
\sup_{c \in \mathscr{Y}_0} \sup_{\tau, \eta \in [0, \infty)} \sup_{x, y \in \mathscr{X}} \frac{\left\| F_c(\tau |x| \vee \eta |y|) - \frac{\tau p(\tau) F_c(|x|) \vee \eta p(\eta) F_c(|y|)}{p(\tau) \vee p(\eta)} \right\|}{\|x\|^\alpha + \|y\|^\alpha}
$$

and

$$
\sup_{\tau, \eta \in [0, \infty)} \sup_{x, y \in \mathscr{X}} \frac{\left\| F(\tau |x| \vee \eta |y|) - \frac{\tau p(\tau) F(|x|) \vee \eta p(\eta) F(|y|)}{p(\tau) \vee p(\eta)} \right\|}{\|x\|^\alpha + \|y\|^\alpha}.
$$

If any three of them are simultaneously finite, then the fourth is also finite.

Lemma 1.2 *Let be given a continuous function* $p : [0, \infty) \to (0, \infty)$ *and, two Banach lattices* $\mathscr{X}$ *and* $\mathscr{Y}$. *Consider a cone-related functional* $F : \mathscr{X} \to \mathscr{Y}$ *and define the functional* $F_c : \mathscr{X} \to \mathscr{Y}$ *by* $F_c(x) = F(x) \wedge c$, *where* $c \in \mathscr{Y}^+$. *Let* $\alpha \in [0, 1)$ *be some number and* $\mathscr{Y}_0 \subset \mathscr{Y}^+$ *some non-empty subset and consider the following four quantities:*

$$
\sup_{c \in \mathscr{Y}_0} \sup_{\tau, \eta \in [0, \infty)} \sup_{x, y \in \mathscr{X}} \frac{\| F_c(\tau |x| \vee \eta |y|) - F(\tau |x| \vee \eta |y|) \|}{\|x\|^\alpha + \|y\|^\alpha},
$$

$$\sup_{c\in\mathscr{Y}_0}\ \sup_{\tau,\,\eta\in[0,\infty)}\ \sup_{x,\,y\in\mathscr{X}}\ \frac{\left\|\ \frac{\tau p(\tau)F_c(|x|)\vee\eta p(\eta)F_c(|y|)}{p(\tau)+p(\eta)}-\frac{\tau p(\tau)F(|x|)\vee\eta p(\eta)F(|y|)}{p(\tau)+p(\eta)}\ \right\|}{\|x\|^\alpha+\|y\|^\alpha},$$

$$\sup_{c\in\mathscr{Y}_0}\ \sup_{\tau,\,\eta\in[0,\infty)}\ \sup_{x,\,y\in\mathscr{X}}\ \frac{\left\|\ F_c\left(\tau\,|x|\vee\eta\,|y|\right)-\frac{\tau p(\tau)F_c(|x|)\vee\eta p(\eta)F_c(|y|)}{p(\tau)+p(\eta)}\ \right\|}{\|x\|^\alpha+\|y\|^\alpha},$$

and

$$\sup_{\tau,\,\eta\in[0,\infty)}\ \sup_{x,\,y\in\mathscr{X}}\ \frac{\left\|\ F\left(\tau\,|x|\vee\eta\,|y|\right)-\frac{\tau p(\tau)F(|x|)\vee\eta p(\eta)F(|y|)}{p(\tau)+p(\eta)}\ \right\|}{\|x\|^\alpha+\|y\|^\alpha}.$$

If any three of them are simultaneously finite, then the fourth is also finite.

Lemma 1.3 *Let be given a continuous function $p : [0,\infty) \to (0,\infty)$ and, two Banach lattices $\mathscr{X}$ and $\mathscr{Y}$. Consider a cone-related functional $F : \mathscr{X} \to \mathscr{Y}$ and define the functional $F_c : \mathscr{X} \to \mathscr{Y}$ by $F_c(x) = F(x) \wedge c$, where $c \in \mathscr{Y}^+$. Let $\alpha \in [0,1)$ be some number and $\mathscr{Y}_0 \subset \mathscr{Y}^+$ some non-empty subset and consider the following four quantities:*

$$\sup_{c\in\mathscr{Y}_0}\ \sup_{\tau,\,\eta\in[0,\infty)}\ \sup_{x,\,y\in\mathscr{X}}\ \frac{\left\|\ F_c\left(\frac{\tau p(\tau)|x|\vee\eta p(\eta)|y|}{p(\tau)\vee p(\eta)}\right)-F\left(\frac{\tau p(\tau)|x|\vee\eta p(\eta)|y|}{p(\tau)\vee p(\eta)}\right)\ \right\|}{\|x\|^\alpha+\|y\|^\alpha},$$

$$\sup_{c\in\mathscr{Y}_0}\ \sup_{\tau,\,\eta\in[0,\infty)}\ \sup_{x,\,y\in\mathscr{X}}\ \frac{\left\|\ \frac{\tau p(\tau)F_c(|x|)\vee\eta p(\eta)F_c(|y|)}{p(\tau)\vee p(\eta)}-\frac{\tau p(\tau)F(|x|)\vee\eta p(\eta)F(|y|)}{p(\tau)\vee p(\eta)}\ \right\|}{\|x\|^\alpha+\|y\|^\alpha},$$

$$\sup_{c\in\mathscr{Y}_0}\ \sup_{\tau,\,\eta\in[0,\infty)}\ \sup_{x,\,y\in\mathscr{X}}\ \frac{\left\|\ F_c\left(\frac{\tau p(\tau)|x|\vee\eta p(\eta)|y|}{p(\tau)\vee p(\eta)}\right)-\frac{\tau p(\tau)F_c(|x|)\vee\eta p(\eta)F_c(|y|)}{p(\tau)\vee p(\eta)}\ \right\|}{\|x\|^\alpha+\|y\|^\alpha}$$

and

$$\sup_{\tau,\,\eta\in[0,\infty)}\ \sup_{x,\,y\in\mathscr{X}}\ \frac{\left\|\ F\left(\frac{\tau p(\tau)|x|\vee\eta p(\eta)|y|}{p(\tau)\vee p(\eta)}\right)-\frac{\tau p(\tau)F(|x|)\vee\eta p(\eta)F(|y|)}{p(\tau)\vee p(\eta)}\ \right\|}{\|x\|^\alpha+\|y\|^\alpha}.$$

If any three of them are simultaneously finite, then the fourth is also finite.

Lemma 1.4 *Let be given a continuous function* $p : [0, \infty) \to (0, \infty)$ *and, two Banach lattices* $\mathscr{X}$ *and* $\mathscr{Y}$. *Consider a cone-related functional* $F : \mathscr{X} \to \mathscr{Y}$ *and define the functional* $F_c : \mathscr{X} \to \mathscr{Y}$ *by* $F_c(x) = F(x) \wedge c$, *where* $c \in \mathscr{Y}^+$. *Let* $\alpha \in [0, 1)$ *be some number and* $\mathscr{Y}_0 \subset \mathscr{Y}^+$ *some non-empty subset and consider the following four quantities:*

$$\sup_{c \in \mathscr{Y}_0} \sup_{\tau, \eta \in [0, \infty)} \sup_{x, y \in \mathscr{X}} \frac{\left\| F_c\left(\frac{\tau p(\tau)|x| \vee \eta p(\eta)|y|}{p(\tau)+p(\eta)}\right) - F\left(\frac{\tau p(\tau)|x| \vee \eta p(\eta)|y|}{p(\tau)+p(\eta)}\right) \right\|}{\|x\|^\alpha + \|y\|^\alpha},$$

$$\sup_{c \in \mathscr{Y}_0} \sup_{\tau, \eta \in [0, \infty)} \sup_{x, y \in \mathscr{X}} \frac{\left\| \frac{\tau p(\tau) F_c(|x|) \vee \eta p(\eta) F_c(|y|)}{p(\tau)+p(\eta)} - \frac{\tau p(\tau) F(|x|) \vee \eta p(\eta) F(|y|)}{p(\tau)+p(\eta)} \right\|}{\|x\|^\alpha + \|y\|^\alpha},$$

$$\sup_{c \in \mathscr{Y}_0} \sup_{\tau, \eta \in [0, \infty)} \sup_{x, y \in \mathscr{X}} \frac{\left\| F_c\left(\frac{\tau p(\tau)|x| \vee \eta p(\eta)|y|}{p(\tau)+p(\eta)}\right) - \frac{\tau p(\tau) F_c(|x|) \vee \eta p(\eta) F_c(|y|)}{p(\tau)+p(\eta)} \right\|}{\|x\|^\alpha + \|y\|^\alpha}$$

and

$$\sup_{\tau, \eta \in [0, \infty)} \sup_{x, y \in \mathscr{X}} \frac{\left\| F\left(\frac{\tau p(\tau)|x| \vee \eta p(\eta)|y|}{p(\tau)+p(\eta)}\right) - \frac{\tau p(\tau) F(|x|) \vee \eta p(\eta) F(|y|)}{p(\tau)+p(\eta)} \right\|}{\|x\|^\alpha + \|y\|^\alpha}.$$

If any three of them are simultaneously finite, then the fourth is also finite.

Proof (of Theorem 1.1) We first show by induction that for any fixed $x \in \mathscr{X}$,

$$\frac{\left\| \frac{F(2^n |x|)}{2^n} - F(|x|) \right\|}{\|x\|^\alpha} \leq \vartheta \sum_{j=0}^{n-1} 2^{j(\alpha-1)} \tag{1.9}$$

whenever $n \in \mathbb{N}$. In fact, for $n = 1$ the statement is obvious by choosing $\tau = \eta = 2$ and $x = y$ in inequality (1.4). Suppose the statement is true for $n = k$. Let us prove it for $n = k + 1$. In fact, let $2x$ replace x and $n = k$ in inequality (1.9) and observe that

$$\frac{\left\| \frac{F(2^k 2|x|)}{2^k} - F(2|x|) \right\|}{\|2x\|^\alpha} \leq \vartheta \sum_{l=0}^{k-1} 2^{l(\alpha-1)}.$$

Hence

$$\frac{\left\| \frac{F\left(2^{(k+1)}|x|\right)}{2^{(k+1)}} - \frac{1}{2}F\left(2\,|x|\right) \right\|}{\|x\|^{\alpha}} \leq \vartheta\, 2^{(\alpha-1)} \sum_{l=0}^{k-1} 2^{l(\alpha-1)} \leq \vartheta \sum_{l=1}^{k} 2^{l(\alpha-1)}.$$

The triangle inequality yields

$$\left\| \frac{F\left(2^{(k+1)}\,|x|\right)}{2^{k+1}} - F\left(|x|\right) \right\| \leq \left\| \frac{F\left(2^{(k+1)}\,|x|\right)}{2^{k+1}} - \frac{F\left(2\,|x|\right)}{2} \right\|$$

$$+ \left\| \frac{F\left(2\,|x|\right)}{2} - F\left(|x|\right) \right\|$$

$$\leq \left(\vartheta \sum_{j=1}^{k} 2^{j(\alpha-1)} + \vartheta \right) \|x\|^{\alpha}$$

$$= \vartheta\,\|x\|^{\alpha} \sum_{j=0}^{k} 2^{j(\alpha-1)}.$$

We have just shown the validity of inequality (1.9) for every $n \in \mathbb{N}$. Since the geometric series

$$\sum_{l=0}^{\infty} 2^{l(\alpha-1)} = \frac{2}{2 - 2^{\alpha}},$$

$(0 \leq \alpha < 1)$, we obtain that

$$\frac{\left\| \frac{F(2^n|x|)}{2^n} - F\left(|x|\right) \right\|}{\|x\|^{\alpha}} \leq \frac{2\vartheta}{2 - 2^{\alpha}}, \quad 0 \leq \alpha < 1. \tag{1.10}$$

Next, note that for all $m > n > 0$ and making the change of variable $y = 2^n x$ we have

$$\left\| 2^{-m} F\left(2^m\,|x|\right) - 2^{-n} F\left(2^n\,|x|\right) \right\| = 2^{-n} \left\| 2^{-m+n} F\left(2^m\,|x|\right) - F\left(2^n\,|x|\right) \right\|$$

$$= 2^{-n} \left\| 2^{-m+n} F\left(2^{m-n}\,|y|\right) - F\left(|y|\right) \right\|$$

$$\leq 2^{-n} \frac{2\vartheta}{2 - 2^{\alpha}} \|y\|^{\alpha}$$

$$= 2^{-n(1-\alpha)} \frac{2\vartheta}{2 - 2^{\alpha}} \|x\|^{\alpha}.$$

Consequently, passing to the limit yields,

$$\lim_{n\to\infty} \left\| 2^{-m} F\left(2^m \, |x|\right) - 2^{-n} F\left(2^n \, |x|\right) \right\| = 0.$$

Since $\mathscr{Y}$ is a Banach space, we can thus conclude that the sequence

$$\left(\frac{F\left(2^n \, |x|\right)}{2^n} \right) \subset \mathscr{Y}^+$$

converges in the $\mathscr{Y}$-norm. Now, define the mapping $T : \mathscr{X} \to \mathscr{Y}$ by

$$T\left(|x|\right) := \lim_{n\to\infty} \frac{F\left(2^n \, |x|\right)}{2^n} \tag{1.11}$$

Clearly, T is a cone-related operator. Let us show that T is maximum preserving. In fact, letting $\tau = \eta = 2^n$ in (1.4) leads to

$$\left\| F\left(2^n \left(|x| \vee |y|\right)\right) - 2^n \left(F\left(|x|\right) \vee F\left(|y|\right)\right) \right\| \le \vartheta \left(\|x\|^\alpha + \|y\|^\alpha\right).$$

Substituting x with $2^n x$ and y with $2^n y$ in this last inequality one can get

$$\left\| F\left(4^n \left(|x| \vee |y|\right)\right) - 2^n \left(F\left(2^n \, |x|\right) \vee F\left(2^n \, |y|\right)\right) \right\| \le 2^{n\alpha} \vartheta \left(\|x\|^\alpha + \|y\|^\alpha\right)$$

which implies

$$4^{-n} \left\| F\left(4^n \left(|x| \vee |y|\right)\right) - 2^n \left(F\left(2^n \, |x|\right) \vee F\left(2^n \, |y|\right)\right) \right\|$$
$$= \left\| 4^{-n} F\left(4^n \left(|x| \vee |y|\right)\right) - 2^{-n} \left(F\left(2^n \, |x|\right) \vee F\left(2^n \, |y|\right)\right) \right\|$$
$$\le 2^{n(\alpha-2)} \vartheta \left(\|x\|^\alpha + \|y\|^\alpha\right).$$

Consequently, passing to the limit yields

$$\left\| T\left(|x| \vee |y|\right) - T\left(|x|\right) \vee T\left(|y|\right) \right\| = 0,$$

$x, y \in \mathscr{X}$ or equivalently

$$T\left(|x| \vee |y|\right) = T\left(|x|\right) \vee T\left(|y|\right),$$

$x, y \in \mathscr{X}$, because

$$\lim_{n\to\infty} 4^{-n} F\left(4^n \, |z|\right) = \lim_{m\to\infty} 2^{-m} F\left(2^m \, |z|\right), \quad z \in \mathscr{X}.$$

Next, we show the validity of the identity

$$T\left(\tau \, |x|\right) = \tau T\left(|x|\right)$$

for all $x \in \mathscr{X}$ and every number $\tau \in [0, \infty)$. In fact, in inequality (1.4) choose $\eta = \tau$, $y = 0$ and substitute $2^n \tau$ for τ to observe via Remark 1.1/(2) that

$$\left\| F \left(2^n \tau \, |x| \right) - 2^n \tau F \left(|x| \right) \right\| \leq \vartheta \, \|x\|^\alpha ,$$

for all $x \in \mathscr{X}$ and every number $\tau \in [0, \infty)$. This inequality can be transformed as

$$\left\| F \left(4^n \tau \, |x| \right) - 2^n \tau F \left(2^n \, |x| \right) \right\| \leq \vartheta \, 2^{n\alpha} \, \|x\|^\alpha$$

if we replace x with $2^n x$. Consequently,

$$\left\| \frac{F \left(4^n \tau \, |x| \right)}{4^n} - \tau \frac{F \left(2^n \, |x| \right)}{2^n} \right\| \leq \vartheta \, 2^{-n(2-\alpha)} \, \|x\|^\alpha .$$

Hence on the one hand,

$$\lim_{n \to \infty} \frac{F \left(4^n \tau \, |x| \right)}{4^n} = \tau \lim_{n \to \infty} \frac{F \left(2^n \, |x| \right)}{2^n} = \tau T \left(|x| \right)$$

and on the other hand by changing the variable $z = \tau x$

$$\lim_{n \to \infty} \frac{F \left(4^n \tau \, |x| \right)}{4^n} = \lim_{n \to \infty} \frac{F \left(4^n \, |z| \right)}{4^n} = T \left(|z| \right) = T \left(\tau \, |x| \right) .$$

Therefore, the semi-homogeneity holds true. In the next step taking the limit in (1.10) leads to (1.5). Further, let us show the unicity. In fact, assume the existence of another such cone-related functional G such that

$$\mathscr{S} := \{ x \in \mathscr{X} : G \left(|x| \right) \neq T \left(|x| \right) \} \neq \emptyset.$$

Then (1.5) implies the existence of some $\vartheta_0 > 0$ and $\beta \in [0, 1)$ such that for each $x \in \mathscr{S}$,

$$\| G \left(|x| \right) - T \left(|x| \right) \| \leq \vartheta_0 \, \|x\|^\beta . \tag{1.12}$$

One can easily deduce from the semi-homogeneity of G and T that $kx \in \mathscr{S}$ for every $k \in \mathbb{N}$ whenever $x \in \mathscr{S}$. Taking into account (1.12) one can easily deduce by the triangle inequality and the semi-homogeneity that

$$\| G \left(|x| \right) - T \left(|x| \right) \| = n^{-1} \| G \left(|nx| \right) - T \left(|nx| \right) \| \leq n^{\alpha-1} \vartheta \, \|x\|^\alpha + n^{\beta-1} \vartheta_0 \, \|x\|^\beta$$

which would imply in the limit that

$$\| G \left(|x| \right) - T \left(|x| \right) \| = 0,$$

or equivalently

$$G\left(|x|\right) = T\left(|x|\right).$$

This would mean that $x \in \mathscr{S}$ implies $x \notin \mathscr{S}$, or equivalently $\mathscr{S} \cap \mathscr{S} = \emptyset$, i.e. $\mathscr{S} = \emptyset$, a contradiction, indeed. Finally, let us prove the moreover-part. In fact, assuming that F satisfies property P4, pick arbitrarily an increasing sequence $(|x_k|) \subset \mathscr{X}^+$ with limit $|x| \in \mathscr{X}^+$. Then by (1.11), the monotonicity of T and the continuity from below of F we have

$$\lim_{k\to\infty} \lim_{n\to\infty} \frac{F\left(2^n |x_k|\right)}{2^n} = \lim_{k\to\infty} T\left(|x_k|\right) \le T\left(|x|\right)$$

$$= \lim_{n\to\infty} \frac{F\left(2^n |x|\right)}{2^n} = \lim_{n\to\infty} \lim_{k\to\infty} \frac{F\left(2^n |x_k|\right)}{2^n}.$$

Thus

$$\lim_{k\to\infty} \lim_{n\to\infty} \frac{F\left(2^n |x_k|\right)}{2^n} \le T\left(|x|\right) = \lim_{n\to\infty} \lim_{k\to\infty} \frac{F\left(2^n |x_k|\right)}{2^n}. \tag{1.13}$$

By the conjunction of both inequalities (1.3) and (1.13), it follows that operator T is continuous from below. To end the proof of the moreover-part we simply note that the reverse conditional is trivial. Therefore, we can conclude on the validity of the argument.   □

Remark 1.2 Theorems 1.1–1.4 and Lemmas 1.1–1.4 remain valid for negative values of the norm exponent α.

1.3   Functional Equation Involving Both Lattice Operations

In the sequel $(\mathscr{X}, \wedge_{\mathscr{X}}, \vee_{\mathscr{X}})$ will denote a vector lattice and $(\mathscr{Y}, \wedge_{\mathscr{Y}}, \vee_{\mathscr{Y}})$ a Banach lattice with $\mathscr{X}^+$ and $\mathscr{Y}^+$ their respective positive cones.

Let us consider the functional equation

$$T\left(|x| \Delta_{\mathscr{X}}^* |y|\right) \Delta_{\mathscr{Y}}^* T\left(|x| \Delta_{\mathscr{X}}^{**} |y|\right) = T\left(|x|\right) \Delta_{\mathscr{Y}}^{**} T\left(|y|\right) \tag{1.14}$$

to hold true for all $x, y \in \mathscr{X}$, where $\Delta_{\mathscr{X}}^*, \Delta_{\mathscr{X}}^{**} \in \{\wedge_{\mathscr{X}}, \vee_{\mathscr{X}}\}$ and $\Delta_{\mathscr{Y}}^*, \Delta_{\mathscr{Y}}^{**} \in \{\wedge_{\mathscr{Y}}, \vee_{\mathscr{Y}}\}$ are fixed lattice operations, and where $T : \mathscr{X} \to \mathscr{Y}$.

Note that if in the special case the above four lattice operations are at the same time the supremum (join) or the infimum (meet), then the functional equation (1.14) is just the defining equation of the join (or meet)-homomorphism. Moreover, if operations $\Delta_{\mathscr{X}}^*$ and $\Delta_{\mathscr{X}}^{**}$ are the same, then the left hand side of (1.14) is the maps of the meets or the joins. Next, we formulate a lattice version of Ulam's stability problem whose solution also solves Eq. (1.14).

Problem 1.1 Let be given lattice operations $\Delta_{\mathscr{X}}^*$, $\Delta_{\mathscr{X}}^{**} \in \{\wedge_{\mathscr{X}}, \vee_{\mathscr{X}}\}$ and $\Delta_{\mathscr{Y}}^*$, $\Delta_{\mathscr{Y}}^{**} \in \{\wedge_{\mathscr{Y}}, \vee_{\mathscr{Y}}\}$, a vector lattice G_1, a vector lattice G_2 endowed with a metric $d(\cdot, \cdot)$ and a positive number ε, does there exist some $\delta > 0$ such that, if a mapping $F : G_1 \to G_2$ satisfies the perturbation inequality

$$d \left(F \left(|x| \, \Delta_{\mathscr{X}}^* \, |y| \right) \Delta_{\mathscr{Y}}^* F \left(|x| \, \Delta_{\mathscr{X}}^{**} \, |y| \right), \, F \left(|x| \right) \Delta_{\mathscr{Y}}^{**} F \left(|y| \right) \right) \le \delta$$

for all x, $y \in G_1$, then an operation-preserving functional $T : G_1 \to G_2$ exists with the property that

$$d \left(T(x), F(x) \right) \le \varepsilon$$

for all $x \in G_1$?

Since the respective proofs of the main theorems in Sects. 1.3–1.5 will be based on a Forti's result in [15], known as the direct method, we thought we should recall it here.

Theorem 1.5 (Forti) *Let (X, d) be a complete metric space and S an appropriate set. Assume that $f : S \to X$ is a function satisfying the inequality*

$$d \left(H \left(f \left(G \left(x \right) \right) \right), \, f \left(x \right) \right) \le \delta \left(x \right), \tag{1.15}$$

for all $x \in S$, where $\delta : S \to [0, \infty)$ is some function. If $H : X \to X$ is a continuous function and satisfies the inequality

$$d \left(H \left(u \right), H \left(v \right) \right) \le \varphi \left(d \left(u, v \right) \right), \quad u, v \in X, \tag{1.16}$$

for a certain non-decreasing subadditive function $\varphi : [0, \infty) \to [0, \infty)$ and the series

$$\sum_{j=0}^{\infty} \varphi^j \left(\delta \left(G^j \left(x \right) \right) \right) \tag{1.17}$$

is convergent for every $x \in S$, then there exists a unique function $F : S \to X$ solution of the functional equation

$$H \left(F \left(G \left(x \right) \right) \right) = F \left(x \right), \quad x \in S, \tag{1.18}$$

and satisfying the following inequality:

$$d\left(F\left(x\right), f\left(x\right)\right) \leq \sum_{j=0}^{\infty} \varphi^j\left(\delta\left(G^j\left(x\right)\right)\right).\tag{1.19}$$

The function F is given by

$$F\left(x\right) = \lim_{n \to \infty} H^n\left(f\left(G^n\left(x\right)\right)\right).\tag{1.20}$$

1.3.1 The Main Results of This Section

Theorem 1.6 *Consider a cone-related functional* $F : \mathscr{X} \to \mathscr{Y}$ *for which there are numbers* $\vartheta > 0$ *and* $\alpha \in [0, 1)$ *such that*

$$\left\|\frac{F\left(|x| \Delta_{\mathscr{X}}^* |y|\right) \Delta_{\mathscr{Y}}^* F\left(|x| \Delta_{\mathscr{X}}^{**} |y|\right)}{\tau} - F\left(\frac{|x|}{\tau}\right) \Delta_{\mathscr{Y}}^{**} F\left(\frac{|y|}{\tau}\right)\right\|$$
$$\leq \frac{\vartheta}{4}\left(\|x\|^\alpha + \|y\|^\alpha\right)\tag{1.21}$$

for all $x, y \in \mathscr{X}$ *and* $\tau \in (0, \infty)$, *where* $\Delta_{\mathscr{X}}^*, \Delta_{\mathscr{X}}^{**} \in \{\wedge_{\mathscr{X}}, \vee_{\mathscr{X}}\}$ *and* $\Delta_{\mathscr{Y}}^*, \Delta_{\mathscr{Y}}^{**} \in \{\wedge_{\mathscr{Y}}, \vee_{\mathscr{Y}}\}$ *are fixed lattice operations. Then the sequence* $\left(2^{-n} F\left(2^n |x|\right)\right)_{n \in \mathbb{N}}$ *is a Cauchy sequence for every* $x \in \mathscr{X}$. *Moreover, let the functional* $T : \mathscr{X} \to \mathscr{Y}$ *be defined by*

$$T\left(|x|\right) = \lim_{n \to \infty} 2^{-n} F\left(2^n |x|\right).\tag{1.22}$$

Then

(a) T is semi-homogeneous, i.e.

$$T\left(\tau |x|\right) = \tau T\left(|x|\right),$$

   for all $x \in \mathscr{X}$ *and all* $\tau \in [0, \infty)$;
(b) T is the unique cone-related functional satisfying both identity (1.14) and inequality

$$\|T\left(|x|\right) - F\left(|x|\right)\| \leq \frac{2^\alpha \vartheta}{2 - 2^\alpha} \|x\|^\alpha\tag{1.23}$$

   which holds for every $x \in \mathscr{X}$.

Theorem 1.7 *Consider a cone-related functional* $F : \mathscr{X} \to \mathscr{Y}$ *for which there are numbers* $\vartheta > 0$ *and* $p \in (1, \infty)$ *such that*

$$\left\| \tau \left(F \left(|x| \, \Delta_{\mathscr{X}}^* \, |y| \right) \Delta_{\mathscr{Y}}^* \, F \left(|x| \, \Delta_{\mathscr{X}}^{**} \, |y| \right) \right) - F \left(\tau \, |x| \right) \Delta_{\mathscr{Y}}^{**} F \left(\tau \, |y| \right) \right\|$$
$$\leq \vartheta \left(\|x\|^p + \|y\|^p \right) \tag{1.24}$$

for all $x, y \in \mathscr{X}$ *and* $\tau \in [0, \infty)$, *where* $\Delta_{\mathscr{X}}^*, \Delta_{\mathscr{X}}^{**} \in \{\wedge_{\mathscr{X}}, \vee_{\mathscr{X}}\}$ *and* $\Delta_{\mathscr{Y}}^*, \Delta_{\mathscr{Y}}^{**} \in \{\wedge_{\mathscr{Y}}, \vee_{\mathscr{Y}}\}$ *are fixed lattice operations. Then the sequence* $\left(2^n F \left(2^{-n} \, |x| \right) \right)$ *is a Cauchy sequence for every* $x \in \mathscr{X}$. *Moreover, let the functional* $T : \mathscr{X} \to \mathscr{Y}$ *be defined by*

$$T \left(|x| \right) = \lim_{n \to \infty} 2^n F \left(2^{-n} \, |x| \right). \tag{1.25}$$

Then

(a) *T is semi-homogeneous, i.e.*

$$T \left(\tau \, |x| \right) = \tau T \left(|x| \right),$$

for all $x \in \mathscr{X}$ *and all* $\tau \in [0, \infty)$;

(b) *T is the unique cone-related functional satisfying both identity* (1.14) *and inequality*

$$\| T \left(|x| \right) - F \left(|x| \right) \| \leq \frac{2^p \vartheta}{2^p - 2} \, \|x\|^p \tag{1.26}$$

which holds for every $x \in \mathscr{X}$.

Before we start the proofs the following obvious remarks are worth being mentioned, as they will be used multiple times.

Remark 1.3 If the conditions of Theorem 1.6 or 1.7 hold true, then $F(0) = 0$.

Remark 1.4 Let Z be a set closed under the scalar multiplication, i.e. $bz \in Z$ whenever $b \in \mathbb{R}$ and $z \in Z$. Given a number $c \in \mathbb{R}$ let the function $\gamma : Z \to Z$ be defined by $\gamma(z) = cz$. Then $\gamma^j : Z \to Z$ the j-th iteration of γ is given by

$$\gamma^j (z) = c^j z$$

for every counting number $j \geq 2$.

Proof (of Theorem 1.6) First, if we choose $\tau = 2$, $y = x$ and replace x by $2x$ in inequality (1.21) then we obviously have

$$\left\| \frac{F \left(2 \, |x| \right)}{2} - F \left(|x| \right) \right\| \leq \vartheta 2^{\alpha - 1} \, \|x\|^{\alpha}. \tag{1.27}$$

Next, let us define the following functions:

1. $G : \mathscr{X} \to \mathscr{X}$, $\quad G\left(|x|\right) = 2\,|x|$.
2. $\delta : \mathscr{X} \to [0, \infty)$, $\quad \delta\left(|x|\right) = \vartheta\, 2^{\alpha-1}\, \|x\|^{\alpha}$.
3. $\varphi : [0, \infty) \to [0, \infty)$, $\quad \varphi\left(t\right) = 2^{-1} t$.
4. $H : \mathscr{Y} \to \mathscr{Y}$, $\quad H\left(|y|\right) = 2^{-1}\, |y|$.
5. $d\left(\cdot, \cdot\right) : \mathscr{Y} \times \mathscr{Y} \to [0, \infty)$, $\quad d\left(y_1, y_2\right) = \|y_1 - y_2\|$.

We shall verify the fulfilment of all the three conditions of the Forti's theorem as follows.

(I) From inequality (1.27) we obviously have

$$d\left(H\left(F\left(G\left(|x|\right)\right)\right), F\left(|x|\right)\right) = \left\| \frac{F\left(2\,|x|\right)}{2} - F\left(|x|\right) \right\|$$

$$\leq \vartheta\, 2^{\alpha-1}\, \|x\|^{\alpha} = \delta\left(|x|\right).$$

(II) $d\left(H\left(|y_1|\right), H\left(|y_2|\right)\right) = 2^{-1}\, \|y_1 - y_2\| = \varphi\left(d\left(y_1, y_2\right)\right)$ for all $y_1, y_2 \in \mathscr{Y}$.

(III) Clearly, on the one hand φ is a non-decreasing subadditive function on the positive half line, and on the other hand by applying Remark 1.4 on both the iterations G^j and φ^j of G and φ respectively, one can observe that

$$\sum_{j=0}^{\infty} \varphi^j \left(\delta\left(G^j\left(|x|\right)\right)\right) = \vartheta\, 2^{\alpha-1}\, \|x\|^{\alpha} \sum_{j=0}^{\infty} 2^{(\alpha-1)j} = \vartheta\, \|x\|^{\alpha}\, \frac{2^{\alpha}}{2 - 2^{\alpha}} < \infty.$$

Then in virtue of the above Forti's theorem sequence $\left(H^n\left(F\left(G^n\,|x|\right)\right)\right)_{n \in \mathbb{N}}$ is a Cauchy sequence for every $x \in \mathscr{X}$ and thus so is sequence $\left(2^{-n} F\left(2^n\,|x|\right)\right)_{n \in \mathbb{N}}$ and furthermore, the mapping (1.22) is the unique functional which satisfies inequality (1.23).

Next, we prove the validity of inequality (1.14). In fact, in (1.21) substitute x with $2^n x$ and y with $2^n y$, and also let $\tau = 1$. Then

$$\left\| F\left(2^n\left(|x| \triangle_{\mathscr{X}}^{*}\, |y|\right)\right) \triangle_{\mathscr{Y}}^{**} F\left(2^n\left(|x| \triangle_{\mathscr{X}}^{**}\, |y|\right)\right) - F\left(2^n\,|x|\right) \triangle_{\mathscr{Y}}^{**} F\left(2^n\,|y|\right) \right\|$$

$$\leq \frac{\vartheta}{4}\, 2^{n\alpha}\left(\|x\|^{\alpha} + \|y\|^{\alpha}\right).$$

Dividing both sides of this last inequality by 2^n yields

$$\left\| \frac{F\left(2^n\left(|x| \triangle_{\mathscr{X}}^{*}\, |y|\right)\right) \triangle_{\mathscr{Y}}^{**} F\left(2^n\left(|x| \triangle_{\mathscr{X}}^{**}\, |y|\right)\right)}{2^n} - \frac{F\left(2^n\,|x|\right) \triangle_{\mathscr{Y}}^{**} F\left(2^n\,|y|\right)}{2^n} \right\|$$

$$\leq \frac{\vartheta}{4}\left(\|x\|^{\alpha} + \|y\|^{\alpha}\right) 2^{(\alpha-1)n}.$$

$$(1.28)$$

Taking the limit in (1.28) we have via (1.25) that

$$\left\| T\left(|x|\,\Delta^*_{\mathscr{X}}\,|y|\right)\Delta^*_{\mathscr{Y}}T\left(|x|\,\Delta^{**}_{\mathscr{X}}\,|y|\right) - T\left(|x|\right)\Delta^{**}_{\mathscr{Y}}T\left(|y|\right)\right\| = 0$$

which is equivalent to

$$T\left(|x|\,\Delta^*_{\mathscr{X}}\,|y|\right)\Delta^*_{\mathscr{Y}}T\left(|x|\,\Delta^{**}_{\mathscr{X}}\,|y|\right) = T\left(|x|\right)\Delta^{**}_{\mathscr{Y}}T\left(|y|\right).$$

Because of Remark 1.3 identity $\tau F\left(|x|\right) = F\left(\tau\,|x|\right)$ is trivial on the one hand for $\tau = 0$ and all $x \in \mathscr{X}$, on the other hand for $x = 0$ and all $\tau \in [0, \infty)$. Without loss of generality let us thus fix arbitrarily a number $\tau \neq 0$ and an $x \in \mathscr{X}\setminus\{0\}$. In (1.21) choose $y = x$ and make the changes τ to τ^{-1} and x to $2^n x$. Then

$$\left\| \tau F\left(2^n\,|x|\right) - F\left(\tau 2^n\,|x|\right)\right\| \leq \frac{\vartheta}{2}\,\|x\|^\alpha\,2^{n\alpha}.$$

Divide both sides of this last inequality by 2^n to get

$$\left\| \tau 2^{-n} F\left(2^n\,|x|\right) - 2^{-n} F\left(\tau 2^n\,|x|\right)\right\| \leq \frac{\vartheta}{2}\,\|x\|^\alpha\,2^{(\alpha-1)n}. \tag{1.29}$$

By taking the limit in (1.29) we have via (1.22) that

$$\left\| \tau T\left(|x|\right) - T\left(\tau\,|x|\right)\right\| = 0$$

or equivalently,

$$T\left(\tau\,|x|\right) = \tau T\left(|x|\right)$$

for all $x \in \mathscr{X}$. We have thus shown the semi-homogeneity of operator T. We can conclude on the validity of the argument.   □

Proof (of Theorem 1.7) First, if we choose $\tau = 2$, $y = x$ and replace x by $2^{-1}x$ in inequality (1.24) then we obviously have

$$\left\| 2F\left(2^{-1}\,|x|\right) - F\left(|x|\right)\right\| \leq \vartheta 2^{1-p}\,\|x\|^p. \tag{1.30}$$

Next, let us define the following functions:

1. $G : \mathscr{X} \to \mathscr{X}, \quad G\left(|x|\right) = 2^{-1}\,|x|$.
2. $\delta : \mathscr{X} \to [0, \infty), \quad \delta\left(|x|\right) = \vartheta 2^{1-p}\,\|x\|^p$.
3. $\varphi : [0, \infty) \to [0, \infty), \quad \varphi\left(t\right) = 2t$.
4. $H : \mathscr{Y} \to \mathscr{Y}, \quad H\left(|y|\right) = 2\,|y|$.
5. $d\left(\cdot, \cdot\right) : \mathscr{Y} \times \mathscr{Y} \to [0, \infty), \quad d\left(y_1, y_2\right) = \|y_1 - y_2\|$.

We shall verify the fulfilment of all the three conditions of the Forti's theorem as follows.

(I) From inequality (1.30) we obviously have

$$d\left(H\left(F\left(G\left(|x|\right)\right)\right),\ F\left(|x|\right)\right) = \left\|2F\left(2^{-1}|x|\right) - F\left(|x|\right)\right\| \le \vartheta 2^{1-p}\|x\|^p$$

$$= \delta\left(|x|\right).$$

(II) $d\left(H\left(|y_1|\right),\ H\left(|y_2|\right)\right) = 2\|y_1 - y_2\| = \phi\left(d\left(y_1,\ y_2\right)\right)$ for all $y_1,\ y_2 \in \mathcal{Y}$.

(III) Clearly, on the one hand φ is a non-decreasing subadditive function on the positive half line, and on the other hand by applying Remark 1.4 on both the iterations G^j and φ^j of G and φ respectively, one can observe that

$$\sum_{j=0}^{\infty} \varphi^j\left(\delta\left(G^j\left(|x|\right)\right)\right) = \vartheta 2^{1-p}\|x\|^p \sum_{j=0}^{\infty} 2^{(1-p)j} = \vartheta\|x\|^p \frac{2^p}{2^p - 2} < \infty.$$

Then in virtue of the above Forti's theorem sequence $\left(H^n\left(F\left(G^n|x|\right)\right)\right)_{n\in\mathbb{N}}$ is a Cauchy sequence for every $x \in \mathcal{X}$ and thus so is sequence $\left(2^n F\left(2^{-n}|x|\right)\right)_{n\in\mathbb{N}}$ and furthermore, the mapping (1.25) is the unique functional which satisfies inequality (1.26). Next, we prove the validity of inequality (1.14). In fact, in (1.24) substitute x with $2^{-n}x$ and y with $2^{-n}y$, and also let $\tau = 1$. Then

$$\left\| F\left(2^{-n}\left(|x|\, \Delta_{\mathcal{X}}^*\, |y|\right)\right) \Delta_{\mathcal{Y}}^{**} F\left(2^{-n}\left(|x|\, \Delta_{\mathcal{X}}^{**}\, |y|\right)\right) - F\left(2^{-n}|x|\right) \Delta_{\mathcal{Y}}^{**} F\left(2^{-n}|y|\right) \right\|$$

$$\le \vartheta 2^{-np}\left(\|x\|^p + \|y\|^p\right).$$

Multiplying both sides of this last inequality by 2^n yields

$$\left\| 2^n \left(F\left(2^{-n}\left(|x|\, \Delta_{\mathcal{X}}^*\, |y|\right)\right) \Delta_{\mathcal{Y}}^{**} F\left(2^{-n}\left(|x|\, \Delta_{\mathcal{X}}^{**}\, |y|\right)\right)\right) \right.$$

$$\left. - 2^n \left(F\left(2^{-n}|x|\right) \Delta_{\mathcal{Y}}^{**} F\left(2^{-n}|y|\right)\right) \right\| \le \vartheta\left(\|x\|^p + \|y\|^p\right) 2^{(1-p)n}.$$

$$(1.31)$$

Taking the limit in (1.31) we have via (1.25) that

$$\left\| T\left(|x|\, \Delta_{\mathcal{X}}^*\, |y|\right) \Delta_{\mathcal{Y}}^* T\left(|x|\, \Delta_{\mathcal{X}}^{**}\, |y|\right) - T\left(|x|\right) \Delta_{\mathcal{Y}}^{**} T\left(|y|\right) \right\| = 0$$

which is equivalent to

$$T\left(|x|\, \Delta_{\mathcal{X}}^*\, |y|\right) \Delta_{\mathcal{Y}}^* T\left(|x|\, \Delta_{\mathcal{X}}^{**}\, |y|\right) = T\left(|x|\right) \Delta_{\mathcal{Y}}^{**} T\left(|y|\right).$$

Because of Remark 1.3 identity $\tau F\left(|x|\right) = F\left(\tau|x|\right)$ is trivial on the one hand for $\tau = 0$ and all $x \in \mathcal{X}$, on the other hand for $x = 0$ and all $\tau \in [0,\ \infty)$. Without loss of generality let us thus fix arbitrarily a number $\tau \neq 0$ and an $x \in \mathcal{X} \setminus \{0\}$. In (1.24) choose $y = x$ and change x to $2^{-n}x$. Then

$$\left\| \tau F\left(2^{-n}|x|\right) - F\left(\tau 2^{-n}|x|\right) \right\| \le \vartheta\|x\|^p 2^{-np}.$$

Multiply both sides of this last inequality by 2^n to get

$$\left\| \tau 2^n F\left(2^{-n}|x|\right) - 2^n F\left(\tau 2^{-n}|x|\right)\right\| \leq \vartheta \|x\|^p 2^{(1-p)n}. \tag{1.32}$$

By taking the limit in (1.32) we have via (1.25) that

$$\|\tau T\left(|x|\right) - T\left(\tau |x|\right)\| = 0$$

or equivalently,

$$T\left(\tau |x|\right) = \tau T\left(|x|\right)$$

for all $x \in \mathscr{X}$. We have thus proved the semi-homogeneity of operator T. We can conclude on the validity of the argument.   □

To end the section we shall provide an example showing that if in (1.24) the parameter τ is omitted and the power p of the norms equals the unity, then stability cannot always be guaranteed. We remind that in the addition environments Gajda in [16] and Găvruţa in [18] gave some interesting examples to show how stability fails when the power of the norms is equal to 1.

Example 1.1 Consider the Lipschitz-continuous function

$$F : [0, \infty) \to [0, \infty), \ F(x) = \sqrt{x^2 + 1}.$$

Fix arbitrarily two numbers $x, y \in [0, \infty)$. Since F is an increasing function the very first equality in the chain of relations here below is valid, implying the subsequent relations in the chain:

$$|F(x \vee y) - (F(x) \wedge F(y))| = |F(x \vee y) - F(x \wedge y)|$$

$$= \left| \sqrt{(x \vee y)^2 + 1} - \sqrt{(x \wedge y)^2 + 1} \right|$$

$$= \frac{(x \vee y)^2 - (x \wedge y)^2}{\sqrt{(x \vee y)^2 + 1} + \sqrt{(x \wedge y)^2 + 1}}$$

$$= |x - y| \cdot \frac{(x \vee y) + (x \wedge y)}{\sqrt{(x \vee y)^2 + 1} + \sqrt{(x \wedge y)^2 + 1}}$$

$$\leq |x - y| \leq x + y$$

for all $x, y \in [0, \infty)$. Now, let $T : [0, \infty) \to [0, \infty)$ be a function such that $T(x) = xT(1)$ for all $x \in [0, \infty)$. Then a simple argument shows

$$\sup_{x \in (0, \infty)} \frac{|F(x) - T(x)|}{x} = \sup_{x \in (0, \infty)} \left| \sqrt{1 + x^{-2}} - T(1) \right| = \infty.$$

Remark 1.5 Theorem 1.6 remains valid for negative values of the norm exponent α.

1.4 Schwaiger's Type Functional Equation

As a consequence of the counterexample given by Gajda (see [16]), many definitions of approximately linear mappings have come to light. In this perspective Schwaiger (cf. [31]) also proposed a functional equation similar to the Cauchy's one and suitably perturbed it and obtained a stability result similar to Hyers-Ulam's original one (see [11, 20, 33]). Schwaiger's theorem reads:

Theorem 1.8 (Schwaiger's Stability Theorem) *Given a real vector space E_1 and a real Banach space E_2, let $f : E_1 \to E_2$ be a mapping for which inequality*

$$\| f(x + \alpha y) - f(x) - \alpha f(y) \| \leq b(\alpha) \tag{1.33}$$

is satisfied for all $\alpha \in \mathbb{R}$. Then there exists a unique linear function $g : E_1 \to E_2$ such that $f - g$ is bounded.

In the sequel $(\mathscr{X}, \wedge_{\mathscr{X}}, \vee_{\mathscr{X}})$ will denote a vector lattice and $(\mathscr{Y}, \wedge_{\mathscr{Y}}, \vee_{\mathscr{Y}})$ a Banach lattice with $\mathscr{X}^+$ and $\mathscr{Y}^+$ their respective positive cones.

Given two positive real numbers p and q consider the functional equation

$$T\left((\tau^q |x|) \vee |y|\right) = \left(\tau^p T(|x|)\right) \vee T(|y|) \tag{1.34}$$

for all $x, y \in \mathscr{X}$ and $\tau \in [0, \infty)$, where T maps $\mathscr{X}$ into $\mathscr{Y}$.

The following simple examples show that the functional equation (1.34) has at least one solution.

Example 1.2 The function $T_1 : [0, \infty) \to [0, \infty)$ defined by $T_1(x) = x$ is a solution of (1.34), for all $\tau, q, x, y \in [0, \infty)$ with the choice $p = q$.

Example 1.3 The function $T_2 : [0, \infty) \to [0, \infty)$ defined by $T_2(x) = \sqrt{x}$ is a solution of (1.34), for all $\tau, q, x, y \in [0, \infty)$ with the choice $p = \frac{q}{2} < q$.

Example 1.4 The function $T_3 : [0, \infty) \to [0, \infty)$ defined by $T_3(x) = x^2$ is a solution of (1.34), for all $\tau, q, x, y \in [0, \infty)$ with the choice $p = 2q > q$.

Example 1.5 Let $\mathscr{X} = B(M, \mathbb{R})$ be the space of all bounded real-valued functions defined on M. Then the functional

$$T : \mathscr{X} \to \mathscr{X}, \text{ such that } T(|f|) = |f|^{\alpha}$$

solves (1.34) for arbitrary positive numbers q and α with $p = q\alpha$.

The goal of this section is to prove the stability of the functional equation (1.34) which can be viewed as a counterpart of the Schwaiger type stability theorem in lattice environment.

Remark 1.6 Given two positive real numbers p and q, if a cone-related operator $T : \mathscr{X} \to \mathscr{Y}$ satisfies the functional equation (1.34), then

1. $T (|x| \vee |y|) = T (|x|) \vee T (|y|)$ for all x, $y \in \mathscr{X}$ and $\tau = 1$;
2.

$$T \left(\tau^q |x| \right) = \tau^p T (|x|) \tag{1.35}$$

for all $x \in \mathscr{X}$ and all $\tau \in [0, \infty) \setminus \{1\}$.

Proof Choosing $\tau = 1$ in (1.34) we obviously obtain that T is a join-homomorphism. To show the second part we first prove that

$$T (0) = 0.$$

In fact, take $x = y = 0$ in (1.34). Then

$$T (0) = \left(\tau^p T (0) \right) \vee T (0).$$

But since τ runs over the non-negative real line, by choosing $\tau = 2$ yields

$$T (0) = (2T (0)) \vee T (0),$$

which is possible only if $T (0) = 0$. Consequently, (1.35) follows if we select $y = 0$ in (1.34).   □

Theorem 1.9 *Given a pair of positive real numbers (p, q), consider a cone-related functional $F : \mathscr{X} \to \mathscr{Y}$ for which there are numbers $\vartheta > 0$ and α with $q\alpha \in (0, p)$ such that*

$$\left\| F \left((\tau^q |x|) \vee |y| \right) - \left(\tau^p F (|x|) \right) \vee F (|y|) \right\| \leq 2^{-p} \vartheta \left(\|x\|^\alpha + \|y\|^\alpha \right) \tag{1.36}$$

for all x, $y \in \mathscr{X}$ and all $\tau \in [0, \infty)$. Then the sequence $\left(2^{-np} F \left(2^{nq} |x| \right) \right)_{n \in \mathbb{N}}$ is a Cauchy sequence for every $x \in \mathscr{X}$. Let the functional $T : \mathscr{X} \to \mathscr{Y}$ be defined by

$$T (|x|) = \lim_{n \to \infty} 2^{-np} F \left(2^{nq} |x| \right). \tag{1.37}$$

Then

a. *T is a solution of the functional equation (1.34);*

b. T is the unique cone-related functional which satisfies inequality

$$\|T\,(|x|) - F\,(|x|)\| \leq \frac{2^{q\alpha}\,\vartheta}{2^p - 2^{q\alpha}}\,\|x\|^\alpha \tag{1.38}$$

for every $x \in \mathscr{X}$.

Moreover, assume that $\mathscr{X}$ is a Banach lattice and F is continuous from below on the positive cone $\mathscr{X}^+$. Then in order that the limit operator T be continuous from below on $\mathscr{X}^+$, it is necessary and sufficient that

$$\lim_{n\to\infty}\lim_{k\to\infty}\frac{F\,(2^{nq}x_k)}{2^{np}} \leq \lim_{k\to\infty}\lim_{n\to\infty}\frac{F\,(2^{nq}x_k)}{2^{np}}, \tag{1.39}$$

for any increasing sequence $(x_k)_{k\in\mathbb{N}} \subset \mathscr{X}^+$.

Theorem 1.10 *Given a pair of positive real numbers* (p, q), *consider a cone-related functional* $F : \mathscr{X} \to \mathscr{Y}$ *for which there are numbers* $\vartheta > 0$ *and* α *with* $q\alpha \in (p, \infty)$ *such that*

$$\left\| F\left((\tau^q\,|x|)\vee|y|\right) - \left(\tau^p F\,(|x|)\right)\vee F\,(|y|)\right\| \leq 2^p\vartheta\left(\|x\|^\alpha + \|y\|^\alpha\right) \tag{1.40}$$

for all $x, y \in \mathscr{X}$ *and all* $\tau \in [0, \infty)$. *Then the sequence* $\left(2^{np}F\left(2^{-nq}\,|x|\right)\right)_{n\in\mathbb{N}}$ *is a Cauchy sequence for every* $x \in \mathscr{X}$. *Let the functional* $T : \mathscr{X} \to \mathscr{Y}$ *be defined by*

$$T\,(|x|) = \lim_{n\to\infty} 2^{np} F\left(2^{-nq}\,|x|\right). \tag{1.41}$$

Then

a. T is a solution of the functional equation (1.34);
b. T is the unique cone-related functional which satisfies inequality

$$\|T\,(|x|) - F\,(|x|)\| \leq \frac{2^p\vartheta}{2^{q\alpha} - 2^p}\,\|x\|^\alpha \tag{1.42}$$

for every $x \in \mathscr{X}$.

Moreover, assume that $\mathscr{X}$ is a Banach lattice and F is continuous from below on the positive cone $\mathscr{X}^+$. Then in order that the limit operator T be continuous from below on $\mathscr{X}^+$, it is necessary and sufficient that

$$\lim_{n\to\infty}\lim_{k\to\infty} 2^{np} F\left(2^{-nq}x_k\right) \leq \lim_{k\to\infty}\lim_{n\to\infty} 2^{np} F\left(2^{-nq}x_k\right), \tag{1.43}$$

for any increasing sequence $(x_k)_{k\in\mathbb{N}} \subset \mathscr{X}^+$.

Before we start the proofs the following obvious remark is worth being mentioned, as it will be used multiple times in the sequel.

Remark 1.7 If the conditions of Theorem 1.9 or 1.10 hold true, then $F(0) = 0$.

Proof In (1.36) or (1.40) choose $x = y = 0$ and observe that

$$\left\| F(0) - \left(\tau^P F(0)\right) \vee F(0) \right\| = 0$$

so that

$$F(0) = \left(\tau^P F(0)\right) \vee F(0).$$

But since τ runs over the non-negative real line, by choosing $\tau = 2$ yields

$$F(0) = (2F(0)) \vee F(0),$$

which is possible only if $F(0) = 0$.   □

Proof (of Theorem 1.9) First, we choose $\tau = 2^{-1}$, $y = 0$ and replacing x by $2^q x$ in (1.36) we obviously have

$$\left\| \frac{F(2^q |x|)}{2^p} - F(|x|) \right\| \le \vartheta 2^{q\alpha - p} \|x\|^\alpha. \tag{1.44}$$

Next, let us define the following functions:

1. $G : \mathscr{X} \to \mathscr{X}$, $\quad G(|x|) = 2^q |x|$.
2. $\delta : \mathscr{X} \to [0, \infty)$, $\quad \delta(|x|) = \vartheta 2^{q\alpha - p} \|x\|^\alpha$.
3. $\varphi : [0, \infty) \to [0, \infty)$, $\quad \varphi(t) = 2^{-p} t$.
4. $H : \mathscr{Y} \to \mathscr{Y}$, $\quad H(|y|) = 2^{-p} |y|$.
5. $d(\cdot, \cdot) : \mathscr{Y} \times \mathscr{Y} \to [0, \infty)$, $\quad d(y_1, y_2) = \|y_1 - y_2\|$.

We shall verify the fulfilment of all the three conditions of the Forti's theorem as follows.

(I) From inequality (1.44) we obviously have

$$d(H(F(G(|x|))), F(|x|)) = \left\| \frac{F(2^q |x|)}{2^p} - F(|x|) \right\|$$

$$\le \vartheta 2^{q\alpha - p} \|x\|^\alpha = \delta(|x|).$$

(II) For all $y_1, y_2 \in \mathscr{Y}$,

$$d(H(|y_1|), H(|y_2|)) = 2^{-p} \|y_1 - y_2\| = \varphi(d(y_1, y_2)).$$

(III) Clearly, on the one hand φ is a non-decreasing subadditive function on the positive half line, and on the other hand by applying Remark 1.4 on both the iterations G^j and φ^j of G and φ respectively, one can observe that

$$\sum_{j=0}^{\infty} \varphi^j \left(\delta \left(G^j \left(|x| \right) \right) \right) = \vartheta 2^{(q\alpha - p)} \|x\|^\alpha \sum_{j=0}^{\infty} 2^{(q\alpha - p)j}$$

$$= \vartheta \|x\|^\alpha \frac{2^{q\alpha}}{2^p - 2^{q\alpha}} < \infty.$$

Then in virtue of Forti's theorem sequence $(H^n (F (G^n |x|)))_{n \in \mathbb{N}}$ is a Cauchy sequence for every $x \in \mathcal{X}$ and thus so is sequence $\left(2^{-np} F \left(2^{nq} |x| \right) \right)_{n \in \mathbb{N}}$ and furthermore, the mapping (1.37) is the unique functional which satisfies inequality (1.38).

Next, we prove that the mapping T, defined in (1.37), satisfies the functional equation (1.34). In fact, in (1.36) substitute x with $2^{nq} x$ also y with $2^{nq} y$, and fix arbitrarily $\tau \in [0, \infty)$. Then

$$\left\| F \left(2^{nq} \left(\left(\tau^q |x| \right) \vee |y| \right) \right) - \left(\tau^p F \left(2^{nq} |x| \right) \right) \vee F \left(2^{nq} |y| \right) \right\|$$

$$\leq \vartheta 2^{-p} 2^{q\alpha n} \left(\|x\|^\alpha + \|y\|^\alpha \right).$$

Dividing both sides of this last inequality by 2^{np} yields

$$\left\| \frac{F \left(2^{nq} \left(\left(\tau^q |x| \right) \vee |y| \right) \right)}{2^{np}} - \frac{\left(\tau^p F \left(2^{nq} |x| \right) \right) \vee F \left(2^{nq} |y| \right)}{2^{np}} \right\| \tag{1.45}$$

$$\leq \vartheta 2^{-p} 2^{(q\alpha - p)n} \left(\|x\|^\alpha + \|y\|^\alpha \right).$$

Taking the limit in (1.45) we have via (1.37) that for all $\tau \in [0, \infty)$ and all $x, y \in \mathcal{X}$

$$\left\| T \left(\left(\tau^q |x| \right) \vee |y| \right) - \left(\tau^p T \left(|x| \right) \right) \vee T \left(|y| \right) \right\| = 0$$

which is equivalent to (1.34).

The moreover part can be proved the same way the moreover parts of the theorems in [5] were, after we will have proved that the limits on both sides of (1.39) exist. In fact, on the one hand, the existence of the limit on the left hand side follows from the combination of the monotonicity of F and (1.37). On the other hand, because of (1.37) the inner limit on the right hand side equals $T(x_k)$ for every $k \in \mathbb{N}$. But since the limit operator T is a join-homomorphism, it is also isotonic or increasing. Consequently, $(T(x_k))_{k \in \mathbb{N}}$ is a convergent sequence. We have thus proved that the limits on both sides of (1.39) exist.

Therefore, we can conclude on the validity of the argument. $\square$

Proof (of Theorem 1.10) First, we choose $\tau = 2$, $y = 0$ and replacing x by $2^{-q}x$ in (1.40) we obviously have

$$\left\| 2^p F \left(2^{-q} |x| \right) - F \left(|x| \right) \right\| \leq \vartheta 2^{p-q\alpha} \|x\|^\alpha . \tag{1.46}$$

Next, let us define the following functions:

1. $G : \mathscr{X} \to \mathscr{X}, \quad G(|x|) = 2^{-q}|x|$.
2. $\delta : \mathscr{X} \to [0, \infty), \quad \delta(|x|) = \vartheta 2^{p-q\alpha} \|x\|^\alpha$.
3. $\varphi : [0, \infty) \to [0, \infty), \quad \varphi(t) = 2^p t$.
4. $H : \mathscr{Y} \to \mathscr{Y}, \quad H(|y|) = 2^p |y|$.
5. $d(\cdot, \cdot) : \mathscr{Y} \times \mathscr{Y} \to [0, \infty), \quad d(y_1, y_2) = \|y_1 - y_2\|$.

We shall verify the fulfilment of all the three conditions of the Forti's theorem as follows.

(I) From inequality (1.46) we obviously have

$$d\left(H\left(F\left(G\left(|x|\right)\right)\right), F\left(|x|\right)\right) = \left\| 2^p F \left(2^{-q} |x| \right) - F \left(|x| \right) \right\|$$

$$\leq \vartheta 2^{p-q\alpha} \|x\|^\alpha$$

$$= \delta(|x|).$$

(II) $d\left(H\left(|y_1|\right), H\left(|y_2|\right)\right) = 2^p \|y_1 - y_2\| = \varphi\left(d\left(y_1, y_2\right)\right)$ for all $y_1, y_2 \in \mathscr{Y}$.

(III) Clearly, on the one hand φ is a non-decreasing subadditive function on the positive half line, and on other hand by applying Remark 1.4 on both the iterations G^j and φ^j of G and φ respectively, one can observe that

$$\sum_{j=0}^\infty \varphi^j \left(\delta \left(G^j \left(|x| \right) \right) \right) = \vartheta 2^{(p-q\alpha)} \|x\|^\alpha \sum_{j=0}^\infty 2^{(p-q\alpha)j}$$

$$= \vartheta \|x\|^\alpha \frac{2^p}{2^{q\alpha} - 2^p} < \infty.$$

Then in virtue of Forti's theorem sequence $\left(H^n \left(F \left(G^n |x|\right)\right)\right)_{n \in \mathbb{N}}$ is a Cauchy sequence for every $x \in \mathscr{X}$ and thus so is sequence $\left(2^{np} F \left(2^{-nq} |x|\right)\right)_{n \in \mathbb{N}}$ and furthermore, the mapping (1.41) is the unique functional which satisfies inequality (1.42).

Next, we prove that the mapping T, defined in (1.41), satisfies the functional equation (1.34). In fact, in (1.40) substitute x with $2^{-nq}x$ also y with $2^{-nq}y$, and fix arbitrarily $\tau \in [0, \infty)$. Then

$$\left\| F \left(2^{-nq} \left(\left(\tau^q |x| \right) \vee |y| \right) \right) - \left(\tau^p F \left(2^{-nq} |x| \right) \vee F \left(2^{-nq} |y| \right) \right) \right\|$$

$$\leq \vartheta 2^p 2^{-q\alpha n} \left(\|x\|^\alpha + \|y\|^\alpha \right).$$

Multiply both sides of this last inequality by 2^{np} to obtain

$$\left\| 2^{np} F \left(2^{-nq} \left(\left(\tau^q |x| \right) \vee |y| \right) \right) - 2^{np} \left(\tau^p F \left(2^{-nq} |x| \right) \vee F \left(2^{-nq} |y| \right) \right) \right\|$$
$$\leq \vartheta 2^p 2^{(p-q\alpha)n} \left(\|x\|^\alpha + \|y\|^\alpha \right). \tag{1.47}$$

Taking the limit in (1.47) we have via (1.41) that for all $\tau \in [0, \infty)$ and all $x, y \in \mathscr{X}$

$$\left\| T \left(\left(\tau^q |x| \right) \vee |y| \right) - \left(\tau^p T \left(|x| \right) \right) \vee T \left(|y| \right) \right\| = 0$$

which is equivalent to (1.34).

The moreover part can be proved the same way the moreover parts of the theorems in [5] were, after we will have proved that the limits on both sides of (1.43) exist. In fact, on the one hand, the existence of the limit on the left hand side follows from the combination of the monotonicity of F and (1.41). On the other hand, because of (1.41) the inner limit on the right hand side equals $T(x_k)$ for every $k \in \mathbb{N}$. But since the limit operator T is a join-homomorphism, it is also isotonic or increasing. Consequently, $(T(x_k))_{k \in \mathbb{N}}$ is a convergent sequence. We have thus proved that the limits on both sides of (1.43) exist. Therefore, we can conclude on the validity of the argument. $\qquad\square$

To end the section we shall provide some example showing that if in (1.40) parameter τ does not range over the whole non-negative half-line and the power α of the norms equals the ratio of p and q, then stability cannot always be guaranteed.

Example 1.6 Fix arbitrarily three numbers $p, q, c \in (0, \infty)$ and consider the function

$$F : \mathbb{R} \to \mathbb{R}, \ F(|x|) = c.$$

Then whenever $\tau \in (0, 1]$ we have:

$$\left| F \left(\left(\tau^q |x| \right) \vee |y| \right) - \left(\tau^p F \left(|x| \right) \right) \vee F \left(|y| \right) \right| = \left| c - \left(\tau^p c \right) \vee c \right| = 0 \leq |x|^\alpha + |y|^\alpha,$$

where $\alpha = \frac{p}{q}$. Since $|x| = \left(|x|^{\frac{1}{q}} \right)^q$, for any function $T : \mathbb{R} \to \mathbb{R}$ which solves Eq. (1.34) the following consecutive relations are true:

$$\sup_{|x| \in (0, \infty)} \frac{|F(|x|) - T(|x|)|}{|x|^\alpha} = \sup_{|x| \in (0, \infty)} \frac{\left| c - T \left(\left(|x|^{\frac{1}{q}} \right)^q \right) \right|}{|x|^\alpha}$$
$$= \sup_{|x| \in (0, \infty)} \frac{\left| c - |x|^\alpha T(1) \right|}{|x|^\alpha}$$
$$= \sup_{|x| \in (0, \infty)} \left| \frac{c}{|x|^\alpha} - T(1) \right| = \infty.$$

1.5   A More General Form of Problem 1.1 with Some Solutions

Consider the following functional equation [10]:

$$
\left(T\left(\left(\tau^q\,|x|\right)\Delta^*_{\mathscr{X}}\left(\eta^q\,|y|\right)\right)\right)\Delta^*_{\mathscr{Y}}\left(T\left(\left(\tau^q\,|x|\right)\Delta^{**}_{\mathscr{X}}\left(\eta^q\,|y|\right)\right)\right)
$$
$$
= \left(\tau^p T\left(|x|\right)\right)\Delta^{**}_{\mathscr{Y}}\left(\eta^p T\left(|y|\right)\right) \tag{1.48}
$$

for all $x,\,y\in\mathscr{X}$ and all $\tau,\,\eta\in[0,\,\infty)$, where $\Delta^*_{\mathscr{X}},\,\Delta^{**}_{\mathscr{X}}\in\{\wedge_{\mathscr{X}},\,\vee_{\mathscr{X}}\}$ and $\Delta^*_{\mathscr{Y}},\,\Delta^{**}_{\mathscr{Y}}\in\{\wedge_{\mathscr{Y}},\,\vee_{\mathscr{Y}}\}$ are fixed lattice operations, where $\mathscr{X}$ and $\mathscr{Y}$ are Banach lattices.

Remark 1.8 If we let $\eta=\tau$ and $y=x$ in Eq. (1.48), we obtain the Schwaiger's type functional equation in lattice environment [7], recalled as follows

$$
T\left(\tau^q\,|x|\right)=\tau^p T\left(|x|\right) \tag{1.49}
$$

for all $x\in\mathscr{X}$ and all $\tau\in[0,\,\infty)$.

The results in this section are straight generalizations of Agbeko [5, 6] and Salahi et al. [29].

We note that fetching for the unique solution of Eq. (1.48) in the sense of Ulam-Hyers-Aoki is equivalent to solving the problem hereafter.

Problem 1.2 Given three numbers $\varepsilon,\,p,\,q\in(0,\,\infty)$, two Riesz spaces G_1 and G_2 with G_2 being endowed with a metric $d(\cdot,\,\cdot)$, four lattice operations $\Delta^*_{G_1},\,\Delta^{**}_{G_1}\in\{\wedge_{G_1},\,\vee_{G_1}\}$ and $\Delta^*_{G_2},\,\Delta^{**}_{G_2}\in\{\wedge_{G_2},\,\vee_{G_2}\}$, does there exist some real number $\delta>0$ such that, if a mapping $F:G_1\to G_2$ satisfies the perturbation inequality

$$
d\left(\left(F\left(\left(\tau^q\,|x|\right)\Delta^*_{G_1}\left(\eta^q\,|y|\right)\right)\right)\Delta^*_{G_2}\left(F\left(\left(\tau^q\,|x|\right)\Delta^{**}_{G_1}\left(\eta^q\,|y|\right)\right)\right),\right.
$$
$$
\left.\left(\tau^p F\left(|x|\right)\right)\Delta^{**}_{G_2}\left(\eta^p F\left(|y|\right)\right)\right)\le\delta
$$

for all $x,\,y\in G_1$ and all $\tau,\,\eta\in[0,\,\infty)$, then an operation-preserving functional $T:G_1\to G_2$ exists with the property that

$$
d\left(T(x),\,F(x)\right)\le\varepsilon
$$

for all $x\in G_1$ and all $\tau,\,\eta\in[0,\,\infty)$?

Letting $\tau=\eta=1$, Problem 1.2 reduces to Problem 1.1, indeed.

1.5.1   The Main Results of This Section

Theorem 1.11 *Given a pair of real numbers* $(p, q) \in (0, \infty) \times (0, \infty)$, *consider a cone-related functional* $F : \mathcal{X} \to \mathcal{Y}$ *for which there are numbers* $\vartheta > 0$ *and* α *with* $q\alpha \in (p, \infty)$ *such that*

$$
\begin{aligned}
&\left\| F\left(\left(\tau^q |x|\right) \Delta^*_{\mathcal{X}} \left(\eta^q |y|\right)\right) \Delta^*_{\mathcal{Y}} F\left(\left(\tau^q |x|\right) \Delta^{**}_{\mathcal{X}} \left(\eta^q |y|\right)\right) \right. \\
&\left. \quad - \left(\tau^p F\left(|x|\right)\right) \Delta^{**}_{\mathcal{Y}} \left(\eta^p F\left(|y|\right)\right) \right\| \le 2^{(p-1)} \vartheta \left(\|x\|^\alpha + \|y\|^\alpha\right)
\end{aligned}
\tag{1.50}
$$

for all $x, y \in \mathcal{X}$ *and all* $\tau, \eta \in [0, \infty)$. *Then the sequence* $\left(2^{np} F\left(2^{-nq} |x|\right)\right)_{n \in \mathbb{N}}$ *is a Cauchy sequence for every* $x \in \mathcal{X}$. *Let the functional* $T : \mathcal{X}^+ \to \mathcal{Y}^+$ *be defined by*

$$
T\left(|x|\right) = \lim_{n \to \infty} 2^{np} F\left(2^{-nq} |x|\right)
\tag{1.51}
$$

for all $x \in \mathcal{X}$. *Then* T *both is a solution of* (1.48) *and uniquely satisfies inequality*

$$
\left\| T\left(|x|\right) - F\left(|x|\right)\right\| \le \frac{2^p \vartheta}{2^{q\alpha} - 2^p} \|x\|^\alpha
\tag{1.52}
$$

for every $x \in \mathcal{X}$.

Theorem 1.12 *Given a pair of real numbers* $(p, q) \in (0, \infty) \times (0, \infty)$, *consider a cone-related functional* $F : \mathcal{X} \to \mathcal{Y}$ *for which there are numbers* $\beta \in [0, \infty)$, $\vartheta > 0$ *and* α *with* $q\alpha \in (0, p)$ *such that*

$$
\begin{aligned}
&\left\| F\left(\left(\tau^q |x|\right) \Delta^*_{\mathcal{X}} \left(\eta^q |y|\right)\right) \Delta^*_{\mathcal{Y}} F\left(\left(\tau^q |x|\right) \Delta^{**}_{\mathcal{X}} \left(\eta^q |y|\right)\right) \right. \\
&\left. \quad - \left(\tau^p F\left(|x|\right)\right) \Delta^{**}_{\mathcal{Y}} \left(\eta^p F\left(|y|\right)\right) \right\| \le \beta + \vartheta 2^{-(p+1)} \left(\|x\|^\alpha + \|y\|^\alpha\right)
\end{aligned}
\tag{1.53}
$$

for all $x, y \in \mathcal{X}$ *and all* $\tau, \eta \in [0, \infty)$. *Then the sequence* $\left(2^{-np} F\left(2^{nq} |x|\right)\right)_{n \in \mathbb{N}}$ *is a Cauchy sequence for every fixed* $x \in \mathcal{X}$. *Let the functional* $T : \mathcal{X}^+ \to \mathcal{Y}^+$ *be defined by*

$$
T\left(|x|\right) = \lim_{n \to \infty} 2^{-np} F\left(2^{nq} |x|\right)
\tag{1.54}
$$

for all $x \in \mathcal{X}$. *Then* T *both is a solution of* (1.48) *and uniquely satisfies inequality*

$$
\left\| T\left(|x|\right) - F\left(|x|\right)\right\| \le \frac{\beta 2^p}{2^p - 1} + \frac{\vartheta \|x\|^\alpha 2^{q\alpha}}{2^p - 2^{q\alpha}}
\tag{1.55}
$$

for every $x \in \mathcal{X}$.

Remark 1.9 If the conditions of Theorem 1.11 or 1.12 hold true, then $F(0) = 0$.

Proof The proof is similar to its counterpart in [6, 7] under the conditions of Theorem 1.11 or 1.12 when $\beta = 0$. Under the condition of Theorem 1.12 with $\beta > 0$, we need to prove that $F(0) = 0$. Suppose in the contrary that $F(0) > 0$ were true. Then by letting $x = y = 0$ and $\eta = \tau$ in (1.53), inequality

$$\left\| F(0) - \tau^p F(0) \right\| \leq \beta$$

follows or equivalently

$$\left| \tau^p - 1 \right| \leq \frac{\beta}{\| F(0) \|} < \infty$$

which, as τ tends to infinity, would lead to an absurdity, indeed. Hence the relation $F(0) = 0$ must be true.

1.5.2　Proof of the Main Results

We shall use the technique in [6] to prove the main theorems.

Proof (of Theorem 1.11) First, if we choose $\tau = \eta = 2$, $y = x$ and replace x by $2^{-q}x$ in inequality (1.50) then we obviously have

$$\left\| 2^p F \left(2^{-q} |x| \right) - F(|x|) \right\| \leq \vartheta 2^{p-q\alpha} \| x \|^\alpha. \tag{1.56}$$

Next, let us define the following functions:

$$
\begin{array}{lll}
G : \mathscr{X}^+ \to \mathscr{X}^+, & G(|x|) = 2^{-q} |x|, & \text{for all } x \in \mathscr{X}^+ \\
\delta : \mathscr{X}^+ \to [0, \infty), & \delta(|x|) = \vartheta 2^{p-q\alpha} \| x \|^\alpha, & \text{for all } x \in \mathscr{X}^+ \\
\varphi : [0, \infty) \to [0, \infty), & \varphi(t) = 2^p t, & \text{for all } t \in [0, \infty) \\
H : \mathscr{Y}^+ \to \mathscr{Y}^+, & H(|y|) = 2^p |y|, & \text{for all } y \in \mathscr{Y}^+ \\
d : \mathscr{Y}^+ \times \mathscr{Y}^+ \to [0, \infty), & d(|y_1|, |y_2|) = \| |y_1| - |y_2| \|, & \text{for all } y_1, y_2 \in \mathscr{Y}^+.
\end{array}
$$

We shall verify the fulfilment of all the three conditions of the Forti's theorem as follows.

(I) From inequality (1.56) we obviously have

$$
\begin{aligned}
d\left(H(F(G(|x|))), F(|x|) \right) &= \left\| H(F(G(|x|))) - F(|x|) \right\| \\
&= \left\| 2^p F \left(2^{-q} |x| \right) - F(|x|) \right\| \\
&\leq \vartheta 2^{p-q\alpha} \| x \|^\alpha = \delta(|x|).
\end{aligned}
$$

(II) $d\left(H\left(|y_1|\right), H\left(|y_2|\right)\right) = 2^p \left\|\,|y_1| - |y_2|\,\right\| = \varphi\left(d\left(|y_1|, |y_2|\right)\right)$ for all $y_1, y_2 \in \mathscr{Y}$.

(III) Clearly, on the one hand φ is a non-decreasing subadditive function on the positive half line, and on the other hand by applying Remark 1.4 on both the iterations G^j and φ^j of G and φ respectively, one can observe that

$$\sum_{j=0}^{\infty} \varphi^j\left(\delta\left(G^j\left(|x|\right)\right)\right) = \vartheta\, 2^{p-q\alpha} \|x\|^\alpha \sum_{j=0}^{\infty} 2^{(p-q\alpha)j} = \vartheta\, \|x\|^\alpha \frac{2^p}{2^{q\alpha} - 2^p} < \infty.$$

Then in view of Forti's theorem, sequence $\left(H^n\left(F\left(G^n\,|x|\right)\right)\right)_{n\in\mathbb{N}}$ is a Cauchy sequence for every $x \in \mathscr{X}$ and thus so is sequence $\left(2^{np} F\left(2^{-nq}\,|x|\right)\right)_{n\in\mathbb{N}}$. Furthermore, the mapping (1.51) satisfies inequality (1.52).

Next, we prove that T solves (1.48). In fact, in (1.50) substitute x with $2^{-nq} x$ also y with $2^{-nq} y$, and fix arbitrarily $\tau, \eta \in [0, \infty)$. Then

$$\left\| F\left(\frac{(\tau^q\,|x|)\,\Delta_{\mathscr{X}}^*\,(\eta^q\,|y|)}{2^{nq}}\right) \Delta_{\mathscr{Y}}^* F\left(\frac{(\tau^q\,|x|)\,\Delta_{\mathscr{X}}^{**}\,(\eta^q\,|y|)}{2^{nq}}\right) \right.$$
$$\left. - \left(\tau^p F\left(\frac{|x|}{2^{nq}}\right)\right) \Delta_{\mathscr{Y}}^{**}\left(\eta^p F\left(\frac{|y|}{2^{nq}}\right)\right) \right\| \leq 2^{(p-1)}\vartheta \left(\left\|\frac{x}{2^{nq}}\right\|^\alpha + \left\|\frac{y}{2^{nq}}\right\|^\alpha\right).$$

Multiplying both sides of this last inequality by 2^{np} yields

$$2^{np}\left\| F\left(\frac{(\tau^q\,|x|)\,\Delta_{\mathscr{X}}^*\,(\eta^q\,|y|)}{2^{nq}}\right) \Delta_{\mathscr{Y}}^* F\left(\frac{(\tau^q\,|x|)\,\Delta_{\mathscr{X}}^{**}\,(\eta^q\,|y|)}{2^{nq}}\right) \right.$$
$$\left. - \left(\tau^p F\left(\frac{|x|}{2^{nq}}\right)\right) \Delta_{\mathscr{Y}}^{**}\left(\eta^p F\left(\frac{|y|}{2^{nq}}\right)\right) \right\| \leq \frac{\vartheta}{2^{(1-p)}} \frac{\|x\|^\alpha + \|y\|^\alpha}{2^{n(q\alpha - p)}}. \tag{1.57}$$

Taking the limit in (1.57) we have via (1.51) that

$$\left\| T\left((\tau^q\,|x|)\,\Delta_{\mathscr{X}}^*\,(\eta^q\,|y|)\right) \Delta_{\mathscr{Y}}^* T\left((\tau^q\,|x|)\,\Delta_{\mathscr{X}}^{**}\,(\eta^q\,|y|)\right) \right.$$
$$\left. - \left(\tau^p T\left(|x|\right)\right) \Delta_{\mathscr{Y}}^{**}\left(\eta^p T\left(|y|\right)\right) \right\| = 0$$

for all $\tau, \eta \in [0, \infty)$ and all $x, y \in \mathscr{X}$, which is equivalent to (1.48). Thus T also satisfies (1.49) in Remark 1.8. Finally, we show the uniqueness, using a technique in [29]. In fact, assume that there is another functional $S : \mathscr{X} \to \mathscr{Y}$ which satisfies (1.48) and the inequality

$$\|S\left(|x|\right) - F\left(|x|\right)\| \leq \delta_2 \|x\|^{\alpha_2}$$

for some numbers $\alpha_2, \delta_2 \in (0, \infty)$ with $q\alpha_2 > p$, and for all $x \in \mathscr{X}$. In (1.52) let

$$\delta_1 := \frac{2^p\,\vartheta}{2^{q\alpha} - 2^p}, \quad \alpha_1 := \alpha$$

and by choosing $\tau = 2^{-n}$ in Eq. (1.49) one can observe that for all $x \in \mathscr{X}$

$$
\begin{aligned}
\|S\,(|x|) - T\,(|x|)\| &= 2^{np}\,\left\|S\left(2^{-nq}\,|x|\right) - T\left(2^{-nq}\,|x|\right)\right\| \\
&\leq 2^{np}\,\left\|F\left(2^{-nq}\,|x|\right) - T\left(2^{-nq}\,|x|\right)\right\| \\
&\quad + 2^{np}\,\left\|S\left(2^{-nq}\,|x|\right) - F\left(2^{-nq}\,|x|\right)\right\| \\
&\leq 2^{np}\delta_1 \left\|2^{-nq}x\right\|^{\alpha_1} + 2^{np}\delta_2 \left\|2^{-nq}x\right\|^{\alpha_2} \\
&= 2^{(p-q\alpha_1)n}\delta_1 \|x\|^{\alpha_1} + 2^{(p-q\alpha_2)n}\delta_2 \|x\|^{\alpha_2} .
\end{aligned}
$$

Hence,

$$
\|S\,(|x|) - T\,(|x|)\| \leq 2^{(p-q\alpha_1)n}\delta_1 \|x\|^{\alpha_1} + 2^{(p-q\alpha_2)n}\delta_2 \|x\|^{\alpha_2}
$$

which, in the limit, yields

$$
\|S\,(|x|) - T\,(|x|)\| = 0
$$

or equivalently

$$
S\,(|x|) = T\,(|x|)
$$

for all $x \in \mathscr{X}$.

This was to be proven.

Proof (of Theorem 1.12) First, if we choose $\tau = \eta = 2^{-1}$, $y = x$ and replace x by $2^q x$ in inequality (1.53) then we obviously have

$$
\left\|2^{-p}F\left(2^q\,|x|\right) - F\,(|x|)\right\| \leq \beta + \vartheta 2^{q\alpha-p}\,\|x\|^{\alpha} . \tag{1.58}
$$

Next, let us define the following functions:

$$
\begin{array}{lll}
G : \mathscr{X}^+ \to \mathscr{X}^+, & G\,(|x|) = 2^q\,|x|, & \text{for all } x \in \mathscr{X}^+ \\
\delta : \mathscr{X}^+ \to [0,\infty), & \delta\,(|x|) = \beta + \vartheta 2^{q\alpha-p}\,\|x\|^{\alpha}, & \text{for all } x \in \mathscr{X}^+ \\
\varphi : [0,\infty) \to [0,\infty), & \varphi\,(t) = 2^{-p}t, & \text{for all } t \in [0,\infty) \\
H : \mathscr{Y}^+ \to \mathscr{Y}^+, & H\,(|y|) = 2^{-p}\,|y|, & \text{for all } y \in \mathscr{Y}^+ \\
d : \mathscr{Y}^+ \times \mathscr{Y}^+ \to [0,\infty), & d\,(|y_1|,\,|y_2|) = \||y_1| - |y_2|\|, & \text{for all } y_1, y_2 \in \mathscr{Y}^+.
\end{array}
$$

We shall verify the fulfilment of all the three conditions of the Forti's theorem as follows.

(I) From inequality (1.58) we obviously have

$$d\left(H\left(F\left(G\left(|x|\right)\right)\right),\, F\left(|x|\right)\right) = \left\|H\left(F\left(G\left(|x|\right)\right)\right) - F\left(|x|\right)\right\|$$

$$= \left\|2^{-p}F\left(2^{q}\,|x|\right) - F\left(|x|\right)\right\|$$

$$\leq \beta + \vartheta 2^{q\alpha-p}\,\|x\|^{\alpha} = \delta\left(|x|\right).$$

(II) $d\left(H\left(|y_1|\right),\, H\left(|y_2|\right)\right) = 2^{-p}\left\|\,|y_1| - |y_2|\,\right\| = \varphi\left(d\left(|y_1|,\, |y_2|\right)\right)$ for all $y_1,\, y_2 \in \mathcal{Y}$.

(III) Clearly, on the one hand φ is a non-decreasing subadditive function on the positive half line, and on the other hand by applying Remark 1.4 on both the iterations G^j and φ^j of G and φ respectively, one can observe that

$$\sum_{j=0}^{\infty}\varphi^{j}\left(\delta\left(G^{j}\left(|x|\right)\right)\right) = \beta\sum_{j=0}^{\infty}2^{-pj} + \vartheta 2^{q\alpha-p}\,\|x\|^{\alpha}\sum_{j=0}^{\infty}2^{(q\alpha-p)j}$$

$$= \frac{\beta 2^{p}}{2^{p}-1} + \frac{\vartheta\,\|x\|^{\alpha}\,2^{q\alpha}}{2^{p}-2^{q\alpha}} < \infty.$$

Then in view of Forti's theorem, sequence $\left(H^{n}\left(F\left(G^{n}\,|x|\right)\right)\right)_{n\in\mathbb{N}}$ is a Cauchy sequence for every $x \in \mathcal{X}$ and thus so is sequence $\left(2^{-np}F\left(2^{nq}\,|x|\right)\right)_{n\in\mathbb{N}}$. Furthermore, the mapping (1.54) satisfies inequality (1.55).

Next, we prove that T solves (1.48). In fact, in (1.53) substitute x with $2^{nq}x$ also y with $2^{nq}y$, and fix arbitrarily $\tau,\, \eta \in [0,\,\infty)$. Then

$$\left\|F\left(2^{nq}\left(\left(\tau^{q}\,|x|\right)\Delta_{\mathcal{X}}^{*}\left(\eta^{q}\,|y|\right)\right)\right)\Delta_{\mathcal{Y}}^{*}F\left(2^{nq}\left(\left(\tau^{q}\,|x|\right)\Delta_{\mathcal{X}}^{**}\left(\eta^{q}\,|y|\right)\right)\right)\right.$$

$$\left. - \left(\tau^{p}F\left(2^{nq}\,|x|\right)\right)\Delta_{\mathcal{Y}}^{**}\left(\eta^{p}F\left(2^{nq}\,|y|\right)\right)\right\|$$

$$\leq \beta + 2^{-(p+1)}\vartheta\left(\left\|2^{nq}x\right\|^{\alpha} + \left\|2^{nq}y\right\|^{\alpha}\right).$$

Dividing both sides of this last inequality by 2^{np} yields

$$\left\|\frac{F\left(2^{nq}\left(\left(\tau^{q}\,|x|\right)\Delta_{\mathcal{X}}^{*}\left(\eta^{q}\,|y|\right)\right)\right)\Delta_{\mathcal{Y}}^{*}F\left(2^{nq}\left(\left(\tau^{q}\,|x|\right)\Delta_{\mathcal{X}}^{**}\left(\eta^{q}\,|y|\right)\right)\right)}{2^{np}}\right.$$

$$\left. - \frac{\left(\tau^{p}F\left(2^{nq}\,|x|\right)\right)\Delta_{\mathcal{Y}}^{**}\left(\eta^{p}F\left(2^{nq}\,|y|\right)\right)}{2^{np}}\right\| \tag{1.59}$$

$$\leq \beta 2^{-np} + 2^{-(p+1)}\vartheta\left(\|x\|^{\alpha} + \|y\|^{\alpha}\right)2^{(q\alpha-p)n}.$$

Taking the limit in (1.59) we have via (1.54) that

$$\left\|T\left(\left(\tau^{q}\,|x|\right)\Delta_{\mathcal{X}}^{*}\left(\eta^{q}\,|y|\right)\right)\Delta_{\mathcal{Y}}^{*}T\left(\left(\tau^{q}\,|x|\right)\Delta_{\mathcal{X}}^{**}\left(\eta^{q}\,|y|\right)\right)\right.$$

$$\left. - \left(\tau^{p}T\left(|x|\right)\right)\Delta_{\mathcal{Y}}^{**}\left(\eta^{p}T\left(|y|\right)\right)\right\| = 0$$

for all τ, $\eta \in [0, \infty)$ and all x, $y \in \mathscr{X}$, which is equivalent to (1.48). Thus T satisfies (1.49) in Remark 1.8. Finally, we show the uniqueness, using a technique in [29]. In fact, assume that there is another functional $S : \mathscr{X} \to \mathscr{Y}$ which satisfies (1.48) and the inequality

$$\| S(|x|) - F(|x|) \| \leq \beta_2 + \delta_2 \|x\|^{\alpha_2}$$

for some numbers α_2, $\delta_2 \in (0, \infty)$, $\beta_2 \in [0, \infty)$ with $q\alpha_2 < p$, and for all $x \in \mathscr{X}$. In (1.55) let

$$\beta_1 := \frac{\beta 2^p}{2^p - 1}, \quad \delta_1 := \frac{\vartheta 2^{qa}}{2^p - 2^{qa}}, \quad \alpha_1 := \alpha$$

and by choosing $\tau = 2^n$ in Eq. (1.49) one can observe that for all $x \in \mathscr{X}$

$$
\begin{aligned}
\| S(|x|) - T(|x|) \| &= 2^{-np} \left\| S\left(2^{nq}|x|\right) - T\left(2^{nq}|x|\right) \right\| \\
&\leq 2^{-np} \left\| F\left(2^{nq}|x|\right) - T\left(2^{nq}|x|\right) \right\| \\
&\quad + 2^{-np} \left\| S\left(2^{nq}|x|\right) - F\left(2^{nq}|x|\right) \right\| \\
&\leq 2^{-np} \left(\beta_1 + \delta_1 \left\|2^{nq}x\right\|^{\alpha_1}\right) + 2^{-np} \left(\beta_2 + \delta_2 \left\|2^{nq}x\right\|^{\alpha_2}\right) \\
&= 2^{-np} (\beta_1 + \beta_2) + \delta_1 2^{(q\alpha_1 - p)n} \|x\|^{\alpha_1} + \delta_2 2^{(q\alpha_2 - p)n} \|x\|^{\alpha_2}.
\end{aligned}
$$

Hence

$$\| S(|x|) - T(|x|) \| \leq 2^{-np} (\beta_1 + \beta_2) + \delta_1 2^{(q\alpha_1 - p)n} \|x\|^{\alpha_1} + \delta_2 2^{(q\alpha_2 - p)n} \|x\|^{\alpha_2}$$

which, in the limit, yields

$$\| S(|x|) - T(|x|) \| = 0$$

or equivalently

$$S(|x|) = T(|x|)$$

for all $x \in \mathscr{X}$. This completes the proof. $\qquad\square$

To end the section we give an example showing that stability fails to occur in general.

Example 1.7 Fix arbitrarily τ, $\eta \in (0, 2)$ and consider the function

$$F : [0, \infty) \to [0, \infty), \quad F(x) = x^{\alpha + 1}, \quad \alpha = \frac{p}{q}.$$

Since F is increasing the first equality in the chain below is valid, entailing the subsequent relations:

$$\left| F\left(\left(\tau^q x\right) \vee \left(\eta^q y\right)\right) - \left(\tau^p F\left(x\right)\right) \wedge \left(\eta^p F\left(y\right)\right) \right|$$

$$= \left| \left(\tau^q x\right)^{\alpha+1} \vee \left(\eta^q y\right)^{\alpha+1} - \left(\tau^p x^{\alpha+1}\right) \wedge \left(\eta^p y^{\alpha+1}\right) \right|$$

$$\leq \left(\tau^q x\right)^{\alpha+1} \vee \left(\eta^q y\right)^{\alpha+1} + \left(\tau^p x^{\alpha+1}\right) \wedge \left(\eta^p y^{\alpha+1}\right)$$

$$\leq \left(2^q x\right)^{\alpha+1} \vee \left(2^q y\right)^{\alpha+1} + \left(2^p x^{\alpha+1}\right) \wedge \left(2^p y^{\alpha+1}\right)$$

$$\leq 2^{p+q}\left(x^{\alpha+1} \vee y^{\alpha+1}\right) + 2^{p+q}\left(x^{\alpha+1} \wedge y^{\alpha+1}\right) = 2^{p+q}\left(x^{\alpha+1} + y^{\alpha+1}\right)$$

for all $x, y \in [0, \infty)$. Now, let $T : [0, \infty) \to [0, \infty)$ be a function such that $T\left(\mu^q x\right) = \mu^p T\left(x\right)$ for all $x \in [0, \infty)$ and all $\mu \in [0, \infty)$. Since $x = \left(x^{1/q}\right)^q$, and α is the ratio of p and q, we can then note that $T\left(x\right) = x^\alpha T\left(1\right)$ for every $x \in [0, \infty)$. Now,

$$\sup_{x \in (0, \infty)} \frac{|F\left(x\right) - T\left(x\right)|}{2^{p+q} x^{\alpha+1}} = \sup_{x \in (0, \infty)} \frac{\left| x^{\alpha+1} - T\left(\left(x^{\frac{1}{q}}\right)^q\right) \right|}{2^{p+q} x^{\alpha+1}} =$$

$$= \sup_{x \in (0, \infty)} \frac{\left| x^{\alpha+1} - x^\alpha T\left(1\right) \right|}{2^{p+q} x^{\alpha+1}}$$

$$= \frac{1}{2^{p+q}} \sup_{x \in (0, \infty)} \left| 1 - \frac{T\left(1\right)}{x} \right| = \infty.$$

The above example about the lack of stability on the real line in lattice environments is the counterpart of the example given by Czerwik [12] in the addition environments for quadratic mappings.

1.6 Lattice-Valued Maps Defined on Semigroups

This section is the collection of notions and results in [9]. Order theory plays an important role in many disciplines of computer science and engineering. For example, it has applications in distributed computing (vector clocks, global predicate detection), cryptography, programming language semantics (fixed-point semantics), and data mining (concept analysis). Moreover, it is useful in other disciplines of mathematics such as logic, combinatorics, number theory or measure theory. Also the semigroup theory is an integral part of modern mathematics, with connections and applications across a broad spectrum of areas such as theoretical physics, computer sciences, control engineering, information sciences, coding theory, topological spaces.

Algebraic semigroups are defined in very simple terms: they are algebraic varieties endowed with a composition law which is associative and a morphism of varieties. We omit its formal definition together with its additional properties, as well known. For basic definitions of ordered structures and related notions which are used throughout the paper the reader is referred to Davey and Priestley [13] and Schaefer [30].

We adopt the following notations: endpoints of open and closed, or half-open or half-closed intervals are denoted by round and square brackets, respectively.

Proceeding to the main part let us assume that $(S, \star)$ is a semigroup, $(L, \le)$ is a lattice and $T : S \to L$ is a lattice-valued mapping. We denote by L^+ the positive cone of L. Inspired by the notion of optimal averages we ask about morphisms between the algebraic structure of G and the order structure of L. To be precise, we are interested in the following functional equation

$$T(x \star y) = T(x) \vee T(y), \quad x, y \in S \tag{1.60}$$

and in its related functional inequalities

$$T(x \star y) \ge T(x) \vee T(y), \quad x, y \in S \tag{1.61}$$

and

$$T(x \star y) \le T(x) \vee T(y), \quad x, y \in S. \tag{1.62}$$

Clearly, (1.61) implies

$$T(x \star y) \ge T(x), \quad T(x \star y) \ge T(y), \quad x, y \in S \tag{1.63}$$

whereas (1.62) implies

$$T(x^2) \le T(x), \quad x \in S, \tag{1.64}$$

respectively. In fact, (1.61) and (1.63) are equivalent. However, inequalities (1.63) and (1.64) do not require the lattice structure of the target space, and they make sense if L is a partially ordered set (poset for short).

Note also that from (1.60) it easily follows that

$$T(x^n) = T(x), \quad x \in S, n \in \mathbb{N}, \tag{1.65}$$

where the powers x^n are defined inductively as follows: $x^1 = x$ and $x^{n+1} = x^n \star x$ for $n \in \mathbb{N}$.

1.6.1 Examples and Basic Properties of Solutions of (1.60), (1.61) and (1.62)

Example 1.8 Fix arbitrarily an element $\omega \in L^+$ and let $(S, \star)$ be any semigroup and $T_\omega: S \to L^+$ be the mapping defined by $T_\omega(x) = \omega$. Then T_ω trivially satisfies the functional equation (1.60).

Example 1.9 Let $n \in \mathbb{N}$ be fixed and $\mathbf{P_n} := \{p_0, p_1, p_2, \ldots, p_{n-1}\}$ any set of n elements. Let us define on the set $\mathbf{P_n}$ the binary operation $\star: \mathbf{P_n} \times \mathbf{P_n} \to \mathbf{P_n}$ by $p_k \star p_m = p_{k \vee m}$. Then one can easily see that $\mathbf{P_n}$ is a commutative idempotent semigroup with the unity element p_0, which is also the unique invertible element of $\mathbf{P_n}$. Define the functional $T: \mathbf{P_n} \to \mathbb{N}$ by $T(p_k) = k$. Then

$$T(p_k \star p_m) = T(p_{k \vee m}) = k \vee m = T(p_k) \vee T(p_m)$$

for all $k, m \in \{0, 1, \ldots, n-1\}$, i.e. T satisfies Eq. (1.60).

One can easily extend this example to an infinite set. Let $\mathbf{P_\infty} := \{p_k : k \in \mathbb{N}_0\}$ be any sequence of arbitrary pairwise distinct elements furnished $\mathbf{P_\infty}$ with the binary operation $\star$ defined as above. Then $\mathbf{P_\infty}$ is a commutative idempotent semigroup with a unity element p_0, which is also the only invertible element in $\mathbf{P_\infty}$ and functional T defined as previously solves (1.60).

Since lattice operations are associative, then trivially every lattice is a commutative semigroup with operation "$\star$" taken as "$\vee$". A natural question is to know what relation there is between operations "$\star$" and "$\vee$". A very short partial answer is as follows.

Remark 1.10 Assume that S is a net $S = \{x_i : i \in I\}$, where I is a linearly ordered set of indices, and define on S the following operation $\star: S \times S \to S$ by $x_i \star x_j = x_{i \vee j}$ for $i, j \in I$. Then $(S, \star)$ is a semigroup. Moreover, one can introduce an order $\prec$ on S by $x_i \prec x_j$ if $i < j$. We will show that both operations $\star$ and $\vee$ coincide if and only if S is monotone increasing, i.e. $x_i \prec x_j$ whenever $i < j$. Indeed, if $x_i \prec x_j$ for $i < j$, then $x_i \star x_j = x_{i \vee j} = x_i \vee x_j$, so "$\star$" and "$\vee$" are identical operations. Similarly, if "$\star$" and "$\vee$" coincide, then we get $x_{i \vee j} = x_i \vee x_j$ for all $i, j \in I$, which implies that $x_i \prec x_j$ for $i < j$.

Definition 1.1 Let (S, τ) be a partially ordered set. Consider the mappings $M_\tau: S \to 2^S$ and $N_\tau: S \to 2^S$, where 2^S is the power set of S, which are defined by

$$M_\tau(q) = \{p \in S : (p, q) \in \tau\} \quad \text{and} \quad N_\tau(p) = \{q \in S : (p, q) \in \tau\}.$$

Note that $M_\tau(x) \subseteq M_\tau(y)$, and $N_\tau(y) \subseteq N_\tau(x)$ for all $x, y \in S$ with $(x, y) \in \tau$.

Example 1.10 Let the set $S := \{r \in \mathbb{Q} : r \geq 1\}$ be endowed with the standard multiplication operation and the standard order $\leq$, which we will denote by τ.

Further, let $(p_k)_{k\in\mathbb{N}}$ be a sequence of elements of S containing 1. Then we have easily:

(a) $M_\tau(x) \cup M_\tau(y) \subseteq M_\tau(xy)$ for all x, $y \in S$, where the sets M_τ are defined in Definition 1.1.

(b) If $x > 1$ and $y > 1$, then $xy \in M_\tau(xy)$ and $xy \notin M_\tau(x) \cup M_\tau(y)$. Therefore $M_\tau(x) \cup M_\tau(y)$ is a proper subset of $M_\tau(xy)$.

(c) Let $x \in S$. Then $M_\tau(x^2) = M_\tau(x)$ if and only if $x = 1$.

(d) If x, $y \in S$, then $M_\tau(xy) = M_\tau(x)$ if and only if $y = 1$.

(e) Assume that $(\alpha_k) \subset [0, \infty)$ is a bounded sequence of distinct numbers. Let the functional $T_1 : S \to [0, \infty)$ be defined by

$$T_1(x) = \sup\{\alpha_k : p_k \in M_\tau(x)\}, \quad x \in S.$$

Since the sequence $(p_k)_{k\in\mathbb{N}}$ contains 1, then the set $\{\alpha_k : p_k \in M_\tau(x)\}$ is always non-empty and therefore T_1 is well-defined. Moreover T_1 solves functional inequality (1.61).

(f) Let $(\beta_k) \subset (0, \infty)$ be a bounded sequence of distinct numbers such that

$$\inf\{\beta_k : k \in \mathbb{N}\} > 0.$$

Let us define the functional $T_2 : S \to (0, \infty)$ by

$$T_2(x) = \frac{1}{\inf\{\beta_k : p_k \in M_\tau(x)\}}, \quad x \in S.$$

Then T_2 is well-defined and solves functional inequality (1.62).

One can easily check that analogous statements can be obtained for operators defined analogously to T_1 and T_2 with the sets M_ρ replaced by the sets N_ρ.

One can modify this example using as the order τ the divisibility relation on the set of natural numbers.

Now, we provide some basic properties of the solutions of inequalities (1.63).

Proposition 1.1 *Let $(S, \star)$ be a semigroup with a unit e and let L be a partially ordered set. Further let $T : S \to L$ satisfy (1.63). Then*

(i) $T(x) \geq T(e)$ *for all $x \in S$,*

(ii) $T(x) = T(x^{-1}) = T(e)$ *for all $x \in S$ which are invertible,*

(iii) *in particular, if S is a group, then every solution of the functional equation (1.60) on S is a constant mapping.*

Proof Point (i) follows directly from the second part of (1.63) applied for $y = e$.

Next, assume that $x \in S$ is invertible. Then one can easily see that with the aid of (i) we have

$$T(x) \geq T(e) = T(x \star x^{-1}) \geq T(x),$$

which proves (ii). Note that the third part is an immediate consequence of part (ii). □

Let Ω be a nonempty set. We will consider the space $\mathbb{R}^{\Omega}$ of real functions defined on Ω with the pointwise multiplication. Let also $(L, \leq)$ be a poset and $(L^+, \leq)$ its positive cone. Further, denote

$$\ker (f) := \{\omega \in \Omega : f(\omega) = 0\}$$

whenever $f \in \mathbb{R}^{\Omega}$. We introduce on $\mathbb{R}^{\Omega}$ two relations "$\leq_{\ker}$" and "$=_{\ker}$" as follows. For $f, g \colon \Omega \to \mathbb{R}$ let us define:

(i) $f \leq_{\ker} g$ if and only if $\ker (f) \subseteq \ker (g)$,
(ii) $f =_{\ker} g$ if and only if $\ker (f) = \ker (g)$.

The next remark is straightforward.

Remark 1.11 For all $f, g, h \colon \Omega \to \mathbb{R}$ we have:

(i) $f \leq_{\ker} f$,
(ii) if $f \leq_{\ker} g$ and $g \leq_{\ker} f$, then $f =_{\ker} g$,
(iii) if $f \leq_{\ker} g$ and $g \leq_{\ker} h$, then $f \leq_{\ker} h$.

Lemma 1.5 *For an arbitrary mapping $\mu \colon 2^{\Omega} \to L^+$ the following two assertions are equivalent:*

(i) $\mu (\ker (f)) \leq \mu (\ker (g))$ *for all* $f, g \in \mathbb{R}^{\Omega}$ *with* $f \leq_{\ker} g$,
(ii) $\mu (A) \leq \mu (B)$ *for all* $A, B \subseteq \Omega$ *with* $A \subseteq B$.

Proof Note that the implication (ii) $\implies$ (i) is obvious. To show the reverse implication (i) $\implies$ (ii), let us assume that $\mu (\ker (f)) \leq \mu (\ker (g))$ for all $f, g \in \mathbb{R}^{\Omega}$ with $f \leq_{\ker} g$. Pick arbitrarily two sets $A, B \subseteq \Omega$ with $A \subseteq B$. Then (since $B^c \subseteq A^c$, i.e. the complement of B is a subset of the complement of A) it can be easily seen that

$$A = \ker \left(\chi_{A^c}\right) \subseteq \ker \left(\chi_{B^c}\right) = B,$$

where χ denotes the characteristic function of a set. Consequently,

$$\mu (A) = \mu \left(\ker \left(\chi_{A^c}\right)\right) \leq \mu \left(\ker \left(\chi_{B^c}\right)\right) = \mu (B),$$

which completes the proof. □

Proposition 1.2 *Endow the set $\mathbb{R}^{\Omega}$ with the relation $\leq_{\ker}$ and the pointwise multiplication operation. Fix $\mu \colon 2^{\Omega} \to L^+$ and let $T \colon \mathbb{R}^{\Omega} \to L^+$ be a mapping defined by*

$$T (f) = \mu (\ker (f)), \quad f \in \mathbb{R}^{\Omega}.$$

Then the following two assertions are equivalent:

(i) $T(f) \leq T(g)$ *for all* f, $g \in \mathbb{R}^{\Omega}$ *with* $f \leq_{\mathrm{ker}} g$,

(i) $\mu(A) \leq \mu(B)$ *for all* A, $B \subseteq \Omega$ *with* $A \subseteq B$.

Moreover, T solves the functional inequality (1.61).

Proof Note that the first part can be easily derived from Lemma 1.5, and the second part is also immediate from the elementary fact that

$$\ker(f \cdot g) = \ker(f) \cup \ker(g).$$

$\square$

1.6.2 Separation Theorems in Lattice Environments

The main results of the present section (cf. [8]) essentially involve the topic of separation (Theorems 1.13 and 1.14) and the Ulam-type stability problem (Theorem 1.16).

Separation theorems have been studied by several authors. A classical result is the Mazur-Orlicz Theorem [26], which was generalized by Kaufman [21] and by Kranz [23]. In 1978 Rodé [28] proved a far reaching generalization of the Hahn-Banach Theorem, which presently is a powerful tool in the theory of functional equations and inequalities. König [22] found a simpler proof of the Rodé's Theorem. Gajda and Kominek [17] presented another approach, which motivated our next two theorems.

In the first two results of this section we deal with the separation problem for inequalities (1.61) and (1.62). To be precise, we ask whether given two solutions of the reverse inequalities can be separated by a solution of the equation.

Theorem 1.13 *Let us be given a σ-continuous lattice L and a multiplicative semigroup $(S, \star)$ which is commutative and has no elements of finite order, i.e. if $x \in S$, then there is no number $n \geq 2$ for which $x^n = x$. Further, let f, $g: S \rightarrow L$ be functionals for which*

$$g(x \star y) \geq g(x) \vee g(y)$$

and

$$f(x \star y) \leq f(x) \vee f(y)$$

for all x, $y \in S$. Suppose that $g(x) \leq f(x)$ and

$$\lim_{n \to \infty} g\left(x^{2^n}\right) = \lim_{n \to \infty} f\left(x^{2^n}\right)$$

for every $x \in S$. Then there is a functional $a: G \to L$ such that

(i) $g(x) \le a(x) \le f(x)$, *for all $x \in S$,*
(ii) $a(x \star y) = a(x) \vee a(y)$, *for all $x, y \in S$.*

Moreover, the functional $a: S \to L$ which meets conditions (i) and (ii) is unique.

First, we prove the two separation theorems. The following two lemmas are crucial in their proofs.

Lemma 1.6 *Let $(S, \star)$ be a commutative multiplicative semigroup and L a σ-continuous lattice. Let $g: S \to L$ be a functional for which*

$$g(x \star y) \ge g(x) \vee g(y)$$

for all $x, y \in S$. Then, for every $x \in G$ the sequence $\left(g\left(x^{2^n}\right)\right)_{n \in \mathbb{N}}$ is increasing, so that the limit

$$\lim_{n \to \infty} g\left(x^{2^n}\right) = \bigvee_{n=1}^{\infty} g\left(x^{2^n}\right)$$

exists.

Lemma 1.7 *Let $(S, \star)$ be a commutative multiplicative semigroup and L a σ-continuous lattice. Let $f: S \to L$ be a functional for which*

$$f(x \star y) \le f(x) \vee f(y)$$

for all $x, y \in S$. Then, for every $x \in S$ the sequence $\left(f\left(x^{2^n}\right)\right)_{n \in \mathbb{N}}$ is decreasing, so that the limit

$$\lim_{n \to \infty} f\left(x^{2^n}\right) = \bigwedge_{n=1}^{\infty} f\left(x^{2^n}\right)$$

exists.

The proofs of Lemmas 1.6 and 1.7 are omitted because they can be easily carried out.

Proof (of Theorem 1.13) Combine the three functional inequalities from assumption of Theorem 1.13 to see that

$$g(x) \vee g(y) \le g(x \star y) \le f(x \star y) \le f(x) \vee f(y) \tag{1.66}$$

for all $x, y \in S$. One can thus easily observe that

$$g(x) \le g\left(x^{2^n}\right) \le f\left(x^{2^n}\right) \le f(x)$$

for all $x \in S$ and $n \in \mathbb{N}$. Then applying Lemmas 1.6 and 1.7 it ensues that

$$g\left(x\right) \leq \lim_{n \to \infty} g\left(x^{2^n}\right) \leq \lim_{n \to \infty} f\left(x^{2^n}\right) \leq f\left(x\right) \tag{1.67}$$

for all $x \in S$. Now, due to our assumptions a mapping $a: S \to L$ can be defined by

$$a\left(x\right) := \lim_{n \to \infty} g\left(x^{2^n}\right) = \lim_{n \to \infty} f\left(x^{2^n}\right)$$

for all $x \in S$. Then (1.67) yields

$$g\left(x\right) \leq a\left(x\right) \leq f\left(x\right)$$

for all $x \in S$. Next, fix arbitrarily x, $y \in S$ and $n \in \mathbb{N}$. Then in (1.66) replace x with x^{2^n} and y with y^{2^n} to observe that

$$g\left(x^{2^n}\right) \vee g\left(y^{2^n}\right) \leq g\left(x^{2^n} \star y^{2^n}\right) \leq f\left(x^{2^n} \star y^{2^n}\right) \leq f\left(x^{2^n}\right) \vee f\left(y^{2^n}\right).$$

By applying the commutativity and passing to the limit in the above chain of inequalities, both Lemmas 1.6 and 1.7 entail

$$a\left(x\right) \vee a\left(y\right) \leq a\left(x \star y\right) \leq a\left(x\right) \vee a\left(y\right).$$

Therefore,

$$a\left(x \star y\right) = a\left(x\right) \vee a\left(y\right)$$

whenever x, $y \in S$.

To end the proof suppose that b is another mapping which satisfies conditions (i) and (ii) of the theorem and fix an arbitrary $x \in S$. Since b also satisfies (1.65) it is easy to see that

$$g\left(x^{2^n}\right) \leq b\left(x\right) = b\left(x^{2^n}\right) \leq f\left(x^{2^n}\right)$$

for all $n \in \mathbb{N}$. Then passing to the limit entails that

$$\lim_{n \to \infty} g\left(x^{2^n}\right) \leq b\left(x\right) \leq \lim_{n \to \infty} f\left(x^{2^n}\right).$$

Consequently, by the assumption, $b\left(x\right) = a\left(x\right)$, which completes the proof.   □

Theorem 1.14 *Let $(S, \star)$ be an Abelian group and $(L, \leq)$ a lattice. Further, suppose $g: S \to L$ and $f: G \to L$ are two mappings for which inequalities (1.61) and (1.62) are met respectively. Then*

(i) $g(x) = g(e)$ *for every* $x \in S$,

(ii) $f(e) \le f(x) \vee f(x^{-1})$ *for every* $x \in S$. *Moreover, given any* $x \in S$, *if* $f(x) = f(x^{-1})$, *then* $f(e) \le f(x)$,

(iii) $f(e) \ge f(x) \vee f(x^{-1})$ *for all* $x \in S$ *if and only if* $f(x) = f(e)$ *for all* $x \in S$.

Furthermore, suppose that $g(x) \le f(x)$ *and* $f(x) = f(x^{-1})$ *for every* $x \in S$. *Then the functionals* f *and* g *can be separated by a constant function, i.e. there exist some* $\beta \in L$ *such that*

$$g(x) \le \beta \le f(x)$$

for all $x \in S$.

Proof In (1.61) replace T by g and simultaneously y by e and x^{-1} to get that

$$g(x) \vee g(e) \le g(x),$$

respectively

$$g(x) \vee g(x^{-1}) \le g(e).$$

These two inequalities lead to the identity $g(x) = g(e)$, $x \in S$. Next, in (1.62) replace T with f and y with x^{-1} to observe that

$$f(e) \le f(x) \vee f(x^{-1}),$$

whenever $x \in S$. To show the biconditional in part (**iii**) we just note that the necessity follows from part (**ii**) and that the sufficiency is obvious. To end the proof, assume that $f(x) = f(x^{-1})$ and $g(x) \le f(x)$ for all $x \in S$. If $\beta \in L$ is defined by

$$\beta := \frac{1}{2}[g(e) + f(e)],$$

then β separates g and f.   □

An easy remark here below is worth being pointed out.

Remark 1.12 Let S, L and f be as in Theorem 1.14. Then the following three conditions are equivalent:

(i) $f(x) = f(x^{-1})$, $x \in S$,

(ii) $f(x) \ge f(x^{-1})$, $x \in S$,

(iii) $f(x) \le f(x^{-1})$, $x \in S$.

1.6.3 Stability Theorems

Now, we will focus on the stability of Eq. (1.60). One can say that Eq. (1.60) is stable in a sense of Ulam if every perturbed solution of (1.60) is close to an exact solution. In view of Proposition 1.1, which says that all solutions of (1.60) on a group are constant mappings, one can expect that every perturbed solution of (1.60) is bounded, or its norm is estimated by a mapping somehow related to an error function. This however is not the case for (1.60), which is shown in Example 1.12. There exists an unbounded approximate solution which satisfied the stability problem with a bounded error function. Instead, we prove that every solution of a perturbed equation vanishes asymptotically. To be more precise, we show that under some assumptions all Hyers sequences of a particular type converges to zero.

We begin with a result, which states that the following functional equation

$$T(x^2) = T(x), \quad x \in S$$

possesses some stability behaviour for mappings defined on a commutative semi-group and taking values in a Banach lattice.

Theorem 1.15 *Let $(S, \star)$ be an arbitrary semigroup and let $\mathscr{B}$ be a Banach lattice. Further, assume that $(\alpha_n)_{n \in \mathbb{N}}$ is a sequence of positive real numbers converging to zero and $\Phi \colon S \to [0, +\infty)$ satisfies*

$$\lim_{n \to +\infty} \alpha_n \sum_{k=0}^{n-1} \Phi(x^{2^k}) = 0, \quad x \in S. \tag{1.68}$$

If $F \colon S \to \mathscr{B}$ is a mapping such that

$$\|F(x^2) - F(x)\| \le \Phi(x), \quad x \in S, \tag{1.69}$$

then for every $x \in S$ the sequence $(\alpha_n F(x^{2^n}))_{n \in \mathbb{N}}$ converges to zero in $\mathscr{B}$.

Theorem 1.16 *Let $(S, \star)$ be a commutative semigroup and $\mathscr{B}$ be a Banach lattice, $F \colon S \to \mathscr{B}$ and $\Psi \colon S \times S \to [0, +\infty)$ satisfy*

$$\|F(x \star y) - F(x) \vee F(y)\| \le \Psi(x, y), \quad x, y \in S. \tag{1.70}$$

If $(\alpha_n)_{n \in \mathbb{N}}$ is a sequence of positive real numbers converging to zero and $\Phi \colon S \to [0, +\infty)$ defined by

$$\Phi(x) = \Psi(x, x), \quad x, y \in S \tag{1.71}$$

satisfies (1.68), *then for every* $x \in S$ *the sequence* $(\alpha_n F(x^{2^n}))_{n \in \mathbb{N}}$ *converges to zero in* $\mathscr{B}$.

Conversely, if there exists a sequence $(\alpha_n)_{n \in \mathbb{N}}$ *of positive real numbers such that for every* $x \in S$ *the sequence* $(\alpha_n F(x^{2^n}))_{n \in \mathbb{N}}$ *converges to some* $T(x)$ *and* Ψ *satisfies*

$$\lim_{n \to +\infty} \alpha_n \Psi(x^{2^n}, y^{2^n}) = 0 \quad x, y \in S, \tag{1.72}$$

then $T : S \to \mathscr{B}$ *is a solution of Eq.* (1.60).

The next example shows that the estimates of Theorem 1.16 are optimal in the sense that the rate of convergence has to be the same in both conditions (1.68) and (1.72).

Example 1.11 Let us take $S = (1, +\infty)$ with $\star$ equal to the standard multiplication in S and $\mathscr{B} = \mathbb{R}$ with the standard order. Next, take

$$F(x) = \log x, \quad x \in (1, +\infty),$$

$$\Psi(x, y) = \min\{\log x, \log y\}, \quad x, y \in (1, +\infty).$$

Then clearly

$$|\log(xy) - \log x \vee \log y| = \min\{\log x, \log y\}$$

for every $x, y \in (1, +\infty)$, thus estimate (1.70) is satisfied. Moreover, it is easy to see that for every sequence $(\alpha_n)_{n \in \mathbb{N}}$ such that

$$\lim_{n \to +\infty} 2^n \alpha_n = 0,$$

both conditions (1.68) and (1.72) hold true.

Next, we will modify the previous example in order to show that there exists an unbounded mapping which satisfies estimate (1.70) with a bounded error function Ψ. Therefore, a direct analogue of Hyers Theorem is not true for functional equation (1.60).

Example 1.12 Let us take $S = (e, +\infty)$, where e is the base of the natural logarithm, with $\star$ equal to the standard multiplication in S and $\mathscr{B} = \mathbb{R}$ with the standard order. Next, take

$$F(x) = \log(\log x), \quad x \in (e, +\infty),$$

and $\Psi(x, y) = \log 2$ for $x, y \in (e, +\infty)$. Fix $x, y \in (e, +\infty)$. Without loss of generality we can assume that $x \leq y$. Then we have

$$| \log(\log(xy)) - \log(\log x) \vee \log(\log y)| = \left| \log \left(\frac{\log x + \log y}{\log y} \right) \right| \leq \log 2.$$

Thus estimate (1.70) is satisfied with bounded Ψ. Thus both conditions (1.68) and (1.72) hold true with every sequence $(\alpha_n)_{n \in \mathbb{N}}$ for which $\lim_{n \to \infty} n\alpha_n = 0$.

Proof (of Theorem 1.15) Fix $x \in S$ and $n \in \mathbb{N}$ and apply (1.69) repeatedly for x, x^2, $\ldots, x^{2^{n-1}}$. Summing up the inequalities obtained and using the triangle inequality we arrive at

$$\|F(x^{2^n}) - F(x)\| \leq \sum_{k=0}^{n-1} \|F(x^{2^{k+1}}) - F(x^{2^k})\| \leq \sum_{k=0}^{n-1} \Phi(x^{2^k}).$$

Next, multiply both sides of this estimate by α_n to see

$$\|\alpha_n F(x^{2^n}) - \alpha_n F(x)\| \leq \alpha_n \sum_{k=0}^{n-1} \Phi(x^{2^k}).$$

Clearly, sequence $(\alpha_n F(x))_{n \in \mathbb{N}}$ tends to zero as n tends to $+\infty$. Thus, it follows from (1.68) that the sequence $\alpha_n F(x)$ converges to zero. $\quad\square$

Proof (of Theorem 1.16) The first part follows immediately from Theorem 1.15 after substitution $y = x$ in (1.70).

To prove the second part, fix $x, y \in S$ and a positive integer n and apply (1.70) with x replaced by x^{2^n} and y replaced by y^{2^n}. Note that by the commutativity of operation $\star$ we have

$$F(x^{2^n} \star y^{2^n}) = F((x \star y)^{2^n}),$$

and after multiplying both sides by α_n we get

$$\|\alpha_n F((x \star y)^{2^n}) - \alpha_n F(x^{2^n}) \vee \alpha_n F(y^{2^n})\| \leq \alpha_n \Psi(x^{2^n}, y^{2^n}).$$

Let n tend to $+\infty$ and use assumption (1.72) to get the assertion. $\quad\square$

Remark 1.13 Let us note that in the proofs of Theorems 1.15 and 1.16 no completeness of the target set $\mathscr{B}$ was used. This is a substantial difference between our approach and a vast majority of other stability results, where completeness of the target space is essential. Note however, that in the second part of Theorem 1.16 this is at least partially hidden in the assumptions, because we assume that a certain sequence is convergent.

References

1. Agbeko, N.K.: On optimal averages. Acta Math. Hungar. **63**, 1–15 (1994)
2. Agbeko, N.K.: On the structure of optimal measures and some of its applications. Publ. Math. Debr. **46**, 79–87 (1995)
3. Agbeko, N.K.: How to characterize some properties of measurable functions. Math. Notes Miskolc **1/2**, 87–98 (2000)
4. Agbeko, N.K.: Mapping bijectively σ-algebras onto power sets. Miskolc Math. Notes **2**(2), 85–92 (2001)
5. Agbeko, N.K.: Stability of maximum preserving functional equations on banach lattices. Miskolc Math. Notes **13**(2), 187–196 (2012)
6. Agbeko, N.K.: The Hyers-Ulam-Aoki type stability of some functional equation on Banach lattices. Bull. Pol. Acad. Sci. Math. **63**, 177–184 (2015)
7. Agbeko, N.K.: A remark on a result of Schwaiger. Indag. Math. **28**, 268–275 (2017)
8. Agbeko, N.K., Fechner, W., Rak, E.: On lattice-valued maps stemming from the notion of optimal average. Acta Math. Hungar. **152**, 72–83 (2017)
9. Agbeko, N.K.: Studies on some addition-free environments. Habilitation Thesis submitted to the University of Debrecen. http://www.uni-miskolc.hu/~matagbek/Habilitation%20Thesis.pdf
10. Agbeko, N.K., Szokol, P.: A generalization of the Hyers-Ulam-Aoki type stability of some Banach lattice-valued functional equation. Extracta Math. **33**, 1–10 (2018)
11. Aoki, T.: Stability of the linear transformation in Banach spaces. J. Math. Soc. Jpn. **2**, 64–66 (1950)
12. Czerwik, S.: On the stability of the quadratic mapping in normed spaces. Abh. Math. Semin. Univ. Hambg. **62**, 59–64 (1992)
13. Davey, B.A., Priestley, H. A.: Introduction to Lattices and Order. Cambridge University Press, New York (2002)
14. Fechner, W.: On the Hyers-Ulam stability of functional equations connected with additive and quadratic mappings. J. Math. Anal. Appl. **322**, 774–786 (2006)
15. Forti, G.L.: Comments on the core of the direct method for proving Hyers-Ulam stability of functional equations. J. Math. Anal. Appl. **295**, 127–133 (2004)
16. Gajda, Z.: On stability of additive mappings. Internat. J. Math. Sci. **14**, 431–434 (1991)
17. Gajda, Z., Kominek, Z.: On separation theorem for subadditive and superadditive functionals. Stud. Math. **100**, 25–38 (1991)
18. Găvruţa, P.: On a problem of G. Isac and Th. M. Rassias concerning the stability of mappings. J. Math. Anal. Appl. **261**, 543–553 (2001)
19. Ger, R., Šemrl, P.: The stability of the exponential equation. Proc. Am. Math. Soc. **124**, 779–787 (1996)
20. Hyers, D.H.: On the stability of the linear functional equation. Proc. Nat. Acad. Sci. U. S. A. **27**, 222–224 (1941)
21. Kaufman, R.: Interpolation of additive functionals. Stud. Math. **27**, 269–272 (1966)
22. König, H.: On the abstract Hahn-Banach theorem due to Rodé. Aequationes Math. **34**, 89–95 (1987)
23. Kranz, P.: Additive functionals on abelian semigroups. Comment. Math. Prace Mat. **16**, 239–246 (1972)
24. Laczkovich, M.: The local stability of convexity, affinity and of the Jensen equation. Aequationes Math. **58**, 135–142 (1999)
25. Maksa, Gy.: The stability of the entropy of degree alpha. J. Math. Anal. Appl. **346**, 17–21 (2008)
26. Mazur, S., Orlicz, W.: Sur les espaces métriques linéaires, II. Stud. Math. **13**, 137–179 (1953)
27. Páles, Zs.: Hyers-Ulam stability of the Cauchy functional equation on square-symmetric groupoids. Publ. Math. Debr. **58**, 651–666 (2001)

28. Rodé, G.: Eine abstrakte version des Satzes von Hahn-Banach. Arch. Math. (Basel) **31**, 474–481 (1978)
29. Salehi, N., Modarres, S.M.S.: Stablity of maximum preserving quadratic functional equation in Banach lattices. Miskolc Math. Notes **17**, 581–589 (2016)
30. Schaefer, H.H.: Banach lattices and positive operators. In: Die Grundlehren der mathematischen Wissenschaften in Einzeldarstellungen, Band 215. Springer, Berlin (1974)
31. Schwaiger, J.: Remark 10. In: Report of 30th International Symposium on Functional Equations. Aequationes mathematicae, vol. 46, p. 289 (1993)
32. Tarski, A.: A lattice-theoretical Fixpoint theorem and its applications. Pac. J. Math. **5**, 285–309 (1955)
33. Ulam, S.M.: A Collection of the Mathematical Problems. Interscience, New York (1960)

Chapter 2
A Purely Fixed Point Approach to the Ulam-Hyers Stability and Hyperstability of a General Functional Equation

Chaimaa Benzarouala and Lahbib Oubbi

Abstract In this paper, using a purely fixed point approach, we produce a new proof of the Ulam-Hyers stability and hyperstability of the general functional equation:

$$\sum_{i=1}^{m} A_i f(\sum_{j=1}^{n} a_{ij}x_j) + A = 0, \qquad (x_1, x_2, \ldots, x_n) \in X^n,$$

considered in Bahyrycz and Olko (Aequationes Math 89:1461, 2015. https://doi.org/10.1007/s00010-014-0317-z), and in Bahyrycz and Olko (Aequationes Math 90:527, 2016. https://doi.org/10.1007/s00010-016-0418-y). Here m and n are positive integers, f is a mapping from a vector space X into a Banach space $(Y, \| \|)$, $A \in Y$ and, for every $i \in \{1, 2, \ldots, m\}$ and $j \in \{1, \ldots, n\}$, A_i and a_{ij} are scalars.

Keywords Hyers-Ulam stability · Hyperstability · Functional equation · Fixed point theorem

Mathematics Subject Classification (2010) Primary 39B82, 47H14, 47J20; Secondary 39B62, 47H10

C. Benzarouala
Department of Mathematics, Team GrAAF, Laboratory LMSA, Center CeReMar, Faculty of Sciences, Mohammed V University in Rabat, Rabat, Morocco

L. Oubbi (✉)
Department of Mathematics, Team GrAAF, Laboratory LMSA, Center CeReMar, Ecole Normale Supérieure, Mohammed V University in Rabat, Takaddoum, Rabat, Morocco
e-mail: oubbi@daad-alumni.de

© Springer Nature Switzerland AG 2019
J. Brzdęk et al. (eds.), *Ulam Type Stability*,
https://doi.org/10.1007/978-3-030-28972-0_2

2.1 Introduction

The problem of stability of functional equations goes back to 1940, when Ulam
[15] asked whether, for a given group G_1, a metric group (G_2, d) and a positive
number ϵ, it exists a number $\delta > 0$ such that, whenever a function $f : G_1 \rightarrow G_2$
satisfies the inequality $d(f(xy), f(x)f(y)) < \delta$, for every $(x, y) \in G_1^2$, there exists
a group homomorphism $F : G_1 \rightarrow G_2$ such that $d(f(x), F(x)) < \epsilon$, for every
$x \in G_1$. Whenever the answer to this problem is in the affirmative, one says that
the homomorphism equation $f(xy) = f(x)f(y)$ is Ulam-stable, or that the group
homomorphisms are stable with respect to the equation $f(xy) = f(x)f(y)$ and the
Ulam-approximation.

   The first partial answer to this problem was given in 1941 by Hyers [8]. He
namely showed that the Cauchy equation $f(x + y) = f(x) + f(y)$ is Ulam stable,
whenever $G_1 = X$ and $G_2 = Y$ are real Banach spaces. Later, in 1978, Rassias
[14] considered Ulam's problem with a new kind of approximation. He allowed the
Cauchy differences to be unbounded, but dominated in the following way:

$$\|f(x + y) - f(x) - f(y)\| \leq \theta(\|x\|^p + \|y\|^p), \quad x, y \in X,$$

for some $\theta \geq 0$ and some $p \in [0, 1[$. With a similar method as Hyers, he
obtained that additive mappings between Banach spaces are stable with respect
to the Cauchy functional equation and the approximation above. Further, in 2003,
Radu [13] used the alternative fixed point theorem to retrieve Rassias' theorem.
In the same year, L. Cadariu and V. Radu introduced in [4] a new approximation
condition generalizing Rassias' one and, using the alternative fixed point theorem,
they showed that the additive mappings between Banach spaces are stable with
respect to the Jensen functional equation $2f(\frac{x+y}{2}) = f(x) + f(y)$ and their new
approximation condition.

   Instead of The Cauchy and the Jensen functional equations, a large variety
of functional equations have been considered in the literature, such as quadratic
functional equations [6], Euler-Lagrange type equations [9, 11], cubic equations
[5], quartic equations [12] and so on. Several authors have also considered systems
of functional equations and studied their stability [10, 11].

   In 2014, A. Bahyrycz and J. Olko proved in [1] the stability of the general
functional equation

$$\sum_{i=1}^{m} A_i f(\sum_{j=1}^{n} a_{ij}x_j) + A = 0, \tag{2.1}$$

where f is a mapping from a vector space X into a Banach space Y, $A_i \in \mathbb{K}^*, A \in Y$
and $a_{ij} \in \mathbb{K}$. Equation (2.1) generalizes most of the linear functional equations in
the literature. The authors used the direct method, with, in midway, a fixed point
theorem of Brzdęk et al. [3], to show the stability of (2.1). In 2015, Dong [7], using

the same theorem of J. Brzdęk et al., proved the hyperstability of Eq. (2.1) when $A = 0$ with respect to two different approximation functions.

In the same sense of Dong et al., Bahyrycz and Olko in [2], using the same theorem of Brzdęk et al., showed the hyperstability of Eq. (2.1).

In this note, using a proof similar to that of Oubbi [11], relying on the classical Banach contraction theorem, we reprove the stability and the hyperstability of (2.1). Along the way, let us notice that, in [1] and [2], the authors assumed that A is a scalar in (2.1). Actually, A must be an element of Y. Fortunately, this does not alter their results.

2.2 Stability of Eq. (2.1)

In all what follows X and Y will be vector spaces on the field $\mathbb{K} \in \{\mathbb{R}, \mathbb{C}\}$ and $f : X \to Y$ will be a mapping. The space Y will be endowed with a complete norm $\| \ \|$. We will denote by A an element of Y and by n, m positive integers, while A_i and a_{ij} will be scalars, $i = 1, \ldots, m$ and $j = 1, \ldots, n$. We will then be concerned with the functional equation (2.1) and its corresponding homogenous one:

$$\sum_{i=1}^{m} A_i f\left(\sum_{j=1}^{n} a_{ij} x_j\right) = 0 \tag{2.2}$$

Theorem 2.1 *Assume $A = 0$ or $(A \neq 0$ and $\sum_{i=1}^{m} A_i \neq 0)$. Suppose that some mapping $\theta : X^n \to \mathbb{R}_+$ exists such that:*

$$\left\| \sum_{i=1}^{m} A_i f\left(\sum_{j=1}^{n} a_{ij} x_j\right) + A \right\| \leq \theta(x_1, \cdots, x_n), \quad x_1, \cdots, x_n \in X. \tag{2.3}$$

Assume also that there exist a non empty set $I \subsetneq \{1, \ldots, m\}$, $c_1, \cdots, c_n \in \mathbb{K}$, and positive numbers ω_i, $i \notin I$, such that:

(i) $\forall i \in I, \beta_i = 1$, *where, for every* $k = 1, \ldots, m$, $\beta_k := \sum_{j=1}^{n} a_{kj} c_j$,
(ii) $\sum_{i \notin I} |A_i| \omega_i < |\sum_{i \in I} A_i|$,
(iii) $\theta(\beta_i x_1, \cdots, \beta_i x_n) \leq \omega_i \theta(x_1, \cdots, x_n)$, $i \notin I$, $x_1, \cdots, x_n \in X$.

Then there exists a unique solution $G : X \to Y$ of (2.1) satisfying

$$\|f(x) - G(x)\| \leq \frac{\theta(c_1 x, \cdots, c_n x)}{|\sum_{i \in I} A_i| - \sum_{i \notin I} |A_i| \omega_i}, \quad x \in X. \tag{2.4}$$

Before giving our proof, let us denote by $T : Y^X \to Y^X$ an operator of the form:

$$(T\xi)(x) := \sum_{i=1}^{k} \alpha_i \xi(\beta_i x), \qquad \xi \in Y^X, \quad x \in X, \tag{2.5}$$

with $k \in \mathbb{N}$ and $\alpha_1, \cdots, \alpha_k, \beta_1, \cdots, \beta_k \in \mathbb{K}$.

We will make use of the following lemma proven in [1].

Lemma 2.1 *Assume that* $\theta : X^n \to \mathbb{R}_+$ *is a mapping and let* $T : Y^X \to Y^X$ *be given by (2.5). Assume that there exist* $\omega_1, \cdots, \omega_k \in \mathbb{R}_+$ *such that* $\sum_{i=1}^{k} |\alpha_i| \omega_i < 1$ *and*

$$\theta(\beta_i(x_1, \cdots, x_n)) \le \omega_i \theta(x_1, \cdots, x_n), \quad i \in \{1, \cdots, k\}, \quad x_1, \cdots, x_n \in X.$$

If f *satisfies the inequality:*

$$\| \sum_{i=1}^{m} A_i f(\sum_{j=1}^{n} a_{ij} x_j) \| \le \theta(x_1, \cdots, x_n), \quad x_1, \cdots, x_n \in X,$$

and if the limit $G(x) := \lim_{n\to\infty} T^n f(x)$ *exists for every* $x \in X$, *then* $G : X \to Y$ *is a solution of (2.2).*

Now we are in a position to prove Theorem 2.1.

Proof Assume that there exist a non empty set $I \subsetneq \{1, \cdots, m\}$, scalars $c_1, \cdots, c_n$, and positive numbers ω_i, $i \notin I$, enjoying the assumptions (i)–(iii). Note that, due to (ii), $A_I := \sum_{i \in I} A_i \ne 0$.

Case $A = 0$: Let $x \in X$ be arbitrary. Putting $x_j = c_j x$, $j \in \{1, \cdots, n\}$ in (2.3), we get

$$\| f(x) - \sum_{i \notin I} \frac{-A_i}{A_I} f(\sum_{j=1}^{n} a_{ij} c_j x) \| \le \frac{\theta(c_1 x, \cdots, c_n x)}{|A_I|}.$$

Consider the set:

$$M_f := \{g : X \to Y, \exists B > 0 \text{ such that } \| f(t) - g(t) \| \le B\theta(c_1 t, \cdots, c_n t), \quad t \in X\}.$$

Then M_f is non empty, for $f \in M_f$. Now, for every $g, h \in M_f$, put

$$d(g, h) := \inf\{B > 0, \| g(t) - h(t) \| \le B\theta(c_1 t, \cdots, c_n t), t \in X\}.$$

Then d is a distance on M_f. Indeed, since $g, h \in M_f$, there exist $B', B'' > 0$ such that, for every $x \in X$,

$$\| g(x) - f(x) \| \le B'\theta(c_1 x, \cdots, c_n x) \quad \text{and} \quad \| h(x) - f(x) \| \le B''\theta(c_1 x, \cdots, c_n x).$$

Therefore

$$\|g(x) - h(x)\| \leq \|g(x) - f(x)\| + \|f(x) - h(x)\| \leq (B' + B'')\theta(c_1 x, \cdots, c_n x).$$

Thus $d(g, h) \leq B' + B'' < +\infty$ for all $g, h \in M_f$. It is clear that $d(g, h) = 0$ if and only if $g = h$, and that $d(g, h) = d(h, g)$, $g, h \in M_f$. For the triangular inequality, let $g, h, k \in M_f$, and $B, B' > 0$ be given so that $d(g, h) < B$ and $d(h, k) < B'$. Then, for all $t \in X$, we have:

$$\begin{aligned}\|g(t) - k(t)\| &\leq \|g(t) - h(t)\| + \|h(t) - k(t)\| \\ &\leq B\theta(c_1 t, \cdots, c_n t) + B'\theta(c_1 t, \cdots, c_n t) \\ &\leq (B + B')\theta(c_1 t, \cdots, c_n t).\end{aligned}$$

Passing to the infimum, we get $d(g, k) \leq d(g, h) + d(h, k)$. Therefore d is a distance on M_f. Now, let's show that the metric space (M_f, d) is complete. If $(g_n)_n$ is a Cauchy sequence in M_f, as the evaluations $\delta_x : g \mapsto g(x)$ are uniformly continuous from M_f into Y, the sequence $(g_n(x))_n$ is Cauchy in Y, for every $x \in X$. By the completeness of Y, it converges to some $g(x)$. But

$$\forall \epsilon > 0, \exists N_\epsilon \in \mathbb{N}, \forall m > n \geq N_\epsilon : d(g_n, g_m) < \epsilon.$$

Therefore

$$\|g_n(x) - g_m(x)\| \leq \epsilon\theta(c_1 x, \cdots, c_n x), \quad m > n \geq N_\epsilon, \quad x \in X. \tag{2.6}$$

Letting m tend to infinity, we get $\|g_n(x) - g(x)\| \leq \epsilon\theta(c_1 x, \cdots, c_n x)$. Thus

$$\|g(x) - f(x)\| \leq \|g(x) - g_n(x)\| + \|g_n(x) - f(x)\| \leq (\epsilon + d(g_n, f))\theta(c_1 x, \cdots, c_n x).$$

Therefore the so defined mapping g belongs to M_f. Again by (2.6), $(g_n)_n$ converges in M_f to g and then (M_f, d) is complete.

Now, for arbitrary $\xi \in M_f$, define a mapping $T\xi$ from X into Y by:

$$T\xi(x) := \sum_{i \notin I} \frac{-A_i}{A_I} \xi\left(\sum_{j=1}^{n} a_{ij} c_j x\right).$$

Since

$$\left\|f(x) - \sum_{i \notin I} \frac{-A_i}{A_I} f\left(\sum_{j=1}^{n} a_{ij} c_j x\right)\right\| \leq \frac{\theta(c_1 x, \cdots, c_n x)}{|A_I|}, \quad x \in X,$$

$$\|f(x) - Tf(x)\| \leq \frac{\theta(c_1 x, \cdots, c_n x)}{|A_I|}, \quad x \in X.$$

Then $Tf \in M_f$. Actually, T is a self mapping of M_f. Indeed, for every $g, h \in M_f$, it holds $\|g(x) - h(x)\| \leq d(h, g)\theta(c_1 x, \cdots, c_n x)$. Then, for $g \in M_f$,

$$\|Tg(x) - f(x)\| \leq \|Tg(x) - Tf(x)\| + \|Tf(x) - f(x)\|$$

$$\leq \sum_{i \notin I} \frac{|A_i|}{|A_I|} \|g(\sum_{j=1}^{n} a_{ij}c_j x) - f(\sum_{j=1}^{n} a_{ij}c_j x)\| + \|Tf(x) - f(x)\|$$

$$\leq \sum_{i \notin I} |\frac{A_i}{A_I}| d(g, f)\theta(\beta_i c_1 x, \cdots, \beta_i c_n x) + \frac{\theta(c_1 x, \cdots, c_n x)}{|A_I|}$$

$$\leq d(f, g) \sum_{i \notin I} |\frac{A_i}{A_I}| \omega_i \theta(c_1 x, \cdots, c_n x) + \frac{\theta(c_1 x, \cdots, c_n x)}{|A_I|}$$

$$\leq \left(d(f, g) \sum_{i \notin I} |\frac{A_i}{A_I}| \omega_i + \frac{1}{|A_I|} \right) \theta(c_1 x, \cdots, c_n x).$$

Whence $Tg \in M_f$, for all $g \in M_f$.

Now, let us show that T is a strictly contracting mapping. Given g and h in M_f. Then

$$\|Tg(x) - Th(x)\| = \| \sum_{i \notin I} \frac{-A_i}{A_I} g(\sum_{j=1}^{n} a_{ij}c_j x) - \sum_{i \notin I} \frac{-A_i}{A_I} h(\sum_{j=1}^{n} a_{ij}c_j x)\|$$

$$\leq \sum_{i \notin I} |\frac{A_i}{A_I}| \|g(\sum_{j=1}^{n} a_{ij}c_j x) - h(\sum_{j=1}^{n} a_{ij}c_j x)\|$$

$$\leq \sum_{i \notin I} |\frac{A_i}{A_I}| d(g, h)\theta(c_1 \sum_{j=1}^{n} a_{ij}c_j x, \cdots, c_n \sum_{j=1}^{n} a_{ij}c_j x)$$

$$\leq d(g, h) \sum_{i \notin I} |\frac{A_i}{A_I}| \omega_i \theta(c_1 x, \cdots, c_n x).$$

If we put $\gamma := \sum_{i \notin I} |\frac{A_i}{A_I}| \omega_i$, then $\gamma < 1$ and

$$d(Tg, Th) \leq \gamma d(g, h), \quad g, h \in M_f,$$

Therefore T is a strictly contracting mapping. By Banach fixed point theorem, there exists a unique mapping $G \in M_f$ such that $TG = G$ and $\lim_{n \to \infty} T^n f = G$. Thanks to Lemma 2.1, G is a solution of (2.1). Furthermore

$$d(f, G) = d(f, \lim_{n \to \infty} T^n f)$$

$$= \lim_{n \to \infty} d(f, T^n f)$$

$$\leq \lim_{n \to \infty} \sum_{j=0}^{n-1} \gamma^j d(f, Tf)$$

$$\leq \frac{d(f, Tf)}{1 - \gamma}$$

$$\leq \frac{1}{|A_I| - \sum_{i \notin I} |A_i| \omega_i}.$$

Thus (2.4) holds.

Case $A \neq 0$ and $\sum_{i=1}^{m} A_i \neq 0$. Define a new function $g : X \to Y$ by $g(x) := f(x) + \frac{A}{\sum_{i=1}^{m} A_i}$. Since

$$\| \sum_{i=1}^{m} A_i f(\sum_{j=1}^{n} a_{ij} x_j) + A \| \leq \theta(x_1, \cdots, x_n), \quad x_1, \cdots, x_n \in X,$$

we get:

$$\| \sum_{i=1}^{m} A_i g(\sum_{j=1}^{n} a_{ij} x_j) \| \leq \theta(x_1, \cdots, x_n), \quad x_1, \cdots, x_n \in X,$$

By the first part of the proof, there exists a unique solution H of (2.2) such that

$$\|g(x) - H(x)\| \leq \frac{\theta(c_1 x, \cdots, c_n x)}{|\sum_{i \in I} A_i| - \sum_{i \notin I} |A_i| \omega_i}, \quad x \in X.$$

Since H is a solution of (2.2), $G := H - \frac{A}{\sum_{i=1}^{m} A_i}$ is a solution of (2.1). But then

$$\|f(x) - G(x)\| \leq \frac{\theta(c_1 x, \cdots, c_n x)}{|\sum_{i \in I} A_i| - \sum_{i \notin I} |A_i| \omega_i}, \quad x \in X.$$

This finishes the proof. □

2.3 Hyperstability of Eq. (2.1)

In this section, with an additional condition than in Theorem 2.1, we show the hyperstability of Eq. (2.1).

Theorem 2.2 *Assume that $A = 0$ or $A \neq 0$ and $\sum_{i=1}^{m} A_i \neq 0$. Let $\theta : X^n \to \mathbb{R}_+$ satisfy (2.3) and let $\omega : \mathbb{K} \to \mathbb{R}_+$ enjoy:*

$$\theta(\beta x_1, \cdots, \beta x_n) \leq \omega(\beta)\theta(x_1, \cdots, x_n), \quad x_1, \cdots, x_n \in X, \quad \beta \in \mathbb{K}.$$

If there exist a non empty set $I \subsetneq \{1, \cdots, m\}$ and a sequence $(c_{k,1}, \cdots, c_{k,n})_{k \in \mathbb{N}}$ of elements of $\mathbb{K}^n$ such that, with the notation $\beta_{k,i} := \sum_{j=1}^{n} a_{ij}c_{k,j}$, $k \in \mathbb{N}$ and $i \in \{1, 2, \ldots, m\}$:

(i) $\beta_{k,i} = 1$ *for all $i \in I$ and all $k \in \mathbb{N}$.*
(ii) $A_I := \sum_{i \in I} A_i \neq 0$, *and* $\lim_{k \to \infty} \sum_{i \notin I} |\frac{A_i}{A_I}| \omega(\beta_{k,i}) < 1$,
(iii) $\lim_{k \to \infty} \theta(c_{k,1}x, \cdots, c_{k,n}x) = 0$,

then f is a solution of Eq. (2.1).

Proof First assume $A = 0$. It follows from (ii) that there exists $k_0 \in \mathbb{N}$ such that:

$$\gamma_k := \sum_{i \notin I} |\frac{A_i}{A_I}| \omega(\beta_{k,i}) < 1, \ \forall k \geq k_0.$$

For $k \geq k_0$ and arbitrary $x \in X$, if we take in (2.3), $x_j = c_{k,j}x$, $j \in \{1, \cdots, n\}$, we will get:

$$\| f(x) - \sum_{i \notin I} \frac{-A_i}{A_I} f(\beta_{k,i}x) \| \leq \frac{\theta(c_{k,1}x, \cdots, c_{k,n}x)}{|A_I|}, \ x \in X. \tag{2.7}$$

Consider the set M_f^k defined by :

$$M_f^k := \{g : X \to Y; \exists B > 0 \text{ such that } \| f(t) - g(t)\| \leq B\theta(c_{k,1}t, \cdots, c_{k,n}t), \ t \in X\}.$$

As in the proof of Theorem 2.1, (M_f^k, d_k) is a complete metric space, with respect to the distance

$$d_k(g, h) := \inf\{B > 0; \|g(t) - h(t)\| \leq B\theta(c_{k,1}t, \cdots, c_{k,n}t), \ \forall t \in X\}, \quad g, h \in M_f^k.$$

Moreover, the self-mapping T_k of M_f^k defined by:

$$T_k\xi(x) := \sum_{i \notin I} \frac{-A_i}{A_I}\xi(\beta_{k,i}x), \quad \xi \in Y^X, x \in X$$

is a contraction with Lipschitz constant $\gamma_k < 1$. By Banach fixed point theorem, T_k admits a unique fixed point $G_k \in M_f^k$ with $G_k(x) = \lim_{n \to \infty} T_k^n f(x)$, $x \in X$. Again, by Lemma 2.1, we have:

$$\sum_{i=1}^{m} A_i G_k(\sum_{j=1}^{n} a_{ij} x_j) = 0. \tag{2.8}$$

and

$$\|f(x) - G_k(x)\| \leq \frac{\theta(c_{k,1}x, \cdots, c_{k,n}x)}{|A_I|(1 - \gamma_k)}, \quad x \in X. \tag{2.9}$$

Letting k tend to $+\infty$ in (2.9), we obtain

$$\lim_{k \to \infty} G_k(x) = f(x), \quad x \in X.$$

Letting k tend to $+\infty$ in (2.8), we obtain that f is a solution of (2.2) or equivalently of (2.1).

Now, if $A \neq 0$ and $\sum_{i=1}^{m} A_i \neq 0$, define a new function $g : X \to Y$ by

$$g(x) := f(x) + \frac{A}{\sum_{i=1}^{m} A_i}.$$

We have

$$\|\sum_{i=1}^{m} A_i f(\sum_{j=1}^{n} a_{ij} x_j) + A\| \leq \theta(x_1, \cdots, x_n), \quad x_1, \cdots, x_n \in X.$$

Then

$$\|\sum_{i=1}^{m} A_i (f(\sum_{j=1}^{n} a_{ij} x_j) + \frac{A}{\sum_{i=1}^{m} A_i})\| \leq \theta(x_1, \cdots, x_n), \quad x_1, \cdots, x_n \in X.$$

i.e.,

$$\|\sum_{i=1}^{m} A_i g(\sum_{j=1}^{n} a_{ij} x_j)\| \leq \theta(x_1, \cdots, x_n), \quad x_1, \cdots, x_n \in X.$$

By the first part of the proof, g satisfies (2.2). Hence f is a solution of (2.1), which finishes the proof.   □

References

1. Bahyrycz, A., Olko, J.: On stability of the general linear equation. Aequationes Math. **89**, 1461–1474 (2015)
2. Bahyrycz, A., Olko, J.: Hyperstability of general linear functional equation. Aequationes Math. **90**, 527–540 (2016)
3. Brzdęk, J., Chudziak, J., Páles, Z.: A fixed point approach to stability of functional equations. Nonlinear Anal. **(74)**, 6728–6732 (2011)
4. Cădariu, L., Radu, V.: Fixed points and the stability of Jensen's functional equation. J. Inequal. Pure Appl. Math. **4**, Article 4 (2003)
5. Chu, H.Y., Kim, A., Yu, S.K.: On the stability of the generalized cubic set-valued functional equation. Appl. Math. Lett. **37**, 7–14 (2014)
6. Czerwik, S.: On the stability of the quadratic mapping in normed spaces. Abh. Math. Semin. Univ. Hambg. **62**, 59–64 (1992)
7. Dong, Z.: On Hyperstability of generalised linear functional equations in several variables. Bull. Aust. Math. Soc. **92**, 259–267 (2015)
8. Hyers, D.H.: On the stability of the linear functional equation. Proc. Natl. Acad. Sci. U. S. A. **27**, 222–224 (1941)
9. Jun, K.W., Kim, H.M., Rassais, T.M.: Extended Hyers-Ulam stability for Cauchy-Jensen mappings. J. Differ. Equ. Appl. **13**, 1139–1153 (2007)
10. Moslehian, M.S.: Hyers-Ulam-Rassias stability of generalized derivations. Int. J. Math. Math. Sci. **2006**, Article ID 93942, 8 (2006)
11. Oubbi, L.: Ulam-Hyers-Rassias stability problem for several kinds of mappings. Afr. Mat. **24**, 525–542 (2013)
12. Patel, B.M., Patel, A.B.: Stability of Quartic functional equations in 2-Banach space. Int. J. Math. Anal. **7**(23), 1097–1107 (2013)
13. Radu, V.: The fixed point alternative and the stability of functional equations. Fixed Point Theory **4**(1), 91–96 (2003)
14. Rassias, T.M.: On the stability of the linear mapping in Banach spaces. Proc. Am. Math. Soc. **72**, 297–300 (1978)
15. Ulam, S.M.: Problems in Modern Mathematics. Chapter VI, Science Editions. Wiley, New York (1960)

Chapter 3
Birkhoff–James Orthogonality Reversing Property and Its Stability

Jacek Chmieliński and Paweł Wójcik

Abstract For real normed spaces, we consider the class of linear operators, approximately preserving or reversing the Birkhoff–James orthogonality. In particular we deal with stability problems.

Keywords Birkhoff–James orthogonality · Approximate orthogonality · Orthogonality preserving mappings · Orthogonality reversing mappings · Stability

Mathematics Subject Classification (2010) Primary 46B20, 39B82; Secondary 47B49

3.1 Introduction

Linear preservers problems and, in particular, *orthogonality preserving property* have been studied widely in various settings of underlying spaces and with various definitions of the orthogonality—cf., e.g., the survey [9]. In the present paper we remind some of these results to give a context for a similar research concerning an analogous *orthogonality reversing property*. We recall some recently published results in this direction and we also present a few original ones. A significant part of the paper is devoted to questions arising from the stability theory. Namely, along with the considered property of an exact preservation (reversing) of orthogonality we study also its approximate counterpart and we estimate how far these properties are each to other.

Throughout the paper we usually assume that $(X, \|\cdot\|)$ is a real normed space, with $\dim X \geq 2$. By S_X we denote the unit sphere in X and $\mathscr{L}(X)$ stands for the

J. Chmieliński (✉) · P. Wójcik
Department of Mathematics, Pedagogical University of Cracow, Kraków, Poland
e-mail: jacek.chmielinski@up.krakow.pl; pawel.wojcik@up.krakow.pl

© Springer Nature Switzerland AG 2019
J. Brzdęk et al. (eds.), *Ulam Type Stability*,
https://doi.org/10.1007/978-3-030-28972-0_3

space of all continuous operators from X into X. For $T \in \mathscr{L}(X)$ we consider its usual operator norm $\|T\|$ and the "lower norm" $[T]$:

$$\|T\| := \sup\{\|Tx\| : \|x\| = 1\} = \inf\{M \geq 0 : \forall x \in X \quad \|Tx\| \leq M\|x\|\};$$

$$[T] := \inf\{\|Tx\| : \|x\| = 1\} = \sup\{m \geq 0 : \forall x \in X \quad m\|x\| \leq \|Tx\|\}.$$

By X^* we mean the dual space, i.e., the (normed) space of all linear and continuous functionals defined on X. For a fixed $x \in X$ by $J(x)$ we denote the (nonempty) set of supporting functionals:

$$J(x) := \{\varphi \in X^* : \quad \|\varphi\| = 1, \quad \varphi(x) = \|x\|\}.$$

3.2   Birkhoff–James Orthogonality and Approximate Orthogonality

In the case of inner product spaces we have a standard orthogonality relation $x \perp y \Leftrightarrow \langle x|y \rangle = 0$, as well as a natural notion of an *approximate orthogonality*

$$x \perp^{\varepsilon} y \Leftrightarrow |\langle x|y \rangle| \leq \varepsilon \|x\| \|y\|$$

(i.e., $|\cos \angle(x, y)| \leq \varepsilon$) with $\varepsilon \in [0, 1)$. If the given norm is not generated by any inner product, then the notion of orthogonality has to be introduced using solely the notion of the norm and by referring to some desired properties. Among various concepts of such relations we have the *Birkhoff–James orthogonality* which is defined by

$$x \perp_B y \quad \Longleftrightarrow \quad \forall \lambda \in \mathbb{R} : \quad \|x + \lambda y\| \geq \|x\|$$

(cf. [3, 20, 21] or a more recent survey [1]). It is known (cf. [21, Corollary 2.2]) that

$$x \perp_B y \quad \Longleftrightarrow \quad \exists \varphi \in J(x) : \varphi(y) = 0.$$

In [5] the following definition of an *approximate Birkhoff–James orthogonality* (or more specifically *ε-Birkhoff–James orthogonality* with $\varepsilon \in [0, 1)$) was introduced:

$$x \perp_B^{\varepsilon} y \quad \Longleftrightarrow \quad \forall \lambda \in \mathbb{R} : \quad \|x + \lambda y\|^2 \geq \|x\|^2 - 2\varepsilon \|x\| \|\lambda y\|.$$

Obviously, for $\varepsilon = 0$, $\perp_B^0 = \perp_B$ and if the norm comes from an inner product, then $\perp_B^{\varepsilon}$ is equivalent to $\perp^{\varepsilon}$. In a recent paper [16] authors have proved the following two characterizations of the approximate Birkhoff–James orthogonality.

Theorem 3.1 ([16, Theorems 2.2, 2.3]) *For $x, y \in X$ and $\varepsilon \in [0, 1)$*

$$x \perp_{B}^{\varepsilon} y \quad \Longleftrightarrow \quad \exists z \in \text{Lin}\{x, y\} : x \perp_{B} z, \ \|z - y\| \leq \varepsilon \|y\|; \tag{3.1}$$

$$x \perp_{B}^{\varepsilon} y \quad \Longleftrightarrow \quad \exists \varphi \in J(x) : \ |\varphi(y)| \leq \varepsilon \|y\|. \tag{3.2}$$

3.3   Operators Preserving or Reversing Orthogonality

This section is devoted to the main considered property. We are interested in linear operators which do not essentially change orthogonality of arguments. By this we mean that they exactly preserve orthogonality (with the same or changed order) or they do it in some sense approximately.

3.3.1   Exact Preservation or Reversal of the Orthogonality

Let $T : X \rightarrow X$ be a nonzero linear mapping. We say that T is *orthogonality preserving* (OP) if

$$x \perp_{B} y \quad \Longrightarrow \quad Tx \perp_{B} Ty, \qquad x, y \in X. \tag{3.3}$$

It is known that T satisfies (3.3) if and only if it is a similarity (a scalar multiple of a linear isometry), i.e.,

$$\|Tx\| = \gamma \|x\|, \qquad x \in X$$

with some $\gamma > 0$. This result is nontrivial (cf. [22]) and remains true also for complex normed spaces X, Y and $T : X \rightarrow Y$ (cf. [4]). For inner product spaces the property (3.3) is, additionally, equivalent with the condition

$$\langle Tx | Ty \rangle = \gamma^2 \langle x | y \rangle, \qquad x, y \in X$$

and the proof is quite elementary (cf. [6, Theorem 1]).

The Birkhoff–James orthogonality is generally not symmetric. If $\dim X \geq 3$, then symmetry of $\perp_{B}$ characterizes inner product spaces among normed ones. If X is a two-dimensional plane it is possible that $\perp_{B}$ is symmetric even though the norm does not come from an inner product (X is a Radon plane—cf. [23, 28]). Therefore, we can consider the property that T keeps orthogonality but in a reverse order.

We say that T is *orthogonality reversing* (OR) if

$$x \perp_{B} y \quad \Longrightarrow \quad Ty \perp_{B} Tx, \qquad x, y \in X. \tag{3.4}$$

Orthogonality reversing operators have been introduced and studied in [10]. Obviously, if the Birkhoff–James orthogonality relation is symmetric, then the properties (3.3) and (3.4) coincide. We say that T *essentially reverses orthogonality* if it is (OR) but not (OP). It has been proved that such operators may exist only on Minkowski planes (2-dimensional normed spaces). Actually, if dim $X \geq 3$, then X admits (OR) operators if and only if X is an inner product space (it was proved first in [10, Theorem 4.1] for smooth spaces only and then, independently, in [35, Theorem 2.1] and [36, Theorem 5] without this restriction).

Note, that we consider here the same (Birkhoff–James) orthogonality relation for x, y and Tx, Ty. Alternatively, one can consider the property that for orthogonal vectors x, y, their images Tx, Ty are orthogonal but in a different sense. Such properties were considered, e.g., in [31, 36].

Finally, in this section, we estimate the distance between the two considered classes of operators. Denote by $\mathrm{Sim}(X)$ the class of all similarities, i.e., the class of all linear orthogonality preserving operators and by $\mathrm{Rev}(X)$ the class of all linear operators reversing orthogonality. For the sake of convenience assume that zero operator belongs to both classes. By $\mathrm{Isom}(X)$ we denote the class of all linear isometries.

Theorem 3.2 *For $T \in \mathrm{Rev}(X)$ and $U \in \mathrm{Isom}(X)$ it holds that*

$$\mathrm{dist}(T, \mathrm{Sim}(X)) \leq \|T\| \, \mathrm{dist}(U, \mathrm{Rev}(X)). \tag{3.5}$$

Moreover, for each $T \in \mathrm{Rev}(X)$ there exists $V \in \mathrm{Sim}(X)$ such that

$$\|T - V\| \leq \gamma_X \|T\|, \tag{3.6}$$

where $\gamma_X := \inf\{\mathrm{dist}(U, \mathrm{Rev}(X)) : U \in \mathrm{Isom}(X)\}$ is a constant depending on X only.

Proof For $T = 0$, (3.5) and (3.6) are obvious. If $0 \neq T \in \mathrm{Rev}(X)$, then X is an inner product space (whence $\mathrm{Rev}(X) = \mathrm{Sim}(X)$) or dim $X \leq 2$. In the first case again (3.5) and (3.6) follow trivially. Assume that dim $X \leq 2$. It is visible that

$$\mathrm{Rev}(X) \circ \mathrm{Rev}(X) \subset \mathrm{Sim}(X) \quad \text{and} \quad \mathrm{Rev}(X) \circ \mathrm{Sim}(X) = \mathrm{Rev}(X).$$

Thus for an arbitrary similarity $V \in \mathrm{Sim}(X)$ we have $T^2 V \in \mathrm{Sim}(X)$ and $TV \in \mathrm{Rev}(X)$. Moreover, for any $S \in \mathrm{Sim}(X)$ we have $\|TU - SU\| = \|T - S\|$ and (U is invertible) $\{SU : S \in \mathrm{Sim}(X)\} = \mathrm{Sim}(X)$, whence

$$\mathrm{dist}(T, \mathrm{Sim}(X)) = \mathrm{dist}(TU, \mathrm{Sim}(X)) \leq \|TU - T^2 V\| \leq \|T\| \, \|U - TV\|.$$

Since V is an arbitrary similarity, TV is an arbitrary operator reversing orthogonality. Passing to the infimum over $\mathrm{Rev}(X)$ we get (3.5).

Since dim $X < \infty$, $\mathrm{Sim}(X)$ and $\mathrm{Isom}(X)$ are closed whence the distances are attained and (3.5) implies (3.6).

3.3.2 Operators Approximately Preserving or Reversing Orthogonality

In real world applications, usually there is always some error in measurement so we may tell that the respective property (like preservation of some relation) holds, to some extent, approximately only. Thus it may be of some interest to consider mappings which transform orthogonal vectors into approximately orthogonal ones. In other words, we are interested in *approximately orthogonality preserving* (AOP) linear operators and, similarly, *approximately orthogonality reversing* (AOR) ones. We would like do describe how far these approximately preserving (reversing) operators are from those which preserve (or reverse) orthogonality exactly.

3.3.2.1 AOP Operators: Review of Results

Let X and Y be real normed spaces and let $T : X \to Y$ be a nonzero linear mapping. For given $\varepsilon \in [0, 1)$ we say that T is ε-*orthogonality preserving* (ε-OP) if

$$x \perp_{\text{B}} y \quad \Longrightarrow \quad Tx \perp_{\text{B}}^{\varepsilon} Ty, \qquad x, y \in X. \tag{3.7}$$

The class of approximately orthogonality preserving operators for inner product spaces has been introduced and studied in [6].

Theorem 3.3 ([6, Theorem 2]) *Let X and Y be inner product spaces and let $T : X \to Y$ be a nonzero linear mapping satisfying (3.7) for some $\varepsilon \in [0, 1)$. Then T is injective, continuous and there exists $\gamma > 0$ such that*

$$|\langle Tx|Ty \rangle - \gamma \langle x|y \rangle| \leq \delta \min\{\gamma \|x\| \|y\|, \|Tx\| \|Ty\|\}, \qquad x, y \in X \tag{3.8}$$

with

$$\delta = 4\varepsilon \left(\frac{1}{1 - \varepsilon} + \sqrt{\frac{1 + \varepsilon}{1 - \varepsilon}} \right). \tag{3.9}$$

The above estimation can be improved if $\dim X < \infty$; namely we may take $\delta = \varepsilon$ in that case (cf. [34, Theorem 5.5]).

For arbitrary normed spaces, Theorem 3.3 was extended in [25].

Theorem 3.4 ([25], Remark 3.1) *Let X, Y be real normed spaces, $\varepsilon \in [0, \frac{1}{2})$. If $T : X \to Y$ is a linear mapping and satisfies (3.7), then*

$$(1 - 8\varepsilon)\|T\| \|x\| \leq \|Tx\| \leq \|T\| \|x\|, \qquad x \in X.$$

The above result holds true also for complex spaces, however with a worse constant $1 - 16\varepsilon$ (cf. [25, Theorem 3.5]).

Let us only mention that similar problems were also considered for other types of orthogonality, like isosceles orthogonality [11], ρ-orthogonality [12, 13, 30], ρ_*-orthogonality [17], bisectric orthogonality [37], Roberts orthogonality [38, 39] and others. The problem was also studied in other structures like Hilbert modules [19, 26] and for concrete spaces like the space of bounded linear operators [27]. See also [9, 15, 32].

3.3.2.2   AOR Operators

In an analogous manner as in the previous part we define the class of operators which approximately reverse orthogonality. For a given $\varepsilon \in [0, 1)$ we say that T is ε-*orthogonality reversing* (ε-OR) if

$$x \perp_B y \quad \Longrightarrow \quad Ty \perp_B^\varepsilon Tx, \qquad x, y \in X. \tag{3.10}$$

We will show that ε-OR operators (with positive ε) may exists on X, even if $\dim X \geq 3$ and X is not an inner product space (as opposed to the case $\varepsilon = 0$).

Theorem 3.5 *Let X be a real, uniformly convex normed space. Then, each linear operator $T : X \to X$ such that $[T] > 0$ satisfies (3.10) with some $\varepsilon < 1$.*

Proof Let T be a linear operator, $[T] > 0$ and assume that the assertion does not hold. Thus for any increasing sequence $\varepsilon_n \nearrow 1$ there exist sequences of unit vectors $x_n, y_n \in S_X$ such that

$$x_n \perp_B y_n \quad \text{and} \quad \frac{Ty_n}{\|Ty_n\|} \,\, \not\perp_B^{\varepsilon_n} \,\, \frac{Tx_n}{\|Tx_n\|}, \qquad n \in \mathbb{N}$$

(we use homogeneity of $\perp_B^\varepsilon$). Since $x_n \perp_B y_n$, we have in particular

$$\|x_n + y_n\| \geq \|x_n\| = 1, \quad \text{and} \quad \|x_n - y_n\| \geq \|x_n\| = 1, \qquad n \in \mathbb{N}. \tag{3.11}$$

Now, since $\frac{Ty_n}{\|Ty_n\|} \,\, \not\perp_B^{\varepsilon_n} \,\, \frac{Tx_n}{\|Tx_n\|}$, applying (3.2), for any $\varphi \in J\left(\frac{Ty_n}{\|Ty_n\|}\right)$ we have

$$\left| \varphi\left(\frac{Tx_n}{\|Tx_n\|}\right) \right| > \varepsilon_n.$$

Without loss of generality, we may assume $\varphi\left(\frac{Tx_n}{\|Tx_n\|}\right) > \varepsilon_n$ (otherwise we replace x_n by $-x_n$ and use (3.11)). Now, we have

$$1 + \varepsilon_n < \varphi\left(\frac{Ty_n}{\|Ty_n\|}\right) + \varphi\left(\frac{Tx_n}{\|Tx_n\|}\right) = \varphi\left(\frac{Ty_n}{\|Ty_n\|} + \frac{Tx_n}{\|Tx_n\|}\right)$$

$$\leq \left\| \frac{Ty_n}{\|Ty_n\|} + \frac{Tx_n}{\|Tx_n\|} \right\| \leq 2.$$

Letting $n \to \infty$ we get

$$\lim_{n \to \infty} \left\| \frac{T y_n}{\|T y_n\|} + \frac{T x_n}{\|T x_n\|} \right\| = 2.$$

The space X was assumed uniformly convex, thus it follows (cf. [18, Fact 9.5])

$$\lim_{n \to \infty} \left\| \frac{T y_n}{\|T y_n\|} - \frac{T x_n}{\|T x_n\|} \right\| = 0.$$

Since $x_n \perp_{\text{B}} y_n$, there is also $\frac{x_n}{\|T x_n\|} \perp_{\text{B}} \frac{y_n}{\|T y_n\|}$, and hence

$$\frac{1}{\|T\|} \leq \left\| \frac{x_n}{\|T x_n\|} \right\| \leq \left\| \frac{x_n}{\|T x_n\|} - \frac{y_n}{\|T y_n\|} \right\| \leq \frac{1}{[T]} \cdot \left\| T\left(\frac{x_n}{\|T x_n\|} - \frac{y_n}{\|T y_n\|} \right) \right\|$$

$$= \frac{1}{[T]} \cdot \left\| \frac{T x_n}{\|T x_n\|} - \frac{T y_n}{\|T y_n\|} \right\|.$$

So it follows

$$\frac{1}{\|T\|} \leq \frac{1}{[T]} \cdot \left\| \frac{T x_n}{\|T x_n\|} - \frac{T y_n}{\|T y_n\|} \right\| \to 0$$

as $n \to \infty$, a contradiction.

Notice that the above theorem can be applied for the identity operator on the considered space X. It follows then, that in each real uniformly convex normed space the Birkhoff–James orthogonality relation has the property

$$x \perp_{\text{B}} y \implies y \perp_{\text{B}}^{\varepsilon} x, \qquad x, y \in X \tag{3.12}$$

with some $\varepsilon \in [0, 1)$. We call the above property an *approximate symmetry* (or, more precisely, ε-symmetry) of the Birkhoff–James orthogonality $\perp_{\text{B}}$. This notion has been introduced and studied in [14]. Theorem 3.5 yields (cf. also [14, Theorem 4.1]) that the Birkhoff–James orthogonality in a real uniformly convex normed space is approximately symmetric. The same holds true, in particular, for a finite-dimensional real smooth normed space (cf. [14, Theorem 4.2]). But generally it is not true—there are normed spaces (or classes of spaces) for which the condition (3.12) does not hold with any $\varepsilon \in [0, 1)$. A simple example is the plane $\mathbb{R}^2$ with the *maximum* norm. We refer to [14] for detailed discussion on this subject.

3.4 Stability Problems

The stability problem for functional properties may be posed in an analogous way
as for functional equations. Namely, if a mapping satisfies some relation (property)
approximately only, we may ask whether there exists another mapping which is
close to the original one and which exactly satisfies the considered property.

The problem of stability of the orthogonality preserving property, posed in [6],
has been studied quite extensively. The first result obtained in [7, Theorem 4] for
finite-dimensional inner product spaces was generalized in [29] to arbitrary Hilbert
spaces.

Theorem 3.6 ([29], Theorem 2.3) *Let X, Y be Hilbert spaces. Then, for each
linear mapping $f : X \rightarrow Y$ satisfying (3.7) there exists a linear orthogonality
preserving mapping $T : X \rightarrow Y$ such that*

$$\|f - T\| \leq \left(1 - \sqrt{\frac{1 - \varepsilon}{1 + \varepsilon}} \right) \min\{\|f\|, \|T\|\}.$$

Additionally, in the case where $X = Y$, the constant $1 - \sqrt{\frac{1-\varepsilon}{1+\varepsilon}}$ can be replaced by
$\frac{1}{2}\left(1 - \sqrt{\frac{1-\varepsilon}{1+\varepsilon}} \right)$ (cf. [33, Theorem 5.4]).

Later, the considerations were carried on in normed spaces (cf. [8, 11, 25]) and
it is known that for some normed spaces the orthogonality preserving property is
stable. Actually, stability of the orthogonality preserving property is equivalent to
the stability of linear isometries (SLI) property. We say that a pair (X, Y) has got the
(SLI) property if there exists a function $\delta : [0, 1) \rightarrow \mathbb{R}_+$ satisfying $\lim_{\varepsilon \to 0} \delta(\varepsilon) = 0$
such that whenever T is an ε-isometry (i.e., $| \|Tx\| - \|x\| | \leq \varepsilon \|x\|$, $x \in X$), then
there exists an isometry U such that $\|T - U\| \leq \delta(\varepsilon)$. The function δ may depend
on X and Y but not on T.

For the orthogonality reversing property the situation is different since it may
happen that there exists an (ε-OR) operator which cannot be approximated by an
(OR) operator, simply because such an operator does not exist in the considered
space. Indeed, as it was mentioned in Sect. 3.3.1, a normed space X admits an (OR)
operator only if $\dim X = 2$ or X is an inner product space. In the latter case the
problem reduces to stability of the orthogonality preserving property. Therefore it is
reasonable to restrict our (OR)-stability considerations to two-dimensional spaces
only, even though we can formulate a formally correct result for a wider class of
normed spaces.

Let us recall now an auxiliary result which will be used in the proof of our
stability theorem. It was proved by Wójcik [31] who considered a general problem
of transferring one relation into another and the stability of this property.

Theorem 3.7 ([31], Theorem 3) *Let X and Y be finite-dimensional normed spaces
and let $R_1 \subset X \times X$, $R_2 \subset Y \times Y$, $R_2^\varepsilon \subset Y \times Y$. Suppose that for all $\varepsilon \in [0, 1)$, the
relations R_2^ε are weakly homogeneous, i.e.,*

$$x \, R_2^{\varepsilon} \, y \quad \Longrightarrow \quad \alpha x \, R_2^{\varepsilon} \, \alpha y, \qquad x, y \in Y, \ \alpha \in \mathbb{R}.$$

Assume that the family of relations $\{R_2^{\varepsilon}\}_{\varepsilon \in [0,1)}$ *is continuous with respect to the relation* R_2, *i.e., for each sequence* $(\varepsilon_n)_{n \in \mathbb{N}}$ *such that* $0 \leq \varepsilon_n < 1$ *and* $\lim_{n \to \infty} \varepsilon_n = 0$ *and for all sequences* $(a_n)_{n \in \mathbb{N}}$, $(b_n)_{n \in \mathbb{N}}$ *in* Y *such that* $\lim_{n \to \infty} a_n = a$, $\lim_{n \to \infty} b_n = b \, (a, b \in Y)$ *we have*

$$\left(a_n \, R_2^{\varepsilon_n} \, b_n, \ n \in \mathbb{N} \right) \quad \Longrightarrow \quad a \, R_2 \, b.$$

Then, for an arbitrary $\delta > 0$ *there exists* $\varepsilon > 0$ *such that for any linear mapping* $f : X \to Y$ *satisfying*

$$x \, R_1 \, y \quad \Longrightarrow \quad f x \, R_2^{\varepsilon} \, f y, \qquad x, y \in X$$

there exists a linear mapping $g : X \to Y$ *satisfying*

$$x \, R_1 \, y \quad \Longrightarrow \quad g x \, R_2 \, g y, \qquad x, y \in X$$

such that

$$\| f - g \| \leq \delta \min\{\| f \|, \| g \|\}.$$

Applying the above theorem we immediately get the (OR)-stability.

Theorem 3.8 *Let* X *be a finite-dimensional normed space. For each* $\delta > 0$ *there exists* $\varepsilon > 0$ *such that for any linear mapping* $A : X \to X$ *satisfying* (3.10) *there exists a linear mapping* $T : X \to X$ *satisfying* (3.4) *and*

$$\| A - T \| \leq \delta \cdot \min\{\| A \|, \| T \|\}.$$

Proof Define a relation $R_1 \subset X \times X$ by $x \, R_1 \, y \ \Leftrightarrow \ x \perp_B y$. For any $\varepsilon \in [0, 1)$ define $R_2^{\varepsilon} \subset X \times X$ by $x \, R_2^{\varepsilon} \, y \ \Leftrightarrow \ y \perp_B^{\varepsilon} x$. Since the orthogonality relations $\perp_B$ and $\perp_B^{\varepsilon}$ satisfy the homogeneity and continuity properties, the assertion follows from Theorem 3.7.

Although we only assumed that the dimension of X if finite, we have to stress again that this result is essential only for two-dimensional spaces. If dim $X \geq 3$ and X is not an inner product space, then no linear nonzero mapping satisfying (3.4) exists and therefore also a nonzero mapping A satisfying (3.10) cannot exist for all $\varepsilon > 0$.

Corollary 3.1 *Let* X *be a normed space with* $3 \leq \dim X < \infty$ *and which is not an inner product space. Then there exists* $\varepsilon_0 \in (0, 1)$ *such that for all* $\varepsilon \leq \varepsilon_0$ *the set of all nonzero linear mappings* $A : X \to X$ *satisfying (with this* ε*)* (3.10) *is empty.*

Notice that there are also two-dimensional normed spaces which does not admit (OR) operators and the above corollary applies also to them.

Bearing the above remarks in mind, Theorem 3.8 may be reformulated as follows.

Theorem 3.9 *Let X be a finite-dimensional normed space. There exists $\varepsilon_* \in (0, 1]$ (depending on X) such that for a linear mapping $A \colon X \to X$ satisfying (3.10) with $\varepsilon \in [0, \varepsilon_*)$, there exists a linear mapping $T \colon X \to X$ satisfying (3.4) and*

$$\|A - T\| \leq \delta(\varepsilon) \, \min\{\|A\|, \|T\|\},$$

where $\delta(\varepsilon)$ satisfies $\delta(\varepsilon) \to 0$ as $\varepsilon \to 0$.

Now, we consider a reverse problem. Given an operator A which is close to an orthogonality reversing operator T, we show that A is approximately orthogonality reversing. For reasons discussed above, we will assume dim $X = 2$. If X is an inner product space the notions of (approximate) orthogonality reversing and preserving properties coincide and this case will be covered by Theorem 3.12 (see the remark following this theorem).

Let us recall some notions and respective properties. Let ρ_X denote the modulus of smoothness of X, i.e.,

$$\rho_X(t) = \sup \left\{ \frac{\|x + th\| + \|x - th\| - 2}{2} : x, h \in S_X \right\}, \qquad t > 0.$$

It is known that (cf. [2, Proposition A.4])

$$t \mapsto \frac{\rho_X(t)}{t} \text{ is a nondecreasing mapping} \tag{3.13}$$

and (cf. [2, 18]) X is uniformly smooth if and only if

$$\frac{\rho_X(t)}{t} \to 0, \quad \text{as} \quad t \to 0. \tag{3.14}$$

Smoothness of X implies card $J(z) = 1$ for all $z \in X \setminus \{0\}$, i.e., $J(z) = \{\varphi_z\}$. Moreover (cf. [2, Proposition A.5]), if φ_x, φ_y are supporting functionals in x and y, respectively, then

$$\|\varphi_x - \varphi_y\| \leq \frac{\rho_X \left(2 \left\| \frac{x}{\|x\|} - \frac{y}{\|y\|} \right\| \right)}{\left\| \frac{x}{\|x\|} - \frac{y}{\|y\|} \right\|}. \tag{3.15}$$

Now we are able to state and prove the announced result. Note that if X is a finite dimensional normed space, then X is smooth if and only if X is uniformly smooth.

Theorem 3.10 *Let X be a real, two-dimensional and smooth space. Let $0 \neq T \in \mathcal{L}(X)$ satisfy (3.4) and let $A \in \mathcal{L}(X)$ satisfy, with $\gamma \in (0, 1)$,*

$$\|T - A\| \leq \gamma \, [T]. \tag{3.16}$$

Then A satisfies (3.10) *with* $\varepsilon = \varepsilon(\gamma)$ *where the mapping* $\gamma \mapsto \varepsilon(\gamma)$ *depends only on X and satisfies the condition*

$$\varepsilon(\gamma) \to 0, \quad as \ \gamma \to 0. \tag{3.17}$$

Proof Notice that property (3.4) yields that T is injective and since X is finite dimensional, $[T] > 0$. Let $x, y \in X$ and $x \perp_{\mathrm{B}} y$. We may assume that $x, y \in S_X$. For $a, b \in X \setminus \{0\}$ we have the Massera-Schaffer inequality (cf. [24]):

$$\left\| \frac{a}{\|a\|} - \frac{b}{\|b\|} \right\| \leq \frac{2\|a - b\|}{\max\{\|a\|, \|b\|\}},$$

thus, in particular,

$$\left\| \frac{a}{\|a\|} - \frac{b}{\|b\|} \right\| \leq \frac{2\|a - b\|}{\|b\|}. \tag{3.18}$$

Applying the above inequality and (3.16) we get

$$\left\| \frac{Ay}{\|Ay\|} - \frac{Ty}{\|Ty\|} \right\| \leq \frac{2\|Ay - Ty\|}{\|Ty\|} \leq \frac{2\|A - T\|}{[T]} \leq \frac{2\gamma[T]}{[T]} = 2\gamma. \tag{3.19}$$

Moreover,

$$\big| \|Tx\| - \|Ax\| \big| \leq \|Tx - Ax\| \leq \|T - A\| \, \|x\| \leq \gamma[T] \, \|x\| \leq \gamma \|Tx\|,$$

whence $(1 - \gamma)\|Tx\| \leq \|Ax\|$ and

$$\|Tx\| \leq \frac{1}{1 - \gamma} \|Ax\|. \tag{3.20}$$

Combining smoothness with (3.2) we see that $a \perp_{\mathrm{B}}^{\varepsilon} b \Leftrightarrow |\varphi_a(b)| \leq \varepsilon\|b\|$ for $a, b \in X \setminus \{0\}$ and a unique supporting functional φ_a. Since $Ty \perp_{\mathrm{B}} Tx$, we have $\varphi_{Ty}(Tx) = 0$ and

$$\begin{aligned}
\big| \varphi_{Ay}(Ax) \big| &= \big| \varphi_{Ay}(Ax) - \varphi_{Ty}(Tx) \big| \\
&\leq \big| \varphi_{Ay}(Ax) - \varphi_{Ay}(Tx) \big| + \big| \varphi_{Ay}(Tx) - \varphi_{Ty}(Tx) \big| \\
&= \big| \varphi_{Ay}(Ax - Tx) \big| + \big| (\varphi_{Ay} - \varphi_{Ty})(Tx) \big| \\
&\leq \|Ax - Tx\| + \big\| \varphi_{Ay} - \varphi_{Ty} \big\| \cdot \|Tx\|,
\end{aligned}$$

whence, using (3.15),

$$\left|\varphi_{Ay}(Ax)\right| \leq \|Ax - Tx\| + 2 \cdot \frac{\rho_X\left(2\left\|\frac{Ay}{\|Ay\|} - \frac{Ty}{\|Ty\|}\right\|\right)}{2\left\|\frac{Ay}{\|Ay\|} - \frac{Ty}{\|Ty\|}\right\|}\|Tx\|.$$

Now, applying (successively) (3.13), (3.19), (3.16) and (3.20) we obtain

$$\left|\varphi_{Ay}(Ax)\right| \leq \|Ax - Tx\| + 2 \cdot \frac{\rho_X(4\gamma)}{4\gamma} \cdot \|Tx\|$$

$$\leq \gamma[T]\|x\| + 2 \cdot \frac{\rho_X(4\gamma)}{4\gamma} \cdot \frac{1}{1-\gamma}\|Ax\|$$

$$\leq \gamma\|Tx\| + 2 \cdot \frac{\rho_X(4\gamma)}{4\gamma} \cdot \frac{1}{1-\gamma}\|Ax\|.$$

Finally, using (3.20),

$$\left|\varphi_{Ay}(Ax)\right| \leq \frac{\gamma}{1-\gamma}\|Ax\| + 2 \cdot \frac{\rho_X(4\gamma)}{4\gamma} \cdot \frac{1}{1-\gamma}\|Ax\|$$

$$= \left(\frac{\gamma}{1-\gamma} + 2 \cdot \frac{\rho_X(4\gamma)}{4\gamma} \cdot \frac{1}{1-\gamma}\right)\|Ax\|.$$

Define $\varepsilon(\gamma) := \frac{\gamma}{1-\gamma} + 2 \cdot \frac{\rho_X(4\gamma)}{4\gamma} \cdot \frac{1}{1-\gamma}$. Then (3.14) yields (3.17). It follows from the above inequalities that $\left|\varphi_{Ay}(Ax)\right| \leq \varepsilon(\gamma) \cdot \|Ax\|$, whence also, cf. (3.2),

$$Ay \perp_{\mathrm{B}}^{\varepsilon(\gamma)} Ax.$$

Finally, A satisfies (3.10) with $\varepsilon = \varepsilon(\gamma)$.

The following theorem is obtained as a consequence.

Theorem 3.11 *Let X be a real, two-dimensional and smooth normed space. Let a nonzero operator $T \in \mathscr{L}(X)$ satisfy (3.4) and let $A \in \mathscr{L}(X)$ satisfy, with $\alpha \in \left(0, \frac{\|T\|}{\|T\|}\right)$, inequality*

$$\|T - A\| \leq \alpha \|T\|. \tag{3.21}$$

Then A satisfies (3.10) with $\varepsilon = \widehat{\varepsilon}(\alpha)$ where the mapping $\alpha \mapsto \widehat{\varepsilon}(\alpha)$ depends on X only and satisfies the condition $\widehat{\varepsilon}(\alpha) \to 0$, as $\alpha \to 0$.

Proof Consider $\gamma := \alpha \frac{\|T\|}{[T]} \in (0, 1)$. Then (3.21) yields $\|T - A\| \leq \alpha \|T\| = \gamma[T]$. Now, for the mapping $\varepsilon(\cdot)$ from Theorem 3.10 we define $\widehat{\varepsilon}(\alpha) := \varepsilon\left(\alpha \cdot \frac{\|T\|}{[T]}\right)$.

Finally, we goes back to the orthogonality preserving property. The following result is obtained by applying similar methods as in two previous theorems. We will present, however, the whole proof for the reader's convenience. We do not need to require finite-dimensionality of X but its uniform smoothness.

Theorem 3.12 *Let X be a real, uniformly smooth normed space and let $0 \neq T \in \mathcal{L}(X)$ satisfy (3.3). If $A \in \mathcal{L}(X)$ satisfies, with $\gamma \in (0, 1)$, inequality $\|T - A\| \leq \gamma \|T\|$, then A satisfies (3.7) with $\varepsilon = \varepsilon(\gamma)$ such that $\varepsilon(\gamma) \to 0$ as $\gamma \to 0$.*

Proof Let $x, y \in X$ and $x \perp_B y$. We may assume that $x, y \in S_X$ and thus (T is a similarity) $\|Tx\| = \|Ty\| = \|T\|$. Applying (3.18) and $\|T - A\| \leq \gamma \|T\|$ we get

$$\left\| \frac{Ay}{\|Ay\|} - \frac{Ty}{\|Ty\|} \right\| \leq \frac{2\|Ay - Ty\|}{\|Ty\|} \leq \frac{2\|A - T\|}{\|T\|} \leq \frac{2\gamma\|T\|}{\|T\|} = 2\gamma. \qquad (3.22)$$

Moreover, $\big| \|Tx\| - \|Ax\| \big| \leq \|Tx - Ax\| \leq \|T - A\| \leq \gamma\|T\| = \gamma\|Tx\|$, whence (3.20) follows. Since $Tx \perp_B Ty$, we have $\varphi_{Tx}(Ty) = 0$ (uniqueness of the supporting functional is guaranteed by smoothness). Therefore

$$|\varphi_{Ax}(Ay)| = |\varphi_{Ax}(Ay) - \varphi_{Tx}(Ty)|$$
$$\leq |\varphi_{Ax}(Ay - Ty)| + |(\varphi_{Ax} - \varphi_{Tx})(Ty)|$$
$$\leq \|Ay - Ty\| + \|\varphi_{Ax} - \varphi_{Tx}\| \cdot \|Ty\|,$$

and by (3.15),

$$|\varphi_{Ax}(Ay)| \leq \|Ay - Ty\| + 2 \cdot \frac{\rho_X \left(2 \left\| \frac{Ax}{\|Ax\|} - \frac{Tx}{\|Tx\|} \right\| \right)}{2 \left\| \frac{Ax}{\|Ax\|} - \frac{Tx}{\|Tx\|} \right\|} \|Ty\|.$$

Proceeding analogously as in the last part of the proof of Theorem 3.10, applying now (3.22), we get

$$|\varphi_{Ax}(Ay)| \leq \left(\frac{\gamma}{1 - \gamma} + 2 \cdot \frac{\rho_X(4\gamma)}{4\gamma} \cdot \frac{1}{1 - \gamma} \right) \|Ay\|.$$

Finally, with $\varepsilon(\gamma) := \frac{\gamma}{1-\gamma} + 2 \cdot \frac{\rho_X(4\gamma)}{4\gamma} \cdot \frac{1}{1-\gamma}$ we have $|\varphi_{Ax}(Ay)| \leq \varepsilon(\gamma) \cdot \|Ay\|$, whence $Ax \perp_B^{\varepsilon(\gamma)} Ay$. Thus A satisfies (3.7).

Notice, that the above theorem covers in particular the case where $\dim X \geq 3$ and T satisfies (3.4) instead of (3.3). Then X must be an inner product space (which obviously is uniformly convex), (3.4) is equivalent to (3.3) and (3.7) is equivalent to (3.10) so A satisfies (3.10).

References

1. Alonso, J., Martini, H., Wu, S.: On Birkhoff orthogonality and isosceles orthogonality in normed linear spaces. Aequationes Math. **83**, 153–189 (2012)
2. Benyamini, Y., Lindenstrauss, J.: In: Geometric Nonlinear Functional Analysis, vol. 1. Colloquium Publications/AMS 48, Providence (2000)
3. Birkhoff, G.: Orthogonality in linear metric spaces. Duke Math. J. **1**, 169–172 (1935)
4. Blanco, A., Turnšek, A.: On maps that preserve orthogonality in normed spaces. Proc. R. Soc. Edinb. Sect. A Math. **136**, 709–716 (2006)
5. Chmieliński, J.: On an ε-Birkhoff orthogonality. J. Inequal. Pure Appl. Math. **6**, Article 79 (2005)
6. Chmieliński, J.: Linear mappings approximately preserving orthogonality. J. Math. Anal. Appl. **304**, 158–169 (2005)
7. Chmieliński, J.: Stability of the orthogonality preserving property in finite-dimensional inner product spaces. J. Math. Anal. Appl. **318**, 433–443 (2006)
8. Chmieliński, J.: Remarks on orthogonality preserving mappings in normed spaces and some stability problems. Banach J. Math. Anal. **1**, 117–124 (2007)
9. Chmieliński, J.: Orthogonality Preserving Property and its Ulam Stability, Chapter 4. In: Rassias, T.M., Brzdęk, J. (eds.) Functional Equations in Mathematical Analysis. Springer Optimization and Its Applications, vol. 52. Springer Science+Business Media, LLC, pp. 33–58 (2012)
10. Chmieliński, J.: Operators reversing orthogonality in normed spaces. Adv. Oper. Theory **1**(1), 8–14 (2016)
11. Chmieliński, J., Wójcik, P.: Isosceles-orthogonality preserving property and its stability. Nonlinear Anal. **72**, 1445–1453 (2010)
12. Chmieliński, J., Wójcik, P.: On a ρ-orthogonality. Aequationes Math. **80**, 45–55 (2010)
13. Chmieliński, J., Wójcik, P.: ρ-orthogonality and its preservation—revisited. In: Brzdęk, J., et al. (ed.) Recent developments in functional equations and inequalities, vol. 99. Banach Center Publications, pp. 17–30 (2013)
14. Chmieliński, J., Wójcik, P.: Approximate symmetry of Birkhoff orthogonality. J. Math. Anal. Appl. **461**, 625–640 (2018)
15. Chmieliński, J., Łukasik, R., Wójcik, P.: On the stability of the orthogonality equation and the orthogonality-preserving property with two unknown functions. Banach J. Math. Anal. **10**, 828–847 (2016)
16. Chmieliński, J., Stypuła, T., Wójcik, P.: Approximate orthogonality in normed spaces and its applications. Linear Algebra Appl. **531**, 305–317 (2017)
17. Dehghani, M., Zamani, A.: Linear mappings approximately preserving ρ_*-orthogonality. Indag. Math. **28**, 992–1001 (2017)
18. Fabian, M., Habala, P., Hájek, P., Montesinos, V., Zizler, V.: Banach space theory. In: The Basis for Linear and Nonlinear Analysis. CMS Books in Mathematics. Springer, Berlin (2011)
19. Ilišević, D., Turnšek, A.: Approximately orthogonality preserving mappings on C^*-modules. J. Math. Anal. Appl. **341**, 298–308 (2008)
20. James, R.C.: Orthogonality in normed linear spaces. Duke Math. J. **12**, 291–301 (1945)
21. James, R.C.: Orthogonality and linear functionals in normed linear spaces. Trans. Am. Math. Soc. **61**, 265–292 (1947)
22. Koldobsky, A.: Operators preserving orthogonality are isometries. Proc. R. Soc. Edinb. Sect. A Math. **123**, 835–837 (1993)
23. Martini, H., Swanepoel, K.J.: Antinorms and Radon curves. Aequationes Math. **72**(1–2), 110–138 (2006)
24. Massera, J.L., Schaffer, J.J.: Linear differential equations and functional analysis I. Ann. Math. **67**, 517–573 (1958)
25. Mojškerc, B., Turnšek, A.: Mappings approximately preserving orthogonality in normed spaces. Nonlinear Anal. **73**, 3821–3831 (2010)

26. Moslehian, M.S., Zamani, A.: Mappings preserving approximate orthogonality in Hilbert C^*-modules. Math. Scand. **122**, 257–276 (2018)
27. Paul, K., Sain, D., Mal, A.: Approximate Birkhoff–James orthogonality in the space of bounded linear operators. Linear Algebra Appl. **537**, 348–357 (2017)
28. Radon, J.: Über eine besondere art ebener konvexer Kurven. Leipz. Ber. **68**, 123–128 (1916)
29. Turnšek, A.: On mappings approximately preserving orthogonality. J. Math. Anal. Appl. **336**, 625–631 (2007)
30. Wójcik, P.: Linear mappings preserving ρ-orthogonality. J. Math. Anal. Appl. **386**, 171–176 (2012)
31. Wójcik, P.: On mappings approximately transferring relations in finite-dimensional normed spaces. Linear Algebra Appl. **460**, 125–135 (2014)
32. Wójcik, P.: Linear mappings approximately preserving orthogonality in real normed spaces. Banach J. Math. Anal. **9**, 134–141 (2015)
33. Wójcik, P.: On certain basis connected with operator and its applications. J. Math. Anal. Appl. **423**, 1320–1329 (2015)
34. Wójcik, P.: Orthogonality of compact operators. Expo. Math. **35**, 86–94 (2017)
35. Wójcik, P.: Inner product spaces and operators reversing orthogonality. Khayyam J. Math. **3**, 22–24 (2017)
36. Wu, S., He, Ch., Yang, G.: Orthogonalities, linear operators, and characterization of inner product spaces. Aequationes Math. **91**, 969–978 (2017)
37. Zamani, A.: Approximately bisectrix-orthogonality preserving mappings. Comment. Math. **54**, 167–176 (2014)
38. Zamani, A., Moslehian, M.S.: Approximate Roberts orthogonality. Aequationes Math. **89**, 529–541 (2015)
39. Zamani, A., Moslehian, M.S.: Approximate Roberts orthogonality sets and $(\delta - \varepsilon) - (a, b)$-isosceles orthogonality preserving mappings. Aequationes Math. **90**, 647–659 (2016)

Chapter 4
Optimal Forward Contract Design for Inventory: A Value-of-Waiting Analysis

Roy O. Davies and Adam J. Ostaszewski

Abstract A classical inventory problem is studied from the perspective of embedded options, reducing inventory-management to the design of optimal contracts for forward delivery of stock (commodity). Financial option techniques à la Black-Scholes are invoked to value the additional 'option to expand stock'. A simplified approach which ignores distant time effects identifies an optimal 'time to deliver' and an optimal 'amount to deliver' for a production process run in continuous time modelled by a Cobb-Douglas revenue function. Commodity prices, quoted in initial value terms, are assumed to evolve as a geometric Brownian process with positive (inflationary) drift. Expected revenue maximization identifies an optimal 'strike price' for the expansion option to be exercised and reveals the underlying martingale in a truncated (censored) commodity price. The paper establishes comparative statics of the censor, using sensitivity analysis on the related **censor functional equation**; key here is that the censor, as a function of the drift and volatility of price, is the solution of a functional equation. Asymptotic approximation allows a tractable analysis of the optimal timing.

Keywords Value of waiting · Censor functional equation · Optimal forward contract · Optimal exercise price · Optimal timing · Comparative statics · Asymptotic approximation · Martingale

Mathematics Subject Classification (2010) Primary 91B32, 91B38, 39B22; Secondary 91G80, 49J55, 49K40

R. O. Davies
School of Mathematics and Actuarial Science, University of Leicester, Leicester, UK

A. J. Ostaszewski (✉)
Department of Mathematics, London School of Economics, London, UK
e-mail: a.j.ostaszewski@lse.ac.uk

4.1   Problem Formulation and Model

We enhance a classical inventory-management problem by studying its embedded options, reducing the problem to the design of optimal contracts for forward delivery of inventory. The approach borrows much from the Black-Scholes model for valuing financial options (see Musiela and Rutkowski [11, Chapter 5]) and reveals the underlying martingale to be a truncated (right-censored) discounted commodity price.

A production process runs continuously over a unit time interval and the manager is permitted to acquire raw input materials at two dates: initially, at time $t = 0$, and again at one other time $\theta < 1$, selected freely, but committed to at time $t = 0$. This framework is intended as a proxy for a multi-stage inventory management problem, since 'proximal' effects of forward contracting, as represented by the date θ, are more significant than any additional 'distal' dates for forward delivery. Distal dates for additional forward deliveries are thus neglected in this model (see the 'Interpretation' paragraph at the end of Sect. 4.4). Inputs are consumed in a continuous production process which creates an instantaneous revenue rate at time t equal to $f(x_t)$ (quoted in present-value terms), where x_t is the instantaneous input rate of consumed material. To begin with $f(x)$ is, as usual, an **Inada-type** increasing function, viz. twice differentiable, unboundedly increasing from zero, with slope unbounded at the origin and strictly decreasing to zero at infinity; eventually $f(x)$ is specialized to a Cobb-Douglas production function. The revenue from any interval $[a, b]$ is taken to be

$$\int_a^b f(x_t)dt.$$

If the manager decides to use up a proportion θx in the period $[0, \theta]$ then, with θ fixed, the Euler-Lagrange equation implies that a constant instantaneous input rate equal to x is optimal. A further quantity $(1 - \theta)y$ may similarly be consumed in the remaining time interval. If the quantity $(1 - \theta)y$ is made up from a contracted forward delivery of $(1 - \theta)u$ and a possible supplement, purchased at time θ, of a non-negative quantity $(1 - \theta)z$, the revenue from the second interval will be

$$\int_\theta^1 f(x_t)dt = (1 - \theta)f(u + z).$$

Values here and below are quoted in discounted terms, i.e. present-value terms relative to time $t = 0$. (We side-step a discussion of the relevant discount factor. In brief, discounting would be done relative to the required rate of return on capital given the risk-class of the investment project; see Dixit and Pindyck [4, Chapter 4, Section 2].)

Whilst the model of revenue assumes a steady (deterministic) market for the output, the input prices are assumed stochastic. (We prefer this modelling choice to the more general approach of including also a stochastic output price. Indeed, what then determines optimal behaviour is the ratio of the two prices; so, in a sense, the present simpler arrangement subsumes it.) Specifically, we suppose that at time 0 the price of inputs is $b_0 = 1$, and that, as time t progresses, the present value of the spot price, b_t, follows the stochastic differential equation

$$\frac{db_t}{b_t} = \bar{\mu}dt + \bar{\sigma}dw_t, \tag{4.1}$$

with w_t a standard Wiener process. It is assumed that the constant growth rate $\bar{\mu}$ is positive, thus modelling anticipated inflation. The (present-value/discounted) expected price at time t is $e^{\bar{\mu}t}$, and so the price is expected to grow above the initial price of unity. The price b_t is log-normally distributed with a mean which we denote by $v = (\bar{\mu} - \frac{1}{2}\bar{\sigma}^2)t$ and a variance $\sigma^2 = \bar{\sigma}^2 t$. Write $q_t(\cdot) = q(\cdot, \bar{\mu}t, \bar{\sigma}\sqrt{t})$ for the density of b_t. Conditional on the initial choice of θ, the expected future revenue consequent on the choice of x, u and z (with z selected at time θ) is

$$\theta(f(x) - x) + (1 - \theta)\left(\int_0^\infty \{f(z + u) - bz\}q_\theta(b)db - u\right). \tag{4.2}$$

This is a classical inventory problem but amended by the explicit inclusion of the 'option to expand inventory' (choice of z) and of a 'forward' contract (choice of u). We shall evaluate the embedded option in a framework reminiscent of Black-Scholes option-pricing. The 'forward contract' is construed here as a contract signed at the earlier date $t = 0$ with an agreed specified delivered quantity, u, a specified delivery date $t = \theta$, and a price standardized here to *unity* per unit delivered. The latter standardization fixes the unit of money, since, as is well-known, in the absence of arbitrage and storage costs the forward price equals the price of inputs at the initial time of contracting, compounded up to term-value at the required rate of interest. Note that the advance purchase of u has by assumption nil resale value on delivery. This makes the delivered asset a 'non-tradeable' commodity, so that the usual martingale valuation approach applied to a discounted security price is not immediately appropriate; our analysis makes use of dynamical programming, as in Eberly and Van Mieghem [5], and thereby identifies the underlying martingale structure via an appropriately truncated (right-censored) price.

Apart from offering a real-options approach with optimal design in mind, in contrast to the classical inventory literature (see for instance Bensousssan et al. [2], or Scarf [15]), an additional contribution of the current paper is to provide information about the sensitivity in regard to model parameters of the critical 'strike price' for stock expansion (its comparative statics and asymptotics), an issue omitted from consideration in Eberly and Van Mieghem [5].

The current study of profit dependence on timing, drift and variance is motivated by the general discrete-time multi-period model of Gietzmann and Ostaszewski [7], but with the simplifying removal of costly liquidation of inventory. There the latter feature was necessary for a more comprehensive study into the dependence of a firm's 'future value' on accounting data. Such themes are explored in [12] in this volume.

Our option-based analysis is simpler than [3], though similar in spirit. There the (retailer's) inventory control problem studies re-distribution of a storable product; one uses a (long) forward (contract) for delivery combined with an option to dispose of any excess (a put, with a lower salvage price) coupled with an option for additional supply (a call, with a penalty cost for 'emergency supply'); for background on these 'option' terms see e.g. [8]. A similar approach, albeit in discrete time, is taken in [10] using at each date a continuum of puts and calls maturing at the next date taken together with a short (negative) forward.

The rest of the paper is organized as follows. In Sect. 4.2 we study optimality conditions, which identify a threshold price level (the price censor) above which it is not worth purchasing the input. We consider its sensitivity (comparative statics) to price drift and volatility in Sect. 4.3: here we view the censor as a function of these two, identified by a functional equation. Then in Sect. 4.4 we assess the expected revenue and in Sect. 4.5 the optimal timing. Proofs (sensitivity analysis) are spread across Sects. 4.6 and 4.7, some of this in outline with details relegated to an Appendix.

4.2   Optimality: The Censor and Value of Waiting

From (4.2) the optimization problem separates into maximization over $x \geq 0$ of $f(x) - x$ (with solution specified by $f'(x) = 1$) and over $u \geq 0$ and over functions $z(.)$ of the (time $t = 0$) expectation

$$E[f(z(b) + u) - bz(b)] - u. \tag{4.3}$$

Definition For any Inada-type strictly concave function $f(x)$ define the 'indirect profit' (i.e. maximized profit) for a deterministic price b by

$$h(b) = \max_{x>0}[f(x) - bx]. \tag{4.4}$$

Evidently $h(b) = f(I(b)) - bI(b)$, where by tradition I denotes the inverse function to f'.

Theorem 1 (Optimal Forward Delivered Quantity) *In the model setting above, with time θ given, let $\tilde{b} = \tilde{b}(\mu, \sigma, \theta)$ be the scalar solving the equation*

$$E[b_\theta \wedge \tilde{b}] = b_0 = 1, \tag{4.5}$$

where b_θ denotes the random price at time θ. Then the profit-optimizing level of the advance purchase $u = u(\mu, \sigma, \theta)$ for (4.3) satisfies

$$f'(u) = \tilde{b}, \tag{4.6}$$

and the optimal expected profit is given by

$$g(\mu, \sigma) := E[h(b), b \leq \tilde{b}] + h(\tilde{b}) \cdot \Pr[b > \tilde{b}]. \tag{4.7}$$

Proof With β arbitrary, select u with $\beta = f'(u)$. Note that $h(\beta) = f(u) - \beta u$ and $h'(\beta) = -u$. Define the right-censored random variable $B_\theta = B_\theta(\beta)$ by

$$B_\theta = b_\theta \wedge \beta.$$

For given price b, the quantity $z = z(b)$ which maximizes $f(u + z) - bz$ is either zero or satisfies the first-order condition

$$f'(z + u) = b.$$

In view of the monotonicity of f' we thus have $z(b) = 0$, unless $b \leq \beta$. Given that u has been purchased at a price of unity, the profit, when $b_\theta \leq \beta$, is $f(u + z) - (b_\theta z(b_\theta) + u) = h(b_\theta) + (ub_\theta - u)$. Otherwise it is $f(u) - u = h(\beta) + u\beta - u$. Thus the expected profit is

$$\Pi(\beta) := E[h(B_\theta) + uB_\theta - u] = E[h(B_\theta)] + uE[B_\theta] - u.$$

Differentiating Π with respect to β, and noting that

$$dE[h(b_\theta \wedge \beta)]/d\beta = h'(\beta)\Pr[b_\theta \geq \beta],$$

we obtain, after some cancellations in view of $h'(\beta) = -u$, the optimality condition $E[B_\theta] = 1$ on β. The model assumption that $\bar{\mu}$ is positive ensures the existence of a solution of Eq. (4.5). With β set equal to the solution $\tilde{b}$ of Eq. (4.5) we have $\tilde{b} = f'(u)$, i.e. (4.6). □

Definition In view of the right-censoring of the price b occurring under the expectation, we call the solution of (4.5) the **censor** $\tilde{b} = \tilde{b}(\mu, \sigma, \theta)$ at time θ. This definition follows Gietzmann and Ostaszewski [6]. The censored variable is thus a martingale.

Remark It is clear from the proof above that the censor describes the upper limit of those prices which trigger the exercise of the option to expand stock. So evidently, $\tilde{b} > 1$. We return in the next section to a consideration of its behaviour. Whilst this threshold role makes the censor similar to the 'optimal ISD control limit' studied by Eberly and Van Mieghem [5], their thresholds correspond to Investing/Staying-put/Disinvesting and are distinct in respect of the treatment of capital depreciation.

Proposition 1 (Value of Waiting) *The expected profit $g(\mu, \sigma)$ defined in (4.7) obtained by optimal forward contracting is no worse than the indirect profit $h(1)$ obtained by only using purchases at initial prices, that is,*

$$h(1) < g(\mu, \sigma) = E[h(b_\theta \wedge \tilde{b})].$$

Proof This follows from a simple application of Jensen's inequality, as $h(b)$ is strictly convex in b. Indeed, we then have

$$h(1) = h(E[b_\theta \wedge \tilde{b}]) < E[h(b_\theta \wedge \tilde{b})].$$

Of course $-h(b)$ is the Fenchel dual of the strictly concave function f, so $-h(b)$ is strictly concave in b (see [14, Section 12]). In the specific case of $f(x)$ twice differentiable the asserted convexity follows from $h''(b) = -1/f''(I(b))$, where I denotes, as before, the inverse function of f'. □

4.3   Sensitivity and the Censor Functional Equation

Assuming an Inada-type production function for the geometric Brownian model adopted for the price as in (4.1), the censor equation (4.5) which defines $\tilde{b} = \tilde{b}(\mu, \sigma)$ can be re-written as

$$1 = e^\mu \Phi(W - \sigma) + \tilde{b}\Phi(-W). \tag{4.8}$$

Here $\Phi(x) = \int_{-\infty}^x \varphi(w)dw$, with $\varphi(w) = e^{-\frac{1}{2}w^2}/\sqrt{2\pi}$, denotes the standard normal cumulative distribution function, $W = w(\tilde{b})$, and

$$w(b) := \frac{\ln b - \nu}{\sigma}, \quad \text{where } \nu = \mu - \frac{1}{2}\sigma^2. \tag{4.9}$$

This formulation leads naturally to a further definition.

Definition The **normal censor** is the function $W(\mu, \sigma)$ defined for $\mu, \sigma > 0$ implicitly by the **censor functional equation** for W

$$e^{-\mu} = F(W, \sigma), \quad \text{where } F(W, \sigma) := \Phi(W - \sigma) + e^{\sigma W - \frac{1}{2}\sigma^2}\Phi(-W). \tag{4.10}$$

We note that W is well defined since $\partial F/\partial W > 0$. It is helpful to be aware of the hidden connection between the function F and the normal hazard rate $H(x) = \varphi(x)/\Phi(-x)$ (or its reciprocal, the Mills Ratio) and to use properties of this function. We refer to Kendall and Stuart [9, p. 104], or Patel and Read [13] for details. From $\varphi(\sigma - W) = e^{\sigma W - \frac{1}{2}\sigma^2}\varphi(W)$,

$$F(W, \sigma) = \varphi(\sigma - W)\left(\frac{1}{H(\sigma - W)} + \frac{1}{H(W)}\right).$$

From (4.10), $W(\mu, \sigma)$ is decreasing in μ, since $e^{-\mu}$ is decreasing. Less obvious is the fact that $W(\mu, \sigma)$ is increasing in σ since in fact $\partial W/\partial \sigma > 1$. This is shown in

Sect. 4.7, where we deduce the comparative statics of $\tilde{b}(\mu, \sigma)$ from corresponding properties of $W(\mu, \sigma)$. The main results proved there are as follows.

Theorem 5 *The censor $\tilde{b}(\mu, \sigma)$ is decreasing in the drift μ and is increasing in the standard deviation σ.*

These two properties together suggest the following result, obtained by setting $\mu = \bar{\mu}\theta$ and $\sigma = \bar{\sigma}\sqrt{\theta}$, and noting that (4.10) permits arbitrary positive θ.

Theorem 6 *The censor $\bar{b}(\theta) := \tilde{b}(\bar{\mu}\theta, \bar{\sigma}\sqrt{\theta})$ is either unimodal or increasing on the interval $0 < \theta < \infty$, according as $\bar{\mu} \geq \frac{1}{2}\bar{\sigma}^2$ or $\bar{\mu} < \frac{1}{2}\bar{\sigma}^2$.*

4.4 Cobb-Douglas Revenue: Asymptotic Results

We now assume $f(x)$ is Cobb-Douglas, specifically $f(x) = 2\sqrt{x}$, so that the indirect profit defined by (4.4) is $h(b) = b^{-1}$. This choice for the power of x inflicts no loss of generality, because in the presence of a log-normally distributed price any other choice of power is equivalent to a re-scaling of $\bar{\mu}, \bar{\sigma}$. Substituting into the definition (4.7) yields

$$g(\mu, \sigma) = e^{(\sigma^2 - \mu)}\Phi(W + \sigma) + e^{-\mu - \sigma W + \frac{1}{2}\sigma^2}\Phi(-W), \qquad (4.11)$$

as the (optimal expected) profit per unit time arising after the re-stocking date θ. We also define the associated function

$$\bar{g}(\theta) := g(\bar{\mu}\theta, \bar{\sigma}\sqrt{\theta}),$$

for $0 < \theta < \infty$ (with some re-sizing of $\bar{\mu}, \bar{\sigma}$ in mind, as in Proposition 4 of Sect. 4.5). To study these functions we are led to analyse the behaviour of first $W(\mu, \sigma)$ and then $\overline{W}(t) := W(\bar{\mu}t, \bar{\sigma}\sqrt{t})$. The following are derived in Sect. 4.6.

Proposition 2 *For fixed $\mu > 0$,*

$$W(\mu, \sigma) = -\frac{\mu}{\sigma} + \frac{1}{2}o + o(o) \quad (\sigma \to 0+).$$

Proposition 3 *For fixed $\mu > 0$ and with $\hat{\mu} := -\Phi^{-1}(e^{-\mu})$,*

$$W(\mu, \sigma) = \sigma - \hat{\mu} - \frac{1}{\sigma - \hat{\mu}}\{1 + o(1)\} \quad (\sigma \to \infty).$$

From (4.11) and standard asymptotic estimates of $\Phi(x)$ (see Abramowitz and Stegan [1, Section 7]) Theorem A below is immediate. It also turns out that $\overline{W}(t)$ behaves rather like $\pm\sqrt{t}$ (except when $\bar{\mu} = \frac{1}{2}\bar{\sigma}^2$).

Theorem A (Asymptotic Behaviour of the Profit $g(\mu, \sigma)$)

(i) $g = e^{\sigma^2 - \mu} + o(1/\sigma)$, as $\sigma \to \infty$;

(ii) $g = e^{-\mu} + (1 - e^{-\mu})\Phi(\mu/\sigma) + o(\sigma)$, as $\sigma \to 0+$.

Theorem B (Behaviour of the Profit $\bar{g}(\theta)$ at the Origin) *We have $\bar{g}'(0) = \bar{\sigma}^2$ so that*

$$\bar{g}(\theta) = 1 + \bar{\sigma}^2\theta + o(\theta) \quad (\theta \to 0).$$

Theorem C (Asymptotic Behaviour of the Profit $\bar{g}(\theta)$ at Infinity)

(i) *If $\bar{\sigma}^2 < \bar{\mu}$, then*

$$\bar{g}(\theta) = 1 + o(1/\sqrt{\theta}) \to 1+ \quad (\theta \to \infty),$$

and $\bar{g}(\theta)$ has a maximum whose location tends to infinity as $\bar{\sigma}^2 \to \bar{\mu}$.

(ii) *If $\bar{\mu} \leq \bar{\sigma}^2 < 2\bar{\mu}$, then*

$$\bar{g}(\theta) = 1 + e^{(\bar{\sigma}^2 - \bar{\mu})\theta} + o(1/\sqrt{\theta}) \quad (\theta \to \infty).$$

(iii) *If $2\bar{\mu} < \bar{\sigma}^2$, then*

$$\bar{g}(\theta) = e^{(\bar{\sigma}^2 - \bar{\mu})\theta} + o(1/\sqrt{\theta}) \quad (\theta \to \infty).$$

(iv) *If $\bar{\sigma}^2 = 2\bar{\mu}$, then*

$$\bar{g}(\theta) = \frac{1}{4} + e^{\bar{\mu}\theta}\Phi(\sqrt{2\bar{\mu}\theta}) + o(1/\sqrt{\theta}) = \frac{1}{4} + e^{\bar{\mu}\theta} + o(1/\sqrt{\theta}) \quad (\theta \to \infty).$$

For the proofs see Sect. 4.6.

Figures 4.1, 4.2, 4.3 and 4.4 with a parameter value $\bar{\mu} = 0.05$ show the computed graphs of $\bar{g}$ (red/bold) alongside the approximation (green/faint) where relevant; Fig. 4.4 shows the first to the right of the two approximations given in the case (iv).

Interpretation Under 'myopic management', i.e. in the absence of forward contracting, for a given re-stocking date θ the expected profit would be

$$E_0[h(b_\theta)] = E[1/b_\theta] = e^{(\bar{\sigma}^2 - \bar{\mu})\theta}.$$

Theorems A and C thus imply that forward contracting advantages lose significance as variance increases, or as the re-stocking date θ advances. This is ultimately our justification for excluding any additional dates for further forward deliveries.

Fig. 4.1 Typical graph of $\bar{g}(\theta)$ in the case (i) $\bar{\sigma}^2 < \bar{\mu}$; here $\bar{\mu} = 0.05$, $\bar{\sigma} = 0.1$

Fig. 4.2 Typical graph of $\bar{g}(\theta)$ (red) in the case (ii) $\bar{\mu} < \bar{\sigma}^2 < 2\bar{\mu}$; here $\bar{\mu} = 0.05$, $\bar{\sigma} = 0.25$

Fig. 4.3 Graph of $\bar{g}(\theta)$ in the case (iii) $2\bar{\mu} < \bar{\sigma}^2$ (red-green merged); here $\bar{\mu} = 0.05$, $\bar{\sigma} = 0.4$

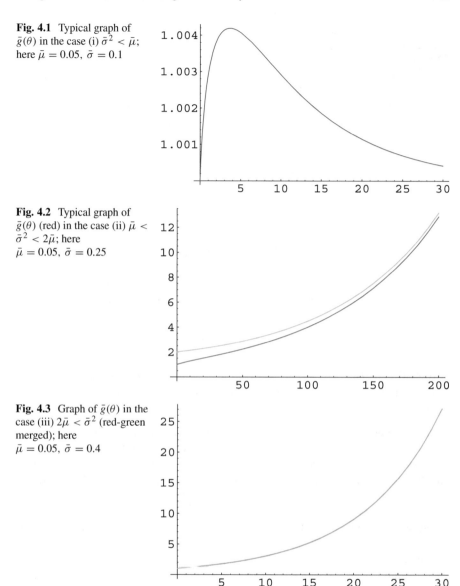

4.5   Cobb-Douglas Optimal Timing: Estimates

Assuming again as above, without much loss of generality, that $f(x) = 2\sqrt{x}$, we turn now to revenue optimization in respect of the time θ to be selected freely in $[0, 1]$. Supposing there are no associated management costs in choosing θ, the optimal revenue $R(\theta)$ for a selected value of θ is, from (4.2), given by

Fig. 4.4 Graph of $\bar{g}(\theta)$ (red) in the case (iv) $\bar{\sigma}^2 = 2\bar{\mu} = 0.1$ and of $\frac{1}{4} + e^{\bar{\mu}\theta}\Phi(\sqrt{2\bar{\mu}\theta})$

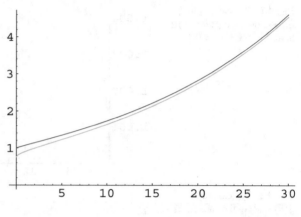

$$R(\theta) := \theta + (1 - \theta)\bar{g}(\theta),$$

the first term being justified by $h(1) = 1$. Since $\bar{g}(0) = 1$, the optimal choice of θ, assuming such exists, is given by the following first-order condition:

$$\frac{\bar{g}(\theta) - \bar{g}(0)}{\bar{g}'(\theta)} = 1 - \theta. \tag{4.12}$$

Proposition 4 *The first-order condition for R in (4.12) is satisfied for some θ with $0 < \theta < 1$. The smallest solution is a local maximum of R. If $\bar{g}$ is concave on $[0, 1]$, then the solution of (4.12) is unique.*

Proof In general, by Proposition 1 on the Value of Waiting (Sect. 4.2), $\bar{g}(1) - \bar{g}(0) > 0$ and so the first assertion is obvious, since the right-hand side of (4.12) is zero at $\theta = 1$ and is positive at $\theta = 0$; indeed, by Theorem B above, the left-hand side has the limiting value zero as $\theta \to 0+$ for $\bar{\sigma} > 0$. If, however, $\bar{g}(1) - \bar{g}(0) = 0$ (i.e. h fails to be strictly convex), then since the function $\bar{g}$ is initially increasing for $\theta > 0$, $\bar{g}$ has an internal local maximum at $\bar{\theta}$ for some $\bar{\theta}$ with $0 < \bar{\theta} < 1$ (by the Mean Value Theorem). In this case the first-order condition for R is satisfied by some $\theta < \bar{\theta}$, since the left-hand side tends to $+\infty$ as $\theta \to \bar{\theta}$.

Any internal solution θ^* to Eq. (4.12) has $\bar{g}'(\theta^*) > 0$ and so the second assertion follows since $R'(\theta^*-) > 0$ and $R'(\theta^*+) < 0$. Observe that if $\bar{g}''(\theta) < 0$, then we have

$$\frac{d}{d\theta}\left(\frac{\bar{g}(\theta) - \bar{g}(0)}{\bar{g}'(\theta)}\right) = 1 - \bar{g}''(\theta)\frac{\bar{g}(\theta) - \bar{g}(0)}{[\bar{g}'(\theta)]^2} > 0,$$

so the third assertion is clear; indeed concavity ensures that the left-hand side of (4.12) is an increasing function of θ. □

One would wish to improve on Proposition 4 to show in more general circumstances (beyond the concavity which can sometimes fail, as Fig. 4.1 shows) that (4.12) has a unique solution, and to study dependence on the two parameters of the problem. This appears analytically intractable. For the purposes of gaining an insight we propose therefore to replace $\bar{g}(t)$ by a function related to it through asymptotic analysis (as t varies), on the grounds that from numerical evaluations the substitute is qualitatively similar. Examination of behaviour for large t may be justified by re-sizing the parameters $\bar{\mu}, \bar{\sigma}$ which enables the termination date to become 'large'. This observation then introduces the advantages of the asymptotic viewpoint.

Guided by Theorems B and C, we are led to a considerably simpler problem obtained by making one of two 'typical' substitutions for $\bar{g}(\theta)$, namely

$$1 + A\theta e^{-\alpha\theta}, \text{ if } \bar{\sigma}^2 < \bar{\mu}, \text{ or } \quad e^{\alpha\theta}, \text{ if } \bar{\mu} < \bar{\sigma}^2,$$

according as variance is low, or high. Here $\alpha := |\bar{\sigma}^2 - \bar{\mu}| > 0$. The substitution in the first of the two situations fits qualitatively with numerical evaluations on the *form* of $\bar{g}$ (see Fig. 4.1); it agrees in the second situation with the general form observed in other figures and also the asymptotic form as $t \to \infty$.

Case (i) $\alpha = \bar{\mu} - \bar{\sigma}^2 > 0$. In this case the optimum time θ is the solution of

$$\theta/(1 - \alpha\theta) = 1 - \theta,$$

a quadratic equation, leading to the explicit formula

$$\theta = \theta(\alpha) := \frac{1}{2} - \frac{1}{\alpha}\left(-1 + \sqrt{1 + \frac{\alpha^2}{4}}\right),$$

so that as α increases from zero the optimal time θ recedes from the mid-point towards the origin. That is, low volatilities move the replenishment timing back.

Case (ii) $\alpha = \bar{\sigma}^2 - \bar{\mu} > 0$. The first-order condition here reduces to

$$(1 - e^{-\alpha\theta})/\alpha = 1 - \theta,$$

with a unique solution in the unit interval. Here we can use a quadratic approximation for the exponential term and solve for θ to obtain, for $\alpha < 2$, the approximation

$$\theta(\alpha) := \frac{1}{1 + \sqrt{1 - \alpha/2}},$$

so that the optimal choice of θ is close to the midpoint $\theta = 1/2$, when α is small, but advances as α increases towards unity (as a direct computation shows). That is, high volatilities bring the replenishment position forward (meaning that waiting longer, beyond the mid-term, is optimal for higher volatilities).

4.6   Asymptotic Analysis: The Proofs

In this section we give outline arguments (for the details, see the Appendix) leading to Propositions 2 and 3 and Theorems B and C of Sect. 4.4.

Lemma 1 *We have for fixed* μ

$$\lim_{\sigma \to 0+} W(\mu, \sigma) = -\infty, \text{ and } \lim_{\sigma \to 0+} \sigma W(\mu, \sigma) = -\mu.$$

This follows directly from the definition of $W(\mu, \sigma)$. We now prove

Proposition 5 *For* $\mu > 0$,

$$W(\mu, \sigma) = -\frac{\mu}{\sigma} + \frac{1}{2}\sigma + o(\sigma) \qquad (\sigma \to 0+).$$

Proof For an intuition, note that for small enough σ we have $e^{-\mu} \sim e^{\sigma W - \frac{1}{2}\sigma^2}$ and so

$$W(\mu, \sigma) \sim -\frac{\mu}{\sigma} + \frac{1}{2}\sigma.$$

This argument can be embellished as follows. For any non-zero ε let

$$W(\varepsilon) := -\frac{\mu}{\sigma} + \frac{1}{2}\sigma + \sigma\varepsilon,$$

so that

$$\sigma - W(\varepsilon) = \frac{\mu}{\sigma} + \frac{1}{2}\sigma - \sigma\varepsilon.$$

We shall prove that for positive ε we have, for small enough σ, that

$$W(-\varepsilon) < W(\mu, \sigma) < W(\varepsilon).$$

This is achieved by showing that for all small enough σ the expression below has the same sign as ε :

$$D(\sigma) = F(W(\varepsilon), \sigma) - F(W(\mu, \sigma), \sigma) = F(W(\varepsilon), \sigma) - e^{-\mu}.$$

This implies the Proposition. Now $D(0+) = 0$ and, since $D(\sigma) = \Phi(W(\varepsilon) - \sigma) + e^{\sigma W(\varepsilon) - \frac{1}{2}\sigma^2}\Phi(-W(\varepsilon))$,

$$D'(\sigma) = e^{-\frac{1}{2}(-W(\varepsilon)+\sigma)^2} \frac{1}{\sqrt{2\pi}} \{-\frac{\mu}{\sigma^2} + \frac{1}{2} - \varepsilon\} + e^{-\mu+\sigma^2\varepsilon}\{2\sigma\varepsilon\}(1+o(\sigma))$$

$$+e^{-\mu+\sigma^2\varepsilon}e^{-\frac{1}{2}W(\varepsilon)^2}\{\frac{\mu}{\sigma^2} + \frac{1}{2} + \varepsilon\}.$$

Note that the first and third terms contain a factor $\sigma \exp[-\mu^2/\sigma^2]$, which is small compared with σ. So for small enough σ the derivative $D'(\sigma)$ has the same sign as ε. So the same is true for $D(\sigma)$.   □

Definitions Recall from (4.10) that $\partial F/\partial W > 0$ and $F(-\infty, \sigma) = 0$, $F(+\infty, \sigma) = 1$. Let m be fixed; for the purposes only of the current section it is convenient to define

$$\overline{\Phi}(m) := 1 - \Phi(m)$$

and to introduce, also as a temporary measure, a variant form $\widehat{W}(m, \sigma)$ of $W(m, \sigma)$ obtained by replacing $e^{-\mu}$ in (4.10) by $\overline{\Phi}(m)$ so that now

$$F(\widehat{W}(m, \sigma), \sigma) = \overline{\Phi}(m) < 1. \tag{4.13}$$

Claim For c any constant

$$\lim_{\sigma\to\infty} F(\sigma - c, \sigma) = \overline{\Phi}(c).$$

The proof is routine.

Conclusion from Claim Notice the consequences for the choices $c = (1 \pm \varepsilon)m$. Since

$$\lim_{\sigma\to\infty} F(\sigma - (1+\varepsilon)m, \sigma) = \overline{\Phi}((1+\varepsilon)m) < \overline{\Phi}(m),$$

for large enough σ we have

$$F(\sigma - (1+\varepsilon)m, \sigma) < F(W, \sigma).$$

Hence for large enough σ we have $W > \sigma - (1+\varepsilon)m$. Similarly, taking $c = (1-\varepsilon)m$ we obtain $W < \sigma - (1-\varepsilon)m$. Thus

$$W(m, \sigma) = \sigma - m\{1 + o(1)\}(\text{as } \sigma \to \infty).$$

This result can be improved by an argument similar to that for Proposition 2 by reference to

$$D(\sigma) = \Phi(\sigma - W) + e^{\sigma W - \frac{1}{2}\sigma^2}\Phi(-W) - \overline{\Phi}(m)$$

to yield the following.

Proposition 6 *With the definition (4.13), for fixed m*

$$\widehat{W}(m, \sigma) = \sigma - m - \frac{1}{\sigma - m}\{1 + o(1)\} \qquad (as\ \sigma \to \infty).$$

Conclusion $\widehat{W}(m, \sigma) = W(\mu, \sigma)$ when $m = \hat{\mu}$ where $e^{-\mu} = \overline{\Phi}(m)$. Restating this equation as

$$e^{-\mu} = 1 - \Phi(\hat{\mu}) = \Phi(-\hat{\mu}),$$

we see that $\hat{\mu} > 0$ if and only if $\mu > \ln 2$, since $\hat{\mu} = -\Phi^{-1}(e^{-\mu})$; in particular for small μ we thus have $\hat{\mu} < 0$.

Lemma 2

$$\lim_{\theta \to 0+} \sqrt{\theta}\overline{W}(\theta) = 0 \ and \ \lim_{\theta \to 0+} \overline{W}(\theta) = +\infty \ for\ fixed\ \bar{\mu}, \bar{\sigma} > 0.$$

This follows again by a routine argument starting from (4.10) but requires the claim below and the definition

$$V := V(\theta) = \overline{W}(\theta) - \bar{\sigma}\sqrt{\theta}.$$

Claim

$$L := \lim_{\theta \to 0+} \sigma V(\theta) = 0.$$

The proof here is by contradiction from (4.10), assuming L non-zero.

Proof of Theorem B Differentiation of (4.10) with respect to θ gives

$$-\bar{\mu}e^{-\bar{\mu}\theta} = \varphi(W - \sigma)(W' - \sigma') + e^{\sigma W - \frac{1}{2}\sigma^2}\varphi(-W)(-W')$$

$$+ \Phi(-W)e^{\sigma W - \frac{1}{2}\sigma^2}[-\frac{1}{2}\bar{\sigma}^2 + (\sigma W)'].$$

Now

$$\varphi(W - \sigma)\sigma' = e^{\sigma W - \frac{1}{2}\sigma^2}\varphi(W)\frac{\bar{\sigma}}{2\sqrt{\theta}}$$

$$= \left(e^{\sigma W - \frac{1}{2}\sigma^2}\frac{\varphi(W)}{W}\right)\frac{1}{\theta}\frac{W\bar{\sigma}\sqrt{\theta}}{2} \to \bar{\mu} \cdot 0 = 0 \qquad (\theta \to 0+),$$

using (4.10) and $\lim_{W \to +\infty} \varphi(W)/(W\Phi(-W)) = 1$ to deal with the bracketed term. Thus

$$-\bar{\mu} = \lim_{\theta \to 0+} [\Phi(-W)(\sigma W)'].$$

Differentiation of (4.11) with respect to θ gives

$$\bar{g}' = [\bar{\sigma}^2 - \bar{\mu}]e^{(\sigma^2-\mu)}\Phi(W+\sigma) + e^{(\sigma^2-\mu)}\varphi(W+\sigma)(W'+\sigma')$$

$$+e^{-\mu-\sigma W+\frac{1}{2}\sigma^2}\varphi(-W)(-W') + e^{-\mu-\sigma W+\frac{1}{2}\sigma^2}\Phi(-W)[\frac{1}{2}\bar{\sigma}^2 - \bar{\mu} - (\sigma W)'].$$

Now

$$\bar{g}'(0) = [\bar{\sigma}^2 - \bar{\mu}] - \lim_{\theta \to 0+} \Phi(-W)[(\sigma W)'] = \bar{\sigma}^2.$$

□

Lemma 3 If $\frac{1}{2}\bar{\sigma}^2 \neq \bar{\mu}$, then

$$\lim_{\theta \to \infty} \overline{W}(\theta) = \pm\infty.$$

Remark This leaves the identification of the appropriate sign as a separate task. The proof is by contradiction from (4.10) by reference to simple properties of the normal hazard rate $H(w) = \varphi(w)/\Phi(-w)$. Lemma 4 below is proved by contradiction. Lemma 5 clarifies the cross-over case.

Lemma 4 $\lim_{\theta \to \infty} \overline{W}(\theta) - \sigma = -\infty$.

Lemma 5 If $\frac{1}{2}\bar{\sigma}^2 = \bar{\mu}$, then $\lim_{\theta \to \infty} \bar{\sigma}\sqrt{\theta}\overline{W}(\theta) = \log 2$.

Conclusion 1 If $\lim_{\theta \to \infty} \overline{W}(\theta) = -\infty$, then

$$\overline{W}(\theta) = -\frac{\bar{\mu} - \frac{1}{2}\bar{\sigma}^2}{\bar{\sigma}}\sqrt{\theta} + o(\sqrt{\theta}), \text{ for } \bar{\mu} > \frac{1}{2}\bar{\sigma}^2.$$

Lemma 6 If $\bar{\sigma}^2 < 2\bar{\mu}$, then

$$\lim_{\theta \to \infty} e^{-\sigma\overline{W}+(\frac{1}{2}\bar{\sigma}^2-\bar{\mu})\theta} = 1.$$

This follows directly from (4.10) and Lemmas 3 and 4.

Proof of Theorem C Lemma 6 establishes case (ii) of Theorem C. Next we note:

Conclusion 2 If $\lim_{\theta \to \infty} \overline{W}(\theta) = +\infty$, then for $\bar{\mu} < \frac{1}{2}\bar{\sigma}^2$

$$\overline{W}(\theta) = (\bar{\sigma} - \sqrt{2\bar{\mu}})\sqrt{\theta} + O(1/\sqrt{\theta}).$$

Case (iii) of Theorem C follows from this estimate. Combining (ii) and (iii) gives (i). Turning to case (iv), if $\bar{\sigma}^2 = 2\bar{\mu}$, then as $\theta \to \infty$ we have $\sigma + \overline{W}(\theta) \to +\infty$, by Lemma 5, so since $\lim_{\theta \to \infty} e^{\sigma \overline{W}(\theta)} = 2$, and appealing to the standard asymptotic estimates of $\Phi(x)$, as $1 - \varphi(x)/x$ for large x,

$$\bar{g}(\theta) = e^{(\bar{\sigma}^2 - \bar{\mu})\theta} \Phi(\sigma + \overline{W}(\theta)) + e^{-\sigma \overline{W} + (\frac{1}{2}\bar{\sigma}^2 - \bar{\mu})\theta} \Phi(-\overline{W}(\theta))$$

$$= e^{\bar{\mu}\theta} \Phi(\sigma + \overline{W}(\theta)) + e^{-\sigma \overline{W}} \Phi(-\overline{W}(\theta))$$

$$= e^{\bar{\mu}\theta} + \frac{1}{4} + o(1/\sqrt{\theta}).$$

This completes the proof of Theorem C.

4.7   Censor Comparative Statics: Reprise

This section considers the sensitivity of $\tilde{b}(\mu, \sigma)$ to μ and σ, and the dependence of $\bar{b}(\theta) = \tilde{b}(\bar{\mu}\theta, \bar{\sigma}\sqrt{\theta})$ on θ as given in Sect. 4.3.

Theorem 2 *The censor $\tilde{b}(\mu, \sigma)$ is decreasing in the drift μ.*

Proof The derivative of $\tilde{b} = \exp(\sigma W + \mu - \frac{1}{2}\sigma^2)$ with respect to μ is positive iff

$$-\sigma \frac{\partial W(\mu, \sigma)}{\partial \mu} > 1. \tag{4.14}$$

But differentiation of (4.10) and

$$\varphi(W(\mu, \sigma) - \sigma) = e^{\sigma W - \frac{1}{2}\sigma^2} \varphi(W(\mu, \sigma))$$

yield

$$1 = \tilde{b}\Phi(-W(\mu, \sigma)) \left(-\sigma \frac{\partial W}{\partial \mu} \right).$$

So (4.14) holds iff $\tilde{b}\Phi(-W(\mu, \sigma)) < 1$. But the latter follows from (4.8).   □

Theorem 3 *The censor $\tilde{b}(\mu, \sigma)$ is increasing in the standard deviation σ.*

Proof Differentiating $\tilde{b} = \exp(\sigma W(\mu, \sigma) + \mu - \frac{1}{2}\sigma^2)$ with respect to σ yields

$$\frac{\partial \tilde{b}}{\partial \sigma} = \tilde{b}(\mu, \sigma) \left\{ \sigma \frac{\partial W}{\partial \sigma} + W(\mu, \sigma) - \sigma \right\}.$$

Differentiating also the censor equation (4.10) with respect to σ, we obtain after some cancellations that

$$\varphi(W(\mu,\sigma) - \sigma) = e^{\sigma W(\mu,\sigma) - \frac{1}{2}\sigma^2} \Phi(-W(\mu,\sigma)) \left\{ W(\mu,\sigma) + \sigma \frac{\partial W}{\partial \sigma} - \sigma \right\}.$$

The bracketed term appearing here and earlier is thus positive, and so

$$\partial \tilde{b}(\mu,\sigma)/\partial \sigma > 0. \qquad \square$$

Using $\varphi(\sigma - W) = e^{\sigma W - \frac{1}{2}\sigma^2} \varphi(W)$ (cf. Sect. 4.2) we note the identity

$$W(\mu,\sigma) + \sigma \frac{\partial W}{\partial \sigma} - \sigma = \frac{\varphi(W(\mu,\sigma))}{\Phi(-W(\mu,\sigma))} = H(W(\mu,\sigma)), \qquad (4.15)$$

where $H(x)$ denotes the normal hazard rate $(\varphi(x)/\Phi(-x))$. Since $H(x) > x$ for all x, Eq. (4.15) gives $\partial W/\partial \sigma > 1$ for $\sigma > 0$. Recalling from Sect. 4.2 that $\partial W/\partial \mu < 0$, we have the following two results:

Theorem 4 *The two functions* $\sigma W(\mu,\sigma) - \frac{1}{2}\sigma^2$, $W(\mu,\sigma) - \sigma$ *are increasing in* σ *for* $\sigma > 0$.

Theorem 5 *The normal censor* $W(\mu,\sigma)$ *is increasing in standard deviation and decreasing with drift.*

Our final result is the following.

Theorem 6 *The censor* $\bar{b}(\theta) := \tilde{b}(\bar{\mu}\theta, \bar{\sigma}\sqrt{\theta})$ *is either unimodal or increasing on the interval* $0 < \theta < \infty$, *according as* $\bar{\mu} \geq \frac{1}{2}\bar{\sigma}^2$ *or* $\bar{\mu} < \frac{1}{2}\bar{\sigma}^2$.

Proof Using $\bar{b}\varphi(W) = e^{\mu}\varphi(W - \sigma)$ and applying the Chain Rule to $\bar{b}(\theta) = \tilde{b}(\bar{\mu}\theta, \bar{\sigma}\sqrt{\theta})$, we obtain

$$\theta \Phi(-W) \frac{d\bar{b}(\theta)}{d\theta} = -\mu \{e^{\mu} \Phi(W - \sigma)\} + \frac{1}{2}\sigma \bar{b}\varphi(W).$$

The stationarity condition for $\bar{b}(\theta)$ can be written using the normal hazard rate $H(x) = \varphi(x)/\Phi(-x)$ as

$$\mu = \frac{1}{2}\sigma H(-W(\mu,\sigma) + \sigma), \qquad (4.16)$$

where $\mu = \bar{\mu}\theta$ and $\sigma = \bar{\sigma}\sqrt{\theta}$, and $W(\mu,\sigma)$ is the normal censor as in (4.10).

We now regard μ and σ as free variables and let $\kappa := \bar{\mu}/\bar{\sigma}^2$ be the dispersion parameter. In this setting we seek a stationary point θ of $\bar{b}(\theta)$ by first finding the values $\mu = \mu^*$ and $\sigma = \sigma^*$ which satisfy Eq. (4.16) simultaneously with the equation:

$$\mu = \kappa\sigma^2. \qquad (4.17)$$

We shall show that this is possible (uniquely) if and only if $\kappa \geq 1/2$ (i.e. $\bar{\mu} \geq \frac{1}{2}\bar{\sigma}^2$). Thus for $\bar{\sigma}^2 > 2\bar{\mu}$ the function $\bar{b}(\theta)$ is increasing, but otherwise has a unique maximum at $\theta = \mu^*/\bar{\mu} = \sigma^{*2}/\bar{\sigma}^2$.

We begin by noting that (4.16) defines an implicit function $\mu = \mu(\sigma)$ for all $\sigma > 0$. Indeed, elimination of μ between (4.10) and (4.16) leads to

$$\exp\left(-\frac{1}{2}\sigma H(-w+\sigma)\right) = F(w,\sigma), \qquad (4.18)$$

and then routine analysis shows that there is a unique solution $w = \omega(\sigma)$ of (4.18). Since $\partial W/\partial \mu < 0$, we may recover $\mu(\sigma) > 0$, for $\sigma > 0$, from $\omega(\sigma) = W(\mu(\sigma), \sigma)$.

Linearization of both sides of (4.18) around $\sigma = 0$ yields the equation

$$H(-w) = 2(\varphi(w) + w\Phi(-w))$$

with unique solution $w = \omega(0) = 0$. Hence $\lim_{\sigma \to 0} W(\mu(\sigma), \sigma) = 0$ and so, for small σ, we have the approximation to (4.16) given by the convex function

$$\mu = \frac{1}{2}\sigma H(\sigma).$$

Numerical investigation of the positive function $w = \omega(\sigma)$ finds its maximum to be 0.051 for σ approximately 2.547. To see why, rewrite (4.18) in the equivalent form:

$$\exp\left(\frac{1}{2}\sigma^2 - \frac{1}{2}\sigma H(-w+\sigma)\right) = \frac{\Phi(w-\sigma)}{\varphi(\sigma)\sqrt{2\pi}} + e^{\sigma w}\Phi(-w).$$

For fixed w with $0 \leq w \leq 1$, and large σ, the left-hand side is close to $e^{\frac{1}{2}(\sigma w - 1)}$, in view of the asymptotic over-approximation $(x + 1/x)$ for $H(x)$ (when x is large), whereas the first term on the right is asymptotic to $1/(\sigma\sqrt{2\pi})$. Neglecting the latter, and replacing $\Phi(-w)$ by $\frac{1}{2}$, the solution for w may be estimated by $(2\log 2 - 1)/\sigma$.

Finally, using the same asymptotic approximation for $H(\sigma)$, we may over-approximate $\frac{1}{2}\sigma H(\sigma - \omega(\sigma))$ by $\frac{1}{2}\sigma^2 + \frac{1}{2}$. From here we may conclude that, for $\kappa > \frac{1}{2}$, Eqs. (4.16) and (4.17) have a solution with a crude over-estimate for σ^* given by

$$(\sigma^*)^2 = \frac{1}{2\kappa - 1}.$$

The supporting line $\mu = \frac{1}{2}H(0)\sigma$ provides the crude under-estimate $\sigma = 1/\kappa\sqrt{2\pi}$. For the special case $\kappa = \frac{1}{2}$ the solution to (4.16) and (4.17) is $\sigma^* = 4.331$. For $\kappa < \frac{1}{2}$ there is no solution, since $\frac{1}{2}\sigma H(\sigma) > \kappa\sigma^2$ for $\sigma > 0$.   □

4.8   Appendix

Proof of Proposition 6 For convenience put

$$R(W, \sigma) := \sqrt{2\pi}\, F(W, \sigma) = \int_{-W+\sigma}^{\infty} e^{-\frac{1}{2}x^2}\, dx + e^{\sigma W - \frac{1}{2}\sigma^2} \int_{W}^{\infty} e^{-\frac{1}{2}x^2}\, dx.$$

Consider an arbitrary non-zero ε; let $W_\varepsilon := \sigma - m - \delta$ and put

$$\delta := \frac{1 - \varepsilon}{\sigma - m}.$$

Now, with D as in the proof of Proposition 5 in Sect. 4.6, as $\delta \to 0$ and $\sigma \to \infty$

$$
\begin{aligned}
D(\sigma) &= \left(\int_{-W+\sigma}^{\infty} e^{-\frac{1}{2}x^2}\, dx + e^{\sigma W - \frac{1}{2}\sigma^2} \int_{W}^{\infty} e^{-\frac{1}{2}x^2}\, dx \right) - \int_{m}^{\infty} e^{-\frac{1}{2}x^2}\, dx \\
&= \left(\int_{m+\delta}^{\infty} e^{-\frac{1}{2}x^2}\, dx - \int_{m}^{\infty} e^{-\frac{1}{2}x^2}\, dx \right) + e^{\sigma(\sigma - m - \delta) - \frac{1}{2}\sigma^2} \int_{\sigma - m - \delta}^{\infty} e^{-\frac{1}{2}x^2}\, dx \\
&= -\delta e^{-\frac{1}{2}(m+\delta)^2} + O(\delta^2) \\
&\quad + e^{\sigma(\sigma - m - \delta) - \frac{1}{2}\sigma^2} \frac{1}{\sigma - m - \delta} e^{-\frac{1}{2}(m+\delta-\sigma)^2} \{1 + O(1/\sigma^2)\} \\
&= -\delta e^{-\frac{1}{2}(m+\delta)^2} + O(\delta^2) + \frac{1}{\sigma - m - \delta} e^{-\frac{1}{2}(m+\delta)^2} \{1 + O(1/\sigma^2)\} \\
&= \left(\frac{1}{\sigma - m - \delta} - \delta \right) e^{-\frac{1}{2}(m+\delta)^2} + O(\delta^2) + O(1/\sigma^2) \\
&= \left(\frac{1}{(\sigma - m) - \frac{1-\varepsilon}{\sigma - m}} - \frac{1-\varepsilon}{\sigma - m} \right) e^{-\frac{1}{2}(m+\delta)^2} + O(\delta^2) + O(1/\sigma^2) \\
&= \left(\frac{(\sigma - m)^2 - (1-\varepsilon)\{(\sigma - m)^2 - (1-\varepsilon)\}}{(\sigma - m)^3 - (1-\varepsilon)(\sigma - m)} \right) e^{-\frac{1}{2}(m+\delta)^2} + O(1/\sigma^2) \\
&= \frac{\varepsilon(\sigma - m)^2 + (1-\varepsilon)^2}{(\sigma - m)^3 - (1-\varepsilon)(\sigma - m)} e^{-\frac{1}{2}(m+\delta)^2} + O(1/\sigma^2) \\
&= \frac{\varepsilon}{\sigma - m} e^{-\frac{1}{2}(m+\delta)^2} + O(1/\sigma^2),
\end{aligned}
$$

and this has the same sign as ε. Thus, for $\varepsilon > 0$,

$$R(W_{-\varepsilon}, \sigma) < R(W(m, \sigma), \sigma) < R(W_\varepsilon, \sigma),$$

and so, since $\partial R(W, \sigma)/\partial W > 0$,

$$W_{-\varepsilon} < W(m, \sigma) < W_\varepsilon. \qquad \square$$

Proof of Lemma 2 We begin with the associated Claim (Sect. 4.6 above), for which we recall that

$$V := V(\theta) = \overline{W}(\theta) - \bar{\sigma}\sqrt{\theta},$$

and then note (by the definition of the normal sensor in Sect. 4.3) that

$$(e^{-\mu} - 1) - \{\Phi(-\sigma - V) - \Phi(-V)\} = [e^{\sigma V + \frac{1}{2}\sigma^2} - 1]\Phi(-\sigma - V). \qquad (4.19)$$

From here, for some V^* between V and $V + \sigma$,

$$(e^{-\mu} - 1) - \sigma\varphi(V^*) = [e^{\sigma V + \frac{1}{2}\sigma^2} - 1]\Phi(-\sigma - V),$$

so that

$$-\bar{\mu}\theta + \sigma\varphi(V^*) \sim [e^{\sigma V + \frac{1}{2}\sigma^2} - 1]\Phi(-\sigma - V).$$

Proof of Claim Put $\bar{V} := \lim_{\theta \to 0+} V(\theta)$, and suppose $L = \lim_{\theta \to 0+} \sigma V(\theta) \neq 0$ along a sequence of values of θ; then

$$V(\theta) \sim L/(\bar{\sigma}\sqrt{\theta}): \qquad \sigma\varphi(V^*) \sim \bar{\sigma}\sqrt{\theta}\exp(-L^2/\bar{\sigma}^2\theta)/\sqrt{2\pi}$$

and so

$$-\bar{\mu}\theta\{1 - (\bar{\sigma}/\bar{\mu}\sqrt{\theta})\exp(-L^2/\bar{\sigma}^2\theta)/\sqrt{2\pi} \sim -\bar{\mu}\theta.$$

So, for small enough θ,

$$[e^{\sigma V + \frac{1}{2}\sigma^2} - 1]\Phi(-\sigma - V) < 0,$$

so that $\bar{V} \leq 0$. Suppose first that $\bar{V} = -\infty$; then $L = 0$, since $\Phi(\infty) = 1$ reduces Eq. (4.19) to

$$0 = (e^L - 1),$$

contradicting $L \neq 0$. Likewise, from (4.19), the finiteness of $\bar{V}$ yields $L = 0$, a final contradiction. $\square_{\text{claim}}$

We turn now to Lemma 2 proper. As above

$$(e^{-\mu} - 1) + \sigma\varphi(V^*) \sim [e^{\sigma V + \frac{1}{2}\sigma^2} - 1]\Phi(-\sigma - V).$$

By the Claim, σV is small; so we may expand the exponential and, dividing by $\sigma = \bar{\sigma}\sqrt{\theta}$, obtain

$$-\frac{\bar{\mu}}{\bar{\sigma}}\sqrt{\theta} + \varphi(V^*) = (V + \frac{1}{2}\sigma)\Phi(-\sigma - V).$$

If $V \to \bar{V}$, a finite limit, then the Mills ratio (hazard rate) defined by

$$H(\bar{V}) := \frac{\varphi(\bar{V})}{\Phi(-\bar{V})}$$

satisfies $H(\bar{V}) = \bar{V}$, a contradiction, since the ratio is always greater than $\bar{V}$. Thus the limit $\bar{V}$ must be infinite, and hence $\varphi(\bar{V}) = 0$. So $\bar{V} = +\infty$, as otherwise $\bar{V} = -\infty$ leads to the contradiction

$$0 = \bar{V}\Phi\left(-\bar{V}\right) = \bar{V} \cdot 1. \qquad \square$$

Proof of Lemma 3 As in the definition of the normal censor

$$e^{-\bar{\mu}\theta} = \Phi(\overline{W}(\theta) - \sigma) + e^{\sigma\overline{W}(\theta) - \frac{1}{2}\sigma^2}\Phi(-\overline{W}(\theta)),$$

or

$$e^{-\bar{\mu}\theta - \sigma\overline{W}(\theta) + \frac{1}{2}\sigma^2} = e^{-\sigma\overline{W}(\theta) + \frac{1}{2}\sigma^2}\Phi(\overline{W}(\theta) - \sigma) + \Phi(-\overline{W}(\theta)), \qquad (4.20)$$

$$e^{-\bar{\mu}\theta - \sigma\overline{W}(\theta) + \frac{1}{2}\sigma^2} = \Phi(-\overline{W}(\theta)) + \varphi(\overline{W}(\theta))/H(\sigma - \overline{W}(\theta)),$$

where, as above, $H(.)$ denotes the hazard rate. Assume that $\overline{W}(\theta) \to \overline{W}$. We are to prove that $\overline{W}$ is not finite. We argue by cases.

Case 1 $\frac{1}{2}\bar{\sigma}^2 > \bar{\mu}$. The left hand side is unbounded, whereas the right-hand side is bounded for large θ by

$$1 + \varphi(\overline{W})/(\bar{\sigma}\sqrt{\theta} - \overline{W}).$$

Case 2 $\frac{1}{2}\bar{\sigma}^2 < \bar{\mu}$. Letting $\theta \to \infty$ gives the contradiction:

$$0 = \Phi(-\overline{W}) + 0. \qquad \square$$

Proof of Lemma 4 As before, if $V := \overline{W}(\theta) - \sigma$, then

$$e^{-\mu} = \Phi(V) + e^{\sigma V + \frac{1}{2}\sigma^2}\Phi(-\sigma - V).$$

Suppose $V \to -\infty$ is false. Then either $V \to \infty$, or $V \to \bar{V}$, a finite limit. In either case we have

$$e^{\sigma V + \frac{1}{2}\sigma^2} \Phi(-\sigma - V) \le e^{-\frac{1}{2}V^2} \varphi(V + \sigma)/(V + \sigma) \to 0,$$

as $\theta \to \infty$ (since $\sigma \to \infty$). This implies that $0 = \Phi(\bar{V})$, a contradiction in either case. So $V \to -\infty$. $\square$

Proof of Lemma 5 As before suppose $\overline{W}(\theta) \to \overline{W}$. If $\overline{W} < 0$ (possibly $-\infty$), then we have in the limit $\Phi(-\overline{W}) = \infty$, a contradiction. If $0 < \overline{W} < \infty$, then by (4.20) above $0 = \Phi(-\overline{W})$, again a contradiction. This leaves two possibilities: either $\overline{W} = \infty$ or $\overline{W} = 0$.

Suppose the former. Noting that

$$1 = \lim_{\theta \to \infty} [e^{\bar{\mu}\theta} \Phi(\overline{W}(\theta) - \sigma) + e^{\sigma \overline{W}(\theta)} \Phi(-\overline{W}(\theta))],$$

then $e^{\sigma \overline{W}(\theta)} \Phi(-\overline{W}(\theta))$ is bounded. But

$$\lim_{\theta \to \infty} e^{\sigma \overline{W}(\theta)} \Phi(-\overline{W}(\theta)) = \lim_{\theta \to \infty} \frac{e^{\sigma \overline{W}(\theta)} e^{-\frac{1}{2}\overline{W}^2}}{\overline{W}(\theta)\sqrt{2\pi}} = \lim_{\theta \to \infty} e^{[\overline{W}(\theta)(\sigma - W)]} \frac{e^{+\frac{1}{2}\overline{W}^2}}{\overline{W}(\theta)\sqrt{2\pi}} = \infty,$$

by Lemma 4 and by our assumption, a contradiction.

Thus after all $\overline{W} = 0$. So

$$1 = \lim_{\theta \to \infty} [e^{\bar{\mu}\theta} \Phi(\overline{W}(\theta) - \sigma) + e^{\sigma \overline{W}(\theta)} \Phi(0)]$$

$$= \lim_{\theta \to \infty} \frac{e^{\frac{1}{2}\bar{\sigma}^2\theta} e^{-\frac{1}{2}\overline{W}^2 - \frac{1}{2}\bar{\sigma}^2\theta + \sigma \overline{W}(\theta)}}{(\sigma - \overline{W}(\theta))\sqrt{2\pi}} + e^{\sigma \overline{W}(\theta)} \Phi(0)]$$

$$= \lim_{\theta \to \infty} e^{\sigma \overline{W}(\theta)} \left(\frac{1}{2} - \frac{1}{\sigma\sqrt{2\pi}} \right) = \frac{1}{2} \lim_{\theta \to \infty} e^{\sigma \overline{W}(\theta)}. \qquad \square$$

Proof of Conclusion 1 Recalling that $\sigma = \bar{\sigma}\sqrt{\theta}$, for any ε, put

$$W_\varepsilon(\theta) := \frac{\bar{\mu} - \frac{1}{2}\bar{\sigma}^2 + \varepsilon}{\bar{\sigma}} \sqrt{\theta} : \qquad e^{-\bar{\mu}\theta + \sigma W_\varepsilon(\theta) + \frac{1}{2}\sigma^2} = e^{\varepsilon\theta}.$$

For $\varepsilon > 0$ and large enough θ,

$$e^{-\bar{\mu}\theta - \sigma \overline{W}(\theta) + \frac{1}{2}\sigma^2} = \Phi(-\overline{W}(\theta)) + \varphi(-\overline{W}(\theta))/H(\sigma - \overline{W}(\theta))$$

$$< e^{\varepsilon\theta} = e^{-\bar{\mu}\theta + \sigma W_\varepsilon(\theta) + \frac{1}{2}\sigma^2} : \qquad -\overline{W}(\theta) < W_\varepsilon(\theta).$$

On the other hand, for $\varepsilon < 0$ and large enough θ

$$e^{-\bar{\mu}\theta - \sigma \overline{W}(\theta) + \frac{1}{2}\sigma^2} > \Phi(-\overline{W}(\theta)) > e^{\varepsilon\theta} = e^{-\bar{\mu}\theta + \sigma W_\varepsilon(\theta) + \frac{1}{2}\sigma^2}.$$

So $W_\varepsilon(\theta) < -\overline{W}(\theta)$. □

Proof of Lemma 6 Here, for large θ,

$$\overline{W}(\theta) = -(\bar{\mu} - \frac{1}{2}\bar{\sigma}^2)\sqrt{\theta} + o(\sqrt{\theta}).$$

So, since $\sigma = \bar{\sigma}\sqrt{\theta}$,

$$\sigma + \overline{W}(\theta) = \frac{\frac{3}{2}\bar{\sigma}^2 - \bar{\mu}}{\bar{\sigma}}\sqrt{\theta} + o(\sqrt{\theta}), \qquad \sigma - \overline{W}(\theta) = \frac{\frac{1}{2}\bar{\sigma}^2 + \bar{\mu}}{\bar{\sigma}}\sqrt{\theta} + o(\sqrt{\theta}).$$

Furthermore, rewriting the normal censor equation,

$$e^{-\bar{\mu}\theta + \frac{1}{2}\sigma^2 - \sigma \overline{W}(\theta)} = e^{-\frac{1}{2}\overline{W}(\theta)^2}\Phi(-\sigma + \overline{W}(\theta))/\varphi(-\overline{W}(\theta) + \sigma) + \Phi(-\overline{W}(\theta)).$$

So by Lemma 4, and since $\overline{W}(\theta) \to -\infty$ (as $\theta \to +\infty$),

$$\lim_{\theta \to \infty} e^{-\bar{\mu}\theta + \frac{1}{2}\sigma^2 - \sigma \overline{W}(\theta)} = 1.$$

In fact, we have $e^{-\bar{\mu}\theta + \frac{1}{2}\sigma^2 - \sigma \overline{W}(\theta)} = 1 + o(1/\sqrt{\theta})$. □

Proof of Conclusion 2 As $2\bar{\mu} > \bar{\sigma}^2$, note that

$$(\bar{\sigma}^2 - \bar{\mu}) - \frac{1}{2}\left(\frac{3}{2}\bar{\sigma} - \frac{\bar{\mu}}{\bar{\sigma}}\right)^2 = (\bar{\sigma}^2 - \bar{\mu}) - \frac{9}{8}\bar{\sigma}^2 - \frac{1}{2}\frac{\bar{\mu}^2}{\bar{\sigma}^2} + \frac{3}{2}\bar{\mu} = -\frac{1}{8}\bar{\sigma}^2 + \frac{1}{2}\bar{\mu} - \frac{1}{2}\frac{\bar{\mu}^2}{\bar{\sigma}^2} < 0;$$

indeed,

$$\bar{\mu}^2 - \bar{\mu}\bar{\sigma}^2 + \frac{1}{4}\bar{\sigma}^4 = \left(\bar{\mu} - \frac{1}{2}\bar{\sigma}^2\right)^2 > 0.$$

From Sect. 4.6,

$$g(\theta) = e^{(\bar{\sigma}^2 - \bar{\mu})\theta}\Phi\left(\sigma + \overline{W}(\theta)\right) + \Phi\left(-\overline{W}(\theta)\right) + o(1/\sqrt{\theta})$$

$$= e^{(\bar{\sigma}^2 - \bar{\mu})\theta}\Phi\left(\frac{\frac{3}{2}\bar{\sigma}^2 - \bar{\mu}}{\bar{\sigma}}\sqrt{\theta}\right) + \Phi\left(\frac{\bar{\mu} - \frac{1}{2}\bar{\sigma}^2}{\bar{\sigma}}\sqrt{\theta}\right) + o(1/\sqrt{\theta}).$$

Applying the asymptotic expansion

$$\Phi(x) \sim 1 - \frac{e^{-x^2/2}}{x\sqrt{2\pi}} \qquad (\text{as } x \to +\infty),$$

for which see [1], yields

$$e^{(\bar{\sigma}^2 - \bar{\mu})\theta} \, \Phi\left(\frac{\frac{3}{2}\bar{\sigma}^2 - \bar{\mu}}{\bar{\sigma}}\sqrt{\theta}\right) = e^{(\bar{\sigma}^2 - \bar{\mu})\theta} + o(1/\sqrt{\theta}). \qquad \square$$

Acknowledgement It is a pleasure to thank Alain Bensoussan for his very helpful advice and encouragement.

Postscript Harold Wilson (1916–1995, British prime minister 1964–1970 and 1974–1976) famously always emphasized the importance of keeping his options open.

References

1. Abramowitz, M.A, Stegan, I.A.: Handbook of Mathematical Functions. Dover, Mineola (1972)
2. Bensoussan, A., Crouhy, M., Proth, J.M.: Mathematical Theory of Production Planning. North-Holland, Amsterdam (1983)
3. Chang, J.S.K, Chang, C., Shi, M.: A market-based martingale valuation approach to optimum inventory control in a doubly stochastic jump-diffusion economy. J. Oper. Res. Soc. **66**, 405–420 (2015)
4. Dixit, A.K., Pindyck, R.S.: Decision Making Under Uncertainty. Princeton University Press, Princeton (1994)
5. Eberly, J.C., Van Mieghem, J.A.: Multi-factor investment under uncertainty. J. Econ. Theory **75**(8), 345–387 (1997)
6. Gietzmann, M.B., Ostaszewski, A.J.: Hedging the purchase of direct inputs in an inflationary environment. Manag. Account. Res. **10**, 61–84 (1999)
7. Gietzmann, M.B., Ostaszewski, A.J.: Predicting firm value: the superiority of q-theory over residual income. Account. Bus. Res. **34**(4), 379–382 (2004)
8. Hull, J.C.: Options, Futures, and Other Derivatives, 10th edn. Pearson Educational Limited, Harlow (2017)
9. Kendall, M.G., Stuart, A.: The Advanced Theory of Statistics, vol. 1. Griffin & Co, London (1963)
10. Kouvelis, P., Zhan, P., Qing D.: Integrated commodity management and financial hedging: a dynamic mean-variance analysis. Prod. Oper. Manag. **27**(6), 1052–1073 (2018)
11. Musiela, M., Rutkowski, M.: Martingale Methods in Financial Modelling. Springer, Berlin (1997)
12. Ostaszewski, A.J.: Subdominant eigenvalue location and the robustness of dividend policy irrelevance. In: J. Brzdęk, D. Popa, T.M. Rassias (eds.), Ulam Type Stability (this volume). https://doi.org/10.1007/978-3-030-28972-0, chapter 13
13. Patel, J.K., Read, C.R.: Handbook of the Normal Distribution. Marcel Dekker, New York (1982)
14. Rockafellar, R.T.: Convex Analysis. Princeton University Press, Princeton (1970)
15. Scarf, H.: The optimality of (S, s) policies in the dynamic inventory problem. In: Arrow, Karli, Suppes (eds.), Mathematical Methods in the Social Sciences, 1959 First Stamford Symposium. University Press, Stanford (1960)

Chapter 5
Ulam-Hyers Stability of Functional Equations in Quasi-β-Banach Spaces

Nguyen Van Dung and Wutiphol Sintunavarat

Abstract In this chapter, we give a survey on Ulam-Hyers stability of functional equations in quasi-β-Banach spaces, in particular in p-Banach spaces, quasi-Banach spaces and (β, p)-Banach spaces.

Keywords Ulam-Hyers stability · b-metric · Quasi-norm · Quasi-β-norm

Mathematics Subject Classification (2010) Primary 39B82; Secondary 39B52

5.1 Quasi-Normed Spaces and Quasi-β-Normed Spaces

In mathematics, more specially in functional analysis, a Banach space is a complete normed vector space. Thus, a Banach space is a vector space with a metric that allows the computation of vector length and distance between vectors and is complete in the sense that a Cauchy sequence of vectors always converges to a well defined limit that is within the space. Banach spaces are named after the Polish mathematician Stefan Banach, who introduced this concept and studied it systematically in 1920–1922 along with Hahn and Helly [69, Chapter 1]. Banach spaces originally grew out of the study of function spaces by Hilbert, Fréchet, and Riesz earlier in the twentieth century. Banach spaces play a central role in functional analysis. In other areas of analysis, the spaces under study are often Banach spaces.

N. Van Dung
Faculty of Mathematics Teacher Education, Dong Thap University, Cao Lanh, Dong Thap, Vietnam
e-mail: nvdung@dthu.edu.vn

W. Sintunavarat (✉)
Department of Mathematics and Statistics, Faculty of Science and Technology, Thammasat University Rangsit Center, Pathumthani, Thailand
e-mail: wutiphol@mathstat.sci.tu.ac.th

© Springer Nature Switzerland AG 2019
J. Brzdęk et al. (eds.), *Ulam Type Stability*,
https://doi.org/10.1007/978-3-030-28972-0_5

Banach spaces were generalized to quasi-Banach spaces. Moreover, there have been very sound reasons to develop understanding these spaces such as the non-normable property of many Banach spaces, and the absence of Hahn-Banach theorem in quasi-Banach spaces, see [44] for fundamental facts in quasi-Banach spaces. Similar to Banach algebras, the notion of quasi-Banach algebras was introduced also.

Definition 5.1 ([39, page 77]; [5, Definition 3]; [45, pages 6–7]; [3, Remark 1.4]; [67, Definition 1.5]) Let A be a vector space over the field $\mathbb{K}$ ($\mathbb{R}$ or $\mathbb{C}$), $\kappa \geq 1$ and $\| \cdot \| : A \to \mathbb{R}_+$ be a function such that for all $x, y \in A$ and all $a \in \mathbb{K}$,

1. $\|x\| = 0$ if and only if $x = 0$.
2. $\|ax\| = |a| \|x\|$.
3. $\|x + y\| \leq \kappa (\|x\| + \|y\|)$.

Then

1. $\| \cdot \|$ is called a *quasi-norm* on A, the smallest possible κ is called the *modulus of concavity* or *quasi-triangle constant*. Without loss of generality we can assume κ is the modulus of concavity. $(A, \| \cdot \|, \kappa)$ is called a *quasi-normed space*
2. $\|.\|$ is called a *p-norm* on A, and $(A, \|.\|, \kappa)$ is called a *p-normed space* if

$$\|x + y\|^p \leq \|x\|^p + \|y\|^p \tag{5.1}$$

for some $0 < p \leq 1$ and for all $x, y \in A$.
3. The sequence $\{x_n\}_n$ is called *convergent* to x if $\lim\limits_{n \to \infty} \|x_n - x\| = 0$, which we denote by $\lim\limits_{n \to \infty} x_n = x$.
4. The sequence $\{x_n\}_n$ is called *Cauchy* if $\lim\limits_{n,m \to \infty} \|x_n - x_m\| = 0$.
5. The quasi-normed space $(A, \| \cdot \|, \kappa)$ is called *quasi-Banach* if each Cauchy sequence is a convergent sequence.
6. The quasi-normed space $(A, \| \cdot \|, \kappa)$ is called *p-Banach* if it is a *p*-normed and quasi-Banach space.
7. The quasi-normed space $(A, \| \cdot \|, \kappa)$ is called a *quasi-normed algebra* if A is an algebra and $\|xy\| \leq C\|x\|\|y\|$ for all $x, y \in A$ and some $C > 0$.
8. The quasi-normed algebra $(A, \| \cdot \|, \kappa)$ is called a *quasi-Banach algebra* if $(A, \| \cdot \|, \kappa)$ is a quasi-Banach space.
9. The quasi-normed algebra $(A, \| \cdot \|, \kappa)$ is called a *p-normed algebra* if $(A, \| \cdot \|, \kappa)$ is a *p*-normed space.
10. The *p*-normed algebra $(A, \| \cdot \|, \kappa)$ is called a *p-Banach algebra* if $(A, \| \cdot \|, \kappa)$ is a *p*-Banach space.

In 1993 Czerwik [21] introduced the notion of a *b-metric* with a coefficient 2. This notion was generalized later with a coefficient $\kappa \geq 1$ [22]. In 2010 Khamsi and Hussain [47] reintroduced the notion of a *b*-metric under the name *metric-type*. Another notion of metric-type, called *s-relaxed_p* metric was introduced in [29, Definition 4.2], see also [46]. A *b*-metric is called *quasi-metric* in [55]. Quasi-metric spaces play an important role in the study of Gromov hyperbolic metric spaces [79, Final remarks], and in the study of optimal transport paths [82]. For convenience the names *b-metric* and *b-metric space* will be used in what follows.

Definition 5.2 ([22, page 263]) Let X be a nonempty set, $\kappa \geq 1$ and $d : X \times X \to \mathbb{R}_+$ be a function satisfying the following conditions for all $x, y, z \in X$:

1. $d(x, y) = 0$ if and only if $x = y$.;
2. $d(x, y) = d(y, x)$;
3. $d(x, z) \leq \kappa \big[d(x, y) + d(y, z) \big]$.

Then

1. d is called a *b-metric* on X and (X, d, κ) is called a *b-metric* space.
2. The sequence $\{x_n\}_n$ is called *convergent* to x in (X, d, κ) if $\lim_{n \to \infty} d(x_n, x) = 0$, written $\lim_{n \to \infty} x_n = x$.
3. The sequence $\{x_n\}_n$ is called *Cauchy* if $\lim_{n,m \to \infty} d(x_n, x_m) = 0$.
4. The space (X, d, κ) is called *complete* if each Cauchy sequence is a convergent sequence.

It is easy to see that every metric space is a b-metric space with $\kappa = 1$, and there exists a b-metric that is not a metric [4, Example 3.9]. Here, we give some known examples of b-metric spaces.

Example 5.1 ([36, page 110]) Let $X = \mathbb{R}$, $p \geq 1$ and the map $d : X \times X \to [0, \infty)$ defined by

$$d(x, y) = |x - y|^p \quad \text{for all } x, y \in X.$$

Then (X, d, κ) is a b-metric space with coefficient $\kappa = 2^{p-1} \geq 1$.

Next example shows the generality of Example 5.1.

Example 5.2 ([10, Example 1.1]) The set $\ell_p(\mathbb{R})$ with $0 < p < 1$, where

$$\ell_p(\mathbb{R}) := \Big\{ \{x_n\} \subset \mathbb{R} : \sum_{n=1}^{\infty} |x_n|^p < \infty \Big\},$$

together with the map $d : \ell_p(\mathbb{R}) \times \ell_p(\mathbb{R}) \to [0, \infty)$ defined by

$$d(x, y) := \Big(\sum_{n=1}^{\infty} |x_n - y_n|^p \Big)^{\frac{1}{p}}$$

for all $x = \{x_n\}$, $y = \{y_n\} \in \ell_p(\mathbb{R})$ is a b-metric space with coefficient $\kappa = 2^{\frac{1}{p}} > 1$. The above result also holds for the general case $\ell_p(X)$ with $0 < p < 1$, where X is a Banach space.

Example 5.3 ([10, Example 1.2]) The set $L_p[0, 1]$ with $0 < p < 1$, where

$$L_p[0, 1] := \left\{ x : [0, 1] \to \mathbb{R} : \int_0^1 |x(t)|^p dt < 1 \right\},$$

together with the map $d : L_p[0, 1] \times L_p[0, 1] \to \mathbb{R}_+$ defined by

$$d(x, y) := \left(\int_0^1 |x(t) - y(t)|^p dt \right)^{1/p}$$

for all $x, y \in L_p[0, 1]$ is a *b*-metric space with constant $\kappa = 2^{\frac{1}{p}} > 1$.

Next we give the result claiming that every *b*-metric space is metrizable.

Theorem 5.1 ([65, Proposition on page 4308]) *Let (Y, d, κ) be a b-metric space,* $\theta = \log_{2\kappa} 2$, *and*

$$D_d(x, y) = \inf \left\{ \sum_{i=1}^n d^\theta (x_i, x_{i+1}) : x_1 = x, x_2, \dots, x_n, x_{n+1} = y \in Y, n \geq 1 \right\}$$

for all $x, y \in Y$. Then D_d is a metric on Y satisfying

$$\frac{1}{4} d^\theta (x, y) \leq D_d(x, y) \leq d^\theta (x, y)$$

for all $x, y \in Y$. In particular, if d is a metric then $\theta = 1$ and $D_d = d$.

It is also easy to see that every normed space is a quasi-normed space with $\kappa = 1$. But, there exist *p*-Banach spaces which are not normable as in following examples.

Example 5.4 ([56, Examples 1–2])

1. Lebesgue spaces L^p with the quasi-norms

$$\|f\|_p = \left(\int_\Omega |f(x)|^p d\mu(x) \right)^{\frac{1}{p}}, \quad f \in L_p$$

are Banach spaces for $1 \leq p < \infty$. For $0 < p < 1$, they are *p*-Banach spaces with the quasi-triangle constant $C = 2^{\frac{1}{p-1}}$.

2. Lorentz spaces $L^{p,q}$ for $0 < p, q < \infty$ and Marcinkiewicz or weak L^p-spaces $L^{p,\infty}$ for $0 < p \leq \infty$ are quasi-Banach spaces determined by the quasi-norms

$$\|f\|_{p,q} = \begin{cases} \left(\int_0^\infty [t^{\frac{1}{p}} f^*(t)]^q \frac{dt}{t} \right)^{\frac{1}{q}} & \text{if } 0 < q < \infty \\ \sup_{t>0} t^{\frac{1}{p}} f^*(t) & \text{if } q = \infty \end{cases}$$

where

$$f^*(t) = \inf\{\lambda > 0 : \mu\big(x \in \Omega : |f(x)| > \lambda\big) \leq t\}.$$

Moreover, the following assertions hold.

a. $\|.\|_{p,q}$ is a norm if and only if either $1 \leq q \leq p$ or $p = q = \infty$.
b. The spaces $L^{p,q}$ are not normable if one of the following conditions holds.

  i. $0 < p < \infty$ and $0 < q < 1$.
  ii. $0 < p < 1$ and $1 \leq q \leq \infty$.
  iii. $p = 1$ and $1 < q \leq \infty$.

However, by Aoki-Rolewicz Theorem, each quasi-norm is equivalent to some p-norm. This theorem plays a very important role in p-normalizing a quasi-normed space.

Theorem 5.2 ([56, Theorem 1/Aoki-Rolewicz Theorem]) *Let* $(Y, \| \cdot \|_Y, \kappa_Y)$ *be a quasi-normed space,* $p = \log_{2\kappa_Y} 2$, *and*

$$|||x|||_Y = \inf\left\{ \left(\sum_{i=1}^{n} \|x_i\|_Y^p \right)^{\frac{1}{p}} : x = \sum_{i=1}^{n} x_i, \, x_i \in Y, n \geq 1 \right\}$$

for all $x \in Y$. *Then* $||| \cdot |||_Y$ *is a quasi-norm on* Y *satisfying*

$$|||x + y|||_Y^p \leq |||x|||_Y^p + |||y|||_Y^p$$

and

$$\frac{1}{2\kappa_Y}\|x\|_Y \leq |||x|||_Y \leq \|x\|_Y \tag{5.2}$$

for all $x, y \in Y$. *In particular, the quasi-norm* $||| \cdot |||_Y$ *is a p-norm, and if* $\| \cdot \|_Y$ *is a norm then* $p = 1$ *and* $||| \cdot |||_Y = \| \cdot \|_Y$.

The main results of [2] were proved by using the following fixed point result of Brzdęk et al. [13], where Y^U is the set of all functions from $U \neq \emptyset$ to $Y \neq \emptyset$.

Theorem 5.3 ([13, Theorem 1]) *Assume that*

1. *U is a nonempty set, Y is a complete metric space, and $\mathcal{T} : Y^U \to Y^U$ is a given function.*
2. *There exist $f_1, \ldots, f_k : U \to U$ and $L_1, \ldots, L_k : U \to \mathbb{R}_+$ such that for all $\xi, \mu \in Y^U$ and $x \in U$,*

$$d\big((\mathcal{T}\xi)(x), (\mathcal{T}\mu)(x)\big) \leq \sum_{i=1}^{k} L_i(x)d\Big(\xi\big(f_i(x)\big), \mu\big(f_i(x)\big)\Big). \tag{5.3}$$

3. *There exist $\varepsilon : U \to \mathbb{R}_+$ and $\varphi : U \to Y$ such that for all $x \in U$,*

$$d\big((\mathscr{T}\varphi)(x), \varphi(x)\big) \le \varepsilon(x)$$

4. *For every $x \in U$,*

$$\varepsilon^*(x) = \sum_{n=0}^{\infty} (\Lambda^n \varepsilon)(x) < \infty$$

where $(\Lambda\delta)(x) = \sum_{i=1}^{k} L_i(x)\delta\big(f_i(x)\big)$ for all $\delta : U \to \mathbb{R}_+$ and $x \in U$.

Then for every $x \in U$, the limit $\lim_{n \to \infty} (\mathscr{T}^n\varphi)(x) = \psi(x)$ exists and the function $\psi : U \to Y$ so defined is a unique fixed point of $\mathscr{T}$ satisfying

$$d\big(\varphi(x), \psi(x)\big) \le \varepsilon^*(x)$$

for all $x \in U$.

The version of Theorem 5.3 in Banach spaces is as follows.

Theorem 5.4 ([2, Theorem 2.1]) *Assume that*

1. *U is a nonempty set, Y is a Banach space, and $\mathscr{T} : Y^U \to Y^U$ is a given function.*
2. *There exist $f_1, \ldots, f_k : U \to U$ and $L_1, \ldots, L_k : U \to \mathbb{R}_+$ such that for all $\xi, \mu \in Y^U$ and $x \in U$,*

$$\|(\mathscr{T}\xi)(x) - (\mathscr{T}\mu)(x)\| \le \sum_{i=1}^{k} L_i(x)\|\xi\big(f_i(x)\big) - \mu\big(f_i(x)\big)\|.$$

3. *There exist $\varepsilon : U \to \mathbb{R}_+$ and $\varphi : U \to Y$ such that for all $x \in U$,*

$$\|(\mathscr{T}\varphi)(x) - \varphi(x)\| \le \varepsilon(x).$$

4. *For every $x \in U$,*

$$\varepsilon^*(x) = \sum_{n=0}^{\infty} (\Lambda^n \varepsilon)(x) < \infty$$

where $(\Lambda\delta)(x) = \sum_{i=1}^{k} L_i(x)\delta\big(f_i(x)\big)$ for all $\delta : U \to \mathbb{R}_+$ and $x \in U$.

Then for every $x \in U$, the limit $\lim_{n \to \infty} (\mathscr{T}^n \varphi)(x) = \psi(x)$ exists and the function $\psi : U \to Y$ so defined is a unique fixed point of $\mathscr{T}$ satisfying for all $x \in U$,

$$\|\varphi(x) - \psi(x)\| \leq \varepsilon^*(x).$$

The next result is an extension of Theorem 5.3 on complete metric spaces to complete b-metric spaces, which was proved by using Theorem 5.1.

Theorem 5.5 ([24, Theorem 2.1]) *Assume that*

1. *U is a nonempty set, (Y, d, κ) is a complete b-metric space, and $\mathscr{T} : Y^U \to Y^U$ is a given function.*
2. *There exist $f_1, \ldots, f_k : U \to U$ and $L_1, \ldots, L_k : U \to \mathbb{R}_+$ such that for all $\xi, \mu \in Y^U$ and $x \in U$,*

$$d\big((\mathscr{T}\xi)(x), (\mathscr{T}\mu)(x)\big) \leq \sum_{i=1}^{k} L_i(x) d\Big(\xi\big(f_i(x)\big), \mu\big(f_i(x)\big)\Big).$$

3. *There exist $\varepsilon : U \to \mathbb{R}_+$ and $\varphi : U \to Y$ such that for all $x \in U$,*

$$d\big((\mathscr{T}\varphi)(x), \varphi(x)\big) \leq \varepsilon(x).$$

4. *For every $x \in U$ and $\theta = \log_{2\kappa} 2$,*

$$\varepsilon^*(x) = \sum_{n=0}^{\infty} (\Lambda^n \varepsilon)^\theta (x) < \infty$$

where $(\Lambda\delta)(x) = \sum_{i=1}^{k} L_i(x)\delta\big(f_i(x)\big)$ for all $\delta : U \to \mathbb{R}_+$ and $x \in U$.

Then we have

1. *For every $x \in U$, the limit $\lim_{n \to \infty} (\mathscr{T}^n \varphi)(x) = \psi(x)$ exists and the function $\psi : U \to Y$ so defined is a fixed point of $\mathscr{T}$ satisfying*

$$d^\theta \big(\varphi(x), \psi(x)\big) \leq 4\varepsilon^*(x) \tag{5.4}$$

for all $x \in U$.
2. *For every $x \in U$, if*

$$\varepsilon^*(x) \leq \Big(M \sum_{n=1}^{\infty} (\Lambda^n \varepsilon)(x)\Big)^\theta < \infty$$

for some positive real number M, then the fixed point of $\mathscr{T}$ satisfying (5.4) is unique.

From Theorem 5.5 we get the following result which is an extension of Theorem 5.4 to quasi-Banach spaces.

Corollary 5.1 ([24, Corollary 2.2]) *Assume that*

1. *U is a nonempty set, $(Y, \|.\|, \kappa)$ is a quasi-Banach space, and $\mathscr{T} : Y^U \to Y^U$ is a given function.*
2. *There exist $f_1, \ldots, f_k : U \to U$ and $L_1, \ldots, L_k : U \to \mathbb{R}_+$ such that for all $\xi, \mu \in Y^U$ and $x \in U$,*

$$\|(\mathscr{T}\xi)(x) - (\mathscr{T}\mu)(x)\| \leq \sum_{i=1}^{k} L_i(x)\|\xi(f_i(x)) - \mu(f_i(x))\|.$$

3. *There exist $\varepsilon : U \to \mathbb{R}_+$ and $\varphi : U \to Y$ such that for all $x \in U$,*

$$\|(\mathscr{T}\varphi)(x) - \varphi(x)\| \leq \varepsilon(x)$$

4. *For every $x \in U$ and $\theta = \log_{2\kappa} 2$,*

$$\varepsilon^*(x) = \sum_{n=0}^{\infty} (\Lambda^n \varepsilon)^\theta(x) < \infty$$

where

$$(\Lambda\delta)(x) = \sum_{i=1}^{k} L_i(x)\delta(f_i(x))$$

for all $\delta : U \to \mathbb{R}_+$ and $x \in U$.

Then we have

1. *For every $x \in U$, the limit $\lim_{n\to\infty} (\mathscr{T}^n\varphi)(x) = \psi(x)$ exists and the function $\psi : U \to Y$ so defined is a fixed point of $\mathscr{T}$ satisfying*

$$\|\varphi(x) - \psi(x)\|^\theta \leq 4\varepsilon^*(x) \tag{5.5}$$

for all $x \in U$.
2. *For every $x \in U$, if*

$$\varepsilon^*(x) \leq \left(M \sum_{n=1}^{\infty} (\Lambda^n \varepsilon)(x)\right)^\theta < \infty$$

for some positive real number M, then the fixed point of $\mathscr{T}$ satisfying (5.5) is unique.

Remark 5.1 ([24, Remark 2.3]) If Y is a complete metric space in Theorem 5.5, then $\kappa = 1$. So $\theta = 1$ and $D_d = d$ and we obtain Theorem 5.3. Moreover the inequality (5.4) becomes better as

$$d\big(\varphi(x), \psi(x)\big) = d^\theta\big(\varphi(x), \psi(x)\big) \leq \varepsilon^*(x).$$

As a generalization of a quasi-normed space, the quasi-β-normed space was introduced as follows.

Definition 5.3 ([40]) Let X be a vector space over the field $\mathbb{K}$ ($\mathbb{R}$ or $\mathbb{C}$), $\kappa \geq 1$, $0 < \beta \leq 1$, and $\| \cdot \| : X \to \mathbb{R}_+$ be a function satisfying the following conditions for all $x, y \in X$ and all $a \in \mathbb{K}$:

1. $\|x\| = 0$ if and only if $x = 0$;
2. $\|ax\| = |a|^\beta \|x\|$;
3. $\|x + y\| \leq \kappa\big(\|x\| + \|y\|\big)$.

Then

1. $\|\cdot\|$ is called a *quasi-β-norm* on X, the smallest possible κ is called the *modulus of concavity* or *quasi-triangle constant*, and $(X, \| \cdot \|, \kappa)$ is called a *quasi-β-normed space*. For a quasi-β-normed space $(X, \| \cdot \|, \kappa)$, without loss of generality we can assume κ is the modulus of concavity.
2. $\|.\|$ is called a *(β,p)-norm* on X, and $(X, \|.\|, \kappa)$ is called a *(β,p)-normed space* if

$$\|x + y\|^p \leq \|x\|^p + \|y\|^p$$

   for some $0 < p \leq 1$ and for all $x, y \in X$.
3. The sequence $\{x_n\}_n$ is called *convergent* to x if $\lim_{n \to \infty} \|x_n - x\| = 0$, which we denote by $\lim_{n \to \infty} x_n = x$.
4. The sequence $\{x_n\}_n$ is called *Cauchy* if $\lim_{n,m \to \infty} \|x_n - x_m\| = 0$.
5. The quasi-β-normed space $(X, \| \cdot \|, \kappa)$ is called *quasi-β-Banach space* if each Cauchy sequence is a convergent sequence.
6. The quasi-β-normed space $(X, \| \cdot \|, \kappa)$ is called *(β, p)-Banach space* if it is a (β, p)-normed and quasi-β-Banach space.

For $\beta = 1$, the quasi-β-norm reduces to a quasi-norm. The following example shows that there exists a quasi-β-norm that is not a quasi-norm.

Example 5.5 ([28, page 333]) Let $X = \mathbb{R}^2$ and for some $0 < p, \beta < 1$, define

$$\|x\|_{p,\beta} = \begin{cases} (|x_1|^{\beta p} + |x_2|^{\beta p})^{\frac{1}{p}} & \text{if } x_2 \neq 0 \\ 2|x_1|^\beta & \text{if } x_2 = 0 \end{cases}$$

for all $x = (x_1, x_2) \in X$. Then $\|.\|_{p,\beta}$ is a (β, p)-norm that is not a quasi-norm.

If X is a quasi-β-normed space with the quasi-β-norm $\|x\|_\beta$, then it is a quasi-normed space with the quasi-norm $\|x\| = \|x\|_\beta^{\frac{1}{\beta}}$. Recall that Aoki-Rolewicz Theorem, see Theorem 5.2, plays a very important role in p-normalizing a quasi-normed space. However, some authors used Aoki-Rolewicz Theorem to p-normalize a quasi-β-normed space, see Sect. 5.3.1. It seems to be not correct. We present here an explicit Aoki-Rolewicz type Theorem to show that we can also p-normalize a quasi-β-normed space as follows.

Theorem 5.6 *Let* $(Y, \| \cdot \|_Y, \kappa_Y)$ *be a quasi-β-normed space with* $0 < \beta \leq 1$, $p = \log_{(2\kappa_Y)^{\frac{1}{\beta}}} 2$, *and*

$$|||x|||_Y = \inf\left\{ \left(\sum_{i=1}^n \|x_i\|_Y^{\frac{p}{\beta}} \right)^{\frac{\beta}{p}} : x = \sum_{i=1}^n x_i, x_i \in Y, n \geq 1 \right\}$$

for all $x \in Y$. *Then* $||| \cdot |||_Y$ *is a quasi-β-norm on* Y *satisfying*

$$|||x + y|||_Y^p \leq |||x|||_Y^p + |||y|||_Y^p \tag{5.6}$$

and

$$\frac{1}{2\kappa_Y}\|x\|_Y^p \leq |||x|||_Y^p \leq \|x\|_Y^p \tag{5.7}$$

for all $x, y \in Y$. *In particular, the quasi-β-norm* $||| \cdot |||_Y$ *is a* (β, p)-*norm, and if* $\| \cdot \|_Y$ *is a norm then* $\beta = p = 1$ *and* $||| \cdot |||_Y = \| \cdot \|_Y$.

Proof Put

$$|||x|||_{Y,\beta} = \inf\left\{ \left(\sum_{i=1}^n \|x_i\|_Y^{\frac{p}{\beta}} \right)^{\frac{1}{p}} : x = \sum_{i=1}^n x_i, x_i \in Y, n \geq 1 \right\}.$$

We find that $\|.\|_Y^{\frac{1}{\beta}}$ is a quasi-norm with the modulus of concavity $\frac{(2\kappa_Y)^{\frac{1}{\beta}}}{2}$. So by Theorem 5.2 we get $|||.|||_{Y,\beta}$ is a quasi-norm on Y satisfying

$$|||x + y|||_{Y,\beta}^p \leq |||x|||_{Y,\beta}^p + |||y|||_{Y,\beta}^p$$

and

$$\frac{1}{(2\kappa_Y)^{\frac{1}{\beta}}}\left(\|x\|_Y^{\frac{1}{\beta}}\right)^p \leq |||x|||_{Y,\beta}^p \leq \left(\|x\|_Y^{\frac{1}{\beta}}\right)^p$$

for all $x, y \in Y$. Since $0 < \beta \leq 1$, we get $\||.\||_Y = (\||.\||_{Y,\beta})^\beta$ is also a quasi-β-norm on Y and

$$
\begin{aligned}
\||x + y\||_Y^p &= \left(\||x + y\||_{Y,\beta}^\beta\right)^p \\
&= \left(\||x + y\||_{Y,\beta}^p\right)^\beta \\
&\leq \left(\||x\||_{Y,\beta}^p + \||y\||_{Y,\beta}^p\right)^\beta \\
&\leq \left(\||x\||_{Y,\beta}^p\right)^\beta + \left(\||y\||_{Y,\beta}^p\right)^\beta \\
&= \||x\||_Y^p + \||y\||_Y^p.
\end{aligned}
$$

Also, we have

$$
\frac{1}{2\kappa_Y}\|x\|_Y^p = \left(\frac{1}{(2\kappa_Y)^{\frac{1}{\beta}}}\left(\|x\|_Y^{\frac{1}{\beta}}\right)^p\right)^\beta \leq \left(\||x\||_{Y,\beta}^p\right)^\beta = \||x\||_Y^p \leq \left(\left(\|x\|_Y^{\frac{1}{\beta}}\right)^p\right)^\beta = \|x\|_Y^p.
$$

In 2005 Baak [7] introduced the so-called *generalized quasi-normed space* as follows.

Definition 5.4 ([7, Definition 2]) Let X be a vector space over the field $\mathbb{K}$ ($\mathbb{R}$ or $\mathbb{C}$), $\kappa > 0$, and $\| \cdot \| : X \to \mathbb{R}_+$ be a function satisfying the following conditions for all $x, x_i \in X$, $i \in \mathbb{N}$, and all $a \in \mathbb{K}$:

1. $\|x\| = 0$ if and only if $x = 0$;
2. $\|ax\| = |a|\|x\|$;
3. $\|\sum_{i=1}^{\infty} x_i\| \leq \kappa \sum_{i=1}^{\infty} \|x_i\|$.

Then $\| \cdot \|$ is called a *generalized quasi-norm* on X, and $(X, \| \cdot \|, \kappa)$ is called a *generalized quasi-normed space*.

A *generalized quasi-Banach space*, a *generalized p-norm* and a *generalized p-Banach space* were defined similarly to the quasi-normed spaces, see [7, page 216]. The author claimed that the generalized quasi-normed space is an extension of a quasi-normed space. However, the class of all generalized quasi-normed spaces is only a special class of quasi-normed spaces. Indeed, by choosing $x_1 = x$, $x_2 = y$ and $x_i = 0$ for all $i \geq 3$ in Definition 5.4 we get $\|x + y\| \leq \kappa(\|x\| + \|y\|)$. Therefore *generalized quasi-normed, generalized p-normed, generalized quasi-Banach, generalized p-Banach* in the sense of Definition 5.4 will be called *strong quasi-normed, strong p-normed, strong quasi-Banach, strong p-Banach* respectively in the next.

5.2   Ulam-Hyers Stability of Functional Equations in Quasi-Banach Spaces

A definition of stability in the case of homomorphisms between groups was suggested by a problem posed by Ulam [80, page 64]. The first answer to Ulam's problem is the result of Hyers. Many authors then studied Ulam-Hyers stability of the following Cauchy equation:

$$f(x + y) = f(x) + f(y), \quad x, y \in X, \tag{5.8}$$

where f is an unknown function from a space X to a space Y endowed with some binary operations and thus forming groups, norm spaces, algebras etc. Next, we give one of the famous results in this direction.

Theorem 5.7 ([41, Theorems 1 and 2]; [6, Theorem on page 64]; [31, Theorem 2]; [53, Theorem 5]; [11, Theorem 1.2]) *Let X, Y be two real normed spaces and $f : X \to Y$ be a function satisfying the inequality*

$$\|f(x + y) - f(x) - f(y)\| \le \alpha(\|x\|^p + \|y\|^p)$$

for all $x, y \in X \setminus \{0\}$, where α and p are real constants with $\alpha > 0$ and $p \ne 1$. Then the following statements hold.

1. *If $p \ge 0$ and Y is complete, then there exists a unique solution $T : X \to Y$ of (5.8) such that*

$$\|f(x) - T(x)\| \le \frac{\alpha}{1 - 2^{p-1}} \|x\|^p \text{ for all } x \in X \setminus \{0\}.$$

2. *If $p < 0$ then f is* additive, *that is, (5.8) holds for all $x, y \in X \setminus \{0\}$.*

In 2006, Maligranda [56] presented an interesting report on the life and work, the books and research of Tosio Aoki who was a Japanese mathematician and published only two papers on functional analysis. Very surprisingly, those two papers have a high impact to Ulam-Hyers stability and quasi-normed spaces. The first paper [5] was on a very useful tool to p-normalize a quasi-normed space which was then reproved by Rolewicz independently [78], see Theorem 5.2. The second paper [6] was on Ulam-Hyers stability of additive map, see the above theorem. Maligranda explained many aspects on Aoki result about Ulam-Hyers stability and even proposed the term *Ulam-Hyers-Aoki stability* instead of *Ulam-Hyers-Rassias stability*, see also [57]. In this chapter we use the term *Ulam-Hyers stability* to merit the asker and the first answerer to the problem.

Ulam-Hyers stability was studied in quasi-Banach spaces where the authors investigated many kinds of functional equations similar to those already considered in Banach spaces. We may distinguish two approaches. In the first one, based on Theorem 5.2, the authors claimed to study Ulam-Hyers stability of functional equations only for p-Banach spaces. In the second, the authors studied Ulam-Hyers stability of functional equations in quasi-Banach spaces which are not assumed to be p-Banach spaces. We must say that the stages in which we study Ulam-Hyers stability of functional equations in quasi-Banach spaces are very similar to those in Banach spaces as follows

1. First stage: Finding a certain kind of functional equations.
2. The second stage: Stating and proving Ulam-Hyers stability of that kind of functional equations in quasi-Banach spaces by using direct method or the fixed point method.

   a. Direct method: the exact solution of the functional equation is explicitly constructed as a limit of a sequence, starting from the given approximate solution [30].
   b. Fixed point method: the exact solution of the functional equation is explicitly constructed as a fixed point of some certain map [14].

5.2.1 Ulam-Hyers Stability of Functional Equations in p-Banach Spaces

The first result in this field belongs to Park et al. [68] published in 2006. In that paper the authors proved Ulam-Hyers stability of the quadratic functional equation of the form

$$f(x + y) + f(x - y) = 2f(x) + 2f(y)$$

in strong p-Banach spaces by the direct method as follows.

Theorem 5.8 ([68, Theorem 3.1]) *Assume that the following conditions hold.*

1. *$(X, \|.\|_X, \kappa_X)$ is a strong quasi-normed space over the field $\mathbb{K}$ and $(Y, \|.\|_Y, \kappa_Y)$ is a strong p-Banach space, and $f : X \to Y$ is a given map.*
2. *There are $r > 2$, $\theta \geq 0$ such that*

$$\|f(x + y) + f(x - y) - 2f(x) - 2f(y)\|_Y \leq \theta\left(\|x\|_X^r + \|y\|_X^r\right)$$

for all $x, y \in X$.

Then there exists a unique quadratic map $Q : X \to Y$ such that

$$\|Q(x) - f(x)\|_Y \leq \frac{2\theta}{(2^{pr} - 4^p)^{\frac{1}{p}}} \|x\|_X^r$$

for all $x \in X$.

After that Park [67] proved Ulam-Hyers stability of homomorphisms in p-Banach algebras. This was also applied to investigate isomorphisms between p-Banach algebras. This problem was continued by Najati and Park [61] but the latter paper was published 1 year earlier. Note that the following statement appeared in [67, page 90] and [59, page 1320] and many papers later.

> By Aoki-Rolewicz Theorem, each quasi-norm is equivalent to some p-norm. Since it is much easier to work with p-norms than quasi-norms, henceforth we restrict our attention mainly to p-norms.

Then, although the titles of many papers are about quasi-Banach spaces, almost all their contents are on p-Banach spaces.

There have been many results on Ulam-Hyers stability of functional equations in p-Banach spaces, and almost all of them were proved by the direct method. The main result of [67] is as follows.

Theorem 5.9 ([67, Theorem 2.1]) *Assume that the following conditions hold.*

1. $(A, \|.\|_A, \kappa_A)$ is a quasi-normed algebra and $(B, \|.\|_B, \kappa_B)$ is a p-Banach algebra over the field $\mathbb{R}$.

2. $r > 1$, $\theta > 0$, and $f : A \to B$ is a map satisfying

$$\|f(x + y) - f(x) - f(y)\|_B \leq \theta \|x\|_A^r \|y\|_A^r$$

$$\|f(xy) - f(x)f(y)\|_B \leq \theta \|x\|_A^r \|y\|_A^r$$

for all $x, y \in A$.

3. For each $x \in A$, $f(tx)$ is continuous in $t \in \mathbb{R}$.

Then there exists a unique homomorphism $H : A \to B$ such that

$$\|f(x) - H(x)\|_B \leq \frac{\theta}{(4^{pr} - 2^p)^{\frac{1}{p}}} \|x\|_A^{2r}, \quad x \in A.$$

Also in [67] Ulam-Hyers stability of homomorphisms in p-Banach algebras associated to Jensen functional equation, isomorphisms between p-Banach algebras associated to Cauchy functional equation and to Jensen functional equation were studied, see [67, Theorem 2.3], [67, Theorem 3.1] and [67, Theorem 3.3], respectively.

In the year 2007 many results on Ulam-Hyers stability of functional equations in p-Banach spaces were published. Park [66] proved Ulam-Hyers stability of homomorphisms in p-Banach algebras and of generalized derivations on p-Banach algebras for the functional equation of the form

$$\sum_{i=1}^{n} f\left(\sum_{j=1}^{n} q(x_i - x_j)\right) + nf\left(\sum_{i=1}^{n} qx_i\right) = nq \sum_{i=1}^{n} f(x_i)$$

see [66, Theorems 2.1 and 2.2] and [66, Theorems 4.2 and 4.3], respectively. These were also applied to investigate isomorphisms between p-Banach algebras, see [66, Theorems 3.1 and 3.2]. Jun and Kim [42] characterized the generalized cubic functional equation, see [42, Theorems 2.1 and 2.3], and then solved the generalized Ulam-Hyers stability problem for Euler-Lagrange type cubic functional equations of the form

$$f(ax + y) + f(x + ay) = (a + 1)(a - 1)^2[f(x) + f(y)] + a(a + 1)f(x + y),$$

where f is an unknown function from a p-norm space to a p-Banach space Y and a is a fixed integer with $a \neq 0, \pm 1$, see [42, Theorems 3.1, 3.3, 3.7 and 3.8], also in p-Banach B-modules, see [42, Theorems 4.1–4.4]. These results were then generalized in [50].

In 2008, Eskandani [27] characterized the following functional equation

$$\sum_{i=1}^{m} f\left(mx_i + \sum_{j=1, j \neq i}^{m} x_j\right) + f\left(\sum_{i=1}^{m} x_i\right) = 2f\left(\sum_{i=1}^{m} mx_i\right)$$

where f is an unknown function from a quasi-norm space to a p-Banach space, $m \in \mathbb{N}$ with $m \geq 2$ [27, Lemma 2.1] and then investigated Ulam-Hyers stability of (5.2.1) in p-Banach spaces [27, Theorems 2.2 and 2.3]. In the same year, Najati and Moghimi [60] established the general solution of the functional equation

$$f(2x + y) + f(2x - y) = f(x + y) + f(x - y) + 2f(2x) - 2f(x),$$

where f is an unknown function from a real vector space to a real vector space [60, Lemmas 2.1 and 2.3, Theorem 2.4], and investigated Ulam-Hyers stability of this equation in p-Banach spaces [60, Theorems 3.2, 3.3, 3.6, 3.7, 3.10 and 3.11]. Also, Najati and Eskandani [59] established the general solution and investigated Ulam-Hyers stability of the following functional equation

$$f(2x + y) + f(2x - y) = 2f(x + y) + 2f(x - y) + 2[f(2x) - 2f(x)]$$

in the p-Banach spaces.

In 2009, Gordji and Khodaei [34] achieved the general solution and the generalized Ulam-Hyers stability of generalized mixed type cubic, quadratic and additive functional equations of the form

$$f(x + ky) + f(x - ky) = k^2 f(x + y) + k^2 f(x - y) + 2(1 - k^2)f(x)$$

for fixed integers k with $k \neq 0, \pm 1$ in the p-Banach spaces.

In 2012, Gao [32] investigated the generalized Ulam-Hyers stability of an n-dimensional quadratic functional equation

$$f\left(\sum_{i=1}^{n} x_i\right) + \sum_{1 \le i \le j \le n} f(x_i - x_j) = n \sum_{i=1}^{n} f(x_i)$$

with $n \ge 2$ in p-Banach spaces by the direct method [32, Theorem 10.2], and by the fixed point method [32, Theorem 10.4]. Zhang and Wang [85] investigated the generalized Ulam-Hyers stability of the quadratic-cubic functional equation of the form

$$6f(x + y) - 6f(x - y) + 4f(3y) = 3f(x + 2y) - 3f(x - 2y) + 9f(2y)$$

in p-Banach spaces [85, Theorems 25.1-25.7].

In 2015, Cho et al. [20] considered the following functional equation

$$\sum_{i=1}^{n} f\left(\sum_{j=1}^{n} q(x_i - x_j)\right) + nf\left(\sum_{i=1}^{n} qx_i\right) = nq \sum_{i=1}^{n} f(x_i)$$

in p-Banach algebras. Then authors proved Ulam-Hyers stability of homomorphisms in p-Banach algebras [20, Subsection 2.3.3] and stability of generalized derivations in p-Banach algebras [20, Subsection 2.3.1].

In 2016 Balamurugan et al. [8] established the general solution and investigated the generalized Ulam-Hyers stability of the following additive-quadratic-cubic-quartic functional equation

$$\begin{aligned}
f(3x &+ 2y + z) + f(3x + 2y - z) + f(3x - 2y + z) + f(3x - 2y - z) \\
&= 48[f(x + y) + f(x - y)] + 24[f(-x + y) + f(-x - y)] \\
&\quad + 12[f(x + z) + f(x - z)] + 6[f(-x + z) + f(-x - z)] \\
&\quad + 4[\tilde{f}(y + z) + \tilde{f}(y - z)] + 20f(2x) + 4f(-2x) - 160f(x) - 80f(-x) \\
&\quad + 2\tilde{f}(2y) - 80\tilde{f}(y) - 24\tilde{f}(z)
\end{aligned}$$

in p-Banach spaces, where $\tilde{f}(x) = f(x) + f(-x)$, see [8, Theorems 12, 13, 16, 19, 22, 25 and 28]. Heidarpour [35] proved the superstability of n-ring homomorphisms on $\mathbb{C}$ [35, Theorems 2.1 and 2.3] and established Ulam-Hyers stability of n-ring homomorphisms in p-Banach algebras [35, Theorems 3.1–3.3, 3.5 and 3.6]. Bodaghi and Kim [9] proved Ulam-Hyers stability for the following mixed quadratic-additive functional equation in p-Banach spaces

$$f(x + my) + f(x - my)$$

$$= \begin{cases} 2f(x) - 2m^2 f(y) + m^2 f(2y) & \text{if } m \text{ is even} \\ f(x + y) + f(x - y) - 2(m^2 - 1)f(y) + (m^2 - 1)f(2y) & \text{if } m \text{ is odd} \end{cases}$$

see [9, Theorems 2.2, 2.6, 2.8 and 2.10].

In 2017 Nikoufar [63] proved the generalized Ulam-Hyers stability of (α_1, α_2)-double Jordan derivations in p-Banach algebras and Isac-Rassias stability of double Jordan derivations in p-Banach algebras [63, Theorems 2.6 and 2.8].

5.2.2 Ulam-Hyers Stability of Functional Equations in Quasi-Banach Spaces Which Are Not Assumed to be p-Banach

As mentioned in Sect. 5.2.1, on studying Ulam-Hyers stability in quasi-Banach spaces, the authors usually restricted their attentions mainly to p-norms. The quotation on page 110 also appeared in the first works [60, page 401], [59, page 1320], and in recent works [20, page 10]. However, quantities relevant to Ulam-Hyers stability of functional equations are not preserved even by equivalent norms in general. Moreover, the inequality (5.1) which may be seen to have the modulus of concavity equal to 1, and the continuity of p-norms were used in many proofs such as in proving the inequalities (3.17) and (3.20) in proof of [60, Theorem 3.2], in proving the inequalities (3.32) and (3.35) in proof of [84, Theorem 3.2].

Inspired by the above facts, some authors were interested in studying Ulam-Hyers stability of functional equations in quasi-Banach spaces where the quasi-norm is not assumed to be a p-norm, and thus, the modulus of concavity is greater than 1 and the quasi-norm is not continuous in general.

To overcome the modulus of concavity greater than 1 and the discontinuity of quasi-norms, we can use the b-metric metrization theorem, see Theorem 5.1. Note that we also use the following squeeze inequalities on b-metric mentioned in Lemma 5.1 to overcome the discontinuity of a b-metric, and thus, of a quasi-norm, in proving Ulam-Hyers stability of functional equations in quasi-normed spaces. Besides that, some authors also used the assumption of strong quasi-Banach spaces to overcome the modulus of concavity greater than 1 and the discontinuity of quasi-norms.

Lemma 5.1 ([1], Lemma 2.1) *Let (X, d, κ) be a b-metric space and $\lim_{n\to\infty} x_n = x$, $\lim_{n\to\infty} y_n = y$. Then we have*

1. $\frac{1}{\kappa^2} d(x, y) \leq \liminf_{n\to\infty} d(x_n, y_n) \leq \limsup_{n\to\infty} d(x_n, y_n) \leq \kappa^2 d(x, y)$.

2. If $x = y$, then $\lim_{n\to\infty} d(x_n, y_n) = 0$.

3. For each $z \in X$, $\frac{1}{\kappa} d(x, z) \leq \liminf_{n\to\infty} d(x_n, z) \leq \limsup_{n\to\infty} d(x_n, z) \leq \kappa d(x, z)$.

The first result in this field belongs to Park et al. [68] published in 2006. In that paper the authors proved Ulam-Hyers stability of the quadratic functional equation of the form

$$f(x + y) + f(x - y) = 2f(x) + 2f(y)$$

in strong quasi-Banach spaces by the direct method as follows.

Theorem 5.10 ([68], Theorem 2.1) *Assume that the following conditions hold.*

1. *X is a strong quasi-normed space over the field $\mathbb{K}$ and $(Y, \|.\|_Y, \kappa_Y)$ is a strong quasi-Banach space, and $f : X \to Y$ is a given map with $f(0) = 0$.*
2. *$\varphi : X \times X \to [0, \infty)$ is a function satisfying*

$$\tilde{\varphi}(x, y) := \sum_{j=1}^{\infty} 4^j \varphi\left(\frac{x}{2^j}, \frac{y}{2^j}\right) < \infty$$

   for all $x, y \in X$.
3. *$\|f(x + y) + f(x - y) - 2f(x) - 2f(y)\|_Y \leq \varphi(x, y)$ for all $x, y \in X$.*

Then there exists a unique quadratic map $Q : X \to Y$ such that

$$\|Q(x) - f(x)\|_Y \leq \frac{K}{4}\tilde{\varphi}(x, x)$$

for all $x \in X$.

In [24] Dung and Hang used the fixed point method to prove an extension of the stability result of Aiemsomboon and Sintunavarat [2, Theorem 2.2] to quasi-Banach spaces by applying Corollary 5.1.

Theorem 5.11 ([24], Theorem 2.5) *Assume that the following conditions hold.*

1. *$(X, \|.\|, \kappa)$ is a quasi-normed space over the field $\mathbb{F}$, $(Y, \|.\|, \kappa)$ is a quasi-Banach space over the field $\mathbb{K}$, and $g : X \to Y$ is a given function.*
2. *There exist $a, b \in \mathbb{F} \setminus \{0\}$, $A, B \in \mathbb{K} \setminus \{0\}$ and $u, v : X \to \mathbb{R}_+$ such that*

$$M_0 = \left\{ n \in \mathbb{N} : \kappa\left(\left|\frac{1}{A}\right| s_1(a + bn)s_2(a + bn) + \left|\frac{B}{A}\right| s_1(n)s_2(n) \right) < 1 \right\}$$

is an infinite set, where $\theta = \log_{2\kappa} 2$,

$$s_1(n) = \inf\{t \in \mathbb{R}_+ : u(nx) \leq tu(x) \text{ for all } x \in X\}$$

$$s_2(n) = \inf\{t \in \mathbb{R}_+ : v(nx) \leq tv(x) \text{ for all } x \in X\}$$

for $n \in \mathbb{F} \setminus \{0\}$, and s_1, s_2 satisfy the following two conditions, where $n \to \infty$ in $\mathbb{F}$ if and only if $|n| \to \infty$,

  a. $\lim\limits_{n \to \infty} s_1(\pm n)s_2(\pm n) = 0$.
  b. $\lim\limits_{n \to \infty} s_1(n) = 0$ *or* $\lim\limits_{n \to \infty} s_2(n) = 0$.

3. *The function* $g : X \to Y$ *satisfies the inequality*

$$\|g(ax + by) - Ag(x) - Bg(y)\| \le u(x)v(y) \, for \, x, y \in X \setminus \{0\}.$$

Then g satisfies the equation

$$g(ax + by) = Ag(x) + Bg(y) \, for \, x, y \in X \setminus \{0\}.$$

By [12, Lemma 3.1] we see that if $g : X \to Y$ satisfies the general linear equation on $X \setminus \{0\}$ then it satisfies the general linear equation on X. From this fact and Theorem 5.11 we get the following result which is an extension of Aiemsomboon and Sintunavarat [2, Theorem 2.3] to quasi-Banach spaces.

Theorem 5.12 ([24], Theorem 2.6) *Assume that the following conditions hold.*

1. $(X, \|.\|, \kappa)$ *is a quasi-normed space over the field* $\mathbb{F}$, $(Y, \|.\|, \kappa)$ *is a quasi-Banach space over the field* $\mathbb{K}$, *and* $g : X \to Y$ *is a given function.*
2. *There exist* $a, b \in \mathbb{F} \setminus \{0\}$, $A, B \in \mathbb{K} \setminus \{0\}$ *and* $u, v : X \to \mathbb{R}_+$ *such that*

$$M_0 = \left\{ n \in \mathbb{N} : \kappa \left(\left| \frac{1}{A} \right| s_1(a + bn)s_2(a + bn) + \left| \frac{B}{A} \right| s_1(n)s_2(n) \right) < 1 \right\}$$

is an infinite set, where

$$s_1(n) = \inf\{t \in \mathbb{R}_+ : u(nx) \le tu(x) \, for \, all \, x \in X\}$$

$$s_2(n) = \inf\{t \in \mathbb{R}_+ : v(nx) \le tv(x) \, for \, all \, x \in X\}$$

for $n \in \mathbb{F} \setminus \{0\}$ *and* s_1, s_2 *satisfy the following two conditions*

a. $\lim_{n \to \infty} s_1(\pm n)s_2(\pm n) = 0.$
b. $\lim_{n \to \infty} s_1(n) = 0 \, or \, \lim_{n \to \infty} s_2(n) = 0.$

3. *The function* $g : X \to Y$ *satisfies the inequality*

$$\|g(ax + by) - Ag(x) - Bg(y)\| \le u(x)v(y) \, for \, x, y \in X \setminus \{0\}.$$

Then g satisfies the equation

$$g(ax + by) = Ag(x) + Bg(y) \, for \, x, y \in X.$$

By using Theorem 5.11 we get an extension of [70, Theorem 2.1] to quasi-Banach spaces as follows.

Corollary 5.2 ([24], Corollary 2.7) *Assume that the following conditions hold.*

1. $(X, \|.\|, \kappa)$ *is a quasi-normed space over the field* $\mathbb{F}$, $(Y, \|.\|, \kappa)$ *is a quasi-Banach space over the field* $\mathbb{K}$, *and* $g : X \to Y$ *is a given function.*

2. *There exist* $a, b \in \mathbb{F} \setminus \{0\}$, $A, B \in \mathbb{K} \setminus \{0\}$, $c \geq 0$, $p, q \in \mathbb{R}$ *with* $p + q < 0$
such that

$$\|g(ax + by) - Ag(x) - Bg(y)\| \leq c\|x\|^p \|y\|^q \text{ for } x, y \in X \setminus \{0\}.$$

Then g satisfies the equation

$$g(ax + by) = Ag(x) + Bg(y) \text{ for } x, y \in X.$$

The next example shows the significance of the obtained results in quasi-Banach spaces.

Example 5.6 ([24, Example 2.8]) Let $X = Y = L^p(0, 1)$ with $0 < p < 1$ and $g : X \to Y$ be defined by $g(x) = \frac{1}{2}x$ for all $x \in X$. Then X, Y are quasi-Banach spaces and all assumptions of Corollary 5.2 are satisfied with $a = A$ and $b = B$. So Corollary 5.2 is applicable to X, Y and g.

However, since $L^p(0, 1)$ with $0 < p < 1$ is not normable [64, page 18], so [2, Theorem 2.2] is not applicable to X, Y and g.

To overcome the modulus of concavity greater than 1 and the discontinuity of quasi-norms, the author of [23] used the squeeze inequality (5.2) presented in an explicit revision of Aoki-Rolewicz, see Theorem 5.2. As illustrations, the authors proved an extension of the main result of [60] in p-Banach spaces to quasi-Banach spaces with better approximation.

Theorem 5.13 ([23, Theorem 2.2]) *Assume that the following conditions hold.*

1. $(X, \| \cdot \|_X, \kappa_X)$ *is a real quasi-normed space and* $(Y, \| \cdot \|_Y, \kappa_Y)$ *is a real quasi-Banach space.*
2. $\varphi : X \times X \to [0, \infty)$ *is a function such that for all* $x, y \in X$,

$$\lim_{n \to \infty} 4^n \varphi\left(\frac{x}{2^n}, \frac{y}{2^n}\right) = 0$$

and for all $x \in X$, *all* $y \in \{0, x, -2x, 3x, 4x\}$, $p = \log_{2\kappa_Y} 2$,

$$\sum_{i=1}^{\infty} 4^{ip} \varphi^p\left(\frac{x}{2^i}, \frac{y}{2^i}\right) < \infty.$$

3. $f : X \to Y$ *is an even function such that* $f(0) = 0$ *and for all* $x, y \in X$,

$$\|f(2x + y) + f(2x - y) - f(x + y) - f(x - y) - 2f(2x) + 2f(x)\|_Y \leq \varphi(x, y).$$

Then for all $x \in X$, *the limit* $Q(x) = \lim_{n \to \infty} 4^n f\left(\frac{x}{2^n}\right)$ *exists, and the function* $Q : X \to Y$ *is a unique quadratic function satisfying*

$$\|f(x) - Q(x)\|_Y \leq \frac{\kappa_Y}{2}\left(\tilde{\psi}(x)\right)^{\frac{1}{p}} \tag{5.9}$$

for all $x \in X$, where

$$\tilde{\psi}(x) = \sum_{i=1}^{\infty} 4^{ip}\left(2^p\varphi^p\left(\frac{x}{3\cdot 2^i}, \frac{x}{3\cdot 2^i}\right) + \varphi^p\left(\frac{x}{3\cdot 2^i}, \frac{x}{2^i}\right) + \varphi^p\left(\frac{x}{3\cdot 2^i}, \frac{4x}{3\cdot 2^i}\right)\right.$$

$$\left. + \varphi^p\left(\frac{x}{3\cdot 2^i}, \frac{-2x}{3\cdot 2^i}\right) + \varphi^p\left(\frac{x}{3\cdot 2^i}, 0\right)\right).$$

The following example illustrates that the approximation defined by (5.9) in Theorem 5.13 can be better than the approximation defined by (3.4) in [60, Theorem 3.2].

Example 5.7 ([23, Example 2.3]) Let $X = Y = L^{\frac{1}{2}}[0, 1]$ and

$$\|x\|_X = \|x\|_Y = \left(\int_0^1 |x(t)|^{\frac{1}{2}} dt\right)^2$$

for all $x \in X$, where

$$L^{\frac{1}{2}}[0, 1] = \left\{f : [0, 1] \to \mathbb{R} : |f|^{\frac{1}{2}} \text{ is Lebesgue integrable}\right\}.$$

Define $f(x) = x^2 + x^4$ for all $x \in X$, $\varphi(x, y) = \left(\int_0^1 6|x(t)y(t)|dt\right)^2$ for all $x, y \in X$. Then

1. All assumptions of Theorem 5.13 and [60, Theorem 3.2] are satisfied with $\kappa_Y = 2$.
2. The approximation defined by (5.9) in Theorem 5.13 is better than the approximation defined by (3.4) in [60, Theorem 3.2].

Remark 5.2 ([23, Remark 2.4]) Similarly to the proof of Theorem 5.13, we can prove extensions of [60, Theorems 3.6, 3.7, 3.10 and 3.11] and many other results in p-Banach spaces to quasi-Banach spaces with better approximation.

As mentioned in Sect. 5.2.1, Ulam-Hyers stability of functional equations in p-Banach algebras was studied by Park [67]. Two of the key techniques in [67] are the proofs of $\mathbb{R}$-linearity of the homomorphism [67, line +14 on page 91], and of the preservation of multiplication [67, line -9 on page 91].

Recall that, to prove the $\mathbb{R}$-linearity of the homomorphism [67, line +14 on page 91], the author used the same reasoning as in [71, page 299], where the dual space of a Banach space was used. However, one of the main differences between Banach spaces and quasi-Banach spaces is the dual space, while the dual space approach is very useful in Banach spaces by Hahn-Banach theorem, the dual space approach fails in quasi-Banach spaces since many quasi-Banach spaces have the trivial dual space, see [44, page 1102].

Also, to prove the preservation of multiplication under the function H [67, line -9 on page 91], the function H was defined by

$$H(x) = \lim_{n \to \infty} 2^n f\left(\frac{x}{2^n}\right)$$

and thus

$$H(xy) = \lim_{n \to \infty} 2^n f\left(\frac{xy}{2^n}\right)$$

but the author used

$$H(xy) = \lim_{n \to \infty} 4^n f\left(\frac{xy}{2^n \cdot 2^n}\right).$$

These limitations were subsequently used in many works, see for example in old one [61, line +8 on page 769] and in recent one [20, Lemma 2.14]. Recall that the quantities relevant to Ulam-Hyers stability are not preserved even under equivalent norms. So Ulam-Hyers stability of homomorphisms in quasi-Banach algebras, where the Banach space is not assumed to be p-Banach, may be different from that in p-Banach algebras.

By the above reasons, some authors were interested in revising the proofs of Ulam-Hyers stability of homomorphisms in p-Banach algebras [67], and interested in extending that to quasi-Banach algebras.

To prove $\mathbb{R}$-linearity of homomorphisms in quasi-Banach algebras, Dung et al. [25] used the following Moore-Osgood Theorem on exchanging limits.

Theorem 5.14 (Moore-Osgood Theorem [33, Theorem 2.1.4.1 and Remark 2.1.4.1]) *Let X, Y be subsets of Hausdorff spaces, $a \in \overline{X} \setminus X$, $b \in \overline{Y} \setminus Y$, Z be a metric space, and $f : X \times Y \to Z$ be a given function such that*

1. $\lim_{x \to a} f(x, y) = h(y)$ *on* $Y \setminus \{b\}$.
2. $\lim_{y \to b} f(x, y) = g(x)$ *uniformly on* $X \setminus \{a\}$.

Then the limits $\lim_{(x,y) \to (a,b)} f(x, y)$, $\lim_{x \to a} \lim_{y \to b} f(x, y)$ *and* $\lim_{y \to b} \lim_{x \to a} f(x, y)$ *exist and*

$$\lim_{(x,y) \to (a,b)} f(x, y) = \lim_{x \to a} \lim_{y \to b} f(x, y) = \lim_{y \to b} \lim_{x \to a} f(x, y).$$

To prove the preservation of multiplication under the function H, Dung et al. [25] used the following, somewhat different, definition of the homomorphism

$$H(x) = \lim_{n \to \infty} 4^n f\left(\frac{x}{4^n}\right). \tag{5.10}$$

These techniques overcome the limitations appeared in many results on Ulam-Hyers stability of homomorphisms in quasi-Banach algebras as mentioned above.

The following result is an extension of [67, Theorem 2.1], where the p-Banach algebra is replaced by a quasi-Banach algebra, and it was proved by using Theorems 5.2, 5.14 and the definition of H stated in (5.10).

Theorem 5.15 ([25], Theorem 4) *Assume that the following conditions hold.*

1. $(A, \|.\|_A, \kappa_A)$ is a quasi-normed algebra and $(B, \|.\|_B, \kappa_B)$ is a quasi-Banach algebra over the field $\mathbb{R}$.
2. $r > 1, \theta > 0$, and $f : A \to B$ is a function satisfying

$$\|f(x + y) - f(x) - f(y)\|_B \leq \theta \|x\|_A^r \|y\|_A^r$$

$$\|f(xy) - f(x)f(y)\|_B \leq \theta \|x\|_A^r \|y\|_A^r$$

for all $x, y \in A$.
3. For each $x \in A$, $f(tx)$ is continuous in $t \in \mathbb{R}$.

Then there exists a unique homomorphism $H : A \to B$ such that

$$\|f(x) - H(x)\|_B \leq \frac{\theta}{(4^{pr} - 2^p)^{\frac{1}{p}}} \|x\|_A^{2r}, \quad x \in A.$$

The techniques used in proof of Theorem 5.15 may be applied to revise and extend other results on Ulam-Hyers stability of homomorphisms in p-Banach algebras to quasi-Banach algebras [25, Remark 5].

5.3   Ulam-Hyers Stability of Functional Equations in Quasi-β-Banach Spaces

Ulam-Hyers stability of functional equations in quasi-β-Banach spaces was first studied in [74] by Rassias and Kim. Then authors investigated Ulam-Hyers stability of many kinds of functional equations similar to that in quasi-Banach spaces. We may distinguish two ways in stating and proving the results. In the first way, based on Theorem 5.2, the authors claimed to study Ulam-Hyers stability of functional equations only for (β, p)-Banach spaces. In fact, we must use Theorem 5.6 to do that. In the second way, the authors studied Ulam-Hyers stability of functional equations in β-Banach spaces directly. We must say that the stages to study Ulam-Hyers stability of functional equations in quasi-β-Banach spaces are very similar to that in quasi-Banach spaces mentioned on page 109.

5.3.1 Ulam-Hyers Stability of Functional Equations in Quasi-β-Banach Spaces

In studying Ulam-Hyers stability of functional equations in quasi-β-Banach spaces, many authors also used the quotation on page 110 to claim that we could replace a quasi-β-Banach space by certain (β, p)-Banach space. However, as mentioned on page 106 we could not use Aoki-Rolewicz Theorem to claim that every quasi-β-norm is equivalent to certain (β, p)-norm. Alternatively we must use Theorem 5.6.

There have been many results on Ulam-Hyers stability of functional equations in (β, p)-Banach spaces. In 2009, Rassias and Kim [74] generalized results obtained for Jensen type maps and established new theorems about Ulam-Hyers stability for general additive functional equations of the form

$$\sum_{1 \leq i < j \leq n} g\left(\frac{x_i + x_j}{2} + \sum_{l=1, k_l \neq i, j}^{n-2} x_{k_l}\right) = \frac{(n-1)^2}{2} \sum_{i=1}^{n} g(x_i) \tag{5.11}$$

in (β, p)-normed spaces. Note that in case $n = 2$, Eq. (5.11) yields Jensen additive equation

$$2g\left(\frac{x + y}{2}\right) = g(x) + g(y)$$

and there are many interesting results concerning Ulam-Hyers stability problems of Jensen equation, see [72] for example. Therefore, Eq. (5.11) is a generalized form of Jensen additive equation. The characterization for additive maps was stated as follows. Note that the case $n \leq 3$ was stated in [62, Lemma 2.2].

Lemma 5.2 ([74, Theorem 2.2]) *Let X and Y be linear spaces, $n \geq 3$ and $f : X \to Y$ be a map. Then f satisfies*

$$\sum_{1 \leq i < j \leq n} f\left(\frac{x_i + x_j}{2} + \sum_{l=1, k_l \neq i, j}^{n-2} x_{k_l}\right) = \frac{(n-1)^2}{2} \sum_{i=1}^{n} f(x_i)$$

for all $x_1, x_2, \ldots, x_n \in X$ if and only if f is additive.

By using the characterization mentioned above, the next result was proved for contractively subadditive equations by using the direct method.

Theorem 5.16 ([74, Theorem 3.1]) *Assume that the following conditions hold.*

1. *X is a linear space over the field $\mathbb{K}$ and $(Y, \|.\|_Y, \kappa_Y)$ is a (β, p)-Banach space with the p-norm $\|.\|_Y$, and $f : X \to Y$ is a given map.*
2. *$\varphi : X^n \to [0, \infty)$ is a contractively subadditive function, that is,*

$$\varphi(x + y) \leq L[\varphi(x) + \varphi(y)]$$

for all $x, y \in X^n$, where the constant L is such that $L > 0$ and $\lambda^{1-\beta} L < 1$.

3. $\|D_1 f(x_1, \ldots, x_n)\| \le \varphi(x_1, \ldots, x_n)$ *for all n-variables* $x_1, \ldots, x_n \in X$, *where*

$$D_1 f(x_1, \ldots, x_n) = \sum_{1 \le i < j \le n} g\left(\frac{x_i + x_j}{2} + \sum_{l=1, k_l \ne i, j}^{n-2} x_{k_l}\right) - \frac{(n-1)^2}{2} \sum_{i=1}^{n} g(x_i).$$

Then there exists a unique additive map $g : X \to Y$ *satisfying* (5.11) *and*

$$\|f(x) - g(x)\|_Y \le \frac{2^\beta \varphi(x, \ldots, x)}{n^\beta \lambda^\beta \sqrt[p]{\lambda^{\beta p} - \lambda^p L^p}}$$

for all $x \in X$ *and* $\lambda = n - 1$.

Similar to Theorem 5.16, the next result is for expansively superadditive equations.

Theorem 5.17 ([74, Theorem 3.2]) *Assume that the following conditions hold.*

1. *X is a linear space over the field* $\mathbb{K}$ *and* $(Y, \|.\|_Y, \kappa_Y)$ *is a* (β, p)-*Banach space with the p-norm* $\|.\|_Y$, *and* $f : X \to Y$ *is a given map.*
2. $\varphi : X^n \to [0, \infty)$ *is a expansively subadditive function, that is,*

$$\varphi(x + y) \ge \frac{1}{L}[\varphi(x) + \varphi(y)]$$

for all $x, y \in X^n$, *where the constant L is such that* $L > 0$ *and* $\lambda^\beta L < 1$.
3. $\|D_1 f(x_1, \ldots, x_n)\| \le \varphi(x_1, \ldots, x_n)$ *for all n-variables* $x_1, \ldots, x_n \in X$.

Then there exists a unique additive map $g : X \to Y$ *satisfying*

$$\sum_{1 \le i < j \le n} g\left(\frac{x_i + x_j}{2} + \sum_{l=1, k_l \ne i, j}^{n-2} x_{k_l}\right) = \frac{(n-1)^2}{2} \sum_{i=1}^{n} g(x_i)$$

and

$$\|f(x) - g(x)\|_Y \le \frac{2^\beta L \varphi(x, \ldots, x)}{n^\beta \lambda^\beta \sqrt[p]{\lambda^p - \lambda^{\beta p} L^p}}$$

for all $x \in X$ *and* $\lambda = n - 1$.

Another result about Ulam-Hyers stability of Eq. (5.11) is as follows.

Theorem 5.18 ([74, Theorem 3.3]) *Assume that the following conditions hold.*

1. *X is a linear space over the field* $\mathbb{K}$ *and* $(Y, \|.\|_Y, \kappa_Y)$ *is a* (β, p)-*Banach space with the p-norm* $\|.\|_Y$, *and* $f : X \to Y$ *is a given map.*
2. $\varphi : X^n \to [0, \infty)$ *is a function satisfying*

$$\phi(x, \ldots, x) := \sum_{i=0}^{\infty} \frac{\kappa_Y^i \varphi(\lambda^i x, \ldots, \lambda^i x)}{\lambda^{\beta i}} < \infty, \quad \lim_{m \to \infty} \frac{\kappa_Y^m \varphi(\lambda^m x_1, \ldots, \lambda^m x_n)}{\lambda^{\beta m}} = 0$$

for all $x, x_1, \ldots, x_n \in X$, where $\lambda = n - 1$.

3. $\|D_1 f(x_1, \ldots, x_n)\|_Y \le \varphi(x_1, \ldots, x_n)$ for all n-variables $x_1, \ldots, x_n \in X$.

Then there exists a unique additive map $g : X \to Y$ satisfying (5.11) and

$$\|f(x) - g(x)\|_Y \le \frac{2^\beta \kappa_Y}{n^\beta \lambda^{2\beta}} \phi(x, \ldots, x)$$

for all $x \in X$.

In 2010, Wang and Liu [81] investigated Ulam-Hyers stability of the following quadratic functional equation

$$2f(2x+y)+2f(2x-y) = 4f(x+y)+4f(x-y)+4f(2x)+f(2y)-8f(x)-8f(y)$$

in (β, p)-normed spaces. They characterized solutions of such quadratic functional equations as follows.

Theorem 5.19 ([81, Theorem 2.1]) *Suppose* $f : X \to Y$ *is a map. Then* f *satisfies (5.3.1) if and only if there are a quadratic map* $Q(x)$ *and a cubic map* $H(x)$ *such that* $f(x) = Q(x) + H(x)$ *for all* $x \in X$.

Then the authors proved Ulam-Hyers stability by using the direct method as follows.

Theorem 5.20 ([81, Theorem 3.1]) *Assume that the following conditions hold.*

1. *X is a linear space over the field* $\mathbb{K}$ *and* $(Y, \|.\|_Y, \kappa_Y)$ *is a* (β, p)-*Banach space with the p-norm* $\|.\|_Y$, *and* $f : X \to Y$ *is a given map with* $f(0) = 0$.
2. $\varphi : X \times X \to [0, \infty)$ *is a function satisfying*

$$\Psi(x) := \sum_{n=1}^{\infty} 4^{\beta n p} \varphi^p\left(0, \frac{x}{2^n}\right) < \infty, \quad \lim_{n \to \infty} 4^{\beta n} \varphi\left(\frac{x}{2^n}, \frac{y}{2^n}\right) = 0$$

   for all $x, y \in X$.
3. $\|Df(x, y)\|_Y \le \varphi(x, y)$ *for all* $x, y \in X$, *where*

$$Df(x, y) = 2f(2x + y) + 2f(2x - y) - 4f(x + y) - 4f(x - y)$$
$$-4f(2x) - f(2y) + 8f(x) + 8f(y).$$

Then there exists a unique quadratic map $Q : X \to Y$ such that

$$\|Q(x) - f(x)\|_Y \le \frac{1}{4^\beta} (\Psi(x))^{\frac{1}{p}}$$

for all $x \in X$.

Some other modifications of Theorem 5.20 were also stated, see [81, Theorems 3.2, 3.4, 3.5, 3.7 and 3.8].

In 2011, Eskandani et al. [28] first characterized quadratic and quartic maps and then established the general solution of the following mixed additive and quadratic functional equation

$$f(\lambda x + y) + f(\lambda x - y) = f(x + y) + f(x - y) + (\lambda - 1)[(\lambda + 2)f(x) + \lambda f(-x)]$$

in (β, p)-normed spaces [28, Lemmas 2.1–2.2, Theorem 2.3]. The authors then investigated the generalized Ulam-Hyers stability of that equation with $\lambda \in \mathbb{N}$ and $\lambda \neq 1$ in (β, p)-normed spaces, see [28, Theorems 3.1, 3.3, 3.4, 3.6, 3.7, 3.9, 3.10, 3.12 and 3.15].

In 2012 Kim et al. [51] studied the following two radical equations

$$f(\sqrt{x^2 + y^2}) = f(x) + f(y)$$

$$f(\sqrt{x^2 + y^2}) + f(\sqrt{|x^2 - y^2|}) = 2f(x) + 2f(y)$$

in (β, p)-Banach spaces, see [51, Lemma 2.1]. Then the authors proved Ulam-Hyers stability by using subadditive and subquadratic functions for radical functional equations in (β, p)-Banach spaces, see [51, Theorems 2.2, 2.3, 2.7 and 2.9–2.12]. Note that the characterizations for quadratic and quartic maps mentioned in that paper were cited from [48].

In 2013, Moradlou and Rassias [58] proved that every generalized additive map of Cauchy-Jensen type of the form

$$2\sum_{j=1}^{n} f\left(\frac{x_j}{2} + \sum_{i=1,i\neq j}^{n} x_i\right) + \sum_{j=1}^{n} f(x_j) = 2nf\left(\sum_{j=1}^{n} x_j\right)$$

is Cauchy additive in (β, p)-normed spaces, see [58, Lemma 2.1]. Then they proved the generalized Ulam-Hyers stability of that functional equation [58, Theorems 2.3–2.4] by using the fixed point method. In the same year, Xu and Rassias also obtained a description of the general solution of the septic and octic functional equations of the form

$$f(x + 4y) - 7f(x + 3y) + 21f(x + 2y) - 35f(x + y) + 35f(x)$$
$$-21f(x - y) + 7f(x - 2y) - f(x - 3y) = 5040f(y)$$

and

$$f(x + 4y) - 8f(x + 3y) + 28f(x + 2y) - 56f(x + y) + 70f(x)$$
$$-56f(x - y) + 28f(x - 2y) - 8f(x - 3y) + f(x - 4y) = 40320f(y)$$

in (β, p)-normed spaces [83, Theorems 2.1 and 2.2] and then proved Ulam-Hyers stability of the septic and octic functional equations in (β, p)-normed spaces [83, Theorems 3.2 3.5] by the direct method.

In 2014, Rassias and Kim [75] characterized the general solution of the (m, n)-Cauchy-Jensen functional equation [75, Theorem 2.1] and established new theorems about Ulam-Hyers stability of general (m, n)-Cauchy-Jensen additive maps in (β, p)-Banach spaces, which generalized results obtained for Cauchy-Jensen type additive maps by the direct method [75, Theorems 3.1–3.4 and 3.7–3.8]. Cho et al. [19] proved the generalized Ulam-Hyers stability results by considering maps satisfying conditions much weaker than Hyers and Rassias conditions for radical quadratic and radical quartic functional equations in (β, p)-Banach spaces.

In 2015, Kim et al. [52] found the general solution of the following Cauchy-Jensen type functional equation of the form

$$f\left(\frac{x+y}{n} + z\right) + f\left(\frac{y+z}{n} + x\right) + f\left(\frac{z+x}{n} + y\right) = \frac{n+2}{n}\left(f(x) + f(y) + f(z)\right)$$

in (β, p)-Banach spaces [52, Lemma 1], and then investigated the generalized Ulam-Hyers stability of the equation in (β, p)-Banach spaces for any fixed nonzero integer n by the direct method, see [52, Theorems 1–4]. Also Hong and Kim [37] considered a modified quadratic functional equation and then investigated its generalized Ulam-Hyers stability in (β, p)-Banach spaces, see [37, Theorems 2.1 and 2.2]. Cădariu et al. [18] used a fixed point theorem [18, Theorem 2.2] to prove some generalized Ulam-Hyers stability theorems for additive Cauchy functional equations as well as for monomial functional equations in (β, p)-Banach spaces, see [17, Theorem 2 on page 104 and Theorem 3 on page 107].

In 2016, Ravi et al. [77] obtained the generalized Ulam-Hyers stability of the functional equation of the form

$$f(x + 6y) - 11f(x + 5y) + 55f(x + 4y) - 165f(x + 3y)$$
$$+330f(x + 2y) - 462f(x + y) + 462f(x) - 330f(x - y)$$
$$+165f(x - 2y) - 55f(x - 3y) + 11f(x - 4y) - f(x - 5y) = 39916800f(y)$$

in (β, p)-Banach spaces by using the fixed point method. The authors also investigated the pertinent stability of the above functional equation using control functions with sum of powers of norms, product of powers of norms and a mixture of products and sums of powers of norms as upper bounds.

Continuing [77], in 2017 Rassias et al. [76] achieved the general solution of the duodecic functional equation of the form

$$f(x+6y) - 12f(x+5y) + 66f(x+4y) - 220f(x+3y)$$
$$+495f(x+2y) - 792f(x+y) + 924f(x) - 792f(x-y)$$
$$+495f(x-2y) - 220f(x-3y) + 66f(x-4y)$$
$$-12f(x-5y) + f(x-6y) = 479001600f(y)$$

in (β, p)-Banach spaces [76, Theorem 3.1], and investigated Ulam-Hyers stability involving a general control function with sum of powers of norms, product of powers of norms and mixed product-sum of powers of norms in (β, p)-Banach spaces via the fixed point method [76, Theorem 4.2].

Recently, EL-Fassi [26] introduced and solved the radical quintic functional equation of the form

$$f\left(\sqrt[5]{x^5 + y^5}\right) = f(x) + f(y)$$

in quasi-β-Banach spaces [26, Theorem 2.1]. The author next established Ulam-Hyers stability results in quasi-β-Banach spaces [26, Theorems 3.1 and 3.2], and then Ulam-Hyers stability by using subadditive and subquadratic functions in (β, p)-Banach spaces for that functional equation [26, Theorems 3.4 and 3.5]. Also Kim and Liang [49] presented general solution of a generalized quadratic functional equation with several variables, and then obtained its generalized Ulam-Hyers stability results in (β, p)-Banach spaces [49, Theorems 3.1–3.5].

Up to now, there has been a few results on Ulam-Hyers stability of functional equations in quasi-β-Banach spaces which are not assumed to be (β, p)-Banach spaces. In 2012 Liguang and Jing [54] investigated Ulam-Hyers stability of a functional equation deriving from additive and quadratic equations of the form

$$f(x+2y) + f(x-2y) = f(x+y) + f(x-y) + 3f(2y) - 6f(y)$$

in quasi-β-Banach spaces without the assumption that they are (β, p)-Banach spaces. The key technique in that paper is to transform the problem to certain generalized metric space, see the proof of [54, Theorem 3.1]. In that proof, the authors claimed without calculation that the function d defined by

$$d(g, h) = \inf\{C : C \in \mathbb{R}, C \geq 0, \|g(x) - h(x)\|_Y \leq C\varphi(0, x), x \in X\}$$

is a generalized metric on the set $\Omega = \{g : X \to Y\}$, where a generalized metric is very similar to a metric except for $d(g, h) \in [0, \infty]$. However, by usual calculation, it is easy to check that d is a generalized b-metric on Ω. It remains an open problem to prove the claim that d is a generalized metric on Ω.

Recently EL-Fassi [26] established Ulam-Hyers stability results for the radical quintic functional equation in quasi-β-Banach spaces as follows.

Theorem 5.21 ([26, Theorem 3.1]) *Assume that the following conditions hold.*

1. $(X, \|.\|, \kappa)$ *is a quasi-β-Banach space.*
2. $f : \mathbb{R} \to X$ *is a φ-approximately radical quintic function, that is,*

$$\left\| f\left(\sqrt[5]{x^5 + y^5}\right) - f(x) - f(y) \right\| \le \varphi(x, y)$$

for all $x, y \in \mathbb{R}$, where $\varphi : \mathbb{R}^2 \to [0, \infty)$ is a function.

3. $\Phi(x, y) := \sum\limits_{j=0}^{\infty} \left(\frac{\kappa}{2^\beta}\right)^j \varphi(2^{\frac{j}{5}}x, 2^{\frac{j}{5}}y) < \infty$ *and* $\lim\limits_{n \to \infty} 2^{-n\beta}\varphi(2^{\frac{n}{5}}x, 2^{\frac{n}{5}}y) = 0$ *for all*
   $x, y \in \mathbb{R}$.

Then there exists a unique quintic map $Q : \mathbb{R} \to X$ satisfying

$$Q\left(\sqrt[5]{x^5 + y^5}\right) = Q(x) + Q(y)$$

and

$$\|f(x) - Q(x)\| \le \frac{\kappa}{2^\beta}\Phi(x, x)$$

for all $x, y \in \mathbb{R}$.

Also see [26, Theorem 3.2] for a similar result. In the proof the author used the following equality

$$\left\| Q\left(\sqrt[5]{x^5 + y^5}\right) - Q(x) - Q(y) \right\| = \lim\limits_{n \to \infty} 2^{-\beta n}\left\| f\left(2^{\frac{n}{5}}\sqrt[5]{x^5 + y^5}\right) - f(2^{\frac{n}{5}}x) - f(2^{\frac{n}{5}}y) \right\|.$$

However, this equality does not hold since the quasi-β-norm $\|.\|$ does not need to be continuous. To overcome this confusion, we must use Lemma 5.1 or Theorem 5.6.

5.3.2 Some Open Problems in Ulam-Hyers Stability of Functional Equations in Quasi-β-Normed Spaces

There have been many results on Ulam-Hyers stability in Banach spaces [73] and Banach algebras [20]. Inspired by these works, we have the following question.

Question 5.1 Generalize Ulam-Hyers stability results in Banach spaces and in Banach algebras to quasi-Banach spaces and quasi-Banach algebras.

Inspired by the work [24], we have the following equation.

Question 5.2 State other fixed point results in b-metric spaces and apply to study the Ulam-Hyers stability of functional equations in quasi-Banach spaces and to study integral equations, see [38] for example; see also [15].

Finally we list some unsolved issues in the literature.

1. Ulam-Hyers stability of Euler-Lagrange type cubic maps in quasi-Banach spaces [42, Remarks 3.2 and 3.5].
2. Ulam-Hyers stability of first-order linear partial differential equations [43, Remark 3] (see also [16]).

Acknowledgements The authors sincerely thank the anonymous reviewers for their helpful comments. The authors also thank members of The Dong Thap Group of Mathematical Analysis and its Applications, Dong Thap University, Vietnam for their discussions on the manuscript. This project was partially completed while the first author visited the Department of Mathematics and Statistics, Faculty of Science and Technology, Thammasat University Rangsit Center, Thailand

The second author would like to thank the Thailand Research Fund and Office of the Higher Education Commission under grant no. MRG6180283 for financial support during the preparation of this manuscript.

References

1. Aghajani, A., Abbas, M., Roshan, J.R.: Common fixed point of generalized weak contractive mappings in partially ordered b-metric spaces. Math. Slovaca **64**(4), 941–960 (2014)
2. Aiemsomboon, L., Sintunavarat, W.: A note on the generalised hyperstability of the general linear equation. Bull. Aust. Math. Soc. **96**(2), 263–273 (2017)
3. Almira, J.M., Luther, U.: Inverse closedness of approximation algebras. J. Math. Anal. Appl. **314**(1), 30–44 (2006)
4. An, T.V., Tuyen, L.Q., Dung, N.V.: Stone-type theorem on b-metric spaces and applications. Topology Appl. **185–186**, 50–64 (2015)
5. Aoki, T.: Locally bounded linear topological spaces. Proc. Imp. Acad. **18**(10), 588–594 (1942)
6. Aoki, T.: On the stability of the linear transformation in Banach spaces. J. Math. Soc. Jpn **2**(1–2), 64–66 (1950)
7. Baak, C.: Generalized quasi-Banach spaces. J. Chungcheong Math. Soc. **18**, 215–222 (2005)
8. Balamurugan, K., Arunkumar, M., Ravindiran, P.: Superstability and stability of approximate n-ring homomorphisms between quasi-Banach algebras. Int. J. Pure Appl. Math. **106**(5), 99–121 (2016)
9. Bodaghi, A., Kim, S.O.: Ulam's type stability of a functional equation deriving from quadratic and additive functions. J. Math. Inequal. **9**(1), 73–84 (2015)
10. Boriceanu, M., Bota, M., Petruşel, A.: Multivalued fractals in b-metric spaces. Cent. Eur. J. Math. **8**(2), 367–377 (2010)
11. Brzdęk, J.: Hyperstability of the Cauchy equation on restricted domains. Acta Math. Hungar. **141**(1–2), 58–67 (2013)
12. Brzdęk, J.: Remarks on stability of some inhomogeneous functional equations. Aequationes Math. **89**(1), 83–96 (2015)
13. Brzdęk, J., Chudziak, J., Páles, Z.: A fixed point approach to stability of functional equations. Nonlinear Anal. **74**(17), 6728–6732 (2011)
14. Brzdęk, J., Cădariu, L., Ciepliński, K.: Fixed Point Theory and the Ulam stability. J. Funct. Spaces **2014**, 1–16 (2014)
15. Brzdęk, J., Karapınar, E., Petruşel, A.: A fixed point theorem and the Ulam stability in generalized d_q-metric spaces. J. Math. Anal. Appl. **467**, 501–520 (2018)
16. Brzdęk, J., Popa, D., Rasa, I., Xu, B.: Ulam stability of operators. In: Mathematical Analysis and Its Applications, 1st edn. Academic Press, Dordrecht (2018)

17. Cădariu, L.: Generalized Ulam–Hyers stability results: a fixed point approach. In: Handbook of Functional Equations, pp. 101–111. Springer, Berlin (2015)
18. Cădariu, L., Găvruţa, L., Găvruţa, P.: Fixed points and generalized Hyers-Ulam stability. Abstr. Appl. Anal. **2012**, 1–10 (2012)
19. Cho, Y.J., Gordji, M.E., Kim, S.S., Yang, Y.: On the stability of radical functional equations in quasi-β-normed spaces. Bull. Korean Math. Soc. **51**(5), 1511–1525 (2014)
20. Cho, Y.J., Park, C., Rassias, T.M., Saadati, R.: Stability of Functional Equations in Banach Algebras. Springer, Cham (2015)
21. Czerwik, S.: Contraction mappings in b-metric spaces. Acta Math. Univ. Ostrav. **1**(1), 5–11 (1993)
22. Czerwik, S.: Nonlinear set-valued contraction mappings in b-metric spaces. Atti Sem. Math. Fis. Univ. Modena **46**, 263–276 (1998)
23. Dung, N.V., Hang, V.T.L.: Stability of a mixed additive and quadratic functional equation in quasi-Banach spaces. J. Fixed Point Theory Appl. **20**, 1–11 (2018)
24. Dung, N.V., Hang, V.T.L.: The generalized hyperstability of general linear equations in quasi-Banach spaces. J. Math. Anal. Appl. **462**, 131–147 (2018)
25. Dung, N.V., Hang, V.T.L., Sintunaravat, W.: Revision and extension on Hyers-Ulam-Rassias stability of homomorphisms in quasi-Banach algebras. Rev. R. Acad. Cienc. Exactas Fís. Nat. Ser. A Mat. **113**(3), 1773–1784 (2018)
26. EL-Fassi, I.-i.: Solution and approximation of radical quintic functional equation related to quintic mapping in quasi-β-Banach spaces. Rev. R. Acad. Cienc. Exactas Fís. Nat. Ser. A Mat. **113**(2), 1–13 (2018)
27. Eskandani, G.Z.: On the Hyers-Ulam–Rassias stability of an additive functional equation in quasi-Banach spaces. J. Math. Anal. Appl. **1**(345), 405–409 (2008)
28. Eskandani, G.Z., Gavruta, P., Rassias, J.M., Zarghami, R.: Generalized Hyers-Ulam stability for a general mixed functional equation in quasi-β-normed spaces. Mediter. J. Math. **8**(3), 331–348 (2011)
29. Fagin, R., Kumar, R., Sivakumar, D.: Comparing top k lists. SIAM J. Discrete Math. **17**(1), 134–160 (2003)
30. Forti, G.-L.: Comments on the core of the direct method for proving Hyers-Ulam stability of functional equations. J. Math. Anal. Appl. **295**(1), 127–133 (2004)
31. Gajda, Z.: On stability of additive mappings. Int. J. Math. Math. Sci. **14**(3), 431–434 (1991)
32. Gao, J.: Generalized Hyers-Ulam stability for general quadratic functional equation in quasi-Banach spaces. In: Functional Equations in Mathematical Analysis, chapter 10, pp. 125–138. Springer, Berlin (2012)
33. Gelbaum, B.R., Olmsted, J.M.: Theorems and Counterexamples in Mathematics. Springer, New York (1990)
34. Gordji, M.E., Khodaei, H.: Solution and stability of generalized mixed type cubic, quadratic and additive functional equation in quasi-Banach spaces. Nonlinear Anal. **71**(11), 5629–5643 (2009)
35. Heidarpour, Z.: Superstability and stability of approximate n-ring homomorphisms between quasi-Banach algebras. Gen. Math. **33**(1), 40–47 (2016)
36. Heinonen, J.: Lectures on analysis on metric spaces. In: Axler, S., Gehring, F.W., Ribet, K.A. (eds.) Universitext. Springer, Berlin (2001)
37. Hong, Y.S., Kim, H.-M.: Approximate quadratic mappings in quasi-β-normed spaces. J. Chungcheong Math. Soc. **28**(2), 311–319 (2015)
38. Hussain, N., Salimi, P., Al-Mezel, S.: Coupled fixed point results on quasi-Banach spaces with application to a system of integral equations. Fixed Point Theory Appl. **2013**, 1–18 (2013)
39. Hyers, D.H.: A note on linear topological spaces. Bull. Am. Math. Soc. **44**(2), 76–80 (1938)
40. Hyers, D.H.: Locally bounded linear topological spaces. Rev. Ci. Lima **41**, 558–574 (1939)
41. Hyers, D.H.: On the stability of the linear functional equation. Proc. Nat. Acad. Sci. **27**(4), 222–224 (1941)
42. Jun, K.-W., Kim, H.-M.: On the stability of Euler-Lagrange type cubic mappings in quasi-Banach spaces. J. Math. Anal. Appl. **2**(332), 1335–1350 (2007)

43. Jung, S.-M.: Hyers-Ulam stability of linear partial differential equations of first order. Appl. Math. Let. **22**(1), 70–74 (2009)
44. Kalton, N.: Quasi-Banach spaces. In: Johnson, W.B., Lindenstrauss, J. (eds.) Handbook of the Geometry of Banach Spaces, vol. 2, pp. 1099–1130. Elsevier, Amsterdam (2003)
45. Kalton, N.J., Peck, N.T., Roberts, J.W.: An F-Space Sampler. London Mathematical Society Lecture Note Series, vol. 89. Cambridge University Press, Cambridge (1984)
46. Khamsi, M.A.: Remarks on cone metric spaces and fixed point theorems of contractive mappings. Fixed Point Theory Appl. **2010**, 1–7 (2010)
47. Khamsi, M.A., Hussain, N.: KKM mappings in metric type spaces. Nonlinear Anal. **73**(9), 3123–3129 (2010)
48. Khodaei, H., Gordji, M.E., Kim, S.S., Cho, Y.J.: Approximation of radical functional equations related to quadratic and quartic mappings. J. Math. Anal. Appl. **395**(1), 284–297 (2012)
49. Kim, H.-M., Liang, H.-M.: Approximate generalized quadratic mappings in (β, p)-Banach spaces. J. Comput. Anal. Appl. **24**(1), 148–160 (2018)
50. Kim, H.-M., Rassias, J.M.: Generalization of Ulam stability problem for Euler-Lagrange quadratic mappings. J. Math. Anal. Appl. **336**(1), 277–296 (2007)
51. Kim, S.S., Cho, Y.J., Gordji, M.E.: On the generalized Hyers-Ulam-Rassias stability problem of radical functional equations. J. Inequal. Appl. **2012**(2012:186), 1–13 (2012)
52. Kim, H.-M., Jun, K.-W., Son, E.: Approximate Cauchy–Jensen type mappings in quasi-β-normed spaces. In: Handbook of Functional Equations, pp. 243–254. Springer, Berlin (2015)
53. Lee, Y.-H.: On the stability of the monomial functional equation. Bull. Korean Math. Soc. **45**(2), 397–403 (2008)
54. Liguang, W., Jing, L.: On the stability of a functional equation deriving from additive and quadratic functions. Adv. Differ. Equ. **2012**(2012:98), 1–12 (2012)
55. Macías, R.A., Segovia, C.: Lipschitz functions on spaces of homogeneous type. Adv. Math. **33**(3), 257–270 (1979)
56. Maligranda, L.: Tosio Aoki (1910–1989). In: International Symposium on Banach and Function Spaces: 14/09/2006-17/09/2006, pp. 1–23. Yokohama Publishers, Yokohama (2008)
57. Maligranda, L.: A result of Tosio Aoki about a generalization of Hyers-Ulam stability of additive functions–a question of priority. Aequationes Math. **75**(3), 289–296 (2008)
58. Moradlou, F., Rassias, T.M.: Generalized Hyers-Ulam-Rassias stability for a general additive functional equation in quasi-β-normed spaces. Bull. Korean Math. Soc. **50**(6), 2061–2070 (2013)
59. Najati, A., Eskandani, G.Z.: Stability of a mixed additive and cubic functional equation in quasi-Banach spaces. J. Math. Anal. Appl. **342**(2), 1318–1331 (2008)
60. Najati, A., Moghimi, M.B.: Stability of a functional equation deriving from quadratic and additive functions in quasi-Banach spaces. J. Math. Anal. Appl. **337**(1), 399–415 (2008)
61. Najati, A., Park, C.: Hyers-Ulam-Rassias stability of homomorphisms in quasi-Banach algebras associated to the pexiderized Cauchy functional equation. J. Math. Anal. Appl. **335**(2), 763–778 (2007)
62. Najati, A., Ranjbari, A.: Stability of homomorphisms for a 3D Cauchy–Jensen type functional equation on C^*-ternary algebras. J. Math. Anal. Appl. **1**(341), 62–79 (2008)
63. Nikoufar, I.: Functions near some (α_1, α_2)-double Jordan derivations in p-Banach algebras. Boll. Unione Mat. Ital. **10**(2), 191–198 (2017)
64. Okada, S. Ricker, W.J., Pérez, E.S.: Optimal domain and integral extension of operators. In: Operator Theory: Advances and Applications, vol. 180. Springer, Berlin (2008)
65. Paluszyński, M., Stempak, K.: On quasi-metric and metric spaces. Proc. Am. Math. Soc. **137**(12), 4307–4312 (2009)
66. Park, C.-G.: Hyers-Ulam-Rassias stability of homomorphisms in quasi-Banach algebras. Banach J. Math. Anal. **1**(1), 23–32 (2007)
67. Park, C.: Hyers-Ulam-Rassias stability of homomorphisms in quasi-Banach algebras. Bull. Sci. Math. **132**, 87–96 (2008)
68. Park, C., Jun, K.-W., Lu, G.: On the quadratic mapping in generalized quasi-Banach spaces. J. Chungcheong Math. Soc. **19**, 263–274 (2006)

69. Pietsch, A.: History of Banach Spaces and Linear Operators. Birkhäuser, Boston (2007)
70. Piszczek, M.: Hyperstability of the general linear functional equation. Bull. Korean Math. Soc. **52**(6), 1827–1838 (2015)
71. Rassias, T.M.: On the stability of the linear mapping in Banach spaces. Proc. Am. Math. Soc. **72**(2), 297–300 (1978)
72. Rassias, J.M.: Refined Hyers-Ulam approximation of approximately Jensen type mappings. Bull. Sci. Math. **131**(1), 89–98 (2007)
73. Rassias, T.M.: Handbook of Functional Equations: Stability Theory, vol. 96. Springer, Berlin (2014)
74. Rassias, J.M., Kim, H.-M.: Generalized Hyers-Ulam stability for general additive functional equations in quasi-β-normed spaces. J. Math. Anal. Appl. **356**(1), 302–309 (2009)
75. Rassias, J.M., Kim, H.-M.: Approximate (m, n)-Cauchy-Jensen mappings in quasi-β-normed spaces. J. Comput. Anal. Appl. **16**, 346–358 (2014)
76. Rassias, J.M., Ravi, K., Kumar, B.V.S.: A fixed point approach to Ulam-Hyers stability of duodecic functional equation in quasi-β-normed spaces. Tbilisi Math. J. **10**(4), 83–101 (2017)
77. Ravi, K., Rassias, J.M., Kumar, B.V.S.: Ulam-Hyers stability of undecic functional equation in quasi-β-normed spaces: fixed point method. Tbilisi Math. J. **9**(2), 83–103 (2016)
78. Rolewicz, S.: On a certain class of linear metric spaces. Bull. Acad. Polon. Sci. **5**, 471–473 (1957)
79. Schroeder, V.: Quasi-metric and metric spaces. Conform. Geom. Dyn. **10**, 355–360 (2006)
80. Ulam, S.M.: Problems in Modern Mathematics. Wiley, New York (1964)
81. Wang, L.G., Liu, B.: The Hyers-Ulam stability of a functional equation deriving from quadratic and cubic functions in quasi-β-normed spaces. Acta Math. Sin. (Eng. Ser.) **26**(12), 2335–2348 (2010)
82. Xia, Q.: The geodesic problem in quasimetric spaces. J. Geom. Anal. **19**, 452–479 (2009)
83. Xu, T.Z., Rassias, J.M.: Approximate septic and octic mappings in quasi-β-normed spaces. J. Comput. Anal. Appl. **15**, 1110–1119 (2013)
84. Xu, T.Z., Rassias, J.M., Xu, W.X.: Generalized Hyers-Ulam stability of a general mixed additive-cubic functional equation in quasi-Banach spaces. Acta Math. Sin. (Eng. Ser.) **28**(3), 529–560 (2012)
85. Zhang, W., Wang, Z.: Stability of the quadratic-cubic functional equation in quasi-Banach spaces. In: Functional Equations in Mathematical Analysis, chapter 25, pp. 319–336. Springer, Berlin (2012)

Chapter 6
On Stability of the Functional Equation of p-Wright Affine Functions in 2-Banach Spaces

El-Sayed El-hady

Abstract We present some stability results for the functional equation of p-Wright affine functions in 2-Banach spaces. In this way we extend several earlier outcomes.

Keywords 2-Norm · 2-Banach space · p-Wright affine function · Functional equation · Ulam stability

Mathematics Subject Classification (2010) Primary 39B82; Secondary 49B62

6.1 Introduction

The subject of functional equations forms a somehow modern branch of mathematics. The importance of functional equations usually comes from their wide range of applications. Functional equations have recent applications in many fields see e.g. [13, 19, 24]. They have applications e.g. in Communication and Network models see [20, 31, 36], in computer graphics [33], in information theory [2, 32], in decision theory [1, 39], and in digital filtering [38]. In this chapter we investigate the stability of the functional equation of the p-Wright affine functions investigated in [4] but in 2-Banach spaces.

Stability is a very important issue with many interesting applications and we refer to, e.g., [9, 11, 12, 14, 15, 26, 34] for more details. Stability can be seen from different points of views see [34] and hundreds of researchers are dealing with such amazing topic. It has applications in optimization theory (see, e.g., [30]), it is related

E.-S. El-hady (✉)
Mathematics Department, College of Science, Jouf University, Sakaka, Kingdom of Saudi Arabia

Basic Science Department, Faculty of Computers and Informatics, Suez Canal University, Ismailia, Egypt
e-mail: elsayed_elhady@ci.suez.edu.eg

© Springer Nature Switzerland AG 2019
J. Brzdęk et al. (eds.), *Ulam Type Stability*,
https://doi.org/10.1007/978-3-030-28972-0_6

to the notion of shadowing (see, e.g., [25]), and it has applications in economics (see [16]). It should be noted that the issue of stability of functional equations was originally motivated by a problem of S.M. Ulam posed in 1940 and Hyers's answer to it published in [26]. Stability is very important because its an efficient tool for evaluating the error people usually face when replacing functions that satisfy some equations only approximately, by the exact solutions to those equations. Roughly speaking, nowadays we say that an equation is stable in some class of functions if any function from that class, satisfying the equation approximately (in some sense), is near (in some way) to an exact solution of the equation. In the last few decades, several stability problems of various (functional, difference, differential, integral) equations have been investigated by many mathematicians (see e.g. [7, 8, 10, 27, 29] for more details), but mainly in classical spaces.

Since the notion of an approximate solution and the idea of nearness of two functions can be understood in many, nonstandard ways, depending on the needs and tools available in a particular situation. One of such non-classical measures of a distance can be introduced by the notion of a 2-norm. As far as we know the concept of linear 2-normed space was introduced first by Gähler in [22], and it seems that the first work on the Hyers-Ulam stability of functional equations in complete 2-normed spaces (that is, 2-Banach spaces) see e.g. [23]. See also [17, 37] for some details in 2-Banach spaces. This chapter is organized as follows: in Sect. 6.2 we recall some definitions and the functional equation of our interest, in Sect. 6.3 we introduce the fixed point theorem used in the stability, in Sect. 6.4 we investigate the stability of the functional equation of the p-Wright affine functions, and in Sect. 6.5 we introduce a simple observation on superstability.

6.2 Preliminaries

Let $0 < p < 1$ be a fixed real number. We say that a function f:

$$f : I \longmapsto \mathbb{R},$$

mapping a real nonempty interval I into the set of reals $\mathbb{R}$ is p-Wright convex provided (see, e.g., [18])

$$f(px_1 + (1 - p)x_2) + f((1 - p)x_1 + px_2) \leq f(x_1) + f(x_2), \qquad x_1, x_2 \in I.$$

If f satisfies the functional equation

$$f(px_1 + (1 - p)x_2) + f((1 - p)x_1 + px_2) = f(x_1) + f(x_2), \tag{6.1}$$

then we say that it is p-Wright affine (see [18]). Note that for $p = 1/2$ Eq. (6.1) becomes the Jensen's functional equation

$$f(\frac{x_1 + x_2}{2}) = \frac{f(x_1) + f(x_2)}{2}.$$

For $p = 1/3$ Eq. (6.1) takes the form

$$f(2x_1 + x_2) + f(x_1 + 2x_2) = f(3x_1) + f(3x_2),$$

which has been investigated by Najati and Park in [35]; in particular, they proved some results on its stability and applied them in the investigation of the generalized (σ, τ)-Jordan derivations on Banach algebras. The cases of more arbitrary p were studied in [18] (see also [28]). We prove some results concerning the Hyers-Ulam stability of (6.1). The method of the proof of the main result corresponds to some observations in [9] and the main tool in it is a fixed point. To present it we need the following three assumptions ($\mathbb{R}_+$ denotes the set of nonnegative reals). Let us recall first (see, for instance, [21]) some definitions.

Definition 6.1 By a linear *2-normed space* we mean a pair $(X, \|., .\|)$ such that X is an at least two-dimensional real linear space and

$$\|\cdot, \cdot\| : X \times X \to \mathbb{R}$$

is a function satisfying the following conditions:

(1) $\|x_1, x_2\| = 0$ if and only if x_1 and x_2 are linearly dependent;
(2) $\|x_1, x_2\| = \|x_2, x_1\|$ for $x_1, x_2 \in X$
(3) $\|x_1, x_2 + x_3\| \leq \|x_1, x_2\| + \|x_1, x_3\|$ for $x_i \in X, i = 1, 2, 3$
(4) $\|\beta x_1, x_2\| = |\beta| \|x_1, x_2\|$ for $\beta \in \mathbb{R}$ and $x_1, x_2 \in X$

Definition 6.2 A sequence $(x_n)n \in \mathbb{N}$ of elements of a linear 2-normed space X is called a *Cauchy sequence* if there are linearly independent $y, z \in X$ such that

$$\lim_{n,m \to \infty} \|x_n - x_m, z\| = 0 = \|x_n - x_m, y\|,$$

whereas $(x_n)n \in \mathbb{N}$ is said to be convergent if there exists an $x \in X$ (called a limit of this sequence and denoted by $\lim_{n \to \infty} X_n$) with

$$\lim_{n,m \to \infty} \|x_n - x, y\| = 0, \qquad y \in X.$$

A linear 2-normed space in which every Cauchy sequence is convergent is called a *2-Banach space*.

Let us also mention that in linear 2-normed spaces, every convergent sequence has exactly one limit and the standard properties of the limit of a sum and a scalar product are valid. Next, it is easily seen that we have the following property.

Lemma 6.1 *If X is a linear 2-normed space, $x, y, z \in X$, y, z are linearly independent, and*

$$\|x, y\| = 0 = \|x, z\|,$$

then $x = 0$.

Let us yet recall a lemma from [37].

Lemma 6.2 *If X is a linear 2-normed space and $(x_n)n \in \mathbb{N}$ is a convergent sequence of elements of X, then*

$$\lim_{n \to \infty} \|x_n, z\| = \|\lim_{n \to \infty} x_n, z\|, \qquad z \in X.$$

It is easy to check that (in view of the Cauchy-Schwarz inequality), if $\langle ., . \rangle$ is a real inner product in a real linear space X, of dimension greater than 1, and

$$\|x_1, x_2\| := \sqrt{\|x_1\|^2 \|x_2\|^2 - \langle x_1, x_2 \rangle^2}, \qquad x_1, x_2 \in X$$

then conditions (1)–(4) are valid.

6.3 Fixed Point Theorem

Let us introduce the following three assumptions:

(A1) S is a nonempty set, $(Y, \|., .\|)$ is a 2-Banach space, Y_0 is a subset of Y containing two linearly independent vectors, $j \in \mathbb{N}$,

$$f_i : S \to S, \qquad g_i : Y_0 \to Y_0, \qquad L_i : S \times Y_0 \to \mathbb{R} \quad \text{for } i = 1, \cdots, j;$$

(A2) $T : Y^S \to Y^S$ is an operator satisfying the inequality

$$\|T\xi(x) - T\mu(x), y\| \le \sum_{i=1}^{j} L_i(x, y) \|\xi(f_i(x)) - \mu(f_i(x)), g_i(y)\|,$$

$$\tag{6.2}$$

$$\xi, \mu \in Y^S, \ x \in S, \ y \in Y_0;$$

(A3) $\Lambda : \mathbb{R}^{S \times Y_0} \to \mathbb{R}^{S \times Y_0}$ is an operator defined by

$$\Lambda\delta(x, y) := \sum_{i=1}^{j} L_i(x, y)\delta(f_i(x), g_i(y)), \ \delta \in \mathbb{R}^{S \times Y_0}, \qquad x \in S, \ y \in Y_0$$

$$\tag{6.3}$$

Now, its the position to present the above mentioned fixed point theorem.

Theorem 6.1 *Let hypotheses (A1)–(A3) hold and functions $\varepsilon : S \times Y_0 \to \mathbb{R}_+$ and $\varphi : S \to Y$ fulfill the following two conditions:*

$$\|T\varphi(x) - \varphi(x), y\| \le \varepsilon(x, y), \qquad x \in S, \ y \in Y_0 \tag{6.4}$$

$$\varepsilon^*(x, y) := \sum_{i=1}^{\infty} (\Lambda^i \varepsilon)(x, y) < \infty, \qquad x \in S, y \in Y_0 \tag{6.5}$$

Then there exists a unique fixed point ψ of T for which

$$\|\varphi(x) - \psi(x), y\| \le \varepsilon^*(x, y), \qquad x \in S, \ y \in Y_0 \tag{6.6}$$

Moreover,

$$\psi(x) = \lim_{l \to \infty} (T^l \varphi)(x), \qquad x \in S. \tag{6.7}$$

6.4   Stability

The next theorem is the main result in this chapter and concerns the stability of Eq. (6.1); it extends the results in [4] and corresponds to some outcomes, e.g., in [3, 5, 6, 9].

Theorem 6.2 *Let (A1) be valid, $p \in \mathbb{R}$, $A, k \in (0, \infty)$,*

$$|p|^k + |1 - p|^k < 1,$$

E be a subset of Y with $0 \in E$ and

$$px_1 + (1 - p)x_2 \in E, \qquad x_1, x_2 \in E, \tag{6.8}$$

and $g : E \to Y$ satisfy

$$\|g(px_1 + (1 - p)x_2) + g((1 - p)x_1 + px_2) - g(x_1) - g(x_2), y\|$$
$$\le A(\|x_1, y\|^k + \|x_2, y\|^k), \qquad x_1, x_2 \in E, y \in Y_0. \tag{6.9}$$

Then there exists a unique solution $G : E \to Y$ of Eq. (6.1) such that

$$\|g(x) - G(x), y\| \le \frac{A\|x, y\|^k}{1 - |p|^k - |1 - p|^k}, \qquad x \in E \tag{6.10}$$

and G is given by:

$$G(x) := g(0) + \lim_{n \to \infty} (T^n g_0)(x), \qquad x \in E, \tag{6.11}$$

where g_0 and T are defined by (6.14) and (6.15). Moreover, G is the unique solution of Eq. (6.1) such that there exists a constant $M \in (0, \infty)$ with

$$\|g(x) - G(x), y\| \le M \|x, y\|^k, \qquad x \in E, y \in Y_0. \tag{6.12}$$

Proof Note that (6.9) with $x_2 = 0$ gives

$$\|g(px_1) + g((1 - p)x_1) - g(x_1) - g(0), y\| \le A(\|x_1, y\|^k + \|y\|^k), \tag{6.13}$$
$$x_1 \in E, y \in Y_0.$$

Write

$$g_0(x_1) = g(x_1) - g(0), \qquad x_1 \in E \tag{6.14}$$

and

$$T\xi(x_1) = \xi(px_1) + \xi((1 - p)x_1), \qquad x_1 \in E, \xi \in Y^E. \tag{6.15}$$

Then (6.13) implies the inequality

$$\|g_0(px_1) + g_0((1 - p)x_1) - g(x_1) - g(0), y\| \le A(\|x_1, y\|^k), \qquad x_1 \in E, \tag{6.16}$$

which means that

$$\|Tg_0(x_1) - g_0(x_1), y\| \le A(\|x_1, y\|^k), \qquad x_1 \in E. \tag{6.17}$$

Further note that (A3) holds with $k = 2$, $f_1(x) = px$, $f_2(x) = (1 - p)x$, $L_i(x) = 1$ for $i = 1, 2, x \in E$. Define Λ as in (A3). Clearly, with $\varepsilon(x) := A(\|x_1, y\|^k)$ for $x \in E$, we have

$$\varepsilon^*(x_1) := \sum_{n=0}^{\infty} (\Lambda^n \varepsilon)(x_1) \tag{6.18}$$

$$\le A(\|x_1, y\|^k) \sum_{n=0}^{\infty} (|p|^k + |1 - p|^k)^n$$

$$= \frac{A(\|x_1, y\|^k)}{1 - |p|^k - |1 - p|^k}, \qquad x_1 \in E.$$

Hence, according to Theorem 6.1, there exists a unique solution $G_0 : X \to Y$ of the equation

$$G_0(x_1) = G_0(px_1) + G_0((1-p)x_1), \qquad x_1 \in E \qquad (6.19)$$

such that

$$\|g_0(x_1) - G_0(x_1)\| \leq \frac{A(\|x_1, y\|^k)}{1 - |p|^k - |1-p|^k}, \qquad x_1 \in E; \qquad (6.20)$$

moreover

$$G_0(x_1) := \lim_{n \to \infty} (T^n g_0)(x_1), \qquad x_1 \in E. \qquad (6.21)$$

Now we show that, for every $x_1, x_2 \in E, n \in \mathbb{N}_0$ (nonnegative integers),

$$\|T^n g_0(px_1 + (1-p)x_2) + T^n g_0((1-p)x_1 + px_2) - T^n g(x_1) - T^n g(x_2), y\| \qquad (6.22)$$

$$\leq A(|p|^k + |1-p|^k)^n (\|x_1, y\|^k + \|x_2, y\|^k), \qquad x_1, x_2 \in E, y \in Y_0$$

It is easy to see that the case $n = 0$ is just (6.9). Next, fix $m \in \mathbb{N}_0$ and assume that (6.22) holds for every $x_1, x_2 \in E$ with $n = m$. Then

$$\|T^{m+1} g_0(px_1 + (1-p)x_2) + T^{m+1} g_0((1-p)x_1 + px_2) \qquad (6.23)$$

$$- T^{m+1} g(x_1) - T^{m+1} g(x_2), y\|$$

$$= \|T^m g_0(p(px_1 + (1-p)x_2)) + T^m g_0((1-p)(px_1 + (1-p)x_2))$$

$$+ T^m g_0(p((1-p)x_1 + px_2)) + T^m g_0((1-p)(1-p)x_1 + px_2))$$

$$- T^m g_0(px_1) - T^m g_0((1-p)x_1) - T^m g_0(px_2) - T^m g_0((1-p)x_2), y\|,$$

$$\leq \|T^m g_0(ppx_1 + (1-p)px_2) + T^m g_0((1-p)px_1 + ppx_2) - T^m g_0(px_1)$$

$$- T^m g_0(px_2), y\|$$

$$+ \|T^m g_0(p(1-p)x_1 + (1-p)(1-p)x_2)$$

$$+ T^m g_0((1-p)(1-p)x_1 + p(1-p)x_2)$$

$$- T^m g_0((1-p)x_1) - T^m g_0(p(1-p)x_2), y\|$$

$$\leq A(|p|^k + |1-p|^k)^m ((p\|x_1, y\|)^k + (p\|x_2, y\|)^k)$$

$$+ (|p|^k + |1-p|^k)^m (((1-p)\|x_1, y\|)^k + ((1-p)\|x_2, y\|)^k)$$

$$= (|p|^k + |1-p|^k)^m ((\|x_1, y\|)^k + (\|x_2, y\|)^k), \qquad x_1, x_2 \in E, y \in Y_0.$$

Thus, by induction we have shown that (6.22) holds for every $x_1, x_2 \in E$ and $n \in \mathbb{N}_0$. Letting $n \to \infty$ in (6.22), we obtain that

$$G_0(px_1 + (1-p)x_2) + G_0((1-p)x_1 + px_2) = G_0(x_1) + G_0(x_2), \qquad (6.24)$$

$$x_1, x_2 \in E.$$

Write $G(x_1) := G_0(x_1) + g(0)$ for $x_1 \in E$. Then it is easily seen that

$$G(px_1 + (1-p)x_2) + G((1-p)x_1 + px_2) = G(x_1) + G(x_2), \qquad x_1, x_2 \in E \tag{6.25}$$

and (6.10) holds. It remains to show the uniqueness of G. So suppose that $M_0 \in (0, \infty)$ and $G_1 : X \to Y$ is a solution to (6.1) with

$$\|g(x_1) - G_1(x_1), y\| \le M_0\|x_1, y\|, \qquad x_1 \in E, y \in Y_0. \tag{6.26}$$

Note that

$$G(0) = g(0) = G_1(0),$$

$$G_1(px_1) + G_1((1-p)x_1) = G_1(x_1) + G_1(0), \qquad x_1 \in E, \tag{6.27}$$

$$G(px_1) + G((1-p)x_1) = G(x_1) + G(0), \qquad x_1 \in E, \tag{6.28}$$

and, by (6.10),

$$\|G(x_1) - G_1(x_1), y\| \le \frac{(M+A)\|x_1, y\|^k}{1 - |p|^k - |1-p|^k} \tag{6.29}$$

$$= (M+A)\|x_1, y\|^k \sum_{n=j}^{\infty} (|p|^k + |1-p|^k)^n, \qquad x_1 \in E.$$

The case $j = 0$ is exactly (6.29). So fix $l \in \mathbb{N}_0$ and assume that (6.29) holds for $j = l$. Then, in view of (6.27) and (6.28),

$$\|G(x_1) - G_1(x_1), y\| \tag{6.30}$$

$$= \|G(px_1) + G((1-p)x_1) - G_1(px_1) - G_1((1-p)x_1), y\|,$$

$$\le \|G(px_1) - G_1(px_1), y\| + \|G((1-p)x_1) - G_1((1-p)x_1), y\|$$

$$\le (M+A)(\|p\|^k\|x_1, y\|^k + \|(1-p)\|^k\|x_1, y\|^k) \sum_{n=l}^{\infty} (|p|^k + |1-p|^k)^n,$$

$$\le (M+A)\|x_1, y\|^k \sum_{n=l+1}^{\infty} (|p|^k + |1-p|^k)^n, \qquad x_1 \in E, y \in Y_0.$$

Thus we have shown (6.29). Now, letting $j \to \infty$ in (6.29) we get $G_1 = G$. $\qquad \square$

6.5 An Observation on Superstability

The following is a very simple observation on the superstability of Eq. (6.1) complements Theorem 6.2.

Theorem 6.3 *Let (A1) be valid, $p \in \mathbb{F}$, $A, k \in (0, \infty)$, $|p|^{2k} + |1 - p|^{2k} < 1$, E be a subset of Y such that $0 \in E$ and (6.8) holds, and $g : E \to Y$ satisfy*

$$\|g(px_1 + (1 - p)x_2) + g((1 - p)x_1 + px_2) - g(x_1) - g(x_2), y\| \qquad (6.31)$$
$$\leq A\|x_1, y\|^k \|x_2, y\|^k$$

for every $x_1, x_2 \in E$, $y \in Y_0$. Then g is a solution to (6.1).

Proof It is easy to see that (6.31) with $x_2 = 0$ gives

$$g(x_1) = g(px_1) + g((1 - p)x_1) - g(0), \qquad x_1 \in E \qquad (6.32)$$

We show that, for every $x_1, x_2 \in E$, $y \in Y_0$, $n \in \mathbb{N}_0$,

$$\|g(px_1 + (1 - p)x_2) + g((1 - p)x_1 + px_2) - g(x_1) - g(x_2), y\| \qquad (6.33)$$
$$\leq A(|p|^{2k} + |1 - p|^{2k})^n \|x_1, y\|^k \|x_2, y\|^k.$$

It is easy to see that the case $n = 0$ is just (6.31). Next, fix $m \in \mathbb{N}_0$ and assume that (6.33) holds for every $x_1, x_2 \in E$, with $n = m$. Then, by (6.32),

$$\|g(px_1 + (1 - p)x_2) + g((1 - p)x_1 + px_2) - g(x_1) - g(x_2), y\| \qquad (6.34)$$
$$= \|g(p(px_1 + (1 - p)x_2)) + g((1 - p)(px_1 + (1 - p)x_2))$$
$$+ g(p((1 - p)x_1 + px_2)) + g((1 - p)((1 - p)x_1 + px_2))$$
$$- g(px_1) - g((1 - p)x_1) - g(px_2) - g((1 - p)x_2), y\|$$
$$\leq A(|p|^{2k} + |1 - p|^{2k})^m \|p\|^k \|x_1, y\|^k \|p\|^k \|x_2, y\|^k$$
$$+ A(|p|^{2k} + |1 - p|^{2k})^m \|1 - p\|^k \|x_1, y\|^k \|1 - p\|^k \|x_2, y\|^k$$
$$= A(|p|^{2k} + |1 - p|^{2k})^{m+1} \|x_1, y\|^k \|x_2, y\|^k$$

for every $x_1, x_2 \in E$, $y \in Y_0$. Therefore, by induction we have shown that (6.33) holds for every $x_1, x_2 \in E$ and $n \in \mathbb{N}_0$. Letting $n \to \infty$ in (6.33), we obtain that g is a solution to (6.1). $\qquad \square$

Acknowledgement This work is funded by Jouf University, Kingdom of Saudi Arabia under the research project number 39/600.

References

1. Abbas, A.E., Aczél, J.: The role of some functional equations in decision analysis. Decision Anal. **7**(2), 215–228 (2010)
2. Aczél, J.: Notes on generalized information functions. Aequationes Math. **22**(1), 97–107 (1981)
3. Bahyrycz, A., Brzdęk, J., Piszczek, M.: Approximately *p*-Wright affine functions, inner product spaces and derivations. Fixed Point Theory **18**, 69–84 (2017)
4. Brzdęk, J.: Stability of the equation of the p-Wright affine functions. Aequationes Math. **85**, 497–503 (2013)
5. Brzdęk, J.: A note on the functions that are approximately *p*-Wright affine. In: Rassias, Th.M. (ed.) Handbook of Functional Equations: Functional Inequalities, Springer Optimization and Its Applications, pp. 43–55. Springer, Berlin (2014)
6. Brzdęk, J., Cădariu, L.: Stability for a family of equations generalizing the equation of *p*-Wright affine functions. Appl. Math. Comput. **276**, 158–171 (2016)
7. Brzdęk, J., Ciepliński, K.: A fixed point approach to the stability of functional equations in non-Archimedean metric spaces. Nonlinear Anal. **74**, 6861–6867 (2011)
8. Brzdęk, J., Ciepliński, K.: Hyperstability and superstability. Abstr. Appl. Anal. **2013**, Article ID 401756 (2013)
9. Brzdęk, J., Ciepliński, K.: On a fixed point theorem in 2-Banach spaces and some of its applications. Acta Math. Sci. **38**, 377–390 (2018)
10. Brzdęk, J., Chudziak, J., Pales, Zs.: A fixed point approach to stability of functional equations. Nonlinear Anal. **74**, 6728–6732 (2011)
11. Brzdęk, J., Cădariu, L., Ciepliński, K.: Fixed point theory and the Ulam stability. J. Funct. Spaces **2014**, 16 pp. (2014). Article ID 829419
12. Brzdęk, J., Ciepliński, K., Leśniak, Z.: On Ulam's type stability of the linear equation and related issues. Discrete Dyn. Nat. Soc. **2014**, 14 pp. (2014). Article ID 536791
13. Brzdęk, J., El-hady, E., Förg-Rob, W., Leśniak, Z.: A note on solutions of a functional equation arising in a queuing model for a LAN gateway. Aequationes Math. **90**(4), 671–681 (2016)
14. Brzdęk, J., Popa, D., Raşa, I.: Hyers–Ulam stability with respect to gauges. J. Math. Anal. Appl. **453**(1), 620–628 (2017)
15. Brzdęk, J., Popa, D., Raşa, I., Xu, B.: Ulam Stability of Operators. Mathematical Analysis and Its Applications, vol. 1. Academic Press/Elsevier, New York (2018)
16. Castillo, E., Ruiz-Cobo, M.R.: Functional Equations and Modelling in Science and Engineering. Marcel Dekker, New York (1992)
17. Chung, S.C., Park, W.G.: Hyers-Ulam stability of functional equations in 2-Banach spaces. Int. J. Math. Anal. (Ruse) **6**(17/20), 951–961 (2012)
18. Daróczy, Z., Lajkó, K., Lovas, R., Maksa, G., Páles, Z.: Functional equations involving means. Acta Math. Hungar. **116**(1–2), 79–87 (2007)
19. El-hady, E., Brzdęk, J., Nassar, H.: On the structure and solutions of functional equations arising from queueing models. Aequationes Math. **91**, 445–477 (2017)
20. El-hady, E., Förg-Rob, W., Nassar, H.: On a functional equation arising from a network model. Appl. Math. **11**(2), 363–372 (2017)
21. Freese, R.W., Cho, Y.J.: Geometry of Linear 2-normed Spaces. Nova Science Publishers, Hauppauge (2001)
22. Gähler, S.: Lineare 2-normierte Räume. Math. Nachr. **28**, 1–43 (1964)
23. Gao, J.: On the stability of the linear mapping in 2-normed spaces. Nonlinear Funct. Anal. Appl. **14**(5), 801–807 (2009)
24. Guillemin, F., Knessl, C., van Leeuwaarden, J.: Wireless three-hop networks with stealing {II}: exact solutions through boundary value problems. Queueing Syst. **74**, 235–272 (2013)
25. Hayes, W., Jackson, K.R.: A survey of shadowing methods for numerical solutions of ordinary differential equations. Appl. Numer. Math. **53**, 299–321 (2005)

26. Hyers, D.H.: On the stability of the linear functional equation. Proc. Nat. Acad. Sci. U. S. A. **27**(4), 222–224 (1941)
27. Hyers, D.H., Isac, G., Rassias, Th.M.: Stability of Functional Equations in Several Variables. Birkhäuser, Boston (1998)
28. Jarczyk, W., Sablik, M.: Duplicating the cube and functional equations. Results Math. **26**, 324–335 (1994)
29. Jung, S.-M.: Hyers-Ulam-Rassias Stability of Functional Equations in Nonlinear Analysis. Springer, New York (2011)
30. Jung, S.-M., Popa, D., Rassias, M.T.: On the stability of the linear functional equation in a single variable on complete metric groups. J. Global Optim. **59**(1), 165–171 (2014)
31. Kindermann, L., Lewandowski, A., Protzel, P.: A framework for solving functional equations with neural networks. In: Neural Information Processing, ICONIP2001 Proceedings, vol. 2, pp. 1075–1078. Fudan University Press, Shanghai (2001)
32. Maksa, G.: The general solution of a functional equation related to the mixed theory of information. Aequationes Math. **22**, 90–96 (1981)
33. Monreal, A., Tomás, M.S.: On some functional equations arising in computer graphics. Aequationes Math. **55**, 61–72 (1998)
34. Moszner, Z.: Stability has many names. Aequationes Math. **90**(5), 983–999 (2016)
35. Najati, A., Park, C.: Stability of homomorphisms and generalized derivations on Banach algebras. J. Inequal. Appl. **2009**, 12 pp. (2009). Article ID 595439
36. Nassar, H., El-Hady, E.: Closed-form solution of a LAN gateway queueing model. In: Pardalos, P., Rassias, Th.M. (eds.) Contributions in Mathematics and Engineering, pp. 393–427. Springer, Cham (2016)
37. Park, W.G.: Approximate additive mappings in 2-Banach spaces and related topics. J. Math. Anal. **376**, 193–202 (2011)
38. Sahoo, P.K., Székelyhidi, L.: On a functional equation related to digital filtering. Aequationes Math. **62**(3), 280–285 (2001)
39. Sundberg, C., Wagner, C.: A functional equation arising in multi-agent statistical decision theory. Aequationes Math. **32**(1), 32–37 (1987)

Chapter 7
On Solutions and Stability of a Functional Equation Arising from a Queueing System

El-Sayed El-hady

Abstract We use the boundary value problems approach to investigate the analytical solution of a two-variable functional equation, which arose from a queueing model. We also provide some remarks on the Ulam stability of such functional equation.

Keywords Boundary value problem · Two-variable functional equation · Queueing model

Mathematics Subject Classification (2010) Primary 30D05, 30E25, 39B32, 60K25, 65Q20; Secondary 39B82

7.1 Introduction

Functional equations have many recent interesting applications in various fields see e.g. [12, 16]. They have applications in Communication models and Network models see e.g. [15, 16, 21], in dynamical systems [3], in information theory [2, 23], in computer graphics [24], decision theory [1, 32], and in digital filtering [31]. In this chapter we are interested in a special case of the interesting class of functional equations surveyed in [12]. It should be noted that so far there is no general solution theory available for such interesting class of equations. In this chapter we investigate the analytical solution of a functional equation arising from a queueing model. This chapter is organized as follows: in Sect. 7.2 we recall the functional equation from

E.-S. El-hady (✉)
Mathematics Department, College of Science, Jouf University, Sakaka, Kingdom of Saudi Arabia

Basic Science Department, Faculty of Computers and Informatics, Suez Canal University, Ismailia, Egypt
e-mail: elsayed_elhady@ci.suez.edu.eg

© Springer Nature Switzerland AG 2019
J. Brzdęk et al. (eds.), *Ulam Type Stability*,
https://doi.org/10.1007/978-3-030-28972-0_7

the original article, in Sect. 7.3 we analyse the kernel which plays a crucial role in the solution, in Sect. 7.4 we introduce a solution of the functional equation using boundary value problem approach, and in Sect. 7.5 we investigate the stability of the functional equation of interest.

7.2 The Functional Equation

The article [22] ends up with the following challenging two-variable functional equation (for functions f of two complex variables)

$$(x(2\rho x + 1) - 2(1 + \rho)xy + y^2)f(x, y) \tag{7.1}$$
$$= (x(2\rho x + 1) - (1 + \rho)xy - \rho xy^2)f(x, 0) + y(y - x)f(0, y),$$

where

$$f(x, y) = \sum_{m,n=0}^{\infty} p_{m,n} x^m y^n, \qquad |x| \le 1, |y| < 1 + 2\rho$$

is the probability generating function (PGF) of the sequence $p_{m,n}$, which is defined in [22],

$$f(x, 0) = \sum_{m=0}^{\infty} p_{m,0} x^m, \qquad |x| \le 1$$

is the generating function of the sequence $p_{m,0}$,

$$f(0, y) = \sum_{n=0}^{\infty} p_{0,n} y^n, \qquad |y| < 1 + 2\rho$$

is the generating function of the sequence $p_{0,n}$, and $0 < \rho < 1$ is some parameter. Equation (7.1) can be written as follows

$$C_1(x, y)f(x, y) = C_2(x, y)f(x, 0) + C_3(x, y)f(0, y), \tag{7.2}$$

where

$$C_1(x, y) = x(2\rho x + 1) - 2(1 + \rho)xy + y^2,$$
$$C_2(x, y) = x(2\rho x + 1) - (1 + \rho)xy - \rho xy^2,$$

and

$$C_3(x, y) = y(y - x).$$

A crucial role in the solution of (7.2) is played by the kernel defined by

$$\{(x, y) : C_1(x, y) = 0\}. \tag{7.3}$$

The solution of (7.2) will be investigated in the next sections. It should be noted that the current functional equation is related to the equations that appear in the literature (see e.g. [10, 27, 30]) as follows:

- It is a special case of the general class of functional equations surveyed in [12].
- It is different from the functional equations solved recently in [6, 13, 29].
- The functions $C_2(x, y), C_3(x, y)$, are not related to each other unlike the case in [15].
- We have only two unknowns namely $f(x, 0)$ and $f(0, y)$ unlike the case in [16].
- The contour L defined below is not a circle unlike the case in [14].
- We have only one system parameter, namely ρ which will simplify the analysis of the kernel unlike the case in [26].
- We have two unknown functions namely $f(x, 0)$ and $f(0, y)$ so we cannot use Rouché's theorem unlike the case in [28].

7.3 Kernel Analysis

The kernel given by (7.3) can be written as

$$\{(x, y) : C_1(x, y) = x(2\rho x + 1) - 2(1 + \rho)xy + y^2 = 0\}. \tag{7.4}$$

It is obvious that (7.4) is a biquadratic equation, i.e. it can be considered as a quadratic equation in x with coefficients in y and also can be considered as a quadratic equation in y with coefficients in x. We have to study the two cases in the following two subsections.

7.3.1 The Kernel as a Function in y

If we consider (7.4) as a quadratic equation in y we can write that

$$2\rho x^2 + x - 2xy - 2\rho xy + y^2 = 0$$

or in the form

$$y^2 + (-2x - 2\rho x)y + 2\rho x^2 + x = 0,$$

which can be rewritten as

$$\alpha(x)y^2 + \beta(x)y + \gamma(x) = 0, \tag{7.5}$$

where

$$\alpha(x) = 1,$$

$$\beta(x) = -2x - 2\rho x,$$

and

$$\gamma(x) = 2\rho x^2 + x.$$

Equation (7.5) has two solutions given by

$$y_{\pm}(x) = \frac{-\beta(x) \pm \sqrt{\beta(x)^2 - 4\alpha(x)\gamma(x)}}{2\alpha(x)}$$

$$= x + \rho x \pm \sqrt{x(x + \rho^2 x - 1)}. \tag{7.6}$$

It is easy to see that the function (7.6) is a local analytic function. That is to say it is an analytic function except at the real zeros of the root which are the two branch points at

$$x_1 = 0, \qquad x_2 = \frac{1}{1 + \rho^2}.$$

This is because when x traverses any small circuit around x_i, $i = 1, 2$, the function

$$x \quad \mapsto \quad x + \rho x \pm \sqrt{x(x + \rho^2 x - 1)}$$

does not return to its original value.

7.3.2   The Kernel as a Function in x

If we consider (7.4) as a quadratic equation in x we can write that

$$2\rho x^2 + x - 2xy - 2\rho xy + y^2 = 0,$$

or in the form

$$2\rho x^2 + (1 - 2y - 2\rho y)x + y^2 = 0,$$

which can be written as

$$\lambda(y)x^2 + \mu(y)x + \nu(y) = 0, \tag{7.7}$$

where

$$\lambda(y) = 2\rho,$$

$$\mu(y) = 1 - 2y - 2\rho y,$$

and

$$\nu(y) = y^2.$$

Equation (7.5) has two solutions given by

$$x_{\pm}(y) = \frac{-\mu(y) \pm \sqrt{\mu(y)^2 - 4\lambda(y)\nu(y)}}{2\lambda(y)}$$

$$= \frac{2y + 2\rho y - 1 \pm \sqrt{4(1 + \rho^2)y^2 - 4(1 + \rho)y + 1}}{4\rho}. \tag{7.8}$$

It is easy to see that the function (7.8) is a local analytic function. This means that it is an analytic function except at the real zeros of the root which are the two branch points at

$$y_1 = \frac{1 + \rho + \sqrt{2\rho}}{2(1 + \rho^2)}, \qquad y_2 = \frac{1 + \rho - \sqrt{2\rho}}{2(1 + \rho^2)}.$$

This is because when y traverses any small circuit around y_j, $j = 1, 2$, the function $x_{\pm}(y)$ defined by (7.8) does not return to its original value.

Lemma 7.1 *For $x \in [x_1, x_2]$ we have $x \in \mathbb{R}$ and the two roots given by*

$$y_+(x) = x + \rho x + \sqrt{x(x + \rho^2 x - 1)},$$

$$y_-(x) = x + \rho x - \sqrt{x(x + \rho^2 x - 1)}$$

are complex conjugates. Hence, the interval (x_1, x_2) is mapped by $x \mapsto y_{\pm}(x)$ onto a contour L. Any point on such a contour satisfies

$$|y(x)|^2 = 2\rho x^2 + x.$$

Proof Follows directly from the fact that the root function in (7.6) is zero for $x = x_1$ and x_2 and negative for $x \in (x_1, x_2)$, which is symmetric with respect to the real line. Using (7.5) and Vieta's formula we can guarantee that any point on that contour satisfies

$$|y(x)|^2 = y_+(x)y_-(x)$$

$$= \frac{\gamma(x)}{\alpha(x)} = 2\rho x^2 + x$$

7.4   Solution of the Functional Equation

Since by definition the main unknown function $f(x, y)$ is an analytic function in the unit disks, this implies that if $C_1(x, y) = 0$ then also

$$C_2(x, y)f(x, 0) + C_3(x, y)f(0, y) = 0. \tag{7.9}$$

Now the solution of the main functional equation is reduced to the solution of the functional equation (7.9) on

$$\{(x, y) : C_1(x, y) = 0\}.$$

It should be noted that it is sufficient to find one unknown of (7.9) and plug it back in (7.9), using the kernel equation i.e. $C_1(x, y) = 0$ one can latter find the other unknown and hence the main unknown $f(x, y)$ will be obtained. Now using the kernel analysis in Sect. 7.3 the main functional equation can be reduced to the following boundary value problem.

Lemma 7.2 *Find a function $f(.)$ which is analytic inside the unit disk and satisfies*

$$\Re(ia(\Upsilon_y(u))f(\Upsilon_y(u))) = 0, \ u \in D$$

for some known function $a(.)$ of a conformal mapping $\Upsilon_y(.)$.

Proof Since the main unknown function $f(x, y)$ is by definition an analytic function in the unit disk, this implies that if $C_1(x, y) = 0$, then also

$$C_2(x, y)f(x, 0) + C_3(x, y)f(0, y) = 0, \tag{7.10}$$

which is equivalent to

$$f(x, 0) = -\frac{C_3(x, y)}{C_2(x, y)}f(0, y). \tag{7.11}$$

Now assume that the function given by (7.6) maps the interval $[x_1, x_2]$ to a closed contour L in the y-domain. That is to say the function $y_+(x)$ maps $[x_1, x_2]$ to a curve in the upper half plane while the function $y_-(x)$ maps $[x_1, x_2]$ to a curve in the lower half plane so that both functions defined by (7.6) map $[x_1, x_2]$ to a closed contour L which is symmetric with respect to the real line. Since for $x \in [x_1, x_2]$ we have $x \in \mathbb{R}$. Then using this interval in the (7.11) we get

$$\Re(i f(x, 0)) = 0 = \Re(-i\frac{C_3(x, y)}{C_2(x, y)} f(0, y)),$$

which can be written as

$$0 = \Re\{-ia(y) f(0, y)\} \tag{7.12}$$

for every $y \in L$. The problem constructed is a Riemann-Hilbert boundary value problem. The classical way to solve is to use some conformal mapping between L^+ and the unit disk:

$$\Pi_y(y) : L^+ \mapsto D^+,$$

with inverse

$$\Upsilon_y(u) : D^+ \mapsto L^+,$$

then the problem (7.12) can be reduced to the following

$$0 = \Re\{-ia(\Upsilon_y(u)) f(\Upsilon_y(u))\}, u \in D.$$

The boundary value problem constructed is a homogenous Riemann-Hilbert boundary value problem. In fact it is a special case of the problem stated, e.g., in [11, 17]. According to [11] the solution of this problem, when it exists, is given by

$$f(0, y) = Q(y)\phi(y), \tag{7.13}$$

where $Q(y)$ is some polynomial, and the function $\phi(y)$ is defined by

$$\phi(y) = \begin{cases} \exp\left(\frac{1}{2i\pi} \int_L \log\left(z^{-\kappa} \frac{\overline{a(z)}}{a(z)}\right) \frac{dz}{z-y}\right) & \text{if } y \in L^+ \\ \frac{1}{y^\kappa} \exp\left(\frac{1}{2i\pi} \int_L \log\left(z^{-\kappa} \frac{\overline{a(z)}}{a(z)}\right) \frac{dz}{z-y}\right) & \text{if } y \in L^- \end{cases}$$

with κ denoting the index of the Riemann–Hilbert problem and $\phi^{(+)}(u)$ being the interior limit of the function $\phi(u)$ on the unit circle. The solution of the Riemann-Hilbert boundary value problem exists when $\kappa < 0$ and is unique if and only if for $k = 0, 1, \cdots, |\kappa| - 1$

$$\int_L \frac{z^k C(z)}{\phi^{(+)}(z)} dz = 0;$$

in that case, the polynomial $Q(u) \equiv 0$. In the case that $\kappa = 0$, the solution is unique and $Q(u)$ is some constant.

7.5   Remarks on Stability of the Functional Equation

Stability of functional equations is an important issue with many interesting applications and we refer to, e.g., [4, 5, 7, 8, 19, 25] for more details. Stability can be seen from different points of views see [25] and hundreds of researchers are dealing with such amazing topic. It can be considered as a branch of optimization theory (see, e.g., [20]), it is related to the notion of shadowing (see, e.g., [18]), and it has applications in economics (see [9]). It should be noted that the issue of stability of functional equations was originally motivated by a problem of S.M. Ulam posed in 1940 and Hyers's answer to it published in [19].

There are many methods illustrated in the literature see e.g. [4] to investigate stability, namely: the direct method, the method of invariant means, the method based on the sandwich theorems, the weighted space method, the fixed point method, and the method of shadowing. The notion of stability of functional equations arises when we replace the functional equation by a functional inequality which can be considered in some sense as a *perturbation* of the equation. The stability question now is:

> How do the solutions of the perturbed equation "the inequality" differ from those of the given functional equation?

It seems that some methods used in such stability could be applied in investigations of solutions to (7.1), or even more general equations of the form

$$C_1(x, y) f(x, y) = C_2(x, y) f(x, 0) + C_3(x, y) f(0, y), \qquad x, y \in D \subset \mathbb{C}. \tag{7.14}$$

It should be noted that the general solution of Eq. (7.14) is a function defined as follows

$$f(\cdot, \cdot): D \subset \mathbb{C} \to \mathbb{C} \tag{7.15}$$

where D is the unit disk in the complex plane, and $\mathbb{C}$ is the set of all complex numbers. For instance, we could use the following classical definition of Ulam-Hyers stability (cf., e.g., [8]): *We say that the functional equation* (7.14) *is Ulam-Hyers stable if there is a $r > 0$ such that for any $\epsilon > 0$ and*

$$g: D \subset \mathbb{C} \to \mathbb{C}$$

with

$$|C_1(x, y)g(x, y) - C_2(x, y)g(x, 0) - C_3(x, y)g(0, y)| \leq \epsilon \qquad (7.16)$$

there exists a solution f to Eq. (7.14) such that

$$|f(z, w) - g(z, w)| \leq r\epsilon, \qquad (z, w) \in D. \qquad (7.17)$$

So, the Ulam-Hyers stability of Eq. (7.14) means that every approximate (in the sense of (7.16)) solution of (7.14) is close (in the sense of (7.17)) to the exact solutions of the equation. Therefore, in some cases, we could use approximate solutions of the equation (which might have a simpler form) knowing that they are close to the functions that solve the equation exactly. This shows that the issue of stability of (7.14) (and various similar equations surveyed in [12]) is of interest and should be investigated.

Acknowledgement This work is funded by Jouf University, Kingdom of Saudi Arabia under the research project number 39/600.

References

1. Abbas, A.E., Aczél, J.: The role of some functional equations in decision analysis. Decision Anal. **7**(2), 215–228 (2010)
2. Aczél, J.: Notes on generalized information functions. Aequationes Math. **22**(1), 97–107 (1981)
3. Balibrea, F., Reich, L., Smítal, J.: Iteration theory: dynamical systems and functional equations. Int. J. Bifurcation Chaos **13**(7), 1627–1647 (2003)
4. Brzdęk, J., Cadariu, L., Ciepliński, K.: Fixed point theory and the Ulam stability. J. Funct. Spaces **2014**, 16 pp. (2014). Article ID 829419
5. Brzdęk, J., Ciepliński, K., Leśniak, Z.: On Ulam's type stability of the linear equation and related issues. Discrete Dyn. Nat. Soc. **2014**, 14 pp. (2014). Article ID 536791
6. Brzdęk, J., El-hady, E., Förg-Rob, W., Leśniak, Z.: A note on solutions of a functional equation arising in a queuing model for a LAN gateway. Aequationes Math. **90**(4), 671–681 (2016)
7. Brzdęk, J., Popa, D., Raşa, I.: Hyers–Ulam stability with respect to gauges. J. Math. Anal. Appl. **453**(1), 620–628 (2017)
8. Brzdęk, J., Popa, D., Raşa, I., Xu, B.: Ulam Stability of Operators. Mathematical Analysis and Its Applications, vol. 1. Academic Press/Elsevier, New York (2018)
9. Castillo, E., Ruiz-Cobo, M.R.: Functional Equations and Modelling in Science and Engineering. Marcel Dekker, New York (1992)
10. Cohen, J.W.: On the asymmetric clocked buffered switch. Queueing Syst. **30**, 385–404 (1998)
11. Dautray, R., Lions, J.: Mathematical Analysis and Numerical Methods for Science and Technology. Springer, Berlin (1991)
12. El-hady, E., Brzdęk, J., Nassar, H.: On the structure and solutions of functional equations arising from queueing models. Aequationes Math. **91**, 445–477 (2017)
13. El-hady, E., Förg-Rob, W., Nassar, H.: On a functional equation arising from a network model. Appl. Math. **11**(2), 363–372 (2017)

14. Fayolle, G., Iasnogorodski, R.: Two coupled processors: the reduction to a Riemann-Hilbert problem. Zeitschrift für Wahrscheinlichkeitstheorie und verwandte Gebiete **47**(3), 325–351 (1979)
15. Guillemin, F., Knessl, C., Van Leeuwaarden, J.: Wireless multihop networks with stealing: large buffer asymptotics via the ray method. SIAM J. Appl. Math. **71**, 1220–1240 (2011)
16. Guillemin, F., Knessl, C., van Leeuwaarden, J.: Wireless three-hop networks with stealing {II}: exact solutions through boundary value problems. Queueing Syst. **74**(2–3), 235–272 (2013)
17. Guillemin, F., Knessl, C., Van Leeuwaarden, J.: First response to letter of G. Fayolle and R. Iasnogorodski. Queueing Syst. **76**, 109–110 (2014)
18. Hayes, W., Jackson, K.R.: A survey of shadowing methods for numerical solutions of ordinary differential equations. Appl. Numer. Math. **53**, 299–321 (2005)
19. Hyers, D.H.: On the stability of the linear functional equation. Proc. Nat. Acad. Sci. U. S. A. **27**(4), 222–224 (1941)
20. Jung, S.-M., Popa, D., Rassias, M.T.: On the stability of the linear functional equation in a single variable on complete metric groups. J. Global Optim. **59**(1), 165–171 (2014)
21. Kindermann, L., Lewandowski, A., Protzel, P.: A framework for solving functional equations with neural networks. In: Neural Information Processing, ICONIP2001 Proceedings, vol. 2, pp. 1075–1078. Fudan University Press, Shanghai (2001)
22. Kingman, J.F.C.: Two similar queues in parallel. Ann. Math. Stat. **32**, 1314–1323 (1961)
23. Maksa, G.: The general solution of a functional equation related to the mixed theory of information. Aequationes Math. **22**, 90–96 (1981)
24. Monreal, A., Tomás, M.S.: On some functional equations arising in computer graphics. Aequationes Math. **55**, 61–72 (1998)
25. Moszner, Z.: Stability has many names. Aequationes Math. **90**(5), 983–999 (2016)
26. Nassar, H.: Two-dimensional queueing model for a LAN gateway. WSEAS Trans. Commun. **5**(9), 437–442 (2006)
27. Nassar, H., Ahmed, M.A.: Performance analysis of an ATM buffered switch transmitting two-class traffic over unreliable channels. AEU Int. J. Electron. Commun. **57**(3), 190–200 (2003)
28. Nassar, H., Al-mahdi, H.: Queueing analysis of an ATM multimedia multiplexer with non-pre-emptive priority. IEEE Proc. Commun. **150**, 189–196 (2003)
29. Nassar, H., El-Hady, E.: Closed-form solution of a LAN gateway queueing model. In: Pardalos, P., Rassias, Th.M. (eds.) Contributions in Mathematics and Engineering, pp. 393–427. Springer, Cham (2016)
30. Resing, J., Örmeci, L.: A tandem queueing model with coupled processors. Oper. Res. Lett. **31**(5), 383–389 (2003)
31. Sahoo, P.K., Székelyhidi, L.: On a functional equation related to digital filtering. Aequationes Math. **62**(3), 280–285 (2001)
32. Sundberg, C., Wagner, C.: A functional equation arising in multi-agent statistical decision theory. Aequationes Math. **32**(1), 32–37 (1987)

Chapter 8
Approximation by Cubic Mappings

Paşc Găvruţa and Laura Manolescu

Abstract Starting with a stability problem posed by Ulam for group homomorphisms, we characterize the functions with values in a Banach space, which can be approximated by cubic mappings with a given error.

Keywords Hyers-Ulam-Rassias stability · Cubic mapping

Mathematics Subject Classification (2010) Primary 39B82; Secondary 39B52

8.1 Introduction

The study of stability problems for various functional equations originated from a question posed by Ulam [38] in 1940 and reads as follows.

Let $(G_1, \circ)$ be a group, $(G_2, *)$ be a metric group with the metric $d(\cdot, \cdot)$ and $\varepsilon > 0$. Does there exits a $\delta > 0$ such that $f : G_1 \to G_2$ satisfies

$$d(f(x \circ y), f(x) * f(y)) \leq \delta, \quad \text{for all } x, y \in G_1$$

then there exists a homomorphism $h : G_1 \to G_2$ with

$$d(f(x), h(x)) \leq \varepsilon, \quad \text{for all } x \in G_1?$$

The first affirmative answer to this question, was the one provided by Hyers [22], who solved the problem for Banach spaces.

P. Găvruţa (✉) · L. Manolescu
Department of Mathematics, Politehnica University of Timişoara, Timişoara, Romania
e-mail: laura.manolescu@upt.ro

© Springer Nature Switzerland AG 2019
J. Brzdęk et al. (eds.), *Ulam Type Stability*,
https://doi.org/10.1007/978-3-030-28972-0_8

Theorem 8.1 (Hyers [22]) *Let $f : E_1 \to E_2$ (E_1, E_2 are Banach spaces) be a function such that*

$$\|f(x + y) - f(x) - f(y)\| \leq \delta$$

for some $\delta > 0$ and for all $x, y \in E_1$. Then the limit

$$T(x) = \lim_{n \to \infty} \frac{f(2^n x)}{2^n}$$

exists for each $x \in E_1$ and $T : E_1 \to E_2$ is the unique additive mapping such that

$$\|f(x) - T(x)\| \leq \delta, \quad \text{for every } x \in E_1.$$

Moreover, if $f(tx)$ is continuous in t for each fixed $x \in E_1$, then the function T is linear.

Another important result was obtained by Rassias [34] for approximately additive mappings, by using the so called *the direct method*.

Theorem 8.2 (Rassias [34]) *Let $f : E_1 \to E_2$ be a function between Banach spaces, such that $f(tx)$ is continuous in t for each fixed x. If f satisfies the functional inequality*

$$\|f(x + y) - f(x) - f(y)\| \leq \theta(\|x\|^p + \|y\|^p)$$

for some $\theta \geq 0$, $0 \leq p < 1$ and for all $x, y \in E_1$, then there exists a unique linear mapping $T : E_1 \to E_2$ such that

$$\|f(x) - T(x)\| \leq \frac{2\theta}{2 - 2^p}\|x\|^p, \quad \text{for each } x \in E_1.$$

A further generalization was obtained by Găvruța [9], by replacing the Cauchy difference by a control mapping φ and also introduced the concept of generalized Hyers-Ulam-Rassias stability in the spirit of Th.M. Rassias' approach.

In [24] was introduced the notion of ψ-additive mapping and was given a generalized solution to Ulam's problem for ψ-additive mappings.

Definition 8.1 Let $\psi : \mathbb{R}_+ \to \mathbb{R}_+$ be a mapping, E_1 and E_2 be normed spaces. A mapping $f : E_1 \to E_2$ is called ψ-additive if there exists $\theta > 0$ such that

$$\|f(x + y) - f(x) - f(y)\| \leq \theta(\psi(\|x\|) + \psi(\|y\|)),$$

for all $x, y \in E_1$.

In [13], Găvruța gave the following characterization of ψ-additive mappings.

Theorem 8.3 *We suppose that ψ verifies the following conditions:*

(i) $\psi(ts) \le \psi(t)\psi(s)$, *for all* $t, s \ge 0$;
(ii) $\psi(t + s) \le \psi(t) + \psi(s)$, *for all* $t, s \ge 0$;
(iii) ψ *is monotone increasing on* $\mathbb{R}_+$;
(iv) *there exists* $t_0 > 0$ *such that* $\psi(t_0) < t_0$.

Let E_1 be a normed space and E_2 a real Banach space, then $f : E_1 \to E_2$ is a ψ-additive mapping if and only if there exists a constant $c > 0$ and an additive mapping $T : E_1 \to E_2$ such that

$$\|f(x) - T(x)\| \le c\psi(\|x\|), \text{ for all } x \in E_1.$$

Other aspects concerning the connection between Ψ-additive mappings and Hyers-Ulam stability were studied in the paper [17].

For basic results on the stability of mappings, one can see the references [6, 7, 23, 27].

For recent results on the *Hyers-Ulam-Rassias stability*, see also [3, 4, 19, 23, 27–30, 32]. Some open problems in this field were solved in the following papers: [2, 10–14, 16, 18, 21].

In the paper [20], we have investigated the approximation of functions by additive and quadratic mappings. We continue that work here by discussing about the approximation of functions by cubic mappings.

8.2 Approximation of Functions by Additive and by Quadratic Mappings

In this section, we present the main results from the paper [20].

We consider S to be an abelian semigroup, X to be a Banach space and the following given functions:

$$f : S \to X \text{ and } \Phi : S \to \mathbb{R}_+.$$

Definition 8.2 We say that f is Φ-approximable by an additive map if there exists $T : S \to X$ additive such that

$$\|f(x) - T(x)\| \le \Phi(x), \quad x \in S.$$

We say that T is the additive Φ-approximation of f.

Problem 8.1 Give conditions on f such that f to be Φ-approximable by an additive map.

We solve this problem by posing minimal conditions on Φ. We denote by

$$\mathscr{A} = \{\Phi : S \to \mathbb{R}_+ : \lim_{n \to \infty} \frac{\Phi(2^n x)}{2^n} = 0, \text{ for any } x \in S\}.$$

Theorem 8.4 *Let be* $\Phi \in \mathscr{A}$. *Then* f *is* Φ-*approximable by an additive map if and only if*

$$\lim_{n \to \infty} \frac{\|f(2^n x + 2^n y) - f(2^n x) - f(2^n y)\|}{2^n} = 0, \quad (\forall) \, x, y \in S$$

and there exists $\Psi \in \mathscr{A}$ *such that*

$$\|f(2^n x) - 2^n f(x)\| \leq \Psi(2^n x) + 2^n \Phi(x), \quad x \in S.$$

In this case, the additive Φ-*approximation of* f *is unique and is given by*

$$T(x) = \lim_{n \to \infty} \frac{f(2^n x)}{2^n}.$$

In the same paper, we give an analogous result for quadratic mappings. The functional equation

$$f(x + y) + f(x - y) = 2f(x) + 2f(y)$$

is called a *quadratic functional equation*. Every solution of the quadratic functional equation is said to be a quadratic mapping. The Hyers-Ulam stability for quadratic functional equation was proved by Skof [37], for mappings acting between a normed space and a Banach space. Cholewa [6] showed that Skof's Theorem remains true when the normed space is replaced with an abelian group.

Theorem 8.5 (Cholewa [6]) *Let* $(G, +)$ *be an abelian group and let* E *be a Banach space. If a function* $f : G \to E$ *satisfies the inequality*

$$\|f(x + y) + f(x - y) - 2f(x) - 2f(y)\| \leq \delta$$

for some $\delta \geq 0$ *and for all* $x, y \in G$, *then there exists a unique quadratic function* $Q : G \to E$ *such that*

$$\|f(x) - Q(x)\| \leq (1/2)\delta,$$

for any $x \in G$.

Let $(G, +)$ be an abelian group and X a Banach space.

Definition 8.3 We say that f is Φ-approximable by a quadratic map if there exists $Q : G \to X$ quadratic mapping such that

$$\|f(x) - Q(x)\| \leq \Phi(x), \quad x \in G.$$

We say that Q is the quadratic Φ approximation of f.

Problem 8.2 Give conditions on f such that f to be Φ-approximable by a quadratic map.

We denote by

$$\mathcal{Q} = \{\Phi : G \to \mathbb{R}_+ : \lim_{n\to\infty} \frac{\Phi(2^n x)}{4^n} = 0, \text{ for any } x \in S\}.$$

The set $\mathcal{Q}$ is the analogous of the set $\mathcal{A}$ from the case of approximation by additive mappings.

In this case, we have the following characterization of functions which can be approximated by quadratic ones.

Theorem 8.6 *Let be $Q \in \mathcal{Q}$. Then f is Φ-approximable by a quadratic map if and only if the following two conditions holds*

(i) $\displaystyle\lim_{n\to\infty} \frac{\|f(2^n x + 2^n y) + f(2^n x - 2^n y) - 2f(2^n x) - 2f(2^n y)\|}{4^n} = 0, \; (\forall) \, x, y \in G$

(ii) *there exists $\Psi \in \mathcal{Q}$ such that*

$$\|f(2^n x) - 4^n f(x)\| \le \Psi(2^n x) + 4^n \Phi(x), x \in G.$$

In this case, the quadratic Φ-approximation of f is unique and is given by

$$Q(x) = \lim_{n\to\infty} \frac{f(2^n x)}{4^n}$$

From this result, we have immediately the result of Borelli and Forti [1] on the stability of quadratic mappings.

8.3 Approximation of Functions by Cubic Mappings

The study of the stability of the cubic functional equation,

$$f(x + 2y) + 3f(x) = 3f(x + y) + f(x - y) + 6f(y), \tag{8.1}$$

in the sense of Ulam, was given by Rassias [35] in 2001. The generalized Hyers-Ulam-Rassias stability of this equation was given by Găvruţa and Cădariu [15] in 2002.

In 2002, Jun and Kim [25], introduced the following form of a cubic functional equation

$$f(2x + y) + f(2x - y) = 2f(x + y) + 2f(x - y) + 12f(x). \tag{8.2}$$

Every solution of this equation is said to be a *cubic function*. They established the general solution and the Hyers-Ulam-Rassias stability for this functional equation. The stability of Eq. (8.2) in fuzzy normed spaces was initiated in [31]. The stability of the cubic functional equation in a non-Archimedean random normed space and intuitionistic Random normed spaces was studied in the paper [36] and in the paper [5], by using the fixed point method.

In the following, we will prove that the functional equations (8.1) and (8.2) are equivalent with two other more functional equations, studied by other authors.

Theorem 8.7 *Let* $(G, +)$ *be an abelian group and* X *be a linear space. The following functional equations are equivalent, for* $f : G \to X$,

(A) $f(x + 2y) + 3f(x) - 3f(x + y) - f(x - y) - 6f(y) = 0$, *for all* $x, y \in G$;

(B) $f(2x + y) + f(2x - y) - 2f(x + y) - 2f(x - y) - 12f(x) = 0$, *for all* $x, y \in G$;

(C) $f(2x + y) + f(x + 2y) - 3f(x) - 3f(y) - 6f(x + y) = 0$, *for all* $x, y \in G$;

(D) $\Delta_y^3 f(x) := \sum_{k=0}^{3} (-1)^{3-k} \binom{3}{k} f(x + ky) - 3!f(y) = 0$, *for all* $x, y \in G$.

Proof $(A) \Rightarrow (B)$

In (A), we take $x = y = 0$ and it follows that $f(0) = 0$.

In (A), we take $y = -x$ and it follows that $-5f(-x) + 3f(x) - f(2x) = 0$.

In (A), we take $x = 0$, $y = x$ and it follows that $-f(-x) - 9f(x) + f(2x) = 0$.

By adding the above relations, we obtain that

$$-6f(-x) - 6f(x) = 0,$$

so f is an odd function.

In (A), $x \to y$ and $y \to x$ and we get:

$$f(2x + y) + 3f(y) - 3f(x + y) - f(y - x) - 6f(x) = 0.$$

We put here instead of y, $-y$ and we use the fact that f is odd

$$(A') \quad f(2x - y) - 3f(y) - 3f(x - y) + f(x + y) - 6f(x) = 0.$$

By adding the previous form of (A) with (A'), we obtain (B).

$(B) \Rightarrow (C)$

In (B), we take $x = y = 0$ and we obtain that $f(0) = 0$.

In (B), we take $x = 0$ and we get $f(y) + f(-y) = 0$, so f is odd.

In (B), we take $y = 0$ it follows $f(2x) - 8f(x) = 0$.

In (B) we replace x with $x + y$ and y with $x - y$ and we obtain:

$$f(3x + y) + f(x + 3y) - 12f(x + y) - 2f(2x) - 2f(2y) = 0$$

and since $f(2x) = 8f(x)$, we have:

$$(B')\ f(3x + y) + f(x + 3y) - 12f(x + y) - 16f(x) - 16f(y) = 0.$$

In (B) we replace x with $x + y$ and y with $2y$ and we obtain:

$$(B'')\ 8f(x + 2y) + 8f(x) - 12f(x + y) - 2f(x + 3y) - 2f(x - y) = 0$$

and by $x \to y$ and $y \to x$

$$(B''')\ 8f(y + 2x) + 8f(y) - 12f(x + y) - 2f(y + 3x) + 2f(x - y) = 0.$$

We add (B'') with (B'''), and using (B'), it follows (C).

$(C) \Rightarrow (D)$

If f verifies (C), then f is odd. Indeed, in (C) we take $x = y = 0$ and it follows that $f(0) = 0$.

In (C) we take $y = -x$ and we get:

$$f(-x) + f(x) - 3[f(x) + f(-x)] = 0,$$

so $f(x) + f(-x) = 0$.
In (C), we take $x = u + 2v$, $y = -u - v$:

$$f(-u) + f(u + 3v) - 3f(u + 2v) - 3f(-u - v) - 6f(v) = 0$$

and since f is odd, we have

$$-f(u) + f(u + 3v) - 3f(u + 2v) + 3f(u + v) - 6f(v) = 0,$$

that is, (D).

$(D) \Rightarrow (A)$

In (D), we replace x with $x - y$ and we obtain:

$$-f(x - y) + 3f(x) - 3f(x + y) + f(x + 2y) - 6f(y) = 0,$$

that is, (A).

Other functional equation equivalent with the ones mentioned above, was studied in the papers [8, 33]. For the stability of equation (D), see Ref. [26].

Let $(G, +)$ be an abelian group and X a Banach space. Let be the functions

$$f : G \to X, \ \Phi : G \to \mathbb{R}_+.$$

We define the following function $C_f : G \times G \to X$,

$$C_f(x, y) = f(2x + y) + f(2x - y) - 2f(x + y) - 2f(x - y) - 12f(x)$$

If F is a cubic map, then

$$F(2x) = 8F(x).$$

Indeed, if $y = 0$:

$$2F(2x) = 4F(x) + 12F(x)$$

that is

$$F(2x) = 8F(x). \tag{8.3}$$

Definition 8.4 We say that f is Φ-approximable by a cubic map if there exists $F : G \to X$ a cubic map such that

$$\|f(x) - F(x)\| \le \Phi(x), \quad \forall x \in G. \tag{8.4}$$

We say that F is the cubic Φ approximation of f.

Problem Give conditions on f such that f to be Φ-approximable by a cubic map.

We denote by

$$\mathscr{C} = \left\{ \Phi : G \to \mathbb{R}_+ : \lim_{n \to \infty} \frac{\Phi(2^n x)}{8^n} = 0, \text{ for all } x \in G \right\}.$$

Theorem 8.8 *Let be $F \in \mathscr{C}$. Then f is Φ-approximable by a cubic map if and only if the following conditions holds*

(i) $\displaystyle \lim_{n \to \infty} \frac{\|C_f(2^n x, 2^n y)\|}{8^n} = 0$, *for all $x, y \in G$;*

(ii) *there exists $\Psi \in \mathscr{C}$ such that*

$$\|f(2^n x) - 8^n f(x)\| \le \Psi(2^n x) + 8^n \Phi(x), x \in G.$$

In this case, the cubic Φ-approximation of f is unique and is given by

$$F(x) = \lim_{n \to \infty} \frac{f(2^n x)}{8^n}.$$

Proof First, we assume that f is Φ-approximable by a cubic map, i.e. there exist

$$F : G \to X$$

such that the condition (8.4) holds. We have, for $x, y \in G$,

$$\|f(2x + y) - F(2x + y)\| \leq \Phi(2x + y)$$

$$\|f(2x - y) - F(2x - y)\| \leq \Phi(2x - y)$$

and also

$$\|f(x + y) - F(x + y)\| \leq \Phi(x + y)$$

$$\|f(x - y) - F(x - y)\| \leq \Phi(x - y)$$

It follows

$$\|C_f(x, y)\| = \|C_f(x, y) - C_F(x, y)\|$$
$$= \|f(2x + y) - F(2x + y) + f(2x - y) - F(2x - y)$$
$$- 2[f(x + y) - F(x + y)] - 2[f(x - y) - F(x - y)]$$
$$- 12[f(x) - F(x)]\|$$
$$\leq \Phi(2x + y) + \Phi(2x - y) + 2\Phi(x + y) + 2\Phi(x - y) + \Phi(x)$$

hence

$$\frac{\|C_f(2^n x, 2^n y)\|}{8^n} \leq \frac{\Phi[2^n(2x + y)]}{8^n} + \frac{\Phi[2^n(2x - y)]}{8^n}$$
$$+ 2\frac{\Phi[2^n(x + y)]}{8^n} + 2\frac{\Phi[2^n(x - y)]}{8^n} + 12\frac{\Phi(2^n x)}{8^n}$$

By letting n go to infinity in the above inequality, we obtain:

$$\lim_{n \to \infty} \frac{\|C_f(2^n x, 2^n y)\|}{8^n} = 0$$

Thus (i) holds.

Now we prove (ii). From (8.4), we have

$$\|f(2^n x) - F(2^n x)\| \leq \Phi(2^n x)$$

So, with (8.3), we obtain

$$\|f(2^n x) - 8^n f(x)\| = \|f(2^n x) - F(2^n x) + 8^n F(x) - 8^n f(x)\|$$
$$\leq \|f(2^n x) - F(2^n x)\| + 8^n \|F(x) - f(x)\|$$
$$\leq 8^n \Phi(x) + \Phi(2^n x)$$

Conversely, from (ii), we get

$$\left\| \frac{f(2^n x)}{8^n} - f(x) \right\| \leq \frac{\Psi(2^n x)}{8^n} + \Phi(x) \tag{8.5}$$

In (8.5), we put instead of x, $2^m x$ to get

$$\left\| \frac{f(2^{n+m} x)}{8^{n+m}} - \frac{f(2^m x)}{8^m} \right\| \leq \frac{\Psi(2^{n+m} x)}{8^{n+m}} + \frac{\Phi(2^m x)}{8^m}$$

By letting n, m go to infinity in the above inequality, we obtain

$$\lim_{n,m\to\infty} \left\| \frac{f(2^{n+m} x)}{8^{n+m}} - \frac{f(2^m x)}{8^m} \right\| = 0$$

Since X is a Banach space, it follows that the limit

$$F(x) = \lim_{n\to\infty} \frac{f(2^n x)}{8^n}$$

exists. And using (8.5)

$$\| F(x) - f(x) \| \leq \Phi(x)$$

From (i) it follows that $C_F(x, y) = 0$, hence F is cubic.
Now we show that F is unique. We suppose that F satisfies (8.4), i.e.

$$\| F(x) - f(x) \| \leq \Phi(x)$$

and exists F' which satisfies

$$\| F'(x) - f(x) \| \leq \Phi(x).$$

By norm inequality, we have $\| F(x) - F'(x) \| \leq 2\Phi(x)$. But, F and F' are cubic mappings and, by putting instead of x, $2^n x$ and we get

$$\| F(2^n x) - F'(2^n x) \| \leq 2\Phi(2^n x)$$

and by dividing the above inequality by 8^n, we get

$$\left\| \frac{F(2^n x)}{8^n} - \frac{F'(2^n x)}{8^n} \right\| \leq 2 \frac{\Phi(2^n x)}{8^n}.$$

But $\lim_{n\to\infty} \frac{\Phi(2^n x)}{8^n} = 0$ so $F(x) = F'(X)$.

As a corollary, we have the result of Jun and Kim [25].

Corollary 8.1 *Let* $\varphi : G \times G \to \mathbb{R}_+$ *be a function such that*

$$\sum_{i=0}^{\infty} \frac{\varphi(2^i x, 0)}{8^i} < \infty \text{ and}$$

$$\lim_{n \to \infty} \frac{\varphi(2^n x, 2^n y)}{8^n} = 0, \text{ for all } x, y \in X.$$

Suppose that a function $f : G \to X$ *satisfies*

$$\|C_f(x, y)\| \leq \varphi(x, y), \text{ for all } x, y \in X. \tag{8.6}$$

Then there exists a unique cubic function $F : X \to Y$ *such that*

$$\|f(x) - F(x)\| \leq \frac{1}{16} \sum_{i=0}^{\infty} \frac{\varphi(2^i x, 0)}{8^i},$$

for all $x \in G$. *The function* F *is given by*

$$F(x) = \lim_{n \to \infty} \frac{f(2^n x)}{8^n}.$$

Proof In (8.6), we take $y = 0$:

$$\|f(2x) - 8f(x)\| \leq \frac{1}{16} 8 \varphi(x, 0)$$

and by induction, we get

$$\|f(2^n x) - 8^n f(x)\| \leq 8^n \frac{1}{16} \sum_{i=0}^{\infty} \frac{\varphi(2^i x, 0)}{8^i}, \quad \text{for all } x \in G.$$

References

1. Borelli, C., Forti, G.L.: On a general Hyers-Ulam stability result. Int. J. Math. Math. Sci. **18**, 229–236 (1995)
2. Brzdęk, J.: Note on stability of the Cauchy equation-an answer to a problem of Th.M. Rassias. Carphtian J. Math. **30**(1), 47–54 (2004)
3. Brzdęk, J., Popa, D., Raşa, I., Xu, B.: Ulam Stability of Opeartors. Academic Press, New York (2018)
4. Brzdęk, J., Karapinar, E., Petruşel, A.: A fixed point theorem and the Ulam stability in generalized dq-metric spaces. J. Math. Anal. Appl. **467**, 501–520 (2018)

5. Cădariu, L., Radu, V.: Fixed points in generalized metric spaces and the stability of a cubic functional equation. In: Fixed Point Theory and Applications, vol. 7, pp. 53–68. Nova Science Publishers, New York (2007)
6. Cholewa, P.W.: Remarks on the stability of functional equations. Aequationes Math. **27**, 76–86 (1984)
7. Czerwik, S.: Stability of Functional Equations of Ulam-Hyers-Rassias Type. Hadronic Press, Palm Harbor (2003)
8. Eskandani, G.Z., Rassias, J.M., Găvruţa, P.: Generalized Hyers-Ulam stability for a general cubic functional equation in quasi-β-normed spaces. Asian Eur. J. Math. **4**(3), 413–425 (2011)
9. Găvruţa, P.: A generalization of the Hyers-Ulam-Rassias stability of approximately additive mappings. J. Math. Anal. Appl. **184**, 431–436 (1994)
10. Găvruţa, P.: An answer to a question of Th.M. Rassias and J. Tabor on mixed stability of mappings. Bul. Stiint. Univ. Politeh. Timis. Ser. Mat. Fiz. **42**(56), 1–6 (1997)
11. Găvruţa, P.: On the Hyers-Ulam-Rassias stability of mappings. In: Milovanovic, G.V. (ed.) Recent Progress in Inequalities, pp. 465–469. Kluwer, Dordrecht (1998)
12. Găvruţa, P.: An answer to a question of John M. Rassias concerning the stability of Cauchy equation. In: Advances in Equations and Inequalities. Hadronic Mathematics Series, pp. 67–71. Hadronic Press, Palm Harbor (1999)
13. Găvruţa, P.: On a problem of G. Isac and Th.M. Rassias concerning the stability of mappings. J. Math. Anal. Appl. **261**, 543–553 (2001)
14. Găvruţa, P.: On the Hyers-Ulam-Rassias stability of the quadratic mappings. Nonlinear Funct. Anal. Appl. **9**(3), 415–428 (2004)
15. Găvruţa, P., Cădariu, L.: General stability of the cubic functional equation. Bul. Ştiinţ. Univ. Politehnica Timişoara Ser. Mat.-Fiz. **47**(61)(1), 59–70 (2002)
16. Găvruţa, P., Cădariu, L.: The generalized stability of a quadratic functional equation. Nonlinear Funct. Anal. Appl. **9**(4), 513–526 (2004)
17. Găvruţa, P., Găvruţa, L.: Ψ-additive mappings and Hyers-Ulam stability. In: Pardalos, P.M., Rassias, Th.M., Khan, A.A. (eds.) Nonlinear Analysis and Variational Problems, pp. 81–86. Springer, New York (2010)
18. Găvruţa, L., Găvruţa, P.: On a problem of John M. Rassias concerning the stability in Ulam sense of Euler-Lagrange equation. In: Rassias, J.M. (ed.) Functionl Equation, Difference Inequalities and Ulam Stability Notions, pp. 47–53. Nova Science Publishers, New York (2010)
19. Găvruţa, L., Găvruţa, P.: Ulam stability problem for frames, Chap.11. In: Rassias, Th.M., Brzdęk, J. (eds.) Functional Equations in Mathematical Analysis. Springer Optimization and Its Applications, pp. 139–152. Springer, New York (2012)
20. Găvruţa, L., Găvruţa, P.: Approximation of functions by additive and by quadratic mappings. In: Rassias, Th.M., Gupta, V. (eds.) Mathematical Analysis, Approximation Theory and Their Applications. Springer Optimization and Its Applications, vol. 111, pp. 281–292. Springer, Cham (2016)
21. Găvruţa, P., Hossu, M., Popescu, D., Căprău, C.: On the stability of mappings and an answer to a problem of Th. M. Rassias. Ann. Math. Blaise Pascal **2**(2), 55–60 (1995)
22. Hyers, D.H.: On the stability of the linear functional equation. Proc. Nat. Acad. Soc. U. S. A. **27**, 222–224 (1941)
23. Hyers, D.H., Isac, G., Rassias, T.H.: Stability of Functional Equations in Several Variables. Birkhäuser, Basel (1998)
24. Isac, G., Rassias, Th.M.: On the Hyers-Ulam stability of ψ-additive mappings. J. Approx. Theory **72**, 131–137 (1993)
25. Jun, K.-W., Kim, H.-M.: The generalized Hyers-Ulam-Rassias stability of a cubic functional equation. J. Math. Anal. Appl. **274**, 867–878 (2002)
26. Jun, K.-W., Lee, Y.-H.: On the stability of a cubic functional equation. J. Chungcheong Math. Soc. **21**(3), 377–384 (2008)
27. Jung, S.-M.: Hyers-Ulam-Rassias Stability of Functional Equations in Nonlinear Analysis. Springer, New York (2011)

28. Jung, S.-M., Rassias, M.Th.: A linear functional equation of third order associated to the Fibonacci numbers. Abstr. Appl. Anal. **2014**, 7 pp. (2014). Article ID 137468
29. Jung, S.-M., Popa, D., Rassias, M.Th.: On the stability of the linear functional equation in a single variable on complete metric groups. J. Glob. Optim. **59**, 165–171 (2014)
30. Lee, Y.-H., Jung, S.-M., Rassias, M.Th.: On an n-dimensional mixed type additive and quadratic functional equation. Appl. Math. Comput. **228**, 13–16 (2014)
31. Mirmostafaee, A.K., Moslehian, M.S.: Fuzzy approximately cubic mappins. Inf. Sci. **178**(19), 3791–3798 (2008)
32. Mortici, C., Rassias, Th.M., Jung, S.-M.: On the stability of a functional equation associated with the Fibbonacci numbers. Abstr. Appl. Anal. **2014**, 6 pp. (2014). Article ID 546046
33. Najati, A., Park, C.: On the stability of a cubic functional equation. Acta Math. Sin. **24**(12), 1953–1964 (2008)
34. Rassias, Th.M: On the stability of the linear mapping in Banach spaces. Proc. Am. Math. Soc. **72**, 297–300 (1978)
35. Rassias, J.M.: Solution of the stability problem for cubic mappings. Glasnik Matematicki **36**(56), 73–84 (2001)
36. Saadati, R., Vaezpour, S.M., Park, C.: The stability of the cubic functional equations in various spaces. Math. Commun. **16**, 131–145 (2011)
37. Skof, F.: Proprietá locali e approssimazione di operatori. Rend. Sem. Mat. Fis. Milani **53**, 113–129 (1983)
38. Ulam, S.M.: A Collection of Mathematical Problems. Interscience Publications, New York (1960)

Chapter 9
Solutions and Stability of Some Functional Equations on Semigroups

Keltouma Belfakih, Elhoucien Elqorachi, and Themistocles M. Rassias

Abstract In this paper we investigate the solutions and the Hyers-Ulam stability of the μ-Jensen functional equation

$$f(xy) + \mu(y)f(x\sigma(y)) = 2f(x), \ x, y \in S,$$

a variant of the μ-Jensen functional equation

$$f(xy) + \mu(y)f(\sigma(y)x) = 2f(x), \ x, y \in S,$$

and the μ-quadratic functional equation

$$f(xy) + \mu(y)f(x\sigma(y)) = 2f(x) + 2f(y), \ x, y \in S,$$

where S is a semigroup, σ is a morphism of S and $\mu \colon S \longrightarrow \mathbb{C}$ is a multiplicative function such that $\mu(x\sigma(x)) = 1$ for all $x \in S$.

Keywords Functional equation · Hyers-Ulam stability · μ-Jensen functional equation · μ-Quadratic functional equation

Mathematics Subject Classification (2010) Primary 49B82; Secondary 39C52, 39C62

K. Belfakih · E. Elqorachi
Department of Mathematics, Faculty of Sciences, University Ibn Zohr, Agadir, Morocco

T. M. Rassias (✉)
Department of Mathematics, National Technical University of Athens, Athens, Greece
e-mail: trassias@math.ntua.gr

9.1   Introduction

In 1940, Ulam [31] delivered a wide ranging talk before the Mathematics Club of the University of Wisconsin in which he posed a number of important unsolved problems. Among those was the question concerning the stability of group homomorphisms: Given a group G_1, a metric group (G_2, d), a number $\epsilon > 0$ and a mapping $f : G_1 \longrightarrow G_2$ which satisfies $d(f(xy), f(x)f(y)) < \epsilon$ for all $x, y \in G_1$, does there exist a homomorphism $g : G_1 \longrightarrow G_2$ and a constant $k > 0$, depending only on G_1 and G_2 such that $d(f(x), g(x)) < k\epsilon$ for all $x \in G_1$?

In the case of a positive answer to this problem, we say that the Cauchy functional equation $f(xy) = f(x)f(y)$ is stable for the pair (G_1, G_2).

The first affirmative partial answer was given in 1941 by Hyers [16] where G_1, G_2 are Banach spaces.

In 1950 Aoki [2] provided a generalization of Hyers' theorem for additive mappings and in 1978 Rassias [22] generalized Hyers' theorem for linear mappings by allowing the Cauchy difference to be unbounded.

Beginning around the year 1980, several results for Hyers-Ulam-Rassias stability of many functional equations have been proved by several mathematicians. For more details, we can refer for example to [3, 8–10, 12–14, 17, 19, 23–26].

Let S be a semigroup with identity element e. Let σ be an involutive morphism of S. That is σ is an involutive homomorphism:

$$\sigma(xy) = \sigma(x)\sigma(y) \text{ and } \sigma(\sigma(x)) = x \text{ for all } x, y \in S,$$

or σ is an involutive anti-homomorphism:

$$\sigma(xy) = \sigma(y)\sigma(x) \text{ and } \sigma(\sigma(x)) = x \text{ for all } x, y \in S.$$

We say that $f : S \longrightarrow \mathbb{C}$ satisfies the Jensen functional equation if

$$f(xy) + f(x\sigma(y)) = 2f(x), \tag{9.1}$$

for all $x, y \in S$.

A complex valued function f defined on a semigroup S is a solution of a variant of the Jensen functional equation if

$$f(xy) + f(\sigma(y)x) = 2f(x), \tag{9.2}$$

for all $x, y \in S$. Equations (9.1) and (9.2) coincide if f is central, and the central solutions are the maps of the form $f = a + c$, where $a : S \longrightarrow \mathbb{C}$ is an additive map such that $a(\sigma(x)) = -a(x)$ and where $c \in \mathbb{C}$ is a constant.

The Jensen functional equation (9.1) takes the form

$$f(xy) + f(xy^{-1}) = 2f(x) \tag{9.3}$$

for all $x, y \in S$ when $\sigma(x) = x^{-1}$ and S is a group. The new equation (9.2) is much simpler than (9.1). For a more general study we refer the reader to Ng's paper [21] and Stetkær's book [26].

The stability in the sense of Hyers-Ulam of the Jensen equations (9.1) and (9.3) has been studied by various authors for the case when S is an abelian group or a vector space. The interested reader is referred to the papers of Jung [18] and Kim [20].

In 2010, Faiziev and Sahoo [11] proved the Hyers-Ulam stability of Eq. (9.3) on some non-commutative groups such as metabelian groups and $T(2, K)$, where K is an arbitrary commutative field with characteristic different from two. They have shown as well that every semigroup can be embedded into a semigroup in which the Jensen equation is stable.

The quadratic functional equation

$$f(x + y) + f(x - y) = 2f(x) + 2f(y), \quad x, y \in S \tag{9.4}$$

has been extensively studied (see for example [1, 17, 26]). It was generalized by Stetkær [25] to the more general equation

$$f(x + y) + f(x + \sigma(y)) = 2f(x) + 2f(y), \quad x, y \in S. \tag{9.5}$$

A stability result for the quadratic functional equation (9.4) was derived by Cholewa [5] and by Czerwik [6]. Bouikhalene et al. [3] stated the stability theorem of Eq. (9.5). In [7] the stability of the quadratic functional equation

$$f(xy) + f(xy^{-1}) = 2f(x) + 2f(y), \quad x, y \in S \tag{9.6}$$

was obtained on amenable groups.

Bouikhalene et al. [4] obtained the stability of the quadratic functional equation

$$f(xy) + f(x\sigma(y)) = 2f(x) + 2f(y), \quad x, y \in S \tag{9.7}$$

on amenable semigroups.

In this paper we consider the following functional equations:
The μ-Jensen functional equation

$$f(xy) + \mu(y)f(x\sigma(y)) = 2f(x), \quad x, y \in S, \tag{9.8}$$

a variant of the μ-Jensen functional equation

$$f(xy) + \mu(y)f(\sigma(y)x) = 2f(x), \quad x, y \in S, \tag{9.9}$$

and the μ-quadratic functional equation

$$f(xy) + \mu(y)f(x\sigma(y)) = 2f(x) + 2f(y), \quad x, y \in S, \tag{9.10}$$

where $\mu \colon S \longrightarrow \mathbb{C}$ is a multiplicative function such that $\mu(x\sigma(x)) = 1$ for all $x \in S$.

Our results are organized as follows. In Sects. 9.2 and 9.3 we give a proof of the Hyers-Ulam stability of the Jensen functional equation (9.1) and a variant of the Jensen functional equation (9.2) on an amenable semigroup. As an application (Sect. 9.4), we prove the Hyers-Ulam stability of the symmetric functional equation

$$f(xy) + f(yx) = 2f(x) + 2f(y), \quad x, y \in G, \tag{9.11}$$

where G is an amenable group.

In Sects. 9.5 and 9.6 we prove that the μ-Jensen equation (9.8), respectively, the μ-quadratic functional equation (9.10) possesses the same solutions as Jensen's functional equation (9.1), respectively, the quadratic functional equation (9.7). Furthermore, we prove the equivalence of their stability theorems on semigroups.

Throughout this paper m denotes a linear functional on the space $B(S, \mathbb{C})$, namely the space of all bounded functions on S.

The linear functional m is called a left, respectively, right invariant mean if and only if

$$\inf_{x \in S} f(x) \leq m(f) \leq \sup_{x \in S} f(x); \quad m(_a f) = m(f); \text{ respectively, } m(f_a) = m(f)$$

for all $f \in B(S, \mathbb{R})$ and $a \in S$, where $_a f$ and f_a are the left and right translates of f defined by $_a f(x) = f(ax); f_a(x) = f(xa), x \in S$.

A semigroup S which admits a left, respectively, right invariant mean on $B(S, \mathbb{C})$ will be called left, respectively, right amenable. If on the space $B(S, \mathbb{C})$ there exists a real linear functional which is simultaneously a left and right invariant mean, then we say that S is two-sided amenable or just amenable. We refer to [15] for the definition and properties of invariant means.

9.2 Stability of a Variant of the Jensen Functional Equation

In this section we investigate the Hyers-Ulam stability of the functional equation (9.2) on amenable semigroups.

Theorem 9.1 *Let S be an amenable semigroup with identity element e. Let σ be an involutive anti-homomorphism, and let $f : G \longrightarrow \mathbb{C}$ be a function. Assume that there exists $\delta \geq 0$ such that*

$$|f(xy) + f(\sigma(y)x) - 2f(x)| \leq \delta \tag{9.12}$$

for all $x, y \in S$. Then, there exists a unique solution $J : S \longrightarrow \mathbb{C}$ of the functional equation (9.2) such that $J(\sigma(x)) = -J(x)$ and

$$|f(x) - J(x) - f(e)| \leq \delta \tag{9.13}$$

for all $x \in S$. Furthermore if S is a group and $\sigma(x) = x^{-1}$ then there exists a unique additive map $a : S \longrightarrow \mathbb{C}$ such that

$$|f(x) - a(x) - f(e)| \leq \delta \tag{9.14}$$

for all $x \in S$.

Proof Let x, y be in S. Replacing x by $\sigma(x)$ in (9.12) we get

$$|f(\sigma(x)y) + f(\sigma(y)\sigma(x)) - 2f(\sigma(x))| \leq \delta \tag{9.15}$$

Adding (9.12) to (9.15), and using the triangle inequality we obtain that

$$|[f(xy) + f(\sigma(y)\sigma(x))] + [f(\sigma(y)x) + f(\sigma(x)y)] - 2[f(x) + f(\sigma(x))]| \leq 2\delta. \tag{9.16}$$

Hence

$$|f^e(xy) + f^e(\sigma(y)x) - 2f^e(x)| \leq \delta, \tag{9.17}$$

where

$$f^e(x) = \frac{f(x) + f(\sigma(x))}{2} \quad \text{for all } x \in S.$$

Subtracting (9.15) from (9.12), and using the triangle inequality we derive that

$$|f^o(xy) + f^o(\sigma(y)x) - 2f^o(x)| \leq \delta \tag{9.18}$$

for all $x, y \in S$, where

$$f^o(x) = \frac{f(x) - f(\sigma(x))}{2} \quad \text{for all } x \in S.$$

Setting $x = e$ in (9.17) we obtain

$$|f^e(y) - f^e(e)| \leq \frac{\delta}{2} \quad \text{for all } x, y \subset S. \tag{9.19}$$

By replacing x by y in (9.18) and by the fact that f^o is odd we get

$$|f^o(yx) - f^o(\sigma(y)x) - 2f^o(y)| \leq \delta. \tag{9.20}$$

This implies that for each y fixed in S, the function $x \longrightarrow f^o(yx) - f^o(\sigma(y)x)$ is bounded. Since S is amenable, then there exists an invariant mean m on the space of complex bounded functions on S and we can define the new mapping on S by

$$\psi(y) = m\{_y f^o -_{\sigma(y)} f^o\}, \quad \text{for all } y \in S. \tag{9.21}$$

Using (9.21) and the fact that m is an invariant mean we get

$$
\begin{aligned}
\psi(yz) + \psi(\sigma(z)y) &= m\{_{yz}f^o -_{\sigma(z)\sigma(y)} f^o\} + m\{_{\sigma(z)y}f^o -_{\sigma(y)z} f^o\} \\
&= m\{_{yz}f^o -_{\sigma(y)z} f^o\} + m\{_{\sigma(z)y}f^o -_{\sigma(z)\sigma(y)} f^o\} \\
&= m\{_z[_yf^o -_{\sigma(y)} f^o]\} + m\{[_yf^o -_{\sigma(y)} f^o]_{\sigma(z)}\} \\
&= m\{_yf^o -_{\sigma(y)} f^o\} + m\{_yf^o -_{\sigma(y)} f^o\} \\
&= \psi(y) + \psi(y) = 2\psi(y)
\end{aligned}
$$

for all $x, y \in S$. The function

$$
J(y) = \frac{\psi(y)}{2}
$$

satisfies the variant of the Jensen functional equation (9.2), $J(\sigma(y)) = -J(y)$ for all $y \in S$, and we have the following inequality

$$
|J(y) - f^o(y)| = |\frac{1}{2}m\{_{yf^o} -_{\sigma(y)} f^o - 2f(y)\}| \tag{9.22}
$$

$$
\leq \frac{1}{2}\sup_{x \in S} |f^o(yx) - f^o(\sigma(y)x) - 2f^o(y)| \leq \frac{\delta}{2}.
$$

Finally, we obtain

$$
\begin{aligned}
|f(y) - J(y) - f(e)| &= |f^e(y) + f^o(y) - J(y) - f(e)| \\
&\leq |f^e(y) - f(e)| + |f^o(y) - J(y)| \leq \delta
\end{aligned}
$$

for all $y \in S$. This proves the first part of Theorem 9.1.

If S is a group and $\sigma(x) = x^{-1}$, then from [26, Proposition 12.29] we have $J = a$, where $a : S \longrightarrow \mathbb{C}$ is an additive map.

Now suppose that there exist two odd functions J_1 and J_2 satisfying the variant of the Jensen functional equation (9.2), and the following inequality

$$
|f(y) - J_i(y) - f(e)| \leq \delta, \text{ with } i = 1, 2. \tag{9.23}
$$

The function $J := J_1 - J_2$ is also a solution of the functional equation (9.2), that is

$$
J(xy) + J(\sigma(y)x) = 2J(x) \text{ for all } x, y \in S. \tag{9.24}
$$

By using the triangle inequality we get $|J(x)| \leq 2\delta$ for all $x \in S$.

Replacing y by x in (9.24) and using that $J(\sigma(x)) = -J(x)$ we get

$$
J(x^2) = 2J(x) \tag{9.25}
$$

and consequently, we get $J(x^{2^n}) = 2^n J(x)$ for all $n \in \mathbb{N}$. Since J is a bounded map then $J(x) = 0$ for all $x \in S$. This completes the proof of Theorem 2.1. $\square$

The stability of Eq. (9.2) has been obtained in [4, Lemma 3.2], on amenable semigroups with identity element and under the condition that σ is an involutive homomorphism. In the following theorem we investigate the Hyers-Ulam stability of the functional equation (9.2) on amenable semigroups without identity element, and where σ is a homomorphism.

Theorem 9.2 *Let S be an amenable semigroup. Let σ be an involutive homomorphism of S and let $f : S \longrightarrow \mathbb{C}$ be a function. Assume that there exists $\delta \geq 0$ such that*

$$|f(xy) + f(\sigma(y)x) - 2f(x)| \leq \delta \tag{9.26}$$

for all $x, y \in S$. Then there exists a unique additive function $a : S \longrightarrow \mathbb{C}$ and $x_0 \in S$ such that

$$|f(x) - a(x) + f(x_0) - f(\sigma(x_0)) - f(x_0^2)| \leq 4\delta \tag{9.27}$$

for all $x \in S$.

Proof In the proof we use some ideas from Stetkær [28].

Let x, y, z be in S. If we replace x by xy and y by z in (9.26) we get

$$|f(xyz) + f(\sigma(z)xy) - 2f(xy)| \leq \delta. \tag{9.28}$$

By replacing x by $\sigma(z)x$ in (9.26) we get

$$|f(\sigma(z)xy) + f(\sigma(y)\sigma(z)x) - 2f(\sigma(z)x)| \leq \delta. \tag{9.29}$$

Replacing y by z in (9.26) and multiplying the result by 2 we get

$$|2f(xz) + 2f(\sigma(z)x) - 4f(x)| \leq 2\delta. \tag{9.30}$$

If we replace y by yz in (9.26) we get

$$|f(xyz) + f(\sigma(y)\sigma(z)x) - 2f(x)| \leq \delta. \tag{9.31}$$

Subtracting (9.31) from (9.29) and using the triangle inequality we get

$$|f(\sigma(z)xy) - 2f(\sigma(z)x) - f(xyz) + 2f(x)| \leq 2\delta. \tag{9.32}$$

Adding (9.30) and (9.32) and using the triangle inequality we obtain

$$|2f(xz) - 2f(x) + f(\sigma(z)xy) - f(xyz)| \leq 4\delta. \tag{9.33}$$

Subtracting (9.33) from (9.28) and applying the triangle inequality we get

$$|2f(xyz) - 2f(xy) - 2f(xz) + 2f(x)| \leq 5\delta, \tag{9.34}$$

which can be written as follows

$$|[2f(xyz) - 2f(x)] - [2f(xy) - 2f(x)] - [2f(xz) - 2f(x)]| \leq 5\delta. \tag{9.35}$$

Now, for each fixed x_0 in S we define on S the function $A_{x_0}(t) = 2f(x_0 t) - 2f(x_0)$. Therefore, the inequality (9.35) can be written as follows

$$|A_{x_0}(yz) - A_{x_0}(y) - A_{x_0}(z)| \leq 5\delta \text{ for all } y, z \in S. \tag{9.36}$$

Since S is an amenable semigroup then by Szekelyhidi [30] there exists a unique additive mapping $b : S \longrightarrow \mathbb{C}$ such that

$$|A_{x_0}(x) - b(x)| \leq 5\delta \text{ for all } x \in S. \tag{9.37}$$

Replacing y in (9.26) by yz we get

$$|f(xyz) + f(\sigma(yz)x) - 2f(x)| \leq \delta. \tag{9.38}$$

If we replace x by $\sigma(y)$ and y by $\sigma(z)x$ in (9.26) we derive

$$|f(\sigma(y)\sigma(z)x) + f(z\sigma(xy)) - 2f(\sigma(y))| \leq \delta. \tag{9.39}$$

Replacing x by z and y by $\sigma(xy)$ in (9.26) we get

$$|f(z\sigma(xy)) + f(xyz) - 2f(z)| \leq \delta. \tag{9.40}$$

Subtracting (9.39) from the sum of (9.38) and (9.40) and applying the triangle inequality we get

$$|2f(xyz) - 2f(x) - 2f(z) + 2f(\sigma(y))| \leq 3\delta. \tag{9.41}$$

By replacing x and y by x_0, and z by x in (9.41) we get

$$|2f(x_0^2 x) - 2f(x_0) - 2f(x) + 2f(\sigma(x_0))| \leq 3\delta, \tag{9.42}$$

which can be expressed as follows

$$|2f(x_0^2 x) - 2f(x_0^2) - 2f(x) - 2f(x_0) + 2f(\sigma(x_0)) + 2f(x_0^2)| \leq 3\delta. \tag{9.43}$$

Since $A_{x_0^2}(x) = 2f(x_0^2 x) - 2f(x_0^2)$, then we have

$$|A_{x_0^2}(x) - 2f(x) - 2f(x_0) + 2f(\sigma(x_0)) + 2f(x_0^2)| \leq 3\delta. \tag{9.44}$$

Subtracting (9.37) from (9.44) and using the triangle inequality we get

$$|f(x) - a(x) + f(x_0) - f(\sigma(x_0)) - f(x_0^2)| \le 4\delta, \tag{9.45}$$

where $a = \frac{1}{2}b$. This completes the proof of Theorem 2.2.   □

9.3   Hyers-Ulam Stability of Eq. (9.1) on Amenable Semigroups

In this section, we investigate the Hyers-Ulam stability of Eq. (9.1) on an amenable semigroup, where σ is an involutive anti-homomorphism.

Theorem 9.3 *Let S be an amenable semigroup with identity element e. Let σ be an involutive anti-homomorphism of S. Let $f : S \longrightarrow \mathbb{C}$ be a function which satisfies the following inequality*

$$|f(xy) + f(x\sigma(y)) - 2f(x)| \le \delta \tag{9.46}$$

for all x, $y \in S$ and for some nonnegative δ. Then there exists a unique solution j of the Jensen equation (9.1) such that $j(\sigma(x)) = -j(x)$ and

$$|f(x) - j(x) - f(e)| \le 3\delta \tag{9.47}$$

for all $x \in S$.

First, we prove the following useful lemma.

Lemma 9.1 *Let S be a semigroup. Let σ be an involutive anti-homomorphism of S. Let $f : S \longrightarrow \mathbb{C}$ be a function such that $f(\sigma(x)) = -f(x)$ for all $x \in S$ and for which there exists a solution g of the Drygas functional equation*

$$g(yx) + g(\sigma(y)x) = 2g(x) + g(y) + g(\sigma(y)), \quad x, y \in S \tag{9.48}$$

such that $|f(x) - g(x)| \le M$, for all $x \in S$ and for some non negative M. Then

$$g(x) = \lim_{n \to +\infty} 2^{-n} f(x^{2^n}) \text{ for all } x \in S. \tag{9.49}$$

Furthermore $g(\sigma(x)) = -g(x)$ for all $x \in S$ and g satisfies the Jensen functional equation

$$g(xy) + g(x\sigma(y)) = 2g(x) \text{ for all } x, y \in S.$$

Proof Replacing y by $x\sigma(x)$ in (9.48) we obtain

$$g((x\sigma(x))^2) + g((x\sigma(x))^2) = 2g(x\sigma(x)) + g(x\sigma(x)) + g(x\sigma(x)), \qquad (9.50)$$

which implies that $g((x\sigma(x))^2) = 2g(x\sigma(x))$ for all $x \in S$.

By applying the induction assumption we get

$$2^n g(x\sigma(x)) = g((x\sigma(x))^{2^n}) \qquad (9.51)$$

for all $n \in \mathbb{N}$ and for all $x \in S$.

Now, by the hypothesis, $f = g + b$ where b is a bounded function. Since f is odd we have $f = g^o + b^o$ and $g^e + b^e = 0$. Using (9.51) and the fact that

$$g((x\sigma(x))^{2^n}) = g^e((x\sigma(x))^{2^n})$$

we get

$$|g(x\sigma(x))| = 2^{-n}|g^e((x\sigma(x))^{2^n})| \le 2^{-n}|b^e(x\sigma(x))^{2^n}|. \qquad (9.52)$$

Letting $n \to +\infty$ in the formula (9.52), we obtain that $g(x\sigma(x)) = 0$ and hence $g(\sigma(x)x) = 0$ for all $x \in S$.

Setting $y = x$ in (9.48) we get

$$g(x^2) = 2g(x) + g(x) + g(\sigma(x)). \qquad (9.53)$$

If we replace x by $\sigma(x)$ in (9.53) we have

$$g(\sigma(x)^2) = 2g(\sigma(x)) + g(x) + g(\sigma(x)). \qquad (9.54)$$

By adding (9.53) and (9.54) we get that $g^e(x^2) = 4g^e(x)$, and by induction it follows that

$$g^e(x^{2^n}) = 2^{2^n} g^e(x) \qquad (9.55)$$

for all $x \in S$ and for all $n \in \mathbb{N}$.

Using (9.55) and the fact that $g^e + b^e = 0$ we have

$$g^e(x) = 2^{-2^n} g^e(x^{2^n}) = -2^{-2^n} b^e(x^{2^n}). \qquad (9.56)$$

Therefore, we get

$$|g^e(x)| = |2^{-2^n} g^e(x^{2^n})| \le 2^{-2^n}|b^e(x^{2^n})|.$$

So by letting $n \to +\infty$ we obtain that $g^e(x) = 0$ for all $x \in S$, which proves that $g(\sigma(x)) = -g(x)$ for all $x \in S$.

Using (9.53) and that g is odd we get that $g(x^2) = 2g(x)$, and by induction we deduce that

$$g(x^{2^n}) = 2^n g(x) \tag{9.57}$$

for all $x \in S$, and for all $n \in \mathbb{N}$.

Using (9.57) we get

$$2^{-n} f(x^{2^n}) = 2^{-n}[g(x^{2^n}) + b^o(x^{2^n})] = g(x) + 2^{-n} b^o(x^{2^n}).$$

Thus

$$|g(x) - 2^{-n} f(x^{2^n})| \leq 2^{-n} |b^o(x^{2^n})|. \tag{9.58}$$

By letting $n \to +\infty$ we obtain

$$g(x) = \lim_{n \to +\infty} 2^{-n} f(x^{2^n}).$$

We will prove that g satisfies the Jensen functional equation (9.1).

Since $g(\sigma(x)) = -g(x)$ for all $x \in S$, the Drygas functional equation (9.48) can be written as follows

$$g(yx) + g(\sigma(y)x) = 2g(x), \quad x, y \in S. \tag{9.59}$$

Replacing x by $\sigma(x)$ in (9.59) we get

$$g(y\sigma(x)) + g(\sigma(y)\sigma(x)) = 2g(\sigma(x)).$$

Using that $g(\sigma(x)) = -g(x)$ for all $x \in S$ we obtain

$$g(x\sigma(y)) + g(xy) = 2g(x), \quad x, y \in S,$$

which means that g satisfies the Jensen functional equation (9.1). This completes the proof of Lemma 9.1. Now, we are ready to prove Theorem 9.3. Setting $x = e$ in (9.46) we get

$$|f^e(y) - f(e)| \leq \frac{\delta}{2} \tag{9.60}$$

for all $y \in S$.

The inequalities (9.46), (9.60) and the triangle inequality yield

$$|f(xy) + f(yx) - 2f(x) - 2f(y) + 2f(e)| \le |f(xy) + f(x\sigma(y)) - 2f(x)|$$
$$+ |f(yx) + f(y\sigma(x) - 2f(y)| + |2f(e) - f(x\sigma(y)) - f(y\sigma(x))| \le 3\delta.$$
$$(9.61)$$

Hence, from (9.46), (9.60) and (9.61) we get

$$|f(yx) + f(\sigma(y)x) - 2f(x)| \le |f(yx) + f(xy) - 2f(y) - 2f(x) + 2f(e)|$$
$$+ |f(\sigma(y)x) + f(x\sigma(y)) - 2f(\sigma(y)) - 2f(x) + 2f(e)|$$
$$+ | - f(xy) - f(x\sigma(y)) + 2f(x)| + |2f(y) + 2f(\sigma(y)) - 4f(e)| \le 9\delta.$$
$$(9.62)$$

From (9.46) and (9.62) we obtain

$$2|f^o(yx) + f^o(y\sigma(x)) - 2f^o(y)| \tag{9.63}$$
$$= |f(yx) - f(\sigma(x)\sigma(y)) + f(y\sigma(x)) - f(x\sigma(y)) - 2f(y) + 2f(\sigma(y))|$$
$$\le |f(yx) + f(y\sigma(x)) - 2f(y)| + |f(x\sigma(y)) + f(\sigma(x)\sigma(y)) - 2f(\sigma(y))|$$
$$\le 10\delta.$$

Consequently we have

$$|f^o(yx) + f^o(y\sigma(x)) - 2f^o(y)| \le 5\delta \tag{9.64}$$

for all $x, y \in S$. Thus for fixed $y \in S$, the functions $x \longrightarrow f^o(yx) - f^o(x\sigma(y))$ and $x \longrightarrow f^o(xy) + f^o(x\sigma(y)) - 2f^o(x)$ are bounded on S.
 Furthermore,

$$m\{f^o_{\sigma(y)\sigma(z)} + f^o_{\sigma(y)z} - 2f^o_{\sigma(y)}\} = m\{(f^o_{\sigma(z)} + f^o_z - 2f^o)_{\sigma(y)}\} \tag{9.65}$$

$$= m\{f^o_{\sigma(z)} + f^o_z - 2f^o\},$$

where m is an invariant mean on S.
 By using (9.62) we get that, for every fixed $y \in S$, the function

$$x \longrightarrow f^o(yx) + f^o(\sigma(y)x) - 2f^o$$

is bounded and

$$m\{_{zy}f^o +_{\sigma(z)y} f^o - 2_y f^o\} = m\{_y(_z f^o +_{\sigma(z)} f^o - 2f^o)\} \tag{9.66}$$

$$= m\{_z f^o +_{\sigma(z)} f^o - 2f^o\}.$$

Now we define the new mapping

$$\phi(y) := m\{_y f^o - f^o_{\sigma(y)}\}, \quad y \in S. \tag{9.67}$$

By using the definition of ϕ and m, the equalities (9.65) and (9.66), we obtain that

$$\phi(zy) + \phi(\sigma(z)y) = m\{_{zy} f^o - f^o_{\sigma(y)\sigma(z)}\} + m\{_{\sigma(z)y} f^o - f^o_{\sigma(y)z}\} \tag{9.68}$$

$$= m\{_{zy} f^o +_{\sigma(z)y} f^o - 2_y f^o\} - m\{f^o_{\sigma(y)\sigma(z)} + f^o_{\sigma(y)z} - 2 f^o_{\sigma(y)}\}$$

$$+ 2m\{_y f^o - f^o_{\sigma(y)}\}$$

$$= m\{_z f^o +_{\sigma(z)} f^o - 2 f^o\} - m\{f^o_{\sigma(z)} + f^o_z - 2 f^o\} + 2m\{_y f^o - f^o_{\sigma(y)}\}$$

$$= m\{_z f^o - f^o_{\sigma(z)}\} + m\{_{\sigma(z)} f^o - f^o_z\} + 2m\{_y f^o - f^o_{\sigma(y)}\}$$

$$= 2\phi(y) + \phi(z) + \phi(\sigma(z)),$$

which implies that ϕ is a solution of the Drygas functional equation (9.48). Furthermore, we have

$$\left|\frac{\phi}{2}(y) - f^o(y)\right| = \frac{1}{2}|\phi(y) - 2 f^o(y)| = \frac{1}{2}|m\{_y f^o - f^o_{\sigma(y)} - 2 f^o(y)\}| \tag{9.69}$$

$$\leq \frac{1}{2} \sup_{x \in S} |f^o(yx) - f^o(x\sigma(y)) - 2 f^o(y)|$$

$$= \frac{1}{2} \sup_{x \in S} |f^o(yx) + f^o(y\sigma(x)) - 2 f^o(y)|$$

$$\leq \frac{5}{2}\delta.$$

By Lemma 9.1, it follows that the function $\frac{\phi}{2}$ is a solution of the Drygas functional equation (9.48) and $\frac{\phi}{2} - f^o$ is a bounded mapping, thus we have

$$\frac{\phi}{2} = \lim_{n \to +\infty} 2^{-n} f^o(x^{2^n}), \tag{9.70}$$

which implies that $\frac{\phi}{2}(\sigma(x)) = -\frac{\phi}{2}(x)$ for all $x \in S$, consequently $\frac{\phi}{2}$ is a solution of the Jensen functional equation (9.1). On the other hand, we have

$$\left|f(x) - \frac{\phi}{2} - f(e)\right| = \left|f^e(x) + f^o(x) - \frac{\phi}{2} - f(e)\right| \tag{9.71}$$

$$\leq |f^e(x) - f(e)| + \left|f^o(x) - \frac{\phi}{2}\right|$$

$$\leq \frac{\delta}{2} + \frac{5\delta}{2} + 3\delta.$$

We can use the same method as in Theorem 9.1 to prove the uniqueness of the derived solution. This completes the proof of Theorem 9.3.   □

9.4   Application: Stability of the Symmetric Functional Equation (9.11)

In this section we use the result obtained in Sect. 9.3 to prove the stability of the symmetric functional equation (9.11).

Theorem 9.4 *Let G be an amenable group, and $f : G \longrightarrow \mathbb{C}$ a function. Assume that there exists a non-negative M such that*

$$|f(xy) + f(yx) - 2f(x) - 2f(y)| \leq M \tag{9.72}$$

for all $x, y \in G$. Then, there exists a unique solution $J : G \longrightarrow \mathbb{C}$ of the symmetric functional equation (9.11) such that

$$|f(x) - J(x) - f(e)| \leq 12M \text{ for all } x \in G. \tag{9.73}$$

Proof In the proof we use some ideas from Stetkær [26, Proposition 2.17].
    Setting $x = y = e$ in (9.72) we get

$$|f(e)| \leq \frac{M}{2}. \tag{9.74}$$

If we replace y by x^{-1} in (9.72) we get

$$|f(e) - f(x) - f(x^{-1})| \leq \frac{M}{2}. \tag{9.75}$$

Subtracting (9.75) from (9.74) and using the triangle inequality we obtain

$$|f(x) + f(x^{-1})| \leq M. \tag{9.76}$$

Replacing x by xy and y by x^{-1} in (9.72) we derive

$$|f(xyx^{-1}) + f(y) - 2f(xy) - 2f(x^{-1})| \leq M. \tag{9.77}$$

Using (9.76), (9.77) and the triangle inequality we deduce that

$$|f(xyx^{-1}) + f(y) - 2f(xy) + 2f(x)| \leq 3M. \tag{9.78}$$

By replacing y by y^{-1} in (9.78) we get that

$$|f(xy^{-1}x^{-1}) + f(y^{-1}) - 2f(xy^{-1}) + 2f(x)| \leq 3M. \tag{9.79}$$

Adding (9.78) to (9.79) and using the triangle inequality we have that

$$|[f(xyx^{-1}) + f((xyx^{-1})^{-1})] + [f(y) + f(y^{-1})] - 2f(xy) \qquad (9.80)$$
$$-2f(xy^{-1}) + 4f(x)| \le 6M.$$

Using (9.76), (9.80) and the triangle inequality we obtain

$$|f(xy) + f(xy^{-1}) - 2f(x)| \le 4M. \qquad (9.81)$$

By applying Theorem 9.3 there exists $J: G \longrightarrow \mathbb{C}$, unique solution of the Jensen functional equation (9.3), that is

$$J(xy) + J(xy^{-1}) = 2J(x), \qquad (9.82)$$

such that $J(x^{-1}) = -J(x)$ and

$$|f(x) - J(x) - f(e)| \le 12M \qquad (9.83)$$

for all $x \in G$. Interchanging x and y in (9.82) we obtain

$$J(yx) + J(yx^{-1}) = 2J(y). \qquad (9.84)$$

Adding (9.82) to (9.84) we get

$$J(xy) + J(yx) + J(xy^{-1}) + J(yx^{-1}) = 2J(x) + 2J(y). \qquad (9.85)$$

Since $J(x^{-1}) = -J(x)$ for all $x \in G$, then we deduce that

$$J(xy) + J(yx) = 2J(x) + 2J(y) \qquad (9.86)$$

for all $x, y \in G$, which means that J satisfies the symmetric functional equation (9.11).

For the uniqueness of the solution J we use that if J is a solution of (9.86) then $J(x^{2^n}) = 2^n J(x)$ for every integer n and for all $x \in G$, and by similar computations to those used above we deduce the rest of the proof.   □

9.5 μ-Jensen Functional Equation

The trigonometric functional equations having a multiplicative function μ in front of terms like $f(x\sigma(y))$ or $f(\sigma(y)x)$ have been studied in many papers. The μ-d'Alembert's functional equation

$$f(xy) + \mu(y)f(xy^{-1}) = 2f(x)f(y), \quad x, y \in S \qquad (9.87)$$

which is an extension of d'Alembert's functional equation

$$f(xy) + f(xy^{-1}) = 2f(x)f(y), \ x, y \in S$$

has been treated systematically by Stetkær [27] on groups. The non-zero solutions of (9.87) on groups with involution are the normalized traces of certain representation of S on $\mathbb{C}^2$. On abelian groups the solutions of (9.87) are

$$f(x) = \frac{\gamma(x) + \mu(x)\gamma(x^{-1})}{2}, \ \text{where } \gamma : S \longrightarrow \mathbb{C}$$

is a multiplicative function (see [27]).

On abelian groups the solutions of μ-Wilson's functional equation

$$f(xy) + \mu(y)f(x\sigma(y)) = 2f(x)g(y), \ x, y \in S$$

are studied in [9] and [29]. We refer also the interested reader to [8] and [10].

In the present section we prove that the μ-Jensen functional equations (9.8), (9.9) have a non-zero solution only if $\mu = 1$. We note that in this case σ is an arbitrary surjective homomorphism which is not necessary involutive.

Theorem 9.5 *Let S be a semigroup, $\sigma : S \longrightarrow S$ be a homomorphism, and μ be a multiplicative function such that $\mu(x\sigma(x)) = 1$ for all $x \in S$. If the functional equation*

$$f(xy) + \mu(y)f(x\sigma(y)) = 2f(x), \ x, y \in S \tag{9.88}$$

has a non-zero solution then $\mu = 1$. That is, the μ-Jensen functional equation (9.88) possesses the same solutions to those of the Jensen functional equation (1.2).

Proof Making the substitutions (xy, z), $(x\sigma(y), z)$ in (9.88) we get respectively

$$f(xyz) + \mu(z)f(xy\sigma(z)) = 2f(xy), \tag{9.89}$$

$$f(x\sigma(y)z) + \mu(z)f(x\sigma(y)\sigma(z)) = 2f(x\sigma(y)). \tag{9.90}$$

Multiplying (9.90) by $\mu(y)$ we obtain

$$\mu(y)f(x\sigma(y)z) + \mu(yz)f(x\sigma(y)\sigma(z)) = 2\mu(y)f(x\sigma(y)). \tag{9.91}$$

Adding (9.89) and (9.91) and applying (9.88) we obtain

$$f(xyz) + \mu(z)f(xy\sigma(z)) + \mu(y)f(x\sigma(y)z) + \mu(yz)f(x\sigma(y)\sigma(z)) = 4f(x). \tag{9.92}$$

By using (9.88), Eq. (9.92) can be written as follows

$$2f(x) + \mu(z)[f(xy\sigma(z)) + \mu(y\sigma(z))f(x\sigma(y)z) = 4f(x). \tag{9.93}$$

Multiplying (9.93) by $\mu(\sigma(z))$ and using the fact that $\mu(z\sigma(z)) = 1$ we get after some simplification that

$$f(xy\sigma(z)) + \mu(y\sigma(z))f(x\sigma(y)z) = 2\mu(\sigma(z))f(x). \tag{9.94}$$

By replacing y in (9.88) by $y\sigma(z)$ we get

$$f(xy\sigma(z)) + \mu(y\sigma(z))f(x\sigma(y)\sigma^2(z)) = 2f(x). \tag{9.95}$$

Subtracting (9.95) from (9.94) we deduce that

$$\mu(y\sigma(z))[f(x\sigma(y)z) - f(x\sigma(y)\sigma^2(z))] = 2[\mu(\sigma(z)) - 1]f(x). \tag{9.96}$$

Multiplying the last identity by $\mu(\sigma(y)z)$ and using the fact that $\mu(z\sigma(z)) = 1$ we obtain that

$$f(x\sigma(y)z) - f(x\sigma(y)\sigma^2(z)) = 2\mu(\sigma(y))[1 - \mu(z)]f(x). \tag{9.97}$$

On the other hand, if we make the substitutions $(x\sigma(y), z)$ and $(x\sigma(y), \sigma(z))$ in (9.88) we deduce respectively

$$f(x\sigma(y)z) + \mu(z)f(x\sigma(y)\sigma(z)) = 2f(x\sigma(y)). \tag{9.98}$$

$$f(x\sigma(y)\sigma(z)) + \mu(\sigma(z))f(x\sigma(y)\sigma^2(z)) = 2f(x\sigma(y)). \tag{9.99}$$

Multiplying (9.99) by $\mu(z)$ and using the fact that $\mu(z\sigma(z)) = 1$ we derive that

$$\mu(z)f(x\sigma(y)\sigma(z)) + f(x\sigma(y)\sigma^2(z)) = 2\mu(z)f(x\sigma(y)). \tag{9.100}$$

Subtracting (9.100) from (9.98) we obtain

$$f(x\sigma(y)z) - f(x\sigma(y)\sigma^2(z)) = 2[1 - \mu(z)]f(x\sigma(y)). \tag{9.101}$$

By comparing (9.101) and (9.97) we deduce that

$$2\mu(\sigma(y))[1 - \mu(z)]f(x) = 2[1 - \mu(z)]f(x\sigma(y)), \tag{9.102}$$

from which we get

$$[1 - \mu(z)][\mu(\sigma(y))f(x) - f(x\sigma(y))] = 0. \tag{9.103}$$

If we suppose that $\mu \neq 1$, then from (9.103) we deduce that

$$f(x\sigma(y)) = \mu(\sigma(y))f(x) \tag{9.104}$$

for all $x, y \in S$. If we combine Eqs. (9.104) and (9.88) we get

$$f(xy) + \mu(y)\mu(\sigma(y))f(x) = 2f(x). \tag{9.105}$$

Since $\mu(y\sigma(y)) = 1$ we deduce that $f(xy) = f(x)$ for all $y \in S$. Therefore (9.88) becomes

$$(\mu(y) - 1)f(x) = 0$$

which means that either $f = 0$ or $\mu = 1$. Since $\mu \neq 1$, then we get $f = 0$, which contradicts the assumption that $f \neq 0$. $\qquad\square$

Theorem 9.6 *Let S be a semigroup, let $\sigma : S \longrightarrow S$ be a homomorphism, and μ be a multiplicative function such that $\mu(x\sigma(x)) = 1$ for all $x \in S$. If the variant of the μ-Jensen functional equation*

$$f(xy) + \mu(y)f(\sigma(y)x) = 2f(x), \quad x, y \in S \tag{9.106}$$

has a non-zero solution, then $\mu = 1$.

Proof The computations used in [10] for $g = 1$ show that for all fixed a in S, the mapping $x \longrightarrow f(ax) - f(a)$ is additive.

On the other hand, by replacing y by yz in (9.106) we get

$$f(xyz) + \mu(yz)f(\sigma(yz)x) = 2f(x). \tag{9.107}$$

If we replace x by $\sigma(y)$ and y by $\sigma(z)x$ in (9.106) and multiply the result obtained by $\mu(yz)$ we deduce that

$$\mu(yz)f(\sigma(yz)x) + \mu(xy)f(z\sigma(xy)) = 2\mu(yz)f(\sigma(y)). \tag{9.108}$$

By replacing x by z and y by $\sigma(xy)$ in (9.106) and multiplying the result obtained by $\mu(xy)$ we get

$$\mu(xy)f(z\sigma(xy)) + f(xyz) = 2\mu(xy)f(z). \tag{9.109}$$

By subtracting the sum of (9.107) and (9.109) from (9.108) we obtain

$$f(xyz) = f(x) + \mu(xy)f(z) - \mu(yz)f(\sigma(y)). \tag{9.110}$$

Since for each fixed a in S the function $x \longrightarrow f(a^2 x) - f(a^2)$ is additive then the new function

$$x \longrightarrow \mu(a^2) f(x) - \mu(a) \mu(x) f(\sigma(a)) + 2f(a) - 2f(a^2)$$
$$= \mu(a)[\mu(a) f(x) - \mu(x) f(\sigma(a))] + 2f(a) - 2f(a^2)$$

is additive. Since $\mu \neq 0$, then we deduce that f is central. That is $f(xy) = f(yx)$ for all $x, y \in S$. For the rest of the proof we use Theorem 9.5. $\qquad\square$

Theorem 9.7 *Let S be a semigroup, $\sigma : S \longrightarrow S$ be an anti-homomorphism which is surjective and $\mu : S \longrightarrow \mathbb{C}$ be a multiplicative function such that $\mu(x\sigma(x)) = 1$ for all $x \in S$. If the μ-Jensen functional equation*

$$f(xy) + \mu(y) f(x\sigma(y)) = 2f(x), \quad x, y \in S \tag{9.111}$$

has a non-zero solution, then $\mu = 1$.

Proof Making the substitutions (xy, z), $(x\sigma(y), z)$ in (9.111) and multiplying the second result by $\mu(y)$ we get respectively

$$f(xyz) + \mu(z) f(xy\sigma(z)) = 2f(xy), \tag{9.112}$$

$$\mu(y) f(x\sigma(y)z) + \mu(yz) f(x\sigma(y)\sigma(z)) = 2\mu(y) f(x\sigma(y)). \tag{9.113}$$

Adding (9.112) to (9.113) and using (9.111) we obtain

$$f(xyz) + \mu(z) f(xy\sigma(z)) + \mu(y) f(x\sigma(y)z) + \mu(yz) f(x\sigma(y)\sigma(z)) = 4f(x). \tag{9.114}$$

If we replace y in (9.111) by yz we get

$$f(xyz) + \mu(yz) f(x\sigma(z)\sigma(y)) = 2f(x). \tag{9.115}$$

Subtracting (9.115) from (9.114) we obtain

$$\mu(yz)[f(x\sigma(y)\sigma(z)) - f(x\sigma(z)\sigma(y))] + \mu(z) f(xy\sigma(z)) + \mu(y) f(x\sigma(y)z) = 2f(x). \tag{9.116}$$

Taking $y = z$ in the last identity we find

$$\mu(y)[f(xy\sigma(y)) + f(x\sigma(y)y)] = 2f(x). \tag{9.117}$$

On the other hand, if we subtract (9.112) from (9.115) and multiply the result by $\mu(\sigma(z))$ and use the fact that $\mu(z\sigma(z)) = 1$ we get

$$\mu(y)f(x\sigma(z)\sigma(y)) - f(xy\sigma(z)) = 2\mu(\sigma(z))f(x) - 2\mu(\sigma(z))f(xy). \tag{9.118}$$

Replacing x in (9.111) by $x\sigma(z)$ implies

$$f(x\sigma(z)y) + \mu(y)f(x\sigma(z)\sigma(y)) = 2f(x\sigma(z)). \tag{9.119}$$

The subtraction of (9.118) from (9.119) yields

$$f(x\sigma(z)y) + f(xy\sigma(z)) = 2f(x\sigma(z)) - 2\mu(\sigma(z))f(x) + 2\mu(\sigma(z))f(xy). \tag{9.120}$$

Since σ is surjective, then by taking $t = \sigma(z)$ in (9.120) we obtain

$$f(xty) + f(xyt) = 2f(xt) + 2\mu(t)f(xy) - 2\mu(t)f(x) \tag{9.121}$$

for all $x, t, y \in S$. Replacing t in (9.121) by y, and y by $\sigma(y)$ and multiplying the resulting formulas obtained by $\mu(y)$ and using the fact that $\mu(y\sigma(y)) = 1$ we get

$$\mu(y)[f(xy\sigma(y)) + f(x\sigma(y)y)] \tag{9.122}$$
$$= 2\mu(y)f(xy) + 2\mu^2(y)f(x\sigma(y)) - 2\mu^2(y)f(x).$$

If we subtract (9.122) from (9.117) we deduce

$$2\mu(y)[f(xy) + \mu(y)f(x\sigma(y))] - 2\mu^2(y)f(x) = 2f(x). \tag{9.123}$$

Using (9.111) we get

$$[\mu(y) - 1]^2 f(x) = 0 \tag{9.124}$$

for all x and y in S. This means that if f is a non-zero solution of (9.121) then $\mu = 1$. $\square$

9.6 Solutions of μ-Quadratic Functional Equation

In this section we consider the μ-quadratic functional equation (1.10), and we prove a similar result as in the precedent section for the μ-quadratic functional equation (9.10).

Theorem 9.8 *Let S be a semigroup, $\sigma : S \longrightarrow S$ be a homomorphism, and μ be a multiplicative function such that $\mu(x\sigma(x)) = 1$ for all $x \in S$. If the μ-quadratic functional equation*

$$f(xy) + \mu(y)f(x\sigma(y)) = 2f(x) + 2f(y), \quad x, y \in S \tag{9.125}$$

has a non-zero solution, then $\mu = 1$. That is, the μ-quadratic functional equation (9.125) possesses the same solutions to those of the quadratic functional equation (1.4)

Proof Making the substitutions (xy, z), $(x\sigma(y), z)$ in (9.125) we get respectively

$$f(xyz) + \mu(z)f(xy\sigma(z)) = 2f(xy) + 2f(z). \tag{9.126}$$

$$f(x\sigma(y)z) + \mu(z)f(x\sigma(y)\sigma(z)) = 2f(x\sigma(y)) + 2f(z). \tag{9.127}$$

Multiplying (9.127) by $\mu(y)$ we get

$$\mu(y)f(x\sigma(y)z) + \mu(yz)f(x\sigma(y)\sigma(z)) = 2\mu(y)f(x\sigma(y)) + 2\mu(y)f(z). \tag{9.128}$$

Adding (9.126) to (9.128) we obtain

$$[f(xyz) + \mu(yz)f(x\sigma(y)\sigma(z))] + [\mu(z)f(xy\sigma(z)) + \mu(y)f(x\sigma(y)z)]$$
$$= 2[f(xy) + \mu(y)f(x\sigma(y))] + 2[1 + \mu(y)]f(z). \tag{9.129}$$

Replacing y by yz in (9.125) we get

$$f(xyz) + \mu(yz)f(x\sigma(y)\sigma(z)) = 2f(x) + 2f(yz). \tag{9.130}$$

Multiplying (9.125) by 2 we derive

$$2[f(xy) + \mu(y)f(x\sigma(y))] = 4f(x) + 4f(y). \tag{9.131}$$

If we subtract (9.130) from the sum of (9.129) and (9.131) we obtain

$$\mu(z)f(xy\sigma(z)) + \mu(y)f(x\sigma(y)z) + 2f(yz) \tag{9.132}$$
$$= 2f(x) + 4f(y) + 2[1 + \mu(y)]f(z).$$

On the other hand, if we replace y by $y\sigma(z)$ in (9.125) we get

$$f(xy\sigma(z)) + \mu(y\sigma(z))f(x\sigma(y)\sigma^2(z)) = 2f(x) + 2f(y\sigma(z)). \tag{9.133}$$

Multiplying the last equality by $\mu(z)$ and using the fact that $\mu(z\sigma(z)) = 1$ we get

$$\mu(z)f(xy\sigma(z)) + \mu(y)f(x\sigma(y)\sigma^2(z)) = 2\mu(z)f(x) + 2\mu(z)f(y\sigma(z)). \tag{9.134}$$

Subtracting (9.134) from (9.132) we deduce that

$$\mu(y)[f(x\sigma(y)z) - f(x\sigma(y)\sigma^2(z))] + 2[f(yz) + \mu(z)f(y\sigma(z))] \tag{9.135}$$
$$= 2[1 - \mu(z)]f(x) + 4f(y) + 2(1 + \mu(y))f(z).$$

If we make the substitution (y, z) in (9.125) and multiply the result obtained by 2 we derive

$$2[f(yz) + \mu(z)f(y\sigma(z))] = 4[f(y) + f(z)]. \tag{9.136}$$

The subtraction of (9.136) from (9.135) implies after some simplification

$$\mu(y)[f(x\sigma(y)z) - f(x\sigma(y)\sigma^2(z))] = 2[1 - \mu(z)]f(x) + 2(\mu(y) - 1)f(z). \tag{9.137}$$

On the other hand, if we make the substitutions $(x\sigma(y), z)$ and $(x\sigma(y), \sigma(z))$ in (9.125) we get respectively

$$f(x\sigma(y)z) + \mu(z)f(x\sigma(y)\sigma(z)) = 2f(x\sigma(y)) + 2f(z). \tag{9.138}$$

$$f(x\sigma(y)\sigma(z)) + \mu(\sigma(z))f(x\sigma(y)\sigma^2(z)) = 2f(x\sigma(y)) + 2f(\sigma(z)). \tag{9.139}$$

Multiplying (9.139) by $\mu(z)$ and using the fact that $\mu(z\sigma(z)) = 1$ we get

$$\mu(z)f(x\sigma(y)\sigma(z)) + f(x\sigma(y)\sigma^2(z)) = 2\mu(z)f(x\sigma(y)) + 2\mu(z)f(\sigma(z)). \tag{9.140}$$

Subtracting (9.140) from (9.138) we obtain

$$f(x\sigma(y)z) - f(x\sigma(y)\sigma^2(z)) \tag{9.141}$$
$$= 2f(x\sigma(y))[1 - \mu(z)] + 2f(z) - 2\mu(z)f(\sigma(z)).$$

Multiplying the last equation by $\mu(y)$ we obtain

$$\mu(y)[f(x\sigma(y)z) - f(x\sigma(y)\sigma^2(z))] = 2\mu(y)[1 - \mu(z)]f(x\sigma(y)) \tag{9.142}$$
$$+ 2\mu(y)f(z) - 2\mu(yz)f(\sigma(z)).$$

Now, if we subtract (9.142) from (9.137) we deduce that

$$2[1 - \mu(z)]f(x) - 2f(z) = 2\mu(y)[1 - \mu(z)]f(x\sigma(y)) - 2\mu(yz)f(\sigma(z)),$$
(9.143)

from which we get

$$[1 - \mu(z)][f(x) - \mu(y)f(x\sigma(y))] = f(z) - \mu(yz)f(\sigma(z)).$$
(9.144)

Taking $y = z$ in (9.144) we obtain

$$[1 - \mu(y)][f(x) - \mu(y)f(x\sigma(y))] = f(y) - \mu(y^2)f(\sigma(y))$$
(9.145)

for all $x, y \in S$.

Setting $\beta(y) = 1 - \mu(y)$ and multiplying (9.125) by $\beta(y)$ and adding the result obtained to (9.145) we derive that

$$\beta(y)[f(xy) - f(x) - 2f(y)] = f(y) - \mu(y^2)f(\sigma(y)).$$
(9.146)

The last equation can be written as follows

$$\beta(y)f(xy) = \beta(y)f(x) + [2\beta(y) + 1]f(y) - \mu(y^2)f(\sigma(y)).$$
(9.147)

Replacing y in (9.147) by $\sigma(y)$, and multiplying the result obtained by $\mu(y^2)$ and using the fact that $\mu(z\sigma(z)) = 1$ we find

$$\mu(y^2)\beta(\sigma(y))f(x\sigma(y))) = \mu(y^2)\beta(\sigma(y))f(x)$$
(9.148)
$$+ \mu(y^2)[2\beta(\sigma(y)) + 1]f(\sigma(y)) - f(\sigma^2(y)).$$

Since $\mu(y\sigma(y)) = 1$ we get that

$$\mu(y)\beta(\sigma(y)) = \mu(y)[1 - \mu(\sigma(y))] = \mu(y) - 1 = -\beta(y),$$

and thus Eq. (9.148) can be written in the form

$$\mu(y)\beta(y)f(x\sigma(y)) = \mu(y)\beta(y)f(x)$$
(9.149)
$$+ [2\mu(y)\beta(y) - \mu(y^2)]f(\sigma(y)) + f(\sigma^2(y)).$$

Adding (9.149) and (9.147) and using (9.125) we get

$$\beta(y)[2f(x) + 2f(y)] = [\beta(y) + \mu(y)\beta(y)]f(x) + [2\beta(y) + 1]f(y)$$
(9.150)
$$+ [2\mu(y)\beta(y) - 2\mu(y^2)]f(\sigma(y)) + f(\sigma^2(y)).$$

Thus

$$\beta(y)f(x) = f(y) + 2\mu(y)[\beta(y) - \mu(y)]f(\sigma(y)) + f(\sigma^2(y)) \qquad (9.151)$$

for all x, y in S.

If $\mu \neq 1$ then there exists $y_0 \in S$ such that $\beta(y_0) \neq 0$ and from (9.151) we deduce that $f(x) = c$, for all $x \in S$, where

$$c = \frac{1}{\beta(y_0)}[f(y_0) + 2\mu(y_0)[\beta(y_0) - \mu(y_0)]f(\sigma(y_0)) + f(\sigma^2(y_0))],$$

which means that f is a constant. From (9.125) we deduce that $f = 0$, which contradicts the assumption that $f \neq 0$. This completes the proof of Theorem 9.8.

$\square$

9.7 Stability of the μ-Jensen Functional Equation

In this section we study the stability of μ-Jensen functional equation (9.8), where σ is a surjective homomorphism, and μ is a bounded multiplicative function such that $\mu(x\sigma(x)) = 1$ for all $x \in S$.

Theorem 9.9 *Let S be a semigroup, $\sigma : S \longrightarrow S$ be a homomorphism, and μ be a bounded multiplicative function such that $\mu(x\sigma(x)) = 1$ for all $x \in S$. If there exists a non-negative scalar δ such that*

$$|f(xy) + \mu(y)f(x\sigma(y)) - 2f(x)| \leq \delta \qquad (9.152)$$

for all x, $y \in S$, then either f is unbounded or $\mu = 1$.

Furthermore, the μ-Jensen functional equation (9.8) is stable if and only if the Jensen functional equation (1.1) is stable.

Proof Making the substitutions (xy, z), $(x\sigma(y), z)$ in (9.152) we get respectively

$$|f(xyz) + \mu(z)f(xy\sigma(z)) - 2f(xy)| \leq \delta, \qquad (9.153)$$

$$|f(x\sigma(y)z) + \mu(z)f(x\sigma(y)\sigma(z)) - 2f(x\sigma(y))| \leq \delta. \qquad (9.154)$$

The multiplicative mapping μ is bounded, thus there exists a nonnegative real M such that $|\mu(x)| \leq M$ for all $x \in S$. Multiplying (9.154) by $\mu(y)$ we get

$$|\mu(y)f(x\sigma(y)z) + \mu(yz)f(x\sigma(y)\sigma(z)) - 2\mu(y)f(x\sigma(y))| \leq M\delta. \qquad (9.155)$$

Adding (9.153) and (9.155) and using the triangle inequality we obtain

$$\|[f(xyz) + \mu(yz)f(x\sigma(y)\sigma(z))] + [\mu(z)f(xy\sigma(z)) + \mu(y)f(x\sigma(y)z)]$$
$$- 2[f(xy) + \mu(y)f(x\sigma(y))]\| \leq (1+M)\delta.$$
$$(9.156)$$

Replacing y by yz in (9.152) we obtain

$$|f(xyz) + \mu(yz)f(x\sigma(y)\sigma(z)) - 2f(x)| \leq \delta. \qquad (9.157)$$

Multiplying (9.152) by 2 we get

$$|2[f(xy) + \mu(y)f(x\sigma(y))] - 4f(x)| \leq 2\delta. \qquad (9.158)$$

If we subtract (9.157) from the sum of (9.156) and (9.158) and use the triangle inequality we obtain

$$|\mu(z)[f(xy\sigma(z)) + \mu(y\sigma(z))f(x\sigma(y)z) - 2f(x)| \leq (4+M)\delta. \qquad (9.159)$$

Multiplying the last inequality by $\mu(\sigma(z))$ and using the fact that $\mu(z\sigma(z)) = 1$ we get after some simplification

$$|f(xy\sigma(z)) + \mu(y\sigma(z))f(x\sigma(y)z) - 2\mu(\sigma(z))f(x)| \leq (4M + M^2)\delta.$$
$$(9.160)$$

On the other hand, if we replace y in (9.152) by $y\sigma(z)$ we get

$$|f(xy\sigma(z)) + \mu(y\sigma(z))f(x\sigma(y)\sigma^2(z)) - 2f(x)| \leq \delta. \qquad (9.161)$$

Subtracting (9.161) from (9.160) we deduce that

$$|\mu(y\sigma(z))[f(x\sigma(y)z) - f(x\sigma(y)\sigma^2(z)) - 2[\mu(\sigma(z)) - 1]f(x)| \qquad (9.162)$$
$$\leq (1 + 4M + M^2)\delta.$$

Multiplying the last identity by $\mu(\sigma(y)z)$ and using the fact that $\mu(z\sigma(z)) = 1$ we obtain

$$|f(x\sigma(y)z) - f(x\sigma(y)\sigma^2(z)) - 2\mu(\sigma(y))[1 - \mu(z)]f(x)| \qquad (9.163)$$
$$\leq (M + 4M^2 + M^3)\delta.$$

On the other hand, if we make the substitutions $(x\sigma(y), z)$ and $(x\sigma(y), \sigma(z))$ in (9.152) we get respectively

$$|f(x\sigma(y)z) + \mu(z)f(x\sigma(y)\sigma(z)) - 2f(x\sigma(y))| \leq \delta, \qquad (9.164)$$

$$|f(x\sigma(y)\sigma(z)) + \mu(\sigma(z))f(x\sigma(y)\sigma^2(z)) - 2f(x\sigma(y))| \leq \delta. \qquad (9.165)$$

Multiplying (9.165) by $\mu(z)$ and using $\mu(z\sigma(z)) = 1$ we derive that

$$|\mu(z)f(x\sigma(y)\sigma(z)) + f(x\sigma(y)\sigma^2(z)) - 2\mu(z)f(x\sigma(y))| \leq M\delta. \qquad (9.166)$$

Subtracting (9.166) from (9.164) and using the triangle inequality we obtain

$$|f(x\sigma(y)z) - f(x\sigma(y)\sigma^2(z)) - 2f(x\sigma(y))[1 - \mu(z)]| \leq (1 + M)\delta. \qquad (9.167)$$

If we subtract (9.167) from (9.163) we deduce that

$$|2[\mu(\sigma(y))[1 - \mu(z)]f(x) - 2f(x\sigma(y)))[1 - \mu(z)]| \qquad (9.168)$$
$$\leq (1 + 2M + 4M^2 + M^3)\delta,$$

from which we get

$$|[1 - \mu(z)][\mu(\sigma(y))f(x) - f(x\sigma(y))]| \leq (1 + 2M + 4M^2 + M^3)\frac{\delta}{2}. \qquad (9.169)$$

If we suppose that $\mu \neq 1$, then there exists $z_0 \in S$ such that $\mu(z_0) \neq 1$. From (9.169) we deduce that

$$|f(x\sigma(y)) - \mu(\sigma(y))f(x)| \leq \phi\delta \qquad (9.170)$$

for all $x, y \in S$, where

$$\phi = \frac{1}{2(1 - \mu(z_0))}(1 + 2M + 4M^2 + M^3).$$

If we multiply (9.170) by $\mu(y)$ and use the fact that $\mu(x\sigma(x)) = 1$, we obtain

$$|\mu(y)f(x\sigma(y)) - f(x)| \leq M\phi\delta. \qquad (9.171)$$

Subtracting (9.152) from (9.171) and using the triangle inequality we get

$$|f(xy) - f(x)| \leq M(\phi + 1)\delta \qquad (9.172)$$

for all $y \in S$. Replacing y by $\sigma(y)$ in (9.172) and multiplying the result by $\sigma(y)$ we obtain

$$|\mu(y)f(x\sigma(y)) - \mu(y)f(x)| \leq M^2(\phi + 1)\delta. \qquad (9.173)$$

Subtracting (9.152) from the sum of (9.172) and (9.173) and using the triangle inequality we deduce

$$|[1 - \mu(y)]f(x)| \le (M^2 + M)(\phi + 1)\delta + \delta. \tag{9.174}$$

Since $\mu \neq 1$ we deduce that f is a bounded function. This completes the proof of Theorem 9.9. $\qquad\square$

9.8 Stability of the μ-Quadratic Functional Equation

In this section we investigate the stability of the μ-quadratic functional equation (1.10).

Theorem 9.10 *Let S be a semigroup, let $\sigma : S \longrightarrow S$ be a homomorphism, and μ be a bounded multiplicative function such that $\mu(x\sigma(x)) = 1$. If there exists a non-negative scalar δ such that*

$$|f(xy) + \mu(y)f(x\sigma(y)) - 2f(x) - 2f(y)| \le \delta, \ x, y \in S, \tag{9.175}$$

then either f is unbounded or $\mu = 1$.

Furthermore, the μ-quadratic functional equation (1.10) is stable if and only if the quadratic functional equation (9.7) is stable.

Proof Making the substitutions (xy, z), $(x\sigma(y), z)$ in (9.175) we get respectively

$$|f(xyz) + \mu(z)f(xy\sigma(z)) - 2f(xy) - 2f(z)| \le \delta. \tag{9.176}$$

$$|f(x\sigma(y)z) + \mu(z)f(x\sigma(y)\sigma(z)) - 2f(x\sigma(y)) - 2f(z)| \le \delta. \tag{9.177}$$

Multiplying (9.177) by $\mu(y)$ we get

$$|\mu(y)f(x\sigma(y)z) + \mu(yz)f(x\sigma(y)\sigma(z)) - 2\mu(y)f(x\sigma(y)) - 2\mu(y)f(z)|$$
$$\le M\delta. \tag{9.178}$$

Adding (9.176) and (9.178) and using the triangle inequality we obtain

$$|[f(xyz) + \mu(yz)f(x\sigma(y)\sigma(z))] + [\mu(z)f(xy\sigma(z)) + \mu(y)f(x\sigma(y)z)]$$
$$- 2[f(xy) + \mu(y)f(x\sigma(y))] - 2[1 + \mu(y)]f(z)| \le (1 + M)\delta. \tag{9.179}$$

Replacing y by yz in (9.175) we get

$$|f(xyz) + \mu(yz)f(x\sigma(y)\sigma(z)) - 2f(x) - 2f(yz)| \leq \delta. \tag{9.180}$$

Multiplying (9.175) by 2 we get

$$|2[f(xy) + \mu(y)f(x\sigma(y))] - 4f(x) - 4f(y)| \leq 2\delta. \tag{9.181}$$

If we subtract (9.180) from the sum of (9.179) and (9.181) and use the triangle inequality we obtain

$$|\mu(z)f(xy\sigma(z)) + \mu(y)f(x\sigma(y)z) + 2f(yz) - 2f(x) - 4f(y) \tag{9.182}$$
$$-2[1 + \mu(y)]f(z)| \leq (4 + M)\delta.$$

On the other hand, if we replace y in (9.175) by $y\sigma(z)$ we deduce that

$$|f(xy\sigma(z)) + \mu(y\sigma(z))f(x\sigma(y)\sigma^2(z)) - 2f(x) - 2f(y\sigma(z))| \leq \delta. \tag{9.183}$$

Multiplying the last inequality by $\mu(z)$ and using the fact that $\mu(z\sigma(z)) = 1$ we get

$$|\mu(z)f(xy\sigma(z)) + \mu(y)f(x\sigma(y)\sigma^2(z)) - 2\mu(z)f(x) - 2\mu(z)f(y\sigma(z))|$$
$$\leq M\delta.$$
$$\tag{9.184}$$

Subtracting (9.184) from (9.182) and using the triangle inequality we obtain that

$$|\mu(y)[f(x\sigma(y)z) - f(x\sigma(y)\sigma^2(z))] + 2[f(yz) + \mu(z)f(y\sigma(z))] \tag{9.185}$$
$$- 2[1 - \mu(z)]f(x) - 4f(y) - 2(1 + \mu(y))f(z)| \leq (4 + 2M)\delta.$$

If we make the substitution (y, z) in (9.175) and multiply the result by 2 we obtain

$$|2[f(yz) + \mu(z)f(y\sigma(z))] - 4[f(y) + f(z)]| \leq 2\delta. \tag{9.186}$$

The subtraction of (9.186) from (9.185) and the triangle inequality provide after some simplification that

$$|\mu(y)[f(x\sigma(y)z) - f(x\sigma(y)\sigma^2(z))] + 2[\mu(z) - 1]f(x) \tag{9.187}$$
$$+2(1 - \mu(y))f(z)| \leq (6 + 2M)\delta.$$

On the other hand, if we make the substitutions $(x\sigma(y), z)$ and $(x\sigma(y), \sigma(z))$ in (9.175) we get respectively

$$|f(x\sigma(y)z) + \mu(z)f(x\sigma(y)\sigma(z)) - 2f(x\sigma(y)) - 2f(z)| \leq \delta. \tag{9.188}$$

$$|f(x\sigma(y)\sigma(z)) + \mu(\sigma(z))f(x\sigma(y)\sigma^2(z)) - 2f(x\sigma(y)) - 2f(\sigma(z))| \leq \delta.$$
(9.189)

Multiplying (9.189) by $\mu(z)$ and using the fact that $\mu(z\sigma(z)) = 1$ we get that

$$|\mu(z)f(x\sigma(y)\sigma(z)) + f(x\sigma(y)\sigma^2(z)) - 2\mu(z)f(x\sigma(y)) - 2\mu(z)f(\sigma(z))|$$
$$\leq M\delta.$$
(9.190)

Subtracting (9.190) from (9.188) and using the triangle inequality we obtain

$$|f(x\sigma(y)z) - f(x\sigma(y)\sigma^2(z)) - 2f(x\sigma(y))[1 - \mu(z)] - 2f(z)$$
(9.191)
$$+ 2\mu(z)f(\sigma(z))| \leq (1 + M)\delta.$$

Multiplying the last identity by $\mu(y)$ we obtain

$$|\mu(y)[f(x\sigma(y)z) - f(x\sigma(y)\sigma^2(z))] - 2\mu(y)[1 - \mu(z)]f(x\sigma(y))$$
(9.192)
$$- 2\mu(y)f(z) + 2\mu(yz)f(\sigma(z))| \leq (M + M^2)\delta.$$

If we subtract (9.192) from (9.187) and use the triangle inequality we obtain that

$$|2[\mu(z) - 1]f(x) + 2(1 - \mu(y))f(z) + 2\mu(y)[1 - \mu(z)]f(x\sigma(y))$$
(9.193)
$$+ 2\mu(y)f(z) - 2\mu(yz)f(\sigma(z))| \leq (6 + 3M + M^2)\delta,$$

from which we get

$$|[\mu(z) - 1][f(x) - \mu(y)f(x\sigma(y))] + f(z) - \mu(yz)f(\sigma(z))|$$
(9.194)
$$\leq (6 + 3M + M^2)\frac{\delta}{2}.$$

Setting $y = z$ in (9.194) we obtain

$$|\beta(y)[f(x) - \mu(y)f(x\sigma(y))] + f(y) - \mu(y^2)f(\sigma(y))| \leq \alpha$$
(9.195)

where $\beta(y) = \mu(y) - 1$ for all $y \in S$, and

$$\alpha = (6 + 3M + M^2)\frac{\delta}{2}.$$

Adding (9.195) to (9.175) multiplied by $\beta(y)$ and using the triangle inequality we obtain

$$|\beta(y)[f(xy) - f(x) - 2f(y)] + f(y) - \mu(y^2)f(\sigma(y))| \leq \alpha + (M + 1)\delta.$$
(9.196)

The last inequality can be written in the form

$$|\beta(y)f(xy) - \beta(y)f(x) - [2\beta(y) - 1]f(y) - \mu(y^2)f(\sigma(y))| \tag{9.197}$$
$$\leq \alpha + (M+1)\delta.$$

Replacing y in (9.197) by $\sigma(y)$, and multiplying the result by $\mu(y^2)$ and using the fact that $\mu(z\sigma(z)) = 1$ we derive

$$|\mu(y^2)\beta(\sigma(y))f(x\sigma(y)) - \mu(y^2)\beta(\sigma(y))f(x) \tag{9.198}$$
$$- \mu(y^2)[2\beta(\sigma(y)) - 1]f(\sigma(y)) - f(\sigma^2(y))| \leq M^2\alpha + (M^2 + M^3)\delta.$$

Since $\mu(y\sigma(y)) = 1$ we get that

$$\mu(y)\beta(\sigma(y)) = \mu(y)[\mu(\sigma(y)) - 1] = 1 - \mu(y) = -\beta(y),$$

and thus inequality (9.197) can be expressed as follows

$$|\mu(y)\beta(y)f(x\sigma(y)) - \mu(y)\beta(y)f(x) - [2\mu(y)\beta(y) + \mu(y^2)]f(\sigma(y))$$
$$+ f(\sigma^2(y))| \leq M^2\alpha + (M^2 + M^3)\delta.$$
$$\tag{9.199}$$

Subtracting (9.175) multiplied by $\beta(y)$ from the sum of (9.199) and (9.197) and using the triangle inequality we get

$$|\beta(y)[2f(x) + 2f(y)] - [\beta(y) + \mu(y)\beta(y)]f(x) - [2\beta(y) - 1]f(y)$$
$$- [2\mu(y)\beta(y) + 2\mu(y^2)]f(\sigma(y)) + f(\sigma^2(y))|$$
$$\leq (1 + M^2)\alpha + (M^3 + M^2 + M + 2)\delta. \tag{9.200}$$

Simplifying the last inequality we obtain

$$|\beta^2(y)f(x) - f(y) - 2\mu(y)f(\sigma(y)) - f(\sigma^2(y))| \tag{9.201}$$
$$\leq (1 + M^2)\alpha + (M^3 + M^2 + M + 2)\delta.$$

Using the triangle inequality we deduce that

$$|\beta^2(y)f(x)| \leq |f(y) + 2\mu(y)f(\sigma(y)) + f(\sigma^2(y))| \tag{9.202}$$
$$+ (M^2 + 1)\alpha + (M^3 + M^2 + M + 2)\delta$$

for all x, y in S.

If $\mu \neq 1$ then there exists $y_0 \in S$ such that $\beta(y_0) \neq 0$. From (9.202) we deduce that f is bounded. This completes the proof of Theorem 9.10. $\qquad\Box$

References

1. Aczél, J., Dhombres, J.: Functional Equations in Several Variables. With Applications to Mathematics, Information Theory and to the Natural and Social Sciences. Encyclopedia of Mathematics and Its Applications, vol. 31. Cambridge University Press, Cambridge (1989)
2. Aoki, T.: On the stability of the linear transformation in Banach spaces. J. Math. Soc. Japan **2**, 64–66 (1950)
3. Bouikhalene, B., Elqorachi, E., Rassias, Th.M.: On the generalized Hyers-Ulam stability of the quadratic functional equation with a general involution. Nonlinear Funct. Anal. Appl. **12**(2), 247–262 (2007)
4. Bouikhalene, B., Elqorachi, E., Redouani, A.: Hyers-Ulam stability of the generalierd quadratic functional equation in Amenables semigroups. J. Inequal. Pure Appl. Math. (JIPAM) **8**(2), 18 pp. (2007). Article 56
5. Cholewa, P.W.: Remarks on the stability of functional equations. Aequationes Math. **27**, 76–86 (1984)
6. Czerwik, S.: On the stability of the quadratic mapping in normed spaces. Abh. Math. Sem. Univ. Hamburg **62**, 59–64 (1992)
7. Dilian, Y.: Contributions to the theory of functional equations. PhD Thesis, University of Waterloo, Waterloo, Ontario, Canada (2006)
8. Elqorachi, E., Rassias M.Th.: Generalized Hyers-Ulam stability of trigonometric functional equations. Mathematics **6**(5), 11 pp. (2018)
9. Elqorachi, E., Manar, Y., Rassias, Th.M.: Hyers-Ulam stability of Wilson's functional equation. In: Pardalos, P.M., Rassias, Th.M. (eds.) Contributions in Mathematics and Engineering: Honor of Constantin Carathéodory. Springer, New York (2016)
10. Elqorachi, E., Redouani, A.: Solutions and stability of a variant of Wilson's functional equation. Proyecciones J. Math. **37**(2), 317–344 (2018)
11. Faiziev, V.A., Sahoo. P.K.: On the stability of Jensen's functional equation on groups. Proc. Indian Acad. Sci. (Math. Sci.) **117**, 31–48 (2007)
12. Forti, G.L.: Hyers-Ulam stability of functional equations in several variables. Aequationes Math. **50**, 143–190 (1995)
13. Gadja, Z.: On stability of additive mapping. Int. J. Math. Math. Sci. **14**, 431–434 (1991)
14. Găvruta, P.: A generalization of the Hyers-Ulam-Rassias stability of approximately additive mappings. J. Math. Anal. Appl. **184**, 431–436 (1994)
15. Greenleaf, F.P.: Invariant Means on topological Groups and their Applications. Van Nostrand, New York (1969)
16. Hyers, D.H.: On the stability of the linear functional equation. Proc. Nat. Acad. Sci. U. S. A. **27**, 222–224 (1941)
17. Hyers, D.H., Isac, G.I., Rassias, Th.M.: Stability of Functional Equations in Several Variables. Birkhäuser, Basel (1998)
18. Jung, S-M.: Hyers-Ulam-Rassias stability of Jensen's equation and its application. Proc. Amer. Math. Soc. **126**, 3137–3134 (1998)
19. Jung, S.-M.: Hyers-Ulam-Rassias Stability of Functional Equations in Nonlinear Analysis, vol. 48. Springer, New York (2011)
20. Kim, J.H., The stability of d'Alembert and Jensen type functional equations. J. Math. Anal. Appl. **325**, 237–248 (2007)
21. Ng, C.T.: Jensen's functional equation on groups, III. Aequationes Math. **62**(1–2), 143–159 (2001)

22. Rassias, Th.M.: On the stability of linear mapping in Banach spaces. Proc. Amer. Math. Soc. **72**, 297–300 (1978)
23. Rassias, J.M.: On approximation of approximately linear mappings by linear mappings. J. Funct. Anal. **46**, 126–120 (1982)
24. Rassias, M.Th.: Solution of a functional equation problem of Steven Butler. Octogon Math. Mag. **12**, 152–153 (2004)
25. Stetkær, H.: Functional equations on abelian groups with involution. Aequationes Math. **54**, 144–172 (1997)
26. Stetkær, H.: Functional Equations on Groups. World Scientific Publishing, New Jersey (2013)
27. Stetkær, H.: D'Alember's functional equations on groups. Banach Cent. Publ. **99**, 173–191 (2013)
28. Stetkær, H.: A variant of d'Alembert's functional equation. Aequationes Math. **89**(3), 657–662 (2015)
29. Stetkær, H.: A note on Wilson's functional equation. Aequationes Math. **91**(5), 945–947 (2017)
30. Szekelyhidi, L.: Note on a stability theorem. Can. Math. Bull. **25**(4), 500–501 (1982)
31. Ulam, S.M.: A Collection of Mathematical Problems. Interscience Publications, New york (1961). Problems in Modern Mathematics (Wiley, New York, 1964)

Chapter 10
Bi-additive s-Functional Inequalities and Quasi-∗-Multipliers on Banach ∗-Algebras

Jung Rye Lee, Choonkil Park, and Themistocles M. Rassias

Abstract Park introduced and investigated the following bi-additive s-functional inequalities

$$\|f(x+y, z+w)+f(x+y, z-w)+f(x-y, z+w)+f(x-y, z-w)-4f(x, z)\|$$
$$\leq \left\|s\left(4f\left(\tfrac{x+y}{2}, z-w\right)+4f\left(\tfrac{x-y}{2}, z+w\right)-4f(x, z)+4f(y, w)\right)\right\|, \quad (10.1)$$

$$\left\|4f\left(\tfrac{x+y}{2}, z-w\right)+4f\left(\tfrac{x-y}{2}, z+w\right)-4f(x, z)+4f(y, w)\right\|$$
$$\leq \|s(f(x+y, z+w)+f(x+y, z-w)+f(x-y, z+w) \quad (10.2)$$
$$+f(x-y, z-w)-4f(x, z))\|,$$

where s is a fixed nonzero complex number with $|s| < 1$. Using the direct method, we prove the Hyers-Ulam stability of quasi-∗-multipliers on Banach ∗-algebras and unital C^*-algebras, associated to the bi-additive s-functional inequalities (10.1) and (10.2).

Keywords Quasi-multiplier on C^*-algebra · Quasi-∗-multiplier on Banach algebra · Hyers-Ulam stability · Direct method · Bi-additive s-functional inequality

J. R. Lee
Daejin University, Pocheon-si, South Korea
e-mail: jrlee@daejin.ac.kr

C. Park
Hanyang University, Seoul, South Korea
e-mail: baak@hanyang.ac.kr

T. M. Rassias (✉)
Department of Mathematics, National Technical University of Athens, Athens, Greece
e-mail: trassias@math.ntua.gr

© Springer Nature Switzerland AG 2019
J. Brzdęk et al. (eds.), *Ulam Type Stability*,
https://doi.org/10.1007/978-3-030-28972-0_10

Mathematics Subject Classification (2010) Primary 39B52, 39B82, 43A22; Secondary 39B62, 46L05

10.1   Introduction and Preliminaries

The stability problem of functional equations originated from a question of Ulam [23] concerning the stability of group homomorphisms. The functional equation

$$f(x + y) = f(x) + f(y)$$

is called the *Cauchy equation*. In particular, every solution of the Cauchy equation is said to be an *additive mapping*. Hyers [13] gave a first affirmative partial answer to the question of Ulam for Banach spaces. Hyers' Theorem was generalized by Aoki [3] for additive mappings and by Rassias [21] for linear mappings by considering an unbounded Cauchy difference. A generalization of the Rassias theorem was obtained by Găvruta [10] by replacing the unbounded Cauchy difference by a general control function in the spirit of Rassias' approach.

Gilányi [11] showed that if f satisfies the functional inequality

$$\|2f(x) + 2f(y) - f(x - y)\| \le \|f(x + y)\| \tag{10.3}$$

then f satisfies the Jordan-von Neumann functional equation

$$2f(x) + 2f(y) = f(x + y) + f(x - y).$$

See also [22]. Fechner [9] and Gilányi [12] proved the Hyers-Ulam stability of the functional inequality (10.3).

Park [17, 18] defined additive ρ-functional inequalities and proved the Hyers-Ulam stability of the additive ρ-functional inequalities in Banach spaces and non-Archimedean Banach spaces. The stability problems of various functional equations and functional inequalities have been extensively investigated by a number of authors (see [2, 4–8, 19]).

The notion of a quasi-multiplier is a generalization of the notion of a multiplier on a Banach algebra, which was introduced by Akemann and Pedersen [1] for C^*-algebras. McKennon [15] extended the definition to a general complex Banach algebra with bounded approximate identity as follows.

Definition 10.1 ([15]) Let A be a complex Banach algebra. A **C**-bilinear mapping $P : A \times A \to A$ is called a *quasi-multiplier* on A if P satisfies

$$P(xy, zw) = x P(y, z)w$$

for all $x, y, z, w \in A$.

Definition 10.2 Let A be a complex Banach $*$-algebra. A bi-additive mapping P : $A \times A \to A$ is called a *quasi-$*$-multiplier* on A if P is **C**-linear in the first variable and satisfies

$$P(x, z) = P(z, x)^*,$$

$$P(xy, z) = x P(y, z)$$

for all $x, y, z \in A$.

It is easy to show that if P is a quasi-$*$-multiplier, then P is conjugate **C**-linear in the second variable and $P(xy, zw) = x P(y, w) z^*$ for all $x, y, z, w \in A$.

This paper is organized as follows: In Sects. 10.2 and 10.3, we prove the Hyers-Ulam stability of the bi-additive s-functional inequalities (10.1) and (10.2) in complex Banach spaces by using the direct method. In Sect. 10.4, we prove the Hyers-Ulam stability and the superstability of quasi-$*$-multipliers on Banach $*$-algebras and unital C^*-algebras associated to the bi-additive s-functional inequalities (10.1) and (10.2).

Throughout this paper, let X be a complex normed space and Y a complex Banach space. Let A be a complex Banach $*$-algebra. Assume that s is a fixed nonzero complex number with $|s| < 1$.

10.2 Bi-additive s-Functional Inequality (10.1)

Park [20] solved the bi-additive s-functional inequality (10.1) in complex normed spaces.

Lemma 10.1 ([20, Lemma 2.1]) *If a mapping* $f : X^2 \to Y$ *satisfies* $f(0, z) = f(x, 0) = 0$ *and*

$$\| f(x+y, z+w) + f(x+y, z-w) + f(x-y, z+w) + f(x-y, z-w) - 4f(x, z) \|$$

$$\leq \left\| s \left(4f\left(\frac{x+y}{2}, z-w\right) + 4f\left(\frac{x-y}{2}, z+w\right) - 4f(x, z) + 4f(y, w) \right) \right\|$$

$$(10.4)$$

for all $x, y, z, w \in X$, *then* $f : X^2 \to Y$ *is bi-additive.*

Using the direct method, we prove the Hyers-Ulam stability of the bi-additive s-functional inequality (10.4) in complex Banach spaces.

Theorem 10.1 *Let* $\varphi : X^2 \to [0, \infty)$ *be a function such that*

$$\Psi(x, y) := \sum_{j=1}^{\infty} 2^j \varphi\left(\frac{x}{2^j}, \frac{y}{2^j}\right) < \infty \qquad (10.5)$$

for all $x, y \in X$. Let $f : X^2 \to Y$ be a mapping satisfying $f(x, 0) = f(0, z) = 0$
and

$$\|f(x+y, z+w)+f(x+y, z-w)+f(x-y, z+w)+f(x-y, z-w)-4f(x, z)\|$$

$$\leq \left\| s\left(4f\left(\frac{x+y}{2}, z-w\right) + 4f\left(\frac{x-y}{2}, z+w\right) - 4f(x, z) + 4f(y, w)\right)\right\|$$

$$+\varphi(x, y)\varphi(z, w) \tag{10.6}$$

for all $x, y, z, w \in X$. Then there exists a unique bi-additive mapping $B : X^2 \to Y$
such that

$$\|f(x, z) - B(x, z)\| \leq \frac{1}{4}\Psi(x, x)\varphi(z, 0) \tag{10.7}$$

for all $x, z \in X$.

Proof Letting $w = 0$ and $y = x$ in (10.6), we get

$$\|2f(2x, z) - 4f(x, z)\| \leq \varphi(x, x)\varphi(z, 0) \tag{10.8}$$

for all $x, z \in X$. So

$$\left\|f(x, z) - 2f\left(\frac{x}{2}, z\right)\right\| \leq \frac{1}{2}\varphi\left(\frac{x}{2}, \frac{x}{2}\right)\varphi(z, 0)$$

for all $x, z \in X$. Hence

$$\left\|2^l f\left(\frac{x}{2^l}, z\right) - 2^m f\left(\frac{x}{2^m}, z\right)\right\| \leq \sum_{j=l}^{m-1} \left\|2^j f\left(\frac{x}{2^j}, z\right) - 2^{j+1} f\left(\frac{x}{2^{j+1}}, z\right)\right\|$$

$$\leq \frac{1}{4} \sum_{j=l}^{m-1} 2^j \varphi\left(\frac{x}{2^j}, \frac{x}{2^j}\right)\varphi(z, 0) \tag{10.9}$$

for all nonnegative integers m and l with $m > l$ and all $x, z \in X$. It follows
from (10.9) that the sequence $\{2^k f(\frac{x}{2^k}, z)\}$ is Cauchy for all $x, z \in X$. Since Y is a
Banach space, the sequence $\{2^k f(\frac{x}{2^k}, z)\}$ converges. So one can define the mapping
$B : X^2 \to Y$ by

$$B(x, z) := \lim_{k \to \infty} 2^k f\left(\frac{x}{2^k}, z\right)$$

for all $x, z \in X$. Moreover, letting $l = 1$ and passing the limit $m \to \infty$ in (10.9),
we get (10.7).

It follows from (10.5) and (10.6) that

$$\|B(x + y, z+w)+B(x+y, z-w)+B(x-y, z+w)+B(x-y, z-w)-4B(x, z)\|$$
$$= \lim_{n\to\infty} \left\| 2^n \left(f\left(\frac{x+y}{2^n}, z+w\right) +f\left(\frac{x+y}{2^n}, z-w\right) +f\left(\frac{x-y}{2^n}, z+w\right) \right.\right.$$
$$\left.\left. +f\left(\frac{x-y}{2^n}, z-w\right) -4f\left(\frac{x}{2^n}, z\right) \right) \right\|$$
$$\leq \lim_{n\to\infty} \left\| 2^n s \left(4f\left(\frac{x+y}{2^{n+1}}, z-w\right) +4f\left(\frac{x-y}{2^{n+1}}, z+w\right) -4f\left(\frac{x}{2^n}, z\right) \right.\right.$$
$$\left.\left. +4f\left(\frac{y}{2^n}, w\right) \right) \right\| + \lim_{n\to\infty} 2^n \varphi\left(\frac{x}{2^n}, \frac{y}{2^n}\right) \varphi(z, w)$$
$$\leq \left\| s \left(4B\left(\frac{x+y}{2}, z-w\right) +4B\left(\frac{x-y}{2}, z+w\right) -4B(x, z)+4B(y, w) \right) \right\|$$

for all $x, y, z, w \in X$. So

$$\|B(x+ y, z+w)+B(x+y, z-w)+B(x-y, z+w)+B(x-y, z-w)-4B(x, z)\|$$
$$\leq \left\| s \left(4B\left(\frac{x+y}{2}, z-w\right) +4B\left(\frac{x-y}{2}, z+w\right) -4B(x, z)+4B(y, w) \right) \right\|$$

for all $x, y, z, w \in X$. By Lemma 10.1, the mapping $B : X^2 \to Y$ is bi-additive.

Now, let $T : X^2 \to Y$ be another bi-additive mapping satisfying (10.7). Then we have

$$\|B(x, z)-T(x, z)\| = \left\| 2^q B\left(\frac{x}{2^q}, z\right) - 2^q T\left(\frac{x}{2^q}, z\right) \right\|$$
$$\leq \left\| 2^q A\left(\frac{x}{2^q}, z\right) - 2^q f\left(\frac{x}{2^q}, z\right) \right\| + \left\| 2^q T\left(\frac{x}{2^q}, z\right) - 2^q f\left(\frac{x}{2^q}, z\right) \right\|$$
$$\leq \frac{2^q}{2} \Psi\left(\frac{x}{2^q}, \frac{x}{2^q}\right) \varphi(z, 0),$$

which tends to zero as $q \to \infty$ for all $x, z \in X$. So we can conclude that $B(x, z) = T(x, z)$ for all $x, z \in X$. This proves the uniqueness of A, as desired.

Corollary 10.1 *Let $r > 1$ and θ be nonnegative real numbers and let $f : X^2 \to Y$ be a mapping satisfying $f(x, 0) = f(0, z) = 0$ and*

$$\|f(x+y, z+w)+f(x+y, z-w)+f(x-y, z+w)+f(x-y, z-w)-4f(x, z)\|$$
$$\leq \left\| s \left(4f\left(\frac{x+y}{2}, z-w\right) +4f\left(\frac{x-y}{2}, z+w\right) -4f(x, z)+4f(y, w) \right) \right\|$$
$$+\theta(\|x\|^r + \|y\|^r)(\|z\|^r + \|w\|^r) \qquad (10.10)$$

for all $x, y, z, w \in X$. *Then there exists a unique bi-additive mapping* $B : X^2 \to Y$ *such that*

$$\| f(x, z) - B(x, z) \| \le \frac{\theta}{2^r - 2} \|x\|^r \|z\|^r$$

for all $x, z \in X$.

Proof The proof follows from Theorem 10.1 by taking $\varphi(x, y) = \sqrt{\theta}(\|x\|^r + \|y\|^r)$ for all $x, y \in X$.

Theorem 10.2 *Let* $\varphi : X^2 \to [0, \infty)$ *be a function such that*

$$\Psi(x, y) := \sum_{j=0}^{\infty} \frac{1}{2^j} \varphi(2^j x, 2^j y) < \infty \tag{10.11}$$

for all $x, y \in X$. *Let* $f : X^2 \to Y$ *be a mapping satisfying (10.6) and* $f(x, 0) = f(0, z) = 0$ *for all* $x, z \in X$. *Then there exists a unique bi-additive mapping* $B : X^2 \to Y$ *such that*

$$\| f(x, z) - B(x, z) \| \le \frac{1}{4} \Psi(x, x) \varphi(z, 0)$$

for all $x, z \in X$.

Proof It follows from (10.8) that

$$\left\| f(x, z) - \frac{1}{2} f(2x, z) \right\| \le \frac{1}{4} \varphi(x, x) \varphi(z, 0)$$

for all $x, z \in X$.

The rest of the proof is similar to the proof of Theorem 10.1.

Corollary 10.2 *Let* $r < 1$ *and* θ *be nonnegative real numbers and let* $f : X^2 \to Y$ *be a mapping satisfying (10.10) and* $f(x, 0) = f(0, z) = 0$ *for all* $x, z \in X$. *Then there exists a unique bi-additive mapping* $B : X^2 \to Y$ *such that*

$$\| f(x, z) - B(x, z) \| \le \frac{\theta}{2 - 2^r} \|x\|^r \|z\|^r$$

for all $x, z \in X$.

Proof The proof follows from Theorem 10.2 by taking $\varphi(x, y) = \sqrt{\theta}(\|x\|^r + \|y\|^r)$ for all $x, y \in X$.

10.3   Bi-additive *s*-Functional Inequality (10.2)

Park [20] solved the bi-additive *s*-functional inequality (10.2) in complex normed spaces.

Lemma 10.2 ([20, Lemma 3.1]) *If a mapping* $f : X^2 \to Y$ *satisfies* $f(0, z) = f(x, 0) = 0$ *and*

$$\left\| 4f\left(\frac{x+y}{2}, z-w\right) + 4f\left(\frac{x-y}{2}, z+w\right) - 4f(x, z) + 4f(y, w) \right\|$$

$$\leq \| s(f(x+y, z+w) + f(x+y, z-w) \tag{10.12}$$

$$+ f(x-y, z+w) + f(x-y, z-w) - 4f(x, z)) \|$$

for all $x, y, z, w \in X$, *then* $f : X^2 \to Y$ *is bi-additive.*

Using the direct method, we prove the Hyers-Ulam stability of the bi-additive *s*-functional inequality (10.12) in complex Banach spaces.

Theorem 10.3 *Let* $\varphi : X^2 \to [0, \infty)$ *be a function satisfying (10.5). Let* $f : X^2 \to Y$ *be a mapping satisfying* $f(x, 0) = f(0, z) = 0$ *and*

$$\left\| 4f\left(\frac{x+y}{2}, z-w\right) + 4f\left(\frac{x-y}{2}, z+w\right) - 4f(x, z) + 4f(y, w) \right\| \tag{10.13}$$

$$\leq \| s(f(x+y, z+w) + f(x+y, z-w) + f(x-y, z+w)$$

$$+ f(x-y, z-w) - 4f(x, z) \| + \varphi(x, y)\varphi(z, w)$$

for all $x, y, z, w \in X$. *Then there exists a unique bi-additive mapping* $B : X^2 \to Y$ *such that*

$$\| f(x, z) - B(x, z) \| \leq \frac{1}{8} \Psi(2x, 0)\varphi(z, 0) \tag{10.14}$$

for all $x, z \in X$.

Proof Letting $y = w = 0$ in (10.13), we get

$$\left\| 8f\left(\frac{x}{2}, z\right) - 4f(x, z) \right\| \leq \varphi(x, 0)\varphi(z, 0) \tag{10.15}$$

for all $x, z \in X$.

The rest of the proof is similar to the proof of Theorem 10.1.

Corollary 10.3 *Let* $r > 1$ *and* θ *be nonnegative real numbers and let* $f : X^2 \to Y$ *be a mapping satisfying* $f(x, 0) = f(0, z) = 0$ *and*

$$\left\| 4f\left(\frac{x+y}{2}, z-w\right) + 4f\left(\frac{x-y}{2}, z+w\right) - 4f(x, z) + 4f(y, w) \right\| \tag{10.16}$$

$$\leq \|s(f(x + y, z + w) + f(x + y, z - w) + f(x - y, z + w)$$
$$+ f(x - y, z - w) - 4f(x, z))\| + \theta(\|x\|^r + \|y\|^r)(\|z\|^r + \|w\|^r)$$

for all $x, y, z, w \in X$. Then there exists a unique bi-additive mapping $B : X^2 \to Y$
such that

$$\|f(x, z) - B(x, z)\| \leq \frac{2^{r-2}\theta}{2^r - 2} \|x\|^r \|z\|^r$$

for all $x, z \in X$.

Proof The proof follows from Theorem 10.3 by taking $\varphi(x, y) = \sqrt{\theta}(\|x\|^r + \|y\|^r)$
for all $x, y \in X$.

Theorem 10.4 *Let* $\varphi : X^2 \to [0, \infty)$ *be a function satisfying (10.11). Let* $f :$
$X^2 \to Y$ *be a mapping satisfying (10.13) and* $f(x, 0) = f(0, z) = 0$ *for all*
$x, z \in X$. *Then there exists a unique bi-additive mapping* $B : X^2 \to Y$ *such that*

$$\|f(x, z) - B(x, z)\| \leq \frac{1}{8}\Psi(2x, 0)\,\varphi(z, 0) \qquad (10.17)$$

for all $x, z \in X$.

Proof It follows from (10.15) that

$$\left\|f(x, z) - \frac{1}{2}f(2x, z)\right\| \leq \frac{1}{8}\varphi(2x, 0)\varphi(z, 0)$$

for all $x, z \in X$.

The rest of the proof is similar to the proofs of Theorems 10.1 and 10.3.

Corollary 10.4 *Let* $r < 1$ *and* θ *be nonnegative real numbers and let* $f : X^2 \to Y$
be a mapping satisfying (10.16) and $f(x, 0) = f(0, z) = 0$ *for all* $x, z \in X$. *Then*
there exists a unique bi-additive mapping $B : X^2 \to Y$ *such that*

$$\|f(x, z) - B(x, z)\| \leq \frac{\theta}{4(2 - 2^r)} \|x\|^r \|z\|^r$$

for all $x, z \in X$.

Proof The proof follows from Theorem 10.4 by taking $\varphi(x, y) = \sqrt{\theta}(\|x\|^r + \|y\|^r)$
for all $x, y \in X$.

10.4   Quasi-∗-Multipliers in *C**-Algebras

In this section, we investigate quasi-∗-multipliers on complex Banach ∗-algebras and unital *C**-algebras associated to the bi-additive *s*-functional inequalities (10.4) and (10.12).

Theorem 10.5 *Let* $\varphi : A^2 \to [0, \infty)$ *be a function such that there exists an* $L < 1$ *with*

$$\Psi(x, y) := \sum_{j=1}^{\infty} 2^j \varphi\left(\frac{x}{2^j}, \frac{y}{2^j}\right) < \infty \tag{10.18}$$

for all $x, y \in A$. *Let* $f : A^2 \to A$ *be a mapping satisfying* $f(x, 0) = f(0, z) = 0$ *and*

$$\|f(\lambda(x + y), z + w) + f(\lambda(x + y), z - w) + f(\lambda(x - y), z + w)$$
$$+ f(\lambda(x - y), z - w) - 4\lambda f(x, z)\| \tag{10.19}$$
$$\leq \left\| s\left(4f\left(\frac{x+y}{2}, z-w\right) + 4f\left(\frac{x-y}{2}, z+w\right) - 4f(x, z) + 4f(y, w)\right)\right\|$$
$$+ \varphi(x, y)\varphi(z, w)$$

for all $\lambda \in \mathbf{T}^1 := \{v \in \mathbf{C} : |v| = 1\}$ *and all* $x, y, z, w \in A$. *Then there exists a unique bi-additive mapping* $B : A^2 \to A$, *which is* **C**-*linear in the first variable, such that*

$$\|f(x, z) - B(x, z)\| \leq \frac{1}{4}\Psi(x, x)\varphi(z, 0) \tag{10.20}$$

for all $x, z \in A$.
  Furthermore, if, in addition, the mapping $f : A^2 \to A$ *satisfies* $f(2x, z) = 2f(x, z)$ *and*

$$\|f(xy, z) - xf(y, z)\| \leq \varphi(x, y)^2 \varphi(z, 0), \tag{10.21}$$
$$\|f(x, z) - f(z, x)^*\| \leq \varphi(x, 0)\varphi(z, 0) \tag{10.22}$$

for all $x, y, z \in A$, *then the mapping* $f : A^2 \to A$ *is a quasi-∗-multiplier.*

Proof Let $\lambda = 1$ in (10.19). By Theorem 10.1, there is a unique bi-additive mapping $B : A^2 \to A$ satisfying (10.20) defined by

$$B(x, z) := \lim_{n \to \infty} 2^n f\left(\frac{x}{2^n}, z\right)$$

for all $x, z \in A$.

If $f(2x, z) = 2f(x, z)$ for all $x, z \in A$, then we can easily show that $B(x, z) = f(x, z)$ for all $x, z \in A$.

Letting $y = x$ and $w = 0$ in (10.19), we get

$$\|2f(2\lambda x, z) - 4\lambda f(x, z)\| \leq \varphi(x, x)\varphi(z, 0)$$

for all $x, z \in A$ and all $\lambda \in \mathbf{T}^1$. So

$$\|2B(2\lambda x, z) - 4\lambda B(x, z)\| = \lim_{n\to\infty} 2^n \left\| 2f\left(2\lambda \frac{x}{2^n}, z\right) - 4\lambda f\left(\frac{x}{2^n}, z\right) \right\|$$

$$\leq \lim_{n\to\infty} 2^n \varphi\left(\frac{x}{2^n}, \frac{x}{2^n}\right) \varphi(z, 0) = 0$$

for all $x, z \in A$ and all $\lambda \in \mathbf{T}^1$. Hence $2B(2\lambda x, z) = 4\lambda B(x, z)$ and so $B(\lambda x, z) = \lambda B(x, z)$ for all $x, z \in A$ and all $\lambda \in \mathbf{T}^1$. By [16, Theorem 2.1], the bi-additive mapping $B : A^2 \to A$ is $\mathbf{C}$-linear in the first variable.

It follows from (10.21) that

$$\|B(xy, z) - xB(y, z)\| = \lim_{n\to\infty} 4^n \left\| f\left(\frac{xy}{2^n \cdot 2^n}, z\right) - \frac{x}{2^n} f\left(\frac{y}{2^n}, z\right) \right\|$$

$$\leq \lim_{n\to\infty} 4^n \varphi\left(\frac{x}{2^n}, \frac{y}{2^n}\right)^2 \varphi(z, 0) = 0$$

for all $x, y, z \in A$. Thus

$$B(xy, z) = xB(y, z)$$

for all $x, y, z \in A$.

It follows from (10.22) that

$$\|B(x, z) - B(z, x)^*\| = \lim_{n\to\infty} 2^n \left\| f\left(x, \frac{z}{2^n}\right) - f\left(\frac{z}{2^n}, x\right)^* \right\|$$

$$\leq \lim_{n\to\infty} 2^n \varphi\left(\frac{x}{2^n}, 0\right) \varphi(z, 0) = 0$$

for all $x, z \in A$. Thus

$$B(x, z) = B(z, x)^*$$

for all $x, z \in A$. Hence the mapping $f : A^2 \to A$ is a quasi-$*$-multiplier.

Corollary 10.5 *Let $r > 2$ and θ be nonnegative real numbers, and let $f : A^2 \to A$ be a mapping satisfying $f(x, 0) = f(0, z) = 0$ and*

$$\|f(\lambda(x+y), z+w) + f(\lambda(x+y), z-w) + f(\lambda(x-y), z+w)$$

$$+f(\lambda(x-y), z-w) - 4\lambda f(x, z)\| \qquad (10.23)$$

$$\leq \left\| s\left(4f\left(\tfrac{x+y}{2}, z-w\right) + 4f\left(\tfrac{x-y}{2}, z+w\right) - 4f(x, z) + 4f(y, w)\right)\right\|$$

$$+\theta(\|x\|^r + \|y\|^r)(\|z\|^r + \|w\|^r)$$

for all $\lambda \in \mathbf{T}^1$ and all $x, y, z, w \in A$. Then there exists a unique bi-additive mapping $B : A^2 \to A$, which is **C**-linear in the first variable, such that

$$\|f(x, z) - B(x, z)\| \leq \frac{\theta}{2^r - 2} \|x\|^r \|z\|^r \qquad (10.24)$$

for all $x, z \in A$.

   If, in addition, the mapping $f : A^2 \to A$ satisfies $f(2x, z) = 2f(x, z)$ and

$$\|f(xy, z) - xf(y, z)\| \leq \theta(\|x\|^r + \|y\|^r)\|z\|^r, \qquad (10.25)$$

$$\|f(x, z) - f(z, x)^*\| \leq \theta \|x\|^r \|z\|^r \qquad (10.26)$$

for all $x, y, z \in A$, then the mapping $f : A^2 \to A$ is a quasi-$*$-multiplier.

Proof The proof follows from Theorem 10.5 by taking $\varphi(x, y) = \sqrt{\theta}(\|x\|^r + \|y\|^r)$ for all $x, y \in A$.

Theorem 10.6 *Let $\varphi : A^2 \to [0, \infty)$ be a function such that*

$$\Psi(x, y) := \sum_{j=0}^{\infty} \frac{1}{2^j}\varphi(2^j x, 2^j y) < \infty \qquad (10.27)$$

*for all $x, y \in A$. Let $f : A^2 \to A$ be a mapping satisfying (10.19) and $f(x, 0) = f(0, z) = 0$ for all $x, z \in A$. Then there exists a unique bi-additive mapping $B : A^2 \to A$, which is **C**-linear in the first variable, such that*

$$\|f(x, z) - B(x, z)\| \leq \frac{1}{4}\Psi(x, x)\,\varphi(z, 0) \qquad (10.28)$$

for all $x, z \in A$.

   If, in addition, the mapping $f : A^2 \to A$ satisfies (10.21), (10.22) and $f(2x, z) = 2f(x, z)$ for all $x, z \in A$, then the mapping $f : A^2 \to A$ is a quasi-$*$-multiplier.

Proof The proof is similar to the proof of Theorem 10.5.

Corollary 10.6 *Let $r < 1$ and θ be nonnegative real numbers, and let $f : A^2 \to A$ be a mapping satisfying (10.23) and $f(x, 0) = f(0, z) = 0$ for all $x, z \in A$. Then* *there exists a unique bi-additive mapping* $B : A^2 \rightarrow A$, *which is* **C**-*linear in the first variable, such that*

$$\|f(x, z) - B(x, z)\| \leq \frac{\theta}{2 - 2^r} \|x\|^r \|z\|^r \tag{10.29}$$

for all $x, z \in A$.

If, in addition, the mapping $f : A^2 \rightarrow A$ *satisfies* (10.25), (10.26) *and* $f(2x, z) = 2f(x, z)$ *for all* $x, z \in A$, *then the mapping* $f : A^2 \rightarrow A$ *is a quasi-*-multiplier.*

Proof The proof follows from Theorem 10.6 by taking $\varphi(x, y) = \sqrt{\theta}(\|x\|^r + \|y\|^r)$ for all $x, y \in A$.

Similarly, we can obtain the following results.

Theorem 10.7 *Let* $\varphi : A^2 \rightarrow [0, \infty)$ *be a function satisfying* (10.18) *and let* $f : A^2 \rightarrow A$ *be a mapping satisfying* $f(x, 0) = f(0, z) = 0$ *and*

$$\left\| 4f\left(\lambda\frac{x+y}{2}, z-w\right) + 4f\left(\lambda\frac{x-y}{2}, z+w\right) - 4\lambda f(x, z) + 4\lambda f(y, w) \right\|$$
$$\leq \|s(f(x+y, z+w) + f(x+y, z-w) + f(x-y, z+w)$$
$$+ f(x-y, z-w) - 4f(x, z))\| + \varphi(x, y)\varphi(z, w) \tag{10.30}$$

for all $\lambda \in \mathbf{T}^1$ *and all* $x, y, z, w \in A$. *Then there exists a unique bi-additive mapping* $B : A^2 \rightarrow A$, *which is* **C**-*linear in the first variable, such that*

$$\|f(x, z) - B(x, z)\| \leq \frac{1}{8}\Psi(2x, 0)\varphi(z, 0) \tag{10.31}$$

for all $x, z \in A$.

If, in addition, the mapping $f : A^2 \rightarrow A$ *satisfies* (10.21), (10.22) *and* $f(2x, z) = 2f(x, z)$ *for all* $x, z \in A$, *then the mapping* $f : A^2 \rightarrow A$ *is a quasi-*-multiplier.*

Corollary 10.7 *Let* $r > 2$ *and* θ *be nonnegative real numbers, and let* $f : A^2 \rightarrow A$ *be a mapping satisfying* $f(x, 0) = f(0, z) = 0$ *and*

$$\left\| 4f\left(\lambda\frac{x+y}{2}, z-w\right) + 4f\left(\lambda\frac{x-y}{2}, z+w\right) - 4\lambda f(x, z) + 4\lambda f(y, w) \right\|$$
$$\leq \|s(f(x+y, z+w) + f(x+y, z-w) + f(x-y, z+w)$$
$$+ f(x-y, z-w) - 4f(x, z))\| + \theta(\|x\|^r + \|y\|^r)(\|z\|^r + \|w\|^r) \tag{10.32}$$

for all $\lambda \in \mathbf{T}^1$ *and all* $x, y, z, w \in A$. *Then there exists a unique bi-additive mapping* $B : A^2 \rightarrow A$, *which is* **C**-*linear in the first variable, such that*

$$\|f(x, z) - B(x, z)\| \leq \frac{2^{r-2}\theta}{2^r - 2}\|x\|^r\|z\|^r \tag{10.33}$$

for all $x, z \in A$.

If, in addition, the mapping $f : A^2 \rightarrow A$ satisfies (10.25), (10.26) and $f(2x, z) = 2f(x, z)$ for all $x, z \in A$, then the mapping $f : A^2 \rightarrow A$ is a quasi-*-multiplier.

Proof The proof follows from Theorem 10.7 by taking $\varphi(x, y) = \sqrt{\theta}(\|x\|^r + \|y\|^r)$ for all $x, y \in A$.

Theorem 10.8 *Let* $\varphi : A^2 \rightarrow [0, \infty)$ *be a function satisfying* (10.27). *Let* $f : A \rightarrow A$ *be a mapping satisfying* (10.30) *and* $f(x, 0) = f(0, z) = 0$ *for all* $x, z \in A$. *Then there exists a unique bi-additive mapping* $B : A^2 \rightarrow A$, *which is* **C**-*linear in the first variable, such that*

$$\|f(x, z) - B(x, z)\| \leq \frac{1}{8}\Psi(2x, 0)\,\varphi(z, 0) \tag{10.34}$$

for all $x, z \in A$.

If, in addition, the mapping $f : A^2 \rightarrow A$ satisfies (10.21), (10.22) and $f(2x, z) = 2f(x, z)$ for all $x, z \in A$, then the mapping $f : A^2 \rightarrow A$ is a quasi-*-multiplier.

Corollary 10.8 *Let* $r < 1$ *and* θ *be nonnegative real numbers, and let* $f : A \rightarrow A$ *be a mapping satisfying* (10.32) *and* $f(x, 0) = f(0, z) = 0$ *for all* $x, z \in A$. *Then there exists a unique bi-additive mapping* $B : A^2 \rightarrow A$, *which is* **C**-*linear in the first variable, such that* (10.29) *holds for all* $x, z \in A$.

If, in addition, the mapping $f : A^2 \rightarrow A$ satisfies (10.25), (10.26) and $f(2x, z) = 2f(x, z)$ for all $x, z \in A$, then the mapping $f : A^2 \rightarrow A$ is a quasi-*-multiplier.

Proof The proof follows from Theorem 10.8 by taking $\varphi(x, y) = \sqrt{\theta}(\|x\|^r + \|y\|^r)$ for all $x, y \in A$.

From now on, assume that A is a unital C^*-algebra with unit e and unitary group $U(A)$.

Theorem 10.9 *Let* $\varphi : A^2 \rightarrow [0, \infty)$ *be a function satisfying* (10.18) *and let* $f : A^2 \rightarrow A$ *be a mapping satisfying* (10.19) *and* $f(x, 0) = f(0, z) = 0$ *for all* $x, z \in A$. *Then there exists a unique bi-additive mapping* $B : A^2 \rightarrow A$, *which is* **C**-*linear in the first variable and satisfies* (10.20).

If, in addition, the mapping $f : A^2 \rightarrow A$ satisfies (10.22), $f(2x, z) = 2f(x, z)$ and

$$\|f(uy, z) - uf(y, z)\| \leq \varphi(u, y)^2\varphi(z, 0), \tag{10.35}$$

for all $u \in U(A)$ and all $x, y, z \in A$, then the mapping $f : A^2 \to A$ is a quasi-∗-multiplier satisfying $f(x, w) = xf(e, e)w^*$ for all $x, w \in A$.

Proof By the same reasoning as in the proof of Theorem 10.5, there is a unique bi-additive mapping $B : A^2 \to A$ satisfying (10.20), which is **C**-linear in the first variable, defined by

$$B(x, z) := \lim_{n \to \infty} 2^n f\left(\frac{x}{2^n}, z\right)$$

for all $x, z \in A$.

If $f(2x, z) = 2f(x, z)$ for all $x, z \in A$, then we can easily show that $B(x, z) = f(x, z)$ for all $x, z \in A$.

By the same reasoning as in the proof of Theorem 10.5, $B(uy, z) = uB(y, z)$ for all $u \in U(A)$ and all $y, z \in A$.

Since B is **C**-linear in the first variable and each $x \in A$ is a finite linear combination of unitary elements (see [14]), i.e., $x = \sum_{j=1}^{m} \lambda_j u_j$ ($\lambda_j \in$ **C**, $u_j \in U(A)$),

$$B(xy, z) = B(\sum_{j=1}^{m} \lambda_j u_j y, z) = \sum_{j=1}^{m} \lambda_j B(u_j y, z) = \sum_{j=1}^{m} \lambda_j u_j B(y, z)$$

$$= (\sum_{j=1}^{m} \lambda_j u_j) B(y, z) = x B(y, z)$$

for all $x, y, z \in A$. So by the same reasoning as in the proof of Theorem 10.5, $B : A^2 \to A$ is a quasi-∗-multiplier and satisfies

$$B(x, w) = B(xe, we) = xB(e, we) = xB(we, e)^* = x(wB(e, e))^* = xB(e, e)^* w^*$$

$$= xB(e, e)w^*$$

for all $x, w \in A$. Thus $f : A^2 \to A$ is a quasi-∗-multiplier and satisfies $f(x, w) = f(xe, we) = xf(e, e)w^*$ for all $x, w \in A$.

Corollary 10.9 *Let $r > 2$ and θ be nonnegative real numbers, and let $f : A^2 \to A$ be a mapping satisfying (10.23) and $f(x, 0) = f(0, z) = 0$ for all $x, z \in A$. Then there exists a unique bi-additive mapping $B : A^2 \to A$, which is **C**-linear in the first variable and satisfies (10.24).*

If, in addition, the mapping $f : A^2 \to A$ satisfies (10.26), $f(2x, z) = 2f(x, z)$ and

$$\|f(uy, z) - uf(y, z)\| \leq \theta(1 + \|y\|^r)\|z\|^r \tag{10.36}$$

for all $u \in U(A)$ and all $x, y, z \in A$, then the mapping $f : A^2 \to A$ is a quasi-∗-multiplier satisfying $f(x, w) = xf(e, e)w^*$ for all $x, w \in A$.

Proof The proof follows from Theorem 10.9 by taking $\varphi(x, y) = \sqrt{\theta}(\|x\|^r + \|y\|^r)$ for all $x, y \in A$.

Theorem 10.10 *Let* $\varphi : A^2 \rightarrow [0, \infty)$ *be a function satisfying* (10.27). *Let* $f : A^2 \rightarrow A$ *be a mapping satisfying* (10.19) *and* $f(x, 0) = f(0, z) = 0$ *for all* $x, z \in A$. *Then there exists a unique bi-additive mapping* $B : A^2 \rightarrow A$, *which is* **C**-*linear in the first variable and satisfies* (10.34).

If, in addition, the mapping $f : A^2 \rightarrow A$ *satisfies* (10.35), (10.22) *and* $f(2x, z) = 2f(x, z)$ *for all* $x, z \in A$, *then the mapping* $f : A^2 \rightarrow A$ *is a quasi-∗-multiplier satisfying* $f(x, w) = xf(e, e)w^*$ *for all* $x, w \in A$.

Proof The proof is similar to the proofs of Theorems 10.6 and 10.9.

Corollary 10.10 *Let* $r < 1$ *and* θ *be nonnegative real numbers, and let* $f : A^2 \rightarrow A$ *be a mapping satisfying* (10.23) *and* $f(x, 0) = f(0, z) = 0$ *for all* $x, z \in A$. *Then there exists a unique bi-additive mapping* $B : A^2 \rightarrow A$, *which is* **C**-*linear in the first variable and satisfies* (10.29).

If, in addition, the mapping $f : A^2 \rightarrow A$ *satisfies* (10.36), (10.26) *and* $f(2x, z) = 2f(x, z)$ *for all* $x, z \in A$, *then the mapping* $f : A^2 \rightarrow A$ *is a quasi-∗-multiplier satisfying* $f(x, w) = xf(e, e)w^*$ *for all* $x, w \in A$.

Proof The proof follows from Theorem 10.10 by taking $\varphi(x, y) = \sqrt{\theta}(\|x\|^r + \|y\|^r)$ for all $x, y \in A$.

Similarly, we can obtain the following results.

Theorem 10.11 *Let* $\varphi : A^2 \rightarrow [0, \infty)$ *be a function satisfying* (10.18) *and let* $f : A^2 \rightarrow A$ *be a mapping satisfying* (10.30) *and* $f(x, 0) = f(0, z) = 0$ *for all* $x, z \in A$. *Then there exists a unique bi-additive mapping* $B : A^2 \rightarrow A$, *which is* **C**-*linear in the first variable and satisfies* (10.31).

If, in addition, the mapping $f : A^2 \rightarrow A$ *satisfies* (10.35), (10.22) *and* $f(2x, z) = 2f(x, z)$ *for all* $x, z \in A$, *then the mapping* $f : A^2 \rightarrow A$ *is a quasi-∗-multiplier satisfying* $f(x, w) = xf(e, e)w^*$ *for all* $x, w \in A$.

Corollary 10.11 *Let* $r > 2$ *and* θ *be nonnegative real numbers, and let* $f : A^2 \rightarrow A$ *be a mapping satisfying* (10.32) *and* $f(x, 0) = f(0, z) = 0$ *for all* $x, z \in A$. *Then there exists a unique bi-additive mapping* $B : A^2 \rightarrow A$, *which is* **C**-*linear in the first variable and satisfies* (10.33).

If, in addition, the mapping $f : A^2 \rightarrow A$ *satisfies* (10.36), (10.26) *and* $f(2x, z) = 2f(x, z)$ *for all* $x, z \in A$, *then the mapping* $f : A^2 \rightarrow A$ *is a quasi-∗-multiplier satisfying* $f(x, w) = xf(e, e)w^*$ *for all* $x, w \in A$.

Proof The proof follows from Theorem 10.11 by taking $\varphi(x, y) = \sqrt{\theta}(\|x\|^r + \|y\|^r)$ for all $x, y \in A$.

Theorem 10.12 *Let* $\varphi : A^2 \rightarrow [0, \infty)$ *be a function satisfying* (10.27). *Let* $f : A^2 \rightarrow A$ *be a mapping satisfying* (10.30) *and* $f(x, 0) = f(0, z) = 0$ *for all* $x, z \in A$. *Then there exists a unique bi-additive mapping* $B : A^2 \rightarrow A$, *which is* **C**-*linear in the first variable and satisfies* (10.34).

If, in addition, the mapping $f : A^2 \to A$ satisfies (10.35), (10.22) *and* $f(2x, z) = 2f(x, z)$ *for all* $x, z \in A$, *then the mapping* $f : A^2 \to A$ *is a quasi-$*$-multiplier satisfying* $f(x, w) = xf(e, e)w^*$ *for all* $x, w \in A$.

Corollary 10.12 *Let* $r < 1$ *and* θ *be nonnegative real numbers, and let* $f : A^2 \to A$ *be a mapping satisfying* (10.32) *and* $f(x, 0) = f(0, z) = 0$ *for all* $x, z \in A$. *Then there exists a unique bi-additive mapping* $B : A^2 \to A$, *which is* **C**-*linear in the first variable and satisfies* (10.29).

If, in addition, the mapping $f : A^2 \to A$ *satisfies* (10.36), (10.26) *and* $f(2x, z) = 2f(x, z)$ *for all* $x, z \in A$, *then the mapping* $f : A^2 \to A$ *is a quasi-$*$-multiplier satisfying* $f(x, w) = xf(e, e)w^*$ *for all* $x, w \in A$.

Proof The proof follows from Theorem 10.12 by taking $\varphi(x, y) = \sqrt{\theta}(\|x\|^r + \|y\|^r)$ for all $x, y \in A$.

Acknowledgement C. Park was supported by Basic Science Research Program through the National Research Foundation of Korea funded by the Ministry of Education, Science and Technology (NRF-2017R1D1A1B04032937).

References

1. Akemann, C.A., Pedersen, G.K.: Complications of semicontinuity in C^*-algebra theory. Duke Math. J. **40**, 785–795 (1973)
2. Amyari, M., Baak, C., Moslehian, M.: Nearly ternary derivations. Taiwan. J. Math. **11**, 1417–1424 (2007)
3. Aoki, T.: On the stability of the linear transformation in Banach spaces. J. Math. Soc. Japan **2**, 64–66 (1950)
4. Brzdęk, J., Popa, D., Raşa, I., Xu, B.: Ulam Stability of Operators. Elsevier, Oxford (2018)
5. El-Fassai, I., Brzdęk, J., Chahbi, A., Kabbaj, S.: On hyperstability of the biadditive functional equation. Acta Math. Sci. B **37**, 1727–1739 (2017)
6. Eshaghi Gordji, M., Ghobadipour, N.: Stability of (α, β, γ)-derivations on Lie C^*-algebras. Int. J. Geom. Meth. Mod. Phys. **7**, 1097–1102 (2010)
7. Eshaghi Gordji, M., Ghaemi, M.B., Alizadeh, B.: A fixed point method for perturbation of higher ring derivationsin non-Archimedean Banach algebras. Int. J. Geom. Meth. Mod. Phys. **8**(7), 1611–1625 (2011)
8. Eshaghi Gordji, M., Fazeli, A., Park, C.: 3-Lie multipliers on Banach 3-Lie algebras. Int. J. Geom. Meth. Mod. Phys. **9**(7), 15 pp. (2012). Article ID 1250052
9. Fechner, W.: Stability of a functional inequalities associated with the Jordan-von Neumann functional equation. Aequationes Math. **71**, 149–161 (2006)
10. Găvruta, P.: A generalization of the Hyers-Ulam-Rassias stability of approximately additive mappings. J. Math. Anal. Appl. **184**, 431–436 (1994)
11. Gilányi, A.: Eine zur Parallelogrammgleichung äquivalente Ungleichung. Aequationes Math. **62**, 303–309 (2001)
12. Gilányi, A.: On a problem by K. Nikodem. Math. Inequal. Appl. **5**, 707–710 (2002)
13. Hyers, D.H.: On the stability of the linear functional equation. Proc. Nat. Acad. Sci. U. S. A. **27**, 222–224 (1941)
14. Kadison, R.V., Ringrose, J.R.: Fundamentals of the Theory of Operator Algebras: Elementary Theory. Academic Press, New York (1983)
15. McKennon, M.: Quasi-multipliers. Trans. Amer. Math. Soc. **233**, 105–123 (1977)

16. Park, C.: Homomorphisms between Poisson JC^*-algebras. Bull. Braz. Math. Soc. **36**, 79–97 (2005)
17. Park, C.: Additive ρ-functional inequalities and equations. J. Math. Inequal. **9**, 17–26 (2015)
18. Park, C.: Additive ρ-functional inequalities in non-Archimedean normed spaces. J. Math. Inequal. **9**, 397–407 (2015)
19. Park, C.: C^*-ternary biderivations and C^*-ternary bihomomorphisms. Mathematics **6**, Art. No. 30 (2018)
20. Park, C.: Bi-additive s-functional inequalities and quasi-*-multipliers on Banach algebras. Mathematics **6**, Art. No. 31 (2018)
21. Rassias, Th.M.: On the stability of the linear mapping in Banach spaces. Proc. Amer. Math. Soc. **72**, 297–300 (1978)
22. Rätz, J.: On inequalities associated with the Jordan-von Neumann functional equation. Aequationes Math. **66**, 191–200 (2003)
23. Ulam, S.M.: A Collection of the Mathematical Problems. Interscience Publications, New York (1960)

Chapter 11
On Ulam Stability of a Generalization of the Fréchet Functional Equation on a Restricted Domain

Renata Malejki

Abstract In this paper we prove the Ulam type stability of a generalization of the Fréchet functional equation on a restricted domain. In the proofs the main tool is a fixed point theorem for some function spaces.

Keywords Ulam type stability · Fixed point theorem · Fréchet equation

Mathematics Subject Classification (2010) Primary 39B82; Secondary 39B52, 39B62, 47H10

11.1 Introduction

Let $\mathbb{K} \in \{\mathbb{R}, \mathbb{C}\}$ ($\mathbb{R}$ and $\mathbb{C}$ denote the fields of real and complex numbers, respectively) and $A_1, \ldots, A_7 \in \mathbb{K}$. Our consideration involve Ulam stability (see, e.g., [15] for more details and suitable references) of the following conditional (i.e., on a restricted domain) functional equation

$$A_1 F(x+y+z) + A_2 F(x) + A_3 F(y) + A_4 F(z) \tag{11.1}$$
$$= A_5 F(x+y) + A_6 F(x+z) + A_7 F(y+z),$$
$$x, y, z \in D, \; x+y+z, x+y, x+z, y+z \in D,$$

in the class of functions $F : D \to Y$, where $D \subset X$ is nonempty, $(X, +)$ is a commutative monoid (i.e., a semigroup with a neutral element denoted by 0) and Y is a Banach space over $\mathbb{K}$.

R. Malejki (✉)
Institute of Mathematics, Pedagogical University of Cracow, Kraków, Poland
e-mail: renata.malejki@up.krakow.pl

© Springer Nature Switzerland AG 2019
J. Brzdęk et al. (eds.), *Ulam Type Stability*,
https://doi.org/10.1007/978-3-030-28972-0_11

It is a natural generalization of several functional equations, stability of which have been already investigated quite intensively (see [1, 3, 4, 6, 8, 13, 17, 22, 29, 30, 33]). Let us mention here the Cauchy functional equation

$$f(x + y) = f(x) + f(y),$$

the Jensen functional equation

$$f(x + y) = \frac{1}{2}\big(f(2x) + f(2y)\big),$$

the Jordan–von Neumann (quadratic) functional equation

$$f(x + y) + f(x - y) = 2f(x) + 2f(y),$$

the Drygas equation

$$f(x + y) + f(x - y) = 2f(x) + f(y) + f(-y),$$

and the Fréchet (see [19, 25]) functional equation

$$f(x + y + z) + f(x) + f(y) + f(z) = f(x + y) + f(x + z) + f(y + z). \quad (11.2)$$

It has been showed in [14] that the set of solutions of Eq. (11.1) is not empty. Moreover, if we assume that at least two coefficients A_i are not equal, then every solution F of this equation, with $F(0) = 0$, is an additive function. Moreover, the condition $A_1 + A_2 + A_3 + A_4 \neq A_5 + A_6 + A_7$ is sufficient to get $F(0) = 0$ for each solution F of (11.1) (see [14]).

11.2 The Main Result

Unless explicitly state otherwise, in what follows we assume that $(X, +)$ is a commutative monoid, $\widehat{X} := X^3 \setminus \{(0, 0, 0)\}$, Y is a Banach space over the field $\mathbb{K} \in \{\mathbb{R}, \mathbb{C}\}$, and $A_1, \ldots, A_7 \in \mathbb{K}$ are fixed.

First we recall another theorem on stability of the generalization of the Fréchet functional equation. It has been proved in [28].

Theorem 11.1 *Let $(X, +)$ be an abelian group, $A_1 \neq 0$ and*

$$A_2 + A_3 + A_4 = A_5 + A_6 + A_7.$$

Assume that $f : X \to Y$, $c : \mathbb{Z} \setminus \{0\} \to [0, \infty)$ and $L : \widehat{X} \to [0, \infty)$ satisfy the following three conditions:

$$\mathscr{M} := \{m \in \mathbb{Z} \setminus \{0\} : |A_7|c(-2m) + |A_5 + A_6|c(m+1)$$
$$+ |A_3 + A_4|c(-m) + |A_2|c(2m+1) < |A_1|\} \neq \emptyset,$$

$$L(kx, ky, kz) \leq c(k)L(x, y, z), \qquad (x, y, z) \in \widehat{X}, m \in \mathscr{M},$$
$$k \in \{-2m, m+1, -m, 2m+1\},$$

$$\|A_1 f(x+y+z) + A_2 f(x) + A_3 f(y) + A_4 f(z) - A_5 f(x+y)$$
$$- A_6 f(x+z) - A_7 f(y+z)\| \leq L(x, y, z), \qquad (x, y, z) \in X^3.$$

Then there is a unique function $F : X \to Y$ satisfying Eq. (11.1) with $D = X$ and such that

$$\|f(x) - F(x)\| \leq \rho_L(x), \qquad x \in X \setminus \{0\},$$

where

$$\rho_L(x) := \inf_{m \in \mathscr{M}} \frac{L((2m+1)x, -mx, -mx)}{|A_1| - \beta_m},$$

$$\beta_m := |A_7|c(-2m) + |A_5 + A_6|c(m+1) + |A_3 + A_4|c(-m) + |A_2|c(2m+1).$$

The following theorem also concerns stability of Eq. (11.1). It complements Theorem 11.1 and generalizes [14, Theorem 13]. It shows that the assumptions of [14, Theorem 13] can be significantly weakened; in particular, that it is still valid on a restricted domain. The proof of it will be provided in the next section. We use in it a fixed point theorem for some function spaces from [12].

Theorem 11.2 *Let $D \subset X, 0 \in D,$*

$$2x, 3x \in D, \qquad x \in D,$$

$\widehat{D} := D^3 \setminus \{(0, 0, 0)\}, A_2 + A_3 + A_4 \neq 0,$

$$\beta_0 := \left| \frac{A_5 + A_6 + A_7 - A_1}{A_2 + A_3 + A_4} \right| < 1,$$

and a function $L : D^3 \to [0, \infty)$ fulfil the condition

$$L(kx, ky, kz) \leq c_k L(x, y, z), \qquad (x, y, z) \in \widehat{D}, k \in \{2, 3\}, \qquad (11.3)$$

with some $c_2, c_3 \in [0, \infty)$ such that $\beta := b_2 c_2 + b_3 c_3 < 1$, where

$$b_2 := \left| \frac{A_5 + A_6 + A_7}{A_2 + A_3 + A_4} \right|, \qquad b_3 := \left| \frac{A_1}{A_2 + A_3 + A_4} \right|. \tag{11.4}$$

If $f : D \to Y$ satisfy the inequality

$$\|A_1 f(x + y + z) + A_2 f(x) + A_3 f(y) + A_4 f(z) \tag{11.5}$$
$$- A_5 f(x + y) - A_6 f(x + z) - A_7 f(y + z)\| \leq L(x, y, z),$$

$$(x, y, z) \in D^3, \ x + y + z, x + y, x + z, y + z \in D.$$

then there exists a unique solution $F : D \to Y$ of Eq. (11.1) such that $F(0) = 0$ and

$$\|f(x) - F(x)\| \leq \rho_L(x), \qquad x \in D, \tag{11.6}$$

where

$$\rho_L(x) := \frac{L(x, x, x)}{|A_2 + A_3 + A_4|(1 - \gamma(x))}, \qquad x \in D, \tag{11.7}$$

with

$$\gamma(x) := \begin{cases} \beta & \text{if} \quad x \neq 0; \\ \beta_0 & \text{if} \quad x = 0. \end{cases}$$

11.3   Proof of Theorem 11.2

In the proof we use the approach initiated in [10] and next applied also in [2, 4, 6, 11, 13, 16, 28, 31, 35]. The main tool in it will be the fixed point theorem for the function spaces proved in [12].

Theorem 11.3 ([12]) *Let the following three hypotheses be valid.*

(H1) S is a nonempty set, E is a Banach space, and functions $f_1, \ldots, f_k : S \to S$ and $l_1, \ldots, l_k : S \to \mathbb{R}_+$ are given.
(H2) $\mathcal{T} : E^S \to E^S$ is an operator satisfying the inequality

$$\|\mathcal{T}\xi(x) - \mathcal{T}\mu(x)\| \leq \sum_{i=1}^{k} l_i(x) \|\xi(f_i(x)) - \mu(f_i(x))\|, \qquad \xi, \mu \in E^S, x \in S.$$

(H3) $\Lambda : \mathbb{R}_+{}^S \to \mathbb{R}_+{}^S$ *is defined by*

$$\Lambda\delta(x) := \sum_{i=1}^{k} l_i(x)\delta(f_i(x)), \qquad \delta \in \mathbb{R}_+{}^S, x \in S.$$

Assume that functions $\varepsilon : S \to \mathbb{R}_+$ *and* $\varphi : S \to E$ *fulfil the following two conditions*

$$\left\| \mathscr{T}\varphi(x) - \varphi(x) \right\| \leq \varepsilon(x), \qquad x \in S, \tag{11.8}$$

$$\varepsilon^*(x) := \sum_{n=0}^{\infty} \Lambda^n \varepsilon(x) < \infty, \qquad x \in S. \tag{11.9}$$

Then there exists a unique fixed point ψ *of* $\mathscr{T}$ *with*

$$\|\varphi(x) - \psi(x)\| \leq \varepsilon^*(x), \qquad x \in S.$$

Moreover,

$$\psi(x) := \lim_{n \to \infty} \mathscr{T}^n\varphi(x), \qquad x \in S.$$

From this theorem we obtain that an appropriately defined operator determines an exact solution of Eq. (11.1) as the limit of a sequence of its iterates on an approximate solution of this equation. Similar results can be found in, e.g., [4–7, 9, 10, 13, 18, 20, 21, 23, 24, 26–28, 31, 32, 34–36].

Proof Inserting x by y and z in condition (11.5) we get following inequality

$$\|A_1 f(3x) + (A_2 + A_3 + A_4)f(x) - (A_5 + A_6 + A_7)f(2x)\| \tag{11.10}$$

$$\leq L(x, x, x), \qquad x \in D.$$

From (11.10) we have

$$\left\| f(x) - \frac{A_5 + A_6 + A_7}{A_2 + A_3 + A_4} f(2x) + \frac{A_1}{A_2 + A_3 + A_4} f(3x) \right\| \leq \varepsilon(x), \quad x \in D, \tag{11.11}$$

where function ε is given by

$$\varepsilon(x) := \frac{L(x, x, x)}{|A_2 + A_3 + A_4|}.$$

Now we define an operator $\mathcal{T}$ and show that properties (H1), (H2) and (H3) of Theorem 11.3 are satisfied. For every $\xi \in Y^D$, $x \in D$ let

$$\mathcal{T}\xi(x) := \frac{A_5 + A_6 + A_7}{A_2 + A_3 + A_4}\xi(2x) - \frac{A_1}{A_2 + A_3 + A_4}\xi(3x). \tag{11.12}$$

Notice that the operator $\mathcal{T}$ has the property

$$\mathcal{T}\xi(0) = \frac{A_5 + A_6 + A_7 - A_1}{A_2 + A_3 + A_4}\xi(0), \qquad \xi \in Y^D. \tag{11.13}$$

It is easy to see that operator $\mathcal{T}$ is linear.

Let $\xi, \mu \in Y^D$. Then by definition of the norm for every $x \in D$ we obtain

$$\|\mathcal{T}\xi(x) - \mathcal{T}\mu(x)\| \le \left|\frac{A_5 + A_6 + A_7}{A_2 + A_3 + A_4}\right| \|\xi(2x) - \mu(2x)\|$$

$$+ \left|\frac{A_1}{A_2 + A_3 + A_4}\right| \|\xi(3x) - \mu(3x)\|.$$

Thus

$$\|\mathcal{T}\xi(x) - \mathcal{T}\mu(x)\| \le b_2 \|\xi(2x) - \mu(2x)\| \tag{11.14}$$

$$+ b_3 \|\xi(3x) - \mu(3x)\|, \qquad x \in D \setminus \{0\}.$$

In case $x = 0$ we have

$$\|\mathcal{T}\xi(0) - \mathcal{T}\mu(0)\| = \left\|\frac{A_5 + A_6 + A_7 - A_1}{A_2 + A_3 + A_4}(\xi(0) - \mu(0))\right\|$$

$$= \left|\frac{A_5 + A_6 + A_7 - A_1}{A_2 + A_3 + A_4}\right| \|\xi(0) - \mu(0)\|.$$

Therefore

$$\|\mathcal{T}\xi(0) - \mathcal{T}\mu(0)\| = \beta_0 \|\xi(0) - \mu(0)\|. \tag{11.15}$$

Consequently,

$$\|\mathcal{T}\xi(x) - \mathcal{T}\mu(x)\| \le \sum_{i=1}^{2} l_i(x)\|\xi(f_i(x)) - \mu(f_i(x))\|, \qquad \xi, \mu \in Y^D, x \in D,$$

where $f_1(x) = 2x$ and $f_2(x) = 3x$, which means that conditions (H1) and (H2) are satisfied, with $k = 2$, $S = X$, $E = Y$, $l_1(x) = b_2$, $l_2(x) = b_3$ for $x \in D \setminus \{0\}$ and $l_1(0) = \beta_0$, $l_2(0) = 0$.

Next we define an operator $\Lambda : \mathbb{R}_+{}^D \rightarrow \mathbb{R}_+{}^D$ as in (H3) by

$$\Lambda \eta(x) := \sum_{i=1}^{2} l_i(x) \eta(f_i(x)), \quad x \in D \tag{11.16}$$

for every $\eta \in \mathbb{R}_+{}^D$. Then for each $\eta \in \mathbb{R}_+{}^D$ we have that

$$\Lambda \eta(x) := b_2 \, \eta(2x) + b_3 \, \eta(3x), \quad x \in D \setminus \{0\}$$

and

$$\Lambda \eta(0) := \beta_0 \eta(0).$$

Let us note that operator Λ is nondecreasing, i.e., $\Lambda \eta \leq \Lambda \zeta$ for all $\eta, \zeta \in \mathbb{R}_+{}^D$ with $\eta \leq \zeta$.

Besides, by (11.14) and (11.15) the relation between operators $\mathscr{T}$ and Λ is following

$$\|\mathscr{T}\xi(x) - \mathscr{T}\mu(x)\| \leq \Lambda(\|\xi - \mu\|)(x), \quad \xi, \mu \in Y^D, x \in D. \tag{11.17}$$

By (11.11) and (11.12) we obtain the estimation

$$\|f(x) - \mathscr{T}f(x)\| \leq \varepsilon(x), \quad x \in D,$$

so, condition (11.8) holds. In the special case when $x = 0$, we get

$$\|f(0) - \mathscr{T}f(0)\| = \left| 1 - \frac{A_5 + A_6 + A_7 - A_1}{A_2 + A_3 + A_4} \right| \|f(0)\| \leq \varepsilon(0).$$

Now we will show that the function series

$$\sum_{n=0}^{\infty} \Lambda^n \varepsilon(x)$$

is convergent for each $x \in D$, i.e., condition (11.9) is satisfied. Fix an $x \in D \setminus \{0\}$. By (11.16) and (11.3), we obtain

$$\Lambda \varepsilon(x) = b_2 \, \varepsilon(2x) + b_3 \, \varepsilon(3x)$$

$$= b_2 \frac{L(2x, 2x, 2x)}{|A_2 + A_3 + A_4|} + b_3 \frac{L(3x, 3x, 3x)}{|A_2 + A_3 + A_4|}$$

$$\leq b_2 c_2 \frac{L(x, x, x)}{|A_2 + A_3 + A_4|} + b_3 c_3 \frac{L(x, x, x)}{|A_2 + A_3 + A_4|}$$

$$= (b_2 c_2 + b_3 c_3) \frac{L(x, x, x)}{|A_2 + A_3 + A_4|}.$$

Thus

$$\Lambda \varepsilon(x) \leq \beta \varepsilon(x).$$ (11.18)

By induction one can show that monotonicity and linearity of Λ implies

$$\Lambda^n \varepsilon(x) \leq \beta^n \varepsilon(x).$$ (11.19)

Consequently, for each $x \in D \setminus \{0\}$ we have the estimate

$$\varepsilon^*(x) = \sum_{n=0}^{\infty} \Lambda^n \varepsilon(x) \leq \varepsilon(x)\left(1 + \sum_{n=1}^{\infty} \beta^n\right)$$

$$= \frac{\varepsilon(x)}{1 - \beta} = \frac{L(x, x, x)}{|A_2 + A_3 + A_4|(1 - \beta)}.$$

In case $x = 0$ we get

$$\Lambda \varepsilon(0) = \beta_0 \varepsilon(0).$$ (11.20)

So, by induction we obtain

$$\Lambda^n \varepsilon(0) = \beta_0^n \varepsilon(0).$$ (11.21)

Hence

$$\varepsilon^*(0) = \sum_{n=0}^{\infty} \Lambda^n \varepsilon(0) = \varepsilon(0)\left(1 + \sum_{n=1}^{\infty} \beta_0^n\right)$$

$$= \frac{\varepsilon(0)}{1 - \beta_0} = \frac{L(0, 0, 0)}{|A_2 + A_3 + A_4|(1 - \beta_0)}.$$

Thus we have shown that

$$\varepsilon^*(x) = \sum_{n=0}^{\infty} \Lambda^n \varepsilon(x) \leq \frac{L(x, x, x)}{|A_2 + A_3 + A_4|(1 - \gamma(x))} < \infty, \qquad x \in D.$$

Because assumptions of Theorem 11.3 are satisfied, in view of this theorem there exists a function $F \colon D \to Y$ satisfying Eq. (11.1) for $x = y = z$, i.e.,

$$F(x) = \frac{A_5 + A_6 + A_7}{A_2 + A_3 + A_4} F(2x) - \frac{A_1}{A_2 + A_3 + A_4} F(3x), \qquad x \in D. \quad (11.22)$$

Moreover,

$$\|f(x) - F(x)\| \le \varepsilon^*(x) \le \frac{L(x,x,x)}{|A_2 + A_3 + A_4|(1 - \gamma(x))}, \qquad x \in D.$$

and

$$F(x) = \lim_{n \to \infty} \mathscr{T}^n f(x), \qquad x \in D.$$

Next we will prove that the function F satisfies Eq. (11.1) for all $x, y, z \in D$, firstly, by induction we will show that for all $(x, y, z) \in D^3$ such that $x + y + z, x + y, x + z, y + z \in D, n \in \mathbb{N}_0 := \mathbb{N} \cup \{0\}$ occurs the condition

$$\|A_1 \mathscr{T}^n f(x + y + z) + A_2 \mathscr{T}^n f(x) + A_3 \mathscr{T}^n f(y) + A_4 \mathscr{T}^n f(z) \qquad (11.23)$$

$$- A_5 \mathscr{T}^n f(x + y) - A_6 \mathscr{T}^n f(x + z) - A_7 \mathscr{T}^n f(y + z)\|$$

$$\le \lambda^n L(x, y, z),$$

where $\lambda := \max\{\beta, \beta_0\}$. For $n = 0$ condition (11.23) follows from (11.5). Now, assume that (11.23) holds for some $n \in \mathbb{N}_0$ and all $(x, y, z) \in D^3$ such that $x + y + z, x + y, x + z, y + z \in D$. Then by (11.12) we obtain

$$\left\| A_1 \mathscr{T}^{n+1} f(x + y + z) + A_2 \mathscr{T}^{n+1} f(x) + A_3 \mathscr{T}^{n+1} f(y) + A_4 \mathscr{T}^{n+1} f(z) \right.$$

$$- A_5 \mathscr{T}^{n+1} f(x + y) - A_6 \mathscr{T}^{n+1} f(x + z) - A_7 \mathscr{T}^{n+1} f(y + z) \Big\|$$

$$= \left\| \frac{A_5 + A_6 + A_7}{A_2 + A_3 + A_4} A_1 \mathscr{T}^n f(2(x + y + z)) - \frac{A_1}{A_2 + A_3 + A_4} A_1 \mathscr{T}^n f(3(x + y + z)) \right.$$

$$+ \frac{A_5 + A_6 + A_7}{A_2 + A_3 + A_4} A_2 \mathscr{T}^n f(2x) - \frac{A_1}{A_2 + A_3 + A_4} A_2 \mathscr{T}^n f(3x)$$

$$+ \frac{A_5 + A_6 + A_7}{A_2 + A_3 + A_4} A_3 \mathscr{T}^n f(2y) - \frac{A_1}{A_2 + A_3 + A_4} A_3 \mathscr{T}^n f(3y)$$

$$+ \frac{A_5 + A_6 + A_7}{A_2 + A_3 + A_4} A_4 \mathscr{T}^n f(2z) - \frac{A_1}{A_2 + A_3 + A_4} A_4 \mathscr{T}^n f(3z)$$

$$- \frac{A_5 + A_6 + A_7}{A_2 + A_3 + A_4} A_5 \mathscr{T}^n f(2(x + y)) + \frac{A_1}{A_2 + A_3 + A_4} A_5 \mathscr{T}^n f(3(x + y))$$

$$- \frac{A_5 + A_6 + A_7}{A_2 + A_3 + A_4} A_6 \mathscr{T}^n f(2(x + z)) + \frac{A_1}{A_2 + A_3 + A_4} A_6 \mathscr{T}^n f(3(x + z))$$

$$- \frac{A_5 + A_6 + A_7}{A_2 + A_3 + A_4} A_7 \mathscr{T}^n f(2(y + z)) + \frac{A_1}{A_2 + A_3 + A_4} A_7 \mathscr{T}^n f(3(y + z)) \Big\|$$

$$\le \left| \frac{A_5 + A_6 + A_7}{A_2 + A_3 + A_4} \right| \lambda^n L(2x, 2y, 2z) + \left| \frac{A_1}{A_2 + A_3 + A_4} \right| \lambda^n L(3x, 3y, 3z)$$

for every $(x, y, z) \in \widehat{D}$ such that $x + y + z, x + y, x + z, y + z \in D$. Hence by (11.3)

$$\|A_1 \mathscr{T}^{n+1} f(x + y + z) + A_2 \mathscr{T}^{n+1} f(x) + A_3 \mathscr{T}^{n+1} f(y) + A_4 \mathscr{T}^{n+1} f(z)$$
$$- A_5 \mathscr{T}^{n+1} f(x + y) - A_6 \mathscr{T}^{n+1} f(x + z) - A_7 \mathscr{T}^{n+1} f(y + z)\|$$
$$\leq \lambda^n (b_2 c_2 + b_3 c_3) L(x, y, z) \leq \lambda^{n+1} L(x, y, z) \tag{11.24}$$

for $(x, y, z) \in \widehat{D}$ such that $x + y + z, x + y, x + z, y + z \in D$. By (11.13),

$$\|(A_1 + A_2 + A_3 + A_4 - A_5 - A_6 - A_7) \mathscr{T}^{n+1} f(0)\|$$
$$= \left\| (A_1 + A_2 + A_3 + A_4 - A_5 - A_6 - A_7) \frac{A_5 + A_6 + A_7 - A_1}{A_2 + A_3 + A_4} \mathscr{T}^n f(0) \right\|$$
$$= \beta_0 \|(A_1 + A_2 + A_3 + A_4 - A_5 - A_6 - A_7) \mathscr{T}^n f(0)\|$$
$$\leq \beta_0 \lambda^n L(0, 0, 0) \leq \lambda^{n+1} L(0, 0, 0),$$

which ends the proof of (11.23). Letting $n \to \infty$ in (11.23), we obtain

$$A_1 F(x + y + z) + A_2 F(x) + A_3 F(y) + A_4 F(z)$$
$$= A_5 F(x + y) + A_6 F(x + z) + A_7 F(y + z), \qquad (x, y, z) \in D^3.$$

Next we will show that $F(0) = 0$. In view of (11.12) we get by induction that

$$\mathscr{T}^n \xi(0) = \left(\frac{A_5 + A_6 + A_7 - A_1}{A_2 + A_3 + A_4} \right)^n \xi(0) = \beta_0^n \xi(0), \qquad \xi \in Y^D, n \in \mathbb{N}.$$

Thus

$$\lim_{n \to \infty} \mathscr{T}^n \xi(0) = 0, \qquad \xi \in Y^D, \tag{11.25}$$

since $\beta_0 < 1$. Consequently, we obtain $F(0) = \lim_{n \to \infty} \mathscr{T}^n f(0) = 0$.

Now, we prove the uniqueness of F. By induction first we show that for all $\xi, \mu \in Y^D, n \in \mathbb{N}$

$$\|\mathscr{T}^n \xi(x) - \mathscr{T}^n \mu(x)\| \leq \Lambda^n (\|\xi - \mu\|)(x), \quad x \in D. \tag{11.26}$$

By (11.17) condition (11.26) holds for $n = 1$. Fix $\xi, \mu \in Y^D$ and let condition (11.26) holds for $n \in \mathbb{N}$. Then by (11.17)

$$\|\mathscr{T}^{n+1} \xi(x) - \mathscr{T}^{n+1} \mu(x)\| = \|\mathscr{T}(\mathscr{T}^n \xi)(x) - \mathscr{T}(\mathscr{T}^n \mu)(x)\|$$
$$\leq \Lambda(\|\mathscr{T}^n \xi - \mathscr{T}^n \mu\|)(x), \quad x \in D.$$

Hence by (11.26) and monotonicity of Λ we obtain

$$\|\mathscr{T}^{n+1}\xi(x) - \mathscr{T}^{n+1}\mu(x)\| \leq \Lambda(\Lambda^n(\|\xi - \mu\|))(x)$$
$$= \Lambda^{n+1}(\|\xi - \mu\|)(x), \quad x \in D.$$

Let $G : X \to Y$ be also a solution of Eq. (11.1) such that $\|f(x) - G(x)\| \leq \rho_L(x)$ for $x \in D$. Then

$$\|G(x) - F(x)\| \leq 2\rho_L(x), \qquad x \in D. \tag{11.27}$$

Hence by (11.26) we get that

$$\|\mathscr{T}^n G(x) - \mathscr{T}^n F(x)\| \leq 2\Lambda^n \rho_L(x) = \frac{2\Lambda^n \varepsilon(x)}{1 - \gamma(x)}, \qquad x \in D,$$

since Λ is a linear operator. Letting $n \to \infty$, by convergence of the series

$$\sum_{n=0}^{\infty} \Lambda^n \varepsilon(x),$$

we obtain

$$\lim_{n \to \infty} \|\mathscr{T}^n G(x) - \mathscr{T}^n F(x)\| = 0, \qquad x \in D.$$

Thus, $\|G(x) - F(x)\| = 0$ for $x \in D$, since G and F are fixed points of $\mathscr{T}$. Finally $G(x) = F(x)$ for every $x \in D$. This completes the proof. $\qquad \square$

For some suitable comments and examples concerning the assumptions used in this paper we refer to [14].

References

1. Alsina, A., Sikorska, J., Tomás, M.S.: Norm Derivatives and Characterizations of Inner Product Spaces. World Scientific Publishing, Singapore (2010)
2. Badora, R., Brzdęk, J.: Fixed points of a mapping and Hyers-Ulam stability. J. Math. Anal. Appl. **413**, 450–457 (2014)
3. Bahyrycz, A., Olko, J.: Hyperstability of general linear functional equation. Aequationes Math. **90**, 527–540 (2016)
4. Bahyrycz, A., Brzdęk, J., Piszczek, M., Sikorska, J.: Hyperstability of the Fréchet equation and a characterization of inner product spaces. J. Funct. Spaces Appl. **2013**, 6 pp. (2013). Article ID 496361
5. Bahyrycz, A., Brzdęk, J., Leśniak, Z.: On approximate solutions of the generalized Volterra integral equation. Nonlinear Anal. RWA **20**, 59–66 (2014)

6. Bahyrycz, A., Brzdęk, J., Jabłońska, E., Malejki, R.: Ulam's stability of a generalization of the Frechet functional equation. J. Math. Anal. Appl. **442**, 537–553 (2016)
7. Bahyrycz, A., Ciepliński, K., Olko, J.: On Hyers-Ulam stability of two functional equations in non-Archimedean spaces. J. Fixed Point Theory Appl. **18**, 433–444 (2016)
8. Brillouët-Belluot, N., Brzdęk, J., Ciepliński, K.: On some recent developments in Ulam's type stability. Abstr. Appl. Anal. **2012**, 41 pp. (2012). Article ID 716936
9. Brzdęk, J.: Remarks on hyperstability of the Cauchy functional equation. Aequationes Math. **86**, 255–267 (2013)
10. Brzdęk, J.: Hyperstability of the Cauchy equation on restricted domains. Acta Math. Hungar. **141**, 58–67 (2013)
11. Brzdęk, J., Ciepliński, K.: A fixed point approach to the stability of functional equations in non-Archimedean metric spaces. Nonlinear Anal. **74**(18), 6861–6867 (2011)
12. Brzdęk, J., Chudziak, J., Páles, Z.: A fixed point approach to stability of functional equations. Nonlinear Anal. **74**, 6728–6732 (2011)
13. Brzdęk, J., Jabłońska, E., Moslehian, M.S., Pacho, P.: On stability of a functional equation of quadratic type. Acta Math. Hungar. **149**, 160–169 (2016)
14. Brzdęk, J., Leśniak, Z., Malejki, R.: On the generalized Fréchet functional equation with constant coefficients and its stability. Aequationes Math. **92**, 355–373 (2018)
15. Brzdęk, J., Popa, D., Raşa, I., Xu, B.: Ulam Stability of Operators. Elsevier, Oxford (2018)
16. Cădariu, L., Găvruţa, L., Găvruţa, P.: Fixed points and generalized Hyers-Ulam stability. Abstr. Appl. Anal. **2012**, 10 pp. (2012). Article ID 712743
17. Dragomir, S.S.: Some characterizations of inner product spaces and applications. Studia Univ. Babes-Bolyai Math. **34**, 50–55 (1989)
18. Fechner, W.: On the Hyers-Ulam stability of functional equations connected with additive and quadratic mappings. J. Math. Anal. Appl. **322**, 774–786 (2006)
19. Fréchet, M.: Sur la définition axiomatique d'une classe d'espaces vectoriels distanciés applicables vectoriellement sur l'espace de Hilbert. Ann. Math. (2) **36**, 705–718 (1935)
20. Gselmann, E.: Hyperstability of a functional equation. Acta Math. Hungar. **124**, 179–188 (2009)
21. Hyers, D.H., Isac, G., Rassias, Th.M.: Stability of Functional Equations in Several Variables. Birkhäuser, Boston (1998)
22. Jordan, P., von Neumann, J.: On inner products in linear, metric spaces. Ann. Math. (2) **36**, 719–723 (1935)
23. Jung, S.-M.: On the Hyers-Ulam stability of the functional equation that have the quadratic property. J. Math. Anal. Appl. **222**, 126–137 (1998)
24. Jung, S.-M., Hyers-Ulam-Rassias Stability of Functional Equations in Nonlinear Analysis. Springer Optimization and Its Applications, vol. 48. Springer, New York (2011)
25. Kannappan, P.: Functional Equations and Inequalities with Applications. Springer Monographs in Mathematics. Springer, New York (2009)
26. Lee, Y.-H.: On the Hyers-Ulam-Rassias stability of the generalized polynomial function of degree 2. J. Chuncheong Math. Soc. **22**, 201–209 (2009)
27. Maksa, G., Páles, Z.: Hyperstability of a class of linear functional equations. Acta Math. Acad. Paedag. Nyìregyháziensis **17**, 107–112 (2001)
28. Malejki, R.: Stability of a generalization of the Fréchet functional equation. Ann. Univ. Paedagog. Crac. Stud. Math. **14**, 69–79 (2015)
29. Moslehian, M.S., Rassias, J.M.: A characterization of inner product spaces concerning an Euler-Lagrange identity. Commun. Math. Anal. **8**, 16–21 (2010)
30. Nikodem, K., Pales, Z.: Characterizations of inner product spaces by strongly convex functions. Banach J. Math. Anal. **5**, 83–87 (2011)
31. Piszczek, M.: Remark on hyperstability of the general linear equation. Aequationes Math. **88**, 163–168 (2014)
32. Popa, D., Raşa, I.: The Fréchet functional equation with application to the stability of certain operators. J. Approx. Theory **164**, 138–144 (2012)

33. Rassias, Th.M.: New characterizations of inner product spaces. Bull Sci. Math. (2) **108**, 95–99 (1984)
34. Sikorska, J.: On a direct method for proving the Hyers-Ulam stability of functional equations. J. Math. Anal. Appl. **372**, 99–109 (2010)
35. Zhang, D.: On hyperstability of generalised linear functional equations in several variables. Bull. Aust. Math. Soc. **92**, 259–267 (2015)
36. Zhang, D.: On Hyers-Ulam stability of generalized linear functional equation and its induced Hyers-Ulam programming problem. Aequationes Math. **90**, 559–568 (2016)

Chapter 12
Miscellanea About the Stability of Functional Equations

Zenon Moszner

Abstract The interesting details about the stability, the superstability, the inverse stability, the absolute stability and the stability in a class for a functional equation, for a system, and the alternation of functional equations, about the approximation of approximation and about the nearness of two approximations are given.

Keywords Stability · Superstability · Inverse stability · Absolute stability · Stability in the class · Stability of the system · Stability of the alternation · Stability of conditional equation · Approximation of approximation · Nearness of two approximations · Translation equation · Geometric concomitant equation · Dynamical system · Uniform b-stability · Inverse b-stability · Stability of difference equation · Questions

Mathematics Subject Classification (2010) Primary 39B82; Secondary 39B62

12.1 Introduction

It is well known that the stability theory of functional equations is inspired by the following S. Ulam's question presented in 1940 and published in [28, p. 63]: when can one assert that the solutions of the inequality lie near to the solutions of the strict equation?

More exactly [28, p. 64]: for what metric groups G with a metric d it is true that for every $\varepsilon > 0$ there exists a k such that for every function $g : G \to G$ with

$$d[g(xy), g(x)g(y)] \leq \varepsilon, \qquad x, y \in G,$$

Z. Moszner (✉)
Institute of Mathematics, Pedagogical University of Cracow, Kraków, Poland
e-mail: zmoszner@up.krakow.pl

© Springer Nature Switzerland AG 2019
J. Brzdęk et al. (eds.), *Ulam Type Stability*,
https://doi.org/10.1007/978-3-030-28972-0_12

there exists a homomorphism $f : G \to G$ for which

$$d[g(x), f(x)] \leq k\varepsilon, \qquad x \in G.$$

It is possible to interpret these questions in various non-equivalent manners [19]. Hyers was the first to give the following interpretation [9]:

Let B_1 and B_2 be the Banach spaces. Does, for every $\varepsilon > 0$, there exist a $\delta > 0$ such that to each function $g : B_1 \to B_2$ with

$$|g(x + y) - g(x) - g(y)| < \delta, \qquad x, y \in B_1, \tag{12.1}$$

there corresponds an additive function $f : B_1 \to B_2$ such that

$$|g(x) - f(x)| \leq \varepsilon, \qquad x \in B_1. \tag{12.2}$$

Hyers proved in [9] (by, so-called, "direct method") that this property is true with $\delta = \varepsilon$ and in this case the equation

$$f(x + y) = f(x) + f(y) \tag{12.3}$$

is said to be stable (in the Hyers sense).

The Hyers result also is true with $\delta = \varepsilon$ if the inequalities in (12.1) and (12.2) are "$< \delta$ and $< \varepsilon$" or "$\leq \delta$ and $\leq \varepsilon$" [15]. On the contrary, it is true with "$\leq \delta$ and $< \varepsilon$" only when $\delta < \varepsilon$ (consider $g(x) = x + \delta$ in (12.1) and (12.2)).

12.2 Stability

(a) The Hyers result has been generalized in different directions.

By a simple modification of the Hyers proof of his theorem we obtain that the Hyers result is valid also when B_1 is a commutative semigroup (this remark is already in [6]) and B_2 is a commutative semigroup divisible by 2 and complete with respect to a metric d for which $d(2a, 2b) \geq 2d(a, b)$ for $a, b \in B_2$ [16].

If B_1 is the semigroup such that

$$\bigwedge_{x, y \in B_1} \bigvee_{n \in \mathbb{N}, \, n \geq 2} n(x + y) = nx + ny,$$

then Eq. (12.3) is stable.

Indeed, in the paper [5] it is proved that, for every function g from this semigroup B_1 to $\mathbb{R}$ such that $|g(x + y) - g(x) - g(y)|$ is bounded, we have $g = f + h$, where f is a solution of (12.3) and the function h is bounded (we do not have any estimate for h in [5]). Thus

$$\frac{g(nx)}{n} = \frac{f(nx)}{n} + \frac{h(nx)}{n} = f(x) + \frac{h(nx)}{n} \to f(x) \qquad \text{for } n \to +\infty.$$

If $|g(x+y)-g(x)-g(y)| \le \varepsilon$, then by the induction $|g(nx)-ng(x)| \le (n-1)\varepsilon$. This implies that $|f(x) - g(x)| \le \varepsilon$.

The supposition that the semigroup B_1 is commutative can be replaced by the assumption that B_1 is left (right) amenable [27].

Moreover, if Eq. (12.1) is stable for the functions f from a semigroup S to the nontrivial Banach space B_1 (i.e., $B_1 \ne \{0\}$), then this equation is stable for the functions f from the semigroup S to the Banach space B_2, too [8].

(b) The above stability is formulated for the other functional equations by the following way. Let

$$L(f) = R(f) \tag{12.4}$$

be a functional equation in which $f : S_1 \to S_2$ is the unknown function, S_1 is an arbitrary nonempty set, S_2 is metric space with a metric d and $L(f)$ and $R(f)$ have their values in S_2. This equation is said to be stable (Ulam-Hyers stable) if for every $\varepsilon > 0$ there exists a $\delta > 0$ such that, for every solution $g : S_1 \to S_2$ of the inequality

$$d[L(g), R(g)] \le \delta, \tag{12.5}$$

there exists a solution f of Eq. (12.4) such that

$$d[g(x), f(x)] \le \varepsilon, \qquad x \in S_1. \tag{12.6}$$

In this case there exists a function $\Phi : (0, +\infty) \to (0, +\infty)$ such that for every $\varepsilon > 0$ if $d[L(g), R(g)] \le \Phi(\varepsilon)$, then we have (12.6) for some solution f of (12.4). If the function Φ is unbounded the stability is called normal.

Comment All functions considered below are from $\mathbb{R}$ to $\mathbb{R}$ and the metric in $\mathbb{R}$ is natural, unless explicitly stated otherwise.

(c) If the inequality (12.5) does not have any solution for some $\delta_0 > 0$, then Eq. (12.4) does not have solutions either and it is stable! It is sufficiently to put $\delta = \delta_0$ for every $\varepsilon > 0$. For instance, the equation $f(x)^2 + 1 = 0$ is of this type. Here the inequality (12.5): $|f(x)^2 + 1| \le \delta$ does not have any solution for every $0 \le \delta < 1$. The inequality (12.5) for the equation $[f(x)]^{-1} = 0$ has a solution for every $\delta > 0$, but this equation is not stable since it does not have any solution.

(d) For two equivalent equations the first may be stable and the second unstable. E.g., the equations $\exp[f(x)+f(y)-f(xy)] = 1$ and $\exp f(xy) = \exp[f(x)+ f(y)]$ are of this type. Indeed, if we have

$$|\exp[g(x) + g(y) - g(xy)] - 1| \le \frac{\exp\varepsilon - 1}{\exp\varepsilon + 1}, \qquad x, y \in \mathbb{R},$$

for some function g, then

$$|\exp g(x) - 1| \le \frac{\exp\varepsilon - 1}{\exp\varepsilon + 1}$$

and so $|g(x) - 0| \le \varepsilon$, whence the stability follows with

$$\delta = (\exp\varepsilon - 1)(\exp\varepsilon + 1)^{-1}.$$

The function $f(x) = 0$ is the unique solution of the second equation. Assume that this equation is stable. For $\varepsilon = 1$ there exists a $\delta > 0$ such that, if

$$|\exp g(xy) - \exp[g(x) + g(y)]| \le \delta,$$

then $|g(x) - 0| \le 1$. Let $n \in \mathbb{N}$ be such that $|\ln\frac{1}{n}| > 1$ and $|\frac{1}{n}(\frac{1}{n} - 1)| \le \delta$. Then the function $g(x) = 0$ for $x \ne 0$ and $g(0) = \ln\frac{1}{n}$ is a solution of the inequality (12.5) and $|g(0) - 0| > 1$, thus a contradiction.

(e) Let $L_i(f) = R_i(f)$ for $i = 1, 2$ be the two functional equations as in (12.4). Assume that for every $\delta > 0$ the inequalities

$$d[L_1(g), R_1(g)] \le \delta \quad \text{and} \quad d[L_2(g), R_2(g)] \le \delta \tag{12.7}$$

are equivalent. Then

a/ the equations $L_1(f) = R_1(f)$ and $L_2(f) = R_2(f)$ are equivalent, too;
b/ if the equation $L_1(f) = R_1(f)$ is stable, then the equation $L_2(f) = R_2(f)$ is stable and vice versa.

Proof Assume that there exists a solution f of the first equation which is not a solution of the second equation. In this case $L_2(f) \ne R_2(f)$ for some values v_0 of the variables in this equation. For $\delta := \frac{1}{2}d[L_2(f), R_2(f)] > 0$, where the variables in $L_2(f)$ and $R_2(f)$ have the values v_0, we obtain that the function f is a solution of the first inequality in (12.7) and it is not the solution for the second inequality in (12.7), thus a contradiction.

Assume that the equation $L_1(f) = R_1(f)$ is stable and let ε and δ be as in the definition of the stability. Let $g : S_1 \to S_2$ be such that $d[L_2(g), R_2(g)] \le \delta$. Thus $d[L_1(g), R_1(g)] \le \delta$. This yields that there exists a solution f of equation $L_1(f) = R_1(f)$ for which $d[g(x), f(x)] \le \varepsilon$. The proof is finished since the function f is the solution of equation $L_2(f) = R_2(f)$, too.

(f) Consider a little stronger version of the Ulam-Hyers stability, namely the s-stability (see, e.g., [11]). Equation (12.4) is said to be s-stable if, for every $\delta > 0$, there exists a $K(\delta)$ such that $\lim_{\delta \to 0} K(\delta) = 0$ and, for every solution $g : S_1 \to S_2$ of the inequality (12.5), there exists a solution f of the equation such that

$$d[g(x), f(x)] \le K(\delta), \qquad x \in S_1.$$

If $\inf K[(0, +\infty)] = 0$, then this stability is said to be normal uniformly b-stable in [20].

The s-stability implies the Ulam-Hyers stability. In fact, for $\varepsilon > 0$ there exists a $\delta > 0$ such that $K(\delta) \le \varepsilon$. We have $d[g(x), f(x)] \le K(\delta) \le \varepsilon$ for this δ. The implication inverse is not true. E.g., Eq. (12.3), where $f : G \to \mathbb{Z}$, $\mathbb{Z}$ is the set of integer numbers with the usual metric and G is the free group generated by two elements, is evidently stable with $\delta < 1$. It is not s-stable since there exists a function $g : G \to \mathbb{Z}$ for which the function $g(x + y) - g(x) - g(y)$ is bounded and the function $g(x) - f(x)$ is unbounded for every homeomorphism $f : G \to \mathbb{Z}$ [6].

12.3   Stability of System

There exist the unstable (stable) system of stable (unstable) functional equations. E.g., the equations

$$(|f(x) - 1| - f(x) + 1) f(x) = 0 \quad \text{and} \quad E(f(x)) = 0,$$

where $E(u)$ denotes the integer part of u, are stable separately and the system of these equations is not stable.

The system of the equations

$$(|f(x) - 1| - f(x) + 1)|f(x)| + |E(f(x))| = 0,$$
$$(|f(x) + 1| + f(x) + 1)|f(x)| + |E(-f(x))| = 0$$

is stable and the equations in this system are unstable (for the proofs see [14]). The above equations are not natural. For the natural equations see Sect. 12.16.

Moreover if one the equation is stable and the other unstable, then their system may be stable (unstable). Indeed, for $f : \mathbb{R} \to \mathbb{R} \setminus \{0\}$ the equation $f^2(x) + 1 = 0$ is stable, the equation $f(x) = 0$ is unstable and their system is stable.

However, for the same f, the equation $f(f(x)) = f(x)$ is stable, the equation $f(x) = 0$ is unstable and their system is unstable.

Indeed, if for $g : \mathbb{R} \to \mathbb{R} \setminus \{0\}$ we have $|g(g(x)) - g(x)| \le \delta$, then for $f(x) = x$ for $x \in g(\mathbb{R})$ and $f(x) = g(x)$ for $x \in \mathbb{R} \setminus g(\mathbb{R})$ we obtain $f(f(x)) = f(x)$ and $|g(x) - f(x)| \le \delta$ (the same proof is good for $f : S_1 \to S_2$ if $S_2 \subset S_1$ and S_2 is a metric space with the metric denoted $|a - b|$). The system of their equations is unstable since for the function $g(x) = \delta$ we have $|g(g(x)) - g(x)| \le \delta$ and $|g(x)| \le \delta$ and the system of the equations in consideration does not have the solution.

By the above, the stability (unstability) of the system gives no information about the stability of the equations in this system.

12.4 Slightly Differing Equations

Ulam has formulated in [28] the following question, too: When is true that the solution of an equation differing slightly from a given one, must of necessity be close to the solution of the given equation? This motivates the following definitions.

The equations $L_1(f) = R_1(f)$ and $L_2(f) = R_2(f)$ are said to be δ-close if $d[L_1(f), L_2(f)] \leq \delta$ and $d[R_1(f), R_2(f)] \leq \delta$. These δ-close equations are called stable for the solutions if there exists a $\varepsilon > 0$ such that for every solution f_1 of the first equation there exists a solution f_2 of the second equation for which $d(f_1, f_2) \leq \varepsilon$ and vice versa. E.g., the equations $f(x + y) = f(x) + f(y)$ and $f(x + y) + \delta = f(x) + f(y)$ for $f : \mathbb{R} \to \mathbb{R}$ are δ-close and stable for the solutions with $\varepsilon = \delta$. The same equations for $f : \mathbb{R} \to \mathbb{R} \setminus \{0\}$ are not stable for the solutions, since $f(x) = \delta$ is the solution of the second equation and the first equation does not have any solution.

12.5 Stability of Alternation

The alternation

$$L_1(f) = R_1(f) \quad \text{or} \quad L_2(f) = R_2(f) \tag{12.8}$$

of two functional equations is said to be stable if, for every $\varepsilon > 0$, there exists a $\delta > 0$ such that, for every function $g : S_1 \to S_2$ for which

$$d[L_1(g), R_1(g)] \leq \delta \quad \text{or} \quad d[L_2(g), R_2(g)] \leq \delta,$$

there exists a solution f of this alternation for which $d[g(x), f(x)] \leq \varepsilon$ for $x \in S_1$.

We have for this stability the same situation as for the system.

Let f be a function from $\mathbb{R}$ to $\mathbb{R} \setminus \{0\}$. The equations $|f(x)| = x$ and $|f(x)| = -x$ are stable and their alternation is not stable.

The equations $f(x) = f(1)x$ and $f(x) = f(0)(1 - x)$ are unstable and their alternation is stable.

From the equations $f(x) = 2$ and $f(x) = x$ the first one is stable, the second one is unstable and their alternation is unstable.

From the equations $f(x) = f(x)$ and $f(x) = f(1)x$ the first is evidently stable, the second is not stable and their alternation is evidently stable [19].

Here we have the same situation as for the system: the stability (unstability) of the alternation gives no information about the stability of the equations in this alternation.

The alternation (12.8) may be written in the form

$$L_1(f) \neq R_1(f) \Rightarrow L_2(f) = R_2(f)$$

and the stability of it is defined as the stability of the alternation.

E.g., in the theory of elections [25] the generalized indicator plurality function $f : \mathbb{R}^n \setminus \{(0, \ldots, 0)\} \to \mathbb{R}^m \setminus \{(0, \ldots, 0)\}$ is considered, which is a solution of following functional equation

$$f(x) \cdot f(y) \neq (0, \ldots, 0) \Rightarrow f(x + y) = f(x) \cdot f(y), \tag{12.9}$$

where $(u_1, \ldots, u_m) \cdot (v_1, \ldots, v_m) = (u_1 v_1, \ldots, u_m v_m)$ and $(x_1, \ldots, x_n) + (y_1, \ldots, y_n) = (x_1 + y_1, \ldots, x_n + y_n)$.

This equation is not stable for $n = m = 1$. In this case our equation has the only solution of the form $f(x) = \exp a(x)$, where $a(x)$ is the additive function. Assume that for $\varepsilon = 1/2$ there exists a $\delta > 0$ such as in the definition of stability. It is possible to suppose that $\delta \leq 1/16$. We have $|g(x)g(y)| \leq \delta$ for $g(x) = \sqrt{\delta}$, thus there exists a additive function $a(x)$ such that $|g(x) - \exp a(x)| = |\sqrt{\delta} - \exp a(x)| \leq 1/2$. For $a(x) \neq 0$ we obtain a contradiction. For $a(x) = 0$ we have

$$\frac{1}{2} \geq |\sqrt{\delta} - 1| = 1 - \sqrt{\delta} \geq 1 - \frac{1}{4} = \frac{3}{4},$$

thus a contradiction, too.

Question Is the situation the same for the other n, m?

12.6 Stability of the Conditional Functional Equation

It is possible to consider the conditional functional equation of the form

$$(C) \Rightarrow L(f) = R(f),$$

where (C) is a condition. The stability of this equation is defined in the following way: for every $\varepsilon > 0$ there exists a $\delta > 0$ such that, for every function $g : \mathbb{R} \to \mathbb{R}$ for which $(C) \Rightarrow d[L(g), R(g)] \leq \delta$ it is true, there exists a solution f of our conditional equation such that $d(g, f) \leq \varepsilon$.

We have here two possibilities: the condition (C) depends on f or not. E.g., the conditional equation

$$(f \text{ is a function differentiable at every point of } \mathbb{R}) \Rightarrow f(f(x)) = f(x)$$

is stable. The functions which are not differentiable at least at one point and the idempotent function are the all solutions of our equation. If the function $g : \mathbb{R} \to \mathbb{R}$ is a solution of the conditional inequality

$$(g \text{ is a function differentiable at every point of } \mathbb{R}) \Rightarrow |g(g(x)) - g(x)| \leq \delta$$

for a $\delta > 0$, then g is not differentiable at least at one point or $|g(g(x)) - g(x)| \leq \delta$. In the first case g is a solution of our equation and $|g(x) - g(x)| = 0 \leq \delta$. In the second case $|g(g(x)) - g(x)| \leq \delta$ and for the function f, given by: $f(x) = x$ for $x \in g(\mathbb{R})$ and $f(x) = g(x)$ for $x \in \mathbb{R} \setminus g(\mathbb{R})$, we have $|g(x) - f(x)| \leq \delta$. Moreover the function f is a solution of our equation since $f(f(x)) = f(x)$.

Equation (12.9) for $n = m = 1$ as the conditional equation is not stable (the proof as in Sect. 12.7 below).

The conditional equation

$$f(x) = f(y) \Rightarrow f(F(x, t)) = f(F(y, t)), \tag{12.10}$$

where the given function F (the transformation law) is a solution of the translation equation with the identity condition, plays the role in the theory of the concomitants of geometric objects. Exactly, the concomitant $f(x) \in S_1$ of a geometric object $x \in S$ with the transformation law F is the geometric object if and only if f is a solution of the above conditional equation [22]. If $f(x)$ is the geometric object, then there exists a transformation law G of this object and we have

$$G(f(x), t)] = f[F(x, t)]. \tag{12.11}$$

This equation is said to be the geometric concomitant equation. This equation implies (12.10) ($\varphi[F(x, t)] = F(\varphi(x), t) = F(\varphi(y), t) = \varphi[F(y, t)] \text{ if } \varphi(x) = \varphi(y)$) but not vice versa even in the case $G = F$. Indeed, e.g., let $(G, +)$ be a semigroup with the neutral element 0 and let $F(x, t) = x + t$ for $x, t \in G$. The injection f from G to G is evidently the solution of (12.10). If moreover f is not the identity and $f(0) = 0$, then it is not a solution of (12.11) since $f(x + t) = f(x) + t$ implies $f(t) = t$ for $t \in G$.

Equations (12.10) and (12.11) are evidently stable if $G(x, t) = F(x, t) = x$. Let $(G, +)$ be as above and let S be the metric space with the metric d such that card $G =$ card S. Equation (12.11) is stable if $G(x, t) = F(x, t) = g^{-1}[g(x) + t]$, where g is a bijection from S to G. In fact, assume that $d[\psi(g^{-1}[g(x) + t]), g^{-1}[[g(\psi(x)) + t]] \leq \varepsilon$ for a function $\psi : S \to S$ and some $\varepsilon > 0$. Let x_0 be such that $g(x_0) = 0$. We have thus that

$$d[\psi(x), g^{-1}[g(\psi(x_0)) + g(x)] \leq \varepsilon, \qquad x \in S.$$

Since the function $g^{-1}[g(\psi(x_0)) + g(x)]$ is a solution of (12.11), so this equation is stable with $\delta = \varepsilon$.

Every transformation law $F : S \times G \to S$, where S is a set and G is a group, i.e., the function F for which

$$F(F(x, t), s) = F(x, t + s), \quad x \in S, \, t, s \in G \quad \text{and} \quad F(x, 0) = x, \quad x \in S,$$

is of the form

$$F(x, t) = h_n^{-1}[h_n(x) + t], \qquad x \in S_n, t \in G,$$

where S_n for $n \in N_1 \subset \mathbb{N}$ are the non-empty sets such that $\bigcup S_n = S$ and, for every $n \in N_1$, there exists a subgroup G_n of G for which $\operatorname{card} S_n = \operatorname{ind} G_n$ and h_n is a bijection from S_n to the family of the right cosets of G by G_n for $n \in N_1$ [13].

Analogously, the transformation law $G : S_1 \times G \to S_1$ is of the form

$$G(x, t) = g_m^{-1}[g_m(x) + t], \qquad x \in S_m^*, t \in G,$$

where S_m^* for $m \in N_2 \subset \mathbb{N}$ are the non-empty sets such that $\bigcup S_m^* = S_1$ and for every $m \in N_2$ there exists a subgroup G_m^* of G for which $\operatorname{card} S = \operatorname{ind} G_m^*$ and g_m is a bijection from S_m^* to the family of right cosets of G by G_m^* for $m \in N_2$.

If the group G is abelian and S_1 is the metric space with the metric d and $G_n = G^*$ for $n \in N_1$ and $G_m^* = G^*$ for $m \in N_2$, then Eq. (12.11) is stable. Indeed, assume that $d[g(F(x, t)), G(g(x), t)] \leq \varepsilon$ for a function $g : S \to S_1$ and some $\varepsilon > 0$. Let $x \in S_k$ and $g(x) \in S_l^*$. We have thus

$$d\left[g[h_k^{-1}(h_k(x) + t)], \, g_l^{-1}\left[g_l(g(x)) + t\right]\right] \leq \varepsilon.$$

There exists x_0 such that $h_k(x_0) = G_k = G^*$. For every $x \in S_k$ there exists $t(x) \in G$ such that $h_k^{-1}[h_k(x_0) + t(x)] = h_k^{-1}[G^* + t(x)] = x$, whence $G^* + t(x) = h_k(x)$. There exists $a_l \in G$ such that $g_l(g(x_0)) = G_l^* + a_l = G^* + a_l$. From here $g_l(g(x_0)) + t(x) = G^* + a_l + t(x) = h_k(x) + a_l$. Thus we obtain

$$d[g(x), g_l^{-1}(h_k(x) + a_l)] \leq \varepsilon.$$

The function $f(x) = g_l^{-1}(h_k(x) + a_l)$ is the solution of (12.11). In fact, for $x \in S_k, g(x) \in S_l^*$ we have $f(x) \in S_l^*$ and this yields that

$$f[F(x, t)] = f[h_k^{-1}(h_k(x) + t)] = g_l^{-1}(h_k(x) + t + a_l) = g_l^{-1}(h_k(x) + a_l + t)$$

$$= g_l^{-1}(g_l(f(x)) + t) = G(f(x), t).$$

Equation (12.11) is thus stable with $\delta = \varepsilon$.

We note that the function g does not occur clearly in the form of the function f. But the dependance of f on g is in "l" by the condition $g(x) \in S_l^*$.

The supposition that F and G are the transformation laws is essential in the above considerations. Indeed, e.g., for the functions $F(x, t) = x + c$ with $c > 0$ and $G(x, t) = x$ from $\mathbb{R} \times \mathbb{R}$ to $\mathbb{R}$, Eq. (12.11) is not stable. It has the form: $f(x + c) = f(x)$ and this equation of periodic function is not stable. In fact, for the function

$$g(x) = \frac{\delta}{c} x$$

we have $|g(x + c) - g(x)| \leq \delta$ and we obtain for every periodic function f with period c that $g(nc) - f(nc) = \delta n - f(0)$, thus the function $g(x) - f(x)$ is unbounded.

Equation (12.11) may be stable for the transformation laws F and G such that the subgroups G_k and G_i^* are not equal. E.g., put $S = [0, +\infty)$, $S_1 = (-\infty, 0)$,

$$F(x, t) = x \exp t = h_n^{-1}[h_n(x) + t], \qquad x \in S, t \in \mathbb{R}, n = 1, 2,$$

where $h_1(x) = [\ln x] : (0, +\infty) \to \mathbb{R}/\{0\}$ and $h_2(x) = [0] : \{0\} \to \mathbb{R}/\mathbb{R}$, $G(x, t) = x \exp t = g_1^{-1}[g_1(x) + t]$ for $x \in S_1$, $t \in \mathbb{R}$, where $g_1(x) = \ln(-x) : (-\infty, 0) \to \mathbb{R}$. Equation (12.11) is stable in this case since the function $g[F(x, t)] - G(g(x), t)$ is unbounded for every function $g : S \to S_1$ (the function $g(0 \cdot \exp t) - g(0) \exp t$ is unbounded). The last example is not very interesting since the inequality (thus the equation, too) does not have any solution.

Problem Give a suitable example without this deficiency.

There exist the transformation laws F and G for which Eq. (12.11) is not stable. Put $S = \mathbb{R}$,

$$S_1 = \left[\bigcup_{n \in \mathbb{N}} \left(\frac{1}{n+1}, \frac{1}{n} \right) \right] \cup \{2\},$$

$G = (\mathbb{R}, +)$, $F(x, t) = x$ for $x, t \in \mathbb{R}$,

$$G(x, t) = g_n^{-1}[g_n(x) + t] \quad \text{for} \quad x \in (1/(n+1), 1/n), t \in \mathbb{R},$$

where g_n is the bijection from $(1/(n+1), 1/n)$ to $\mathbb{R}$, and $G(2, t) = 2$ for $t \in \mathbb{R}$. The function $f(x) = 2$ is the only solution of Eq. (12.11). Indeed, if f is a solution of (12.11), then $f(x) = G(f(x), t)$, thus $f(x)$ is the fixed point of the function G. It is impossibly that

$$f(x) \in \bigcup_{n \in \mathbb{N}} \left(\frac{1}{n+1}, \frac{1}{n} \right),$$

whence $f(x) \in \{2\}$.

Assume that (12.11) is stable. This yields that for $\varepsilon = 1$ there exists a $\delta > 0$ such that, for every function $g : S \to S_1$, if

$$|g[F(x, t)] - G(g(x), t)| \leq \delta, \qquad x, t \in \mathbb{R},$$

then $|g(x) - 2| \leq 1$. Let $n \in \mathbb{N}$ be such that $1/n - 1/(n+1) \leq 2\delta$. For the function $g(x) = (2n+1)/(2n(n+1))$ we obtain

$$|g[F(x,t)] - G(g(x), t)| = |g(x) - G(g(x), t)| \leq 1/(2n(n+1)) \leq \delta,$$

since $g(x)$ is the midpoint of the interval $(1/(n+1), 1/n)$ and consequently we have $G(g(x), t) \in (1/(n+1), 1/n)$, too. Thus

$$1 \geq |g(x) - 2| = |(2n+1)/(2n(n+1)) - 2| > 1.$$

Question For which transformation laws F (for which transformation laws F and G) Eq. (12.10) (Eq. (12.11)) is stable?

We note that the implications

$$g(x) = g(y) \Rightarrow d\{g[F(x,t)], F(g(x), t)\} \leq \delta$$

and

$$d[g(x), g(y)] \leq \delta \Rightarrow d\{g[F(x,t)], F(g(x), t)\} \leq \delta$$

are not equivalent. The second implies evidently the first but non vice versa. Put $F(x,t) = x + t : \mathbb{R} \times \mathbb{R} \to \mathbb{R}$. Let γ be the injection from $\mathbb{R}/\mathbb{Q}$ to $\mathbb{R}$ such that $\gamma([0]) = 0$, $\gamma([\pi]) = \delta$ and $\gamma([2\pi]) = 3\delta$, where $[a]$ means the coset in $\mathbb{R}/\mathbb{Q}$ for which $a \in [a]$. Put $g(x) = \gamma([x])$ for $x \in \mathbb{R}$. Our implications have in this case the following form

$$\gamma([x]) = \gamma([y]) \Rightarrow |\gamma([x+t]) - \gamma([y+t])| \leq \delta$$

and

$$|\gamma([x]) - \gamma([y])| \leq \delta \Rightarrow |\gamma([x+t]) - \gamma([y+t])| \leq \delta.$$

The first of these implications is true since γ is an injection. The second is false, since for $x = 0$ and $y = t = \pi$ it has the form

$$|0 - \delta| \leq \delta \Rightarrow |\delta - 3\delta| \leq \delta.$$

Conclusion The first of our implications has more solution than the second, despite the fact that the family of functions g for which $|g(x) - g(y)| \leq \delta$ is larger than the family of functions g such that $g(x) = g(y)$.

If the condition (C) does not depend on the function f, then the equation is said to be the equation on a restricted domain, too. In fact, it is proved in [10] that the conditional equation of Jensen

$$|x| + |y| \geq a \Rightarrow 2f\left(\frac{x+y}{2}\right) = f(x) + f(y)$$

for f from a normed space to a Banach space and with some $a > 0$, is stable.
   The equation

$$1 + \alpha xy \neq 0 \Rightarrow f\left(\frac{x+y}{1+\alpha xy}\right) = f(x)f(y)(1+\alpha xy)$$

for $f : \mathbb{R} \to \mathbb{R}$ and $\alpha > 0$, which is from the special theory of relativity, is a conditional equation, too.

Question Is this equation stable?

12.7   Superstability

Equation (12.4) is said to be superstable if the inequality $d[L(f), R(f)] \leq \delta$ for some $\delta > 0$ implies that the function f is bounded or it is a solution of Eq. (12.4).
   The superstable equation may be unstable. E.g., by [1] (the first paper on the subject of superstability) the equation $f(x+y) = f(x)f(y)$, for $f : \mathbb{R} \to (0, +\infty)$, is superstable. Assume that this equation is stable. Then for $\varepsilon = 1/2$ there is a $\delta_0 > 0$ such that for every function $g : \mathbb{R} \to (0, +\infty)$ with $|g(x + y) - g(x)g(y)| \leq \delta_0$, there exists a solution f of the equation such that $|g(x) - f(x)| \leq 1/2$. This condition is satisfied for all $\delta \leq \delta_0, \delta > 0$. For

$$g(x) = \frac{1 - \sqrt{1 - 4\delta}}{2},$$

where $0 < \delta \leq \min\{\delta_0, 1/4\}$, we have $g(x + y) - g(x)g(y) = \delta$, thus there exists a solution f of the equation such that $|g(x) - f(x)| \leq 1/2$. The function g is bounded, hence the function f is bounded, too. Thus $f(x) = 1$. So, we have

$$\frac{1}{2} \geq \left|\frac{1 - \sqrt{1 - 4\delta}}{2} - 1\right| \to 1 \quad \text{for} \quad \delta \to 0,$$

which is a contradiction.

12.8   Stability of the Squares of Functional Equations

The square of the functional equation $L(f) = R(f)$ is of the form $[L(f)]^2 = [R(f)]^2$, if this operation is defined. The stabilities (the superstabilities) of the equations of this form are considered in many papers. E.g., in the paper [3], the superstability of the equation of the square of Cauchy equation $[f(x + y)]^2 = [f(x) + f(y)]^2$, (this equation is not equivalent to the Cauchy equation [12]) has been proved for the function f from an abelian semigroup to a finite-dimensional normed algebra, which is commutative (this supposition is omitted in [3]). In the paper [18] is proved that the equation $[f(y)]^2 = [f(x + y) - f(x)]^2$ is superstable if f is mapping an abelian semigroup to the algebra $A \in \{\mathbb{R}, \mathbb{C}\}$ and the equation $[f(x+y)]^2 = [f(x)+f(y)]^2$ is stable for the function f from a semigroup divisible by 2 to the algebra with multiplicative norm.

Attention! The Lemma 2.8 in the paper [18] is not true! It reads as follows:

Let G be a groupoid divisible by 2. If the function $f : G \to A$, *where* $A \in \{\mathbb{R}, \mathbb{C}\}$, *is a solution of the equation* $[f(2x) - 2f(x)]f(2x) = 0$, *then* $f(2x) = 2f(x)$. *Thus if f is bounded, then* $f = 0$.

Indeed, define a function $f : \mathbb{R} \to \mathbb{R}$ by $f(x) = x$ for $x = 2^k$, where $k = 0, -1, -2, \ldots$, and $f(x) = 0$ otherwise. Then f is a bounded solution of the above equation, $0 = f(2) \neq 2f(1) = 2$ and it is not identically equal zero.

This error may affect validity of some results in the paper [18], in which this lemma has been used. These results are: Theorem 2.7, the first part of Proposition 2.26, and the statement that the alternations

$$f(x + y) - f(x) - f(y) = 0 \qquad \text{or} \qquad f(x + y) - f(x) + f(y) = 0$$

and

$$f(2x) - 2f(x) = 0 \qquad \text{or} \qquad f(2x) = 0$$

are unstable.

Question Are the above results true even though the lemma 2.8 is false? Especially, is the equation $[f(x + y) - f(x)]^2 = f(y)^2$ stable?

12.9   Inverse Stability

(a) The Ulam question may be inverted as follows: suppose that g can be approximated by a solution of an equation. Is g in this case an approximate solution of this equation?

More exactly: Eq. (12.4) is said to be inversely stable if for every $\varepsilon > 0$ there exists a $\delta > 0$ such that, for every function $g : S_1 \to S_2$ for which there exists a solution f of Eq. (12.4) such that $d(g, f) \leq \delta$, we have $d[L(g), R(g)] \leq \varepsilon$ [17].

In this case there exists a function $\Psi : (0, +\infty) \to (0, +\infty)$ such that, for every $\varepsilon > 0$, if $d(g, f) \leq \Psi(\varepsilon)$ for some solution f of Eq. (12.4), then $d[L(g), R(g)] \leq \varepsilon$. If the function Ψ is unbounded, then the inverse stability is called normal.

The equation $f(x) = 0$ is evidently stable and inversely stable. The equation $f(f(x)) = f(x)$ is stable (see Sect. 12.3) and it is not inversely stable [17]. The equation of homomorphism $f(xy) = f(x) + f(y)$ from the free group generated by two elements to $\mathbb{R}$ is unstable (see [6] and the remark in Example 1 in [16]) and it is not inversely stable for $f : \mathbb{R} \to \mathbb{R}$ [15]. The equation $f(x + y) = f(x)f(y)$ for $f : \mathbb{R} \to (0, +\infty)$ is unstable (see Sect. 12.7) and it is not inversely stable [17].

By the above, the stability and the inverse stability are independent.

(b) If the equation does not have any solution, then it is evidently inversely stable.

Let $(S_1, +)$ be a groupoid and $(S_2, +)$ be a groupoid with a metric d, which is invariant with respect to the operation "$+$" in S_2. Then Eq. (12.3) is inversely stable for $f : S_1 \to S_2$ by the following inequalities

$$d[g(x + y), g(x) + g(y)] \le d[g(x + y), f(x + y)] + d[f(x + y), f(x) + f(y)]$$
$$+ d[f(x) + f(y), g(x) + f(y)]$$
$$+ d[g(x) + f(y), g(x) + g(y)]$$
$$= d[g(x + y), f(x + y)] + d[f(x + y), f(x) + f(y)]$$
$$+ d[f(x), g(x)] + d[f(y), g(y)],$$

where $f, g : \mathbb{R} \to \mathbb{R}$ and f is a solution of (12.3).

The equation

$$f(x + y)^2 = [f(x) + f(y)]^2 \tag{12.12}$$

is equivalent to Eq. (12.3) if $f : \mathbb{R} \to \mathbb{R}$ [8] and it is inversely unstable. Indeed, for $g(x) = x + \delta$ we have $|g(x) - x| \le \delta$ and the function

$$|g(x + y)^2 - [g(x) + g(y)]^2| = (2x + 2y + 3\delta)\delta$$

is unbounded.

12.10   Inverse Stability of System

The system of two inversely stable equations is evidently inversely stable.

The system of two inversely unstable equations may be inversely stable. E.g., Eq. (12.12) and $f(x)^2 = |x|$ are not inversely stable and their system is inversely stable since it does not have the solution.

The system may be inversely stable if one of the equations in this system is inversely stable and the second is not. It is so if the first equation does not have the solution and the second is an arbitrary inversely unstable equation. The system does not have the solution, thus it is inversely stable.

Moreover the system may be inversely unstable if one of the equations in this system is inversely stable and the second is not. E.g., the system of equations: (12.3) and (12.7) has this property.

Thus for the inverse stability non all cases are possible.

12.11 Inverse Stability for Alternation

The alternation (12.8) is said to be inversely stable if, for every $\varepsilon > 0$, there exists a $\delta > 0$ such that, for every function $g : S_1 \to S_2$ for which $d[g(x), f(x)] \le \delta$ for $x \in S_1$ and for some solution f of the alternation we have

$$d[L_1(g), R_1(g)] \le \varepsilon \quad \text{or} \quad d[L_2(g), R_2(g)] \le \varepsilon.$$

The alternation of two inversely stable equations may be inversely unstable. E.g., the equations $f(x)^2 = x$ and $f(x)^2 = -x$ are inversely stable as the equations which do not have the solutions. On the contrary their alternation is not inversely stable since for $g(x) = \sqrt{|x|} + \delta$ we have

$$|g(x) - \sqrt{|x|}| \le \delta$$

and the functions

$$|g(x)^2 - x| = ||x| + 2\delta|x| + \delta^2 - x|$$

and

$$|g(x)^2 + x| = ||x| + 2\delta|x| + \delta^2 + x|$$

are unbounded.

If the first equation in the alternation is inversely stable and the second is inversely unstable, then this alternation may be inversely stable. E.g., for the alternation of Eqs. (12.3) and (12.12), the first is inversely stable and the second is not. Their alternation is inversely stable since every solution of this alternation is an additive function and the first equation is inversely stable.

The equation $f(x) = 0$ is evidently inversely stable and the equation $f(x)^2 = x^2$ is not inversely stable (take the function $g(x) = x + \delta$). The alternation of these equations is not inversely stable. For the indirect proof assume that this alternation is inversely stable. Thus for $\varepsilon = 1$ there exists a $\delta > 0$ such that, for every function g, if $|g(x) - x| \le \delta$, then $|g(x) - 0| \le 1$ or $|g(x)^2 - x^2| \le 1$. For $g(x) = x + \delta$ we have $|g(x) - x| \le \delta$ and thus we obtain $|x + \delta| \le 1$ or $|(x+\delta)^2 - x^2| = |2\delta x + \delta^2| \le 1$ and this is impossible for

$$x > \max\left[0, 1 - \delta, \frac{1 - \delta^2}{2\delta}\right].$$

The equations $f(x)^2 = x^2$ and $1/f(x) = 1/x$ for $f : (0, +\infty) \to (0, +\infty)$ are not inversely stable and their alternation is inversely stable. For $g(x) = x + \delta$ we have $|g(x) - x| \le \delta$ and the functions

$$|g(x)^2 - x^2| = 2\delta x + \delta^2 \qquad \text{and} \qquad \left| \frac{1}{g(x)} - \frac{1}{x} \right| = \delta x (x + \delta)$$

are unbounded.

The function $f(x) = x$ is the only solution for the alternation of these equations. We remark that for $\varepsilon > 0$ we have

$$\alpha(\delta) := \frac{\varepsilon - \delta^2}{2\delta} \to +\infty$$

and

$$\beta(\delta) := \frac{\varepsilon \delta + \sqrt{\varepsilon^2 \delta^2 + 4\varepsilon \delta}}{2\varepsilon} \to 0$$

if $\delta \to 0+$.

Since the function $\alpha(\delta)$ is increasing for $0 < \delta \le \sqrt{\varepsilon}$ and the function $\beta(\delta)$ is decreasing for $\delta > 0$, so there exists a $\delta_0 > 0$ such that $\alpha(\delta_0) \ge 1$ and $\beta(\delta_0) \le 1$. Let the function $g : (0, +\infty) \to (0, +\infty)$ be such that $|g(x) - x| \le \delta_0$. Let $h(x) = g(x) - x$. Then $|h(x)| \le \delta_0$.

For $0 < x \le 1$ we have $x \le \alpha(\delta_0)$ and this implies that

$$|g(x)^2 - x^2| = |2xh(x) + h(x)^2| \le 2x|h(x)| + h(x)^2 \le 2x\delta_0 + \delta_0^2 \le \varepsilon.$$

For $x \ge 1$ we have

$$\left| \frac{1}{g(x)} - \frac{1}{x} \right| = \frac{|h(x)|}{|x(x + h(x))|} \le \varepsilon,$$

since

$$|x(x + h(x))| \ge x^2 - x|h(x)| \ge x^2 - x\delta_0 \ge \frac{\delta_0}{\varepsilon}$$

and therefore $\varepsilon x^2 - \varepsilon \delta_0 - \delta_0 \ge 0$ for $x \ge \beta(\delta_0)$.

The function $g(x)$ is thus the solution of the alternation

$$|g(x)^2 - x^2| \le \varepsilon \qquad \text{or} \qquad \left| \frac{1}{g(x)} - \frac{1}{x} \right| \le \varepsilon$$

and this yields that the alternation of our equations is inversely stable.

12.12 Absolute Stability

If Eq. (12.4) is stable and inversely stable, then it is said to be absolutely stable.

Equation (12.3) is absolutely stable if the function f is from the commutative semigroup S_1 to the Banach space S_2. It is not absolutely stable if S_1 is the free group generated by two elements, since it is not stable in this case (see Sect. 12.7). The equation $f(x)^2 = a$ for $a \geq 0$ is inversely stable [17]. It is stable, too; thus it is absolutely stable. Indeed, if $|g(x)^2 - a| \leq \varepsilon^2$, then $|g(x) - \sqrt{a}| \leq \varepsilon$ or $|g(x) + \sqrt{a}| \leq \varepsilon$ and consequently $|g(x) - f(x)| \leq \varepsilon$ for a solution f of the equation of the form

$$
f(x) = \begin{cases} \sqrt{a} & \text{if } |g(x) - \sqrt{a}| \leq \varepsilon, \\ -\sqrt{a} & \text{if } |g(x) - \sqrt{a}| > \varepsilon \text{ and } |g(x) + \sqrt{a}| \leq \varepsilon. \end{cases}
$$

The equation $f(f(x)) = f(x)$ is stable (see Sect. 12.6). It is not inversely stable [17], thus non absolutely stable, too. The equation $f(x + y) = f(x)f(y)$ for a function $f : \mathbb{R} \to (0, +\infty)$ is not stable (see Sect. 12.7) nor inversely stable (take the function $g(x) = \exp x + \delta$); thus it is not absolutely stable (double reason).

12.13 Approximation of Approximation

The Ulam question suggests the following question: is the approximation of the approximation of a functional equation an approximation of this equation?

More exactly: Eq. (12.4) is said to be approximately stable (in short: app-stable), if for every $\varepsilon > 0$ there exists a $\delta > 0$ such that, for every functions $g_1, g_2 : S_1 \to S_2$ for which $d[g_1(x), g_2(x)] \leq \delta$ for $x \in S_1$ and $d[L(g_1), R(g_1)] \leq \delta$, we have $d[L(g_2), R(g_2)] \leq \varepsilon$.

Equation (12.3) is app-stable if the function f is from the commutative semigroup S_1 to the Banach space S_2, since it is absolutely stable (see Sect. 12.16 and the following).

If Eq. (12.4) is absolutely stable, then it is app-stable. Assume that this equation is absolutely stable. Thus it is inversely stable. So for an $\varepsilon > 0$ there exists a $\delta_1 > 0$ such that, for every function $g : S_1 \to S_2$, if $d[g(x), f(x)] \leq \delta_1$ for some solution f of (12.4), then $d[L(g), R(g)] \leq \varepsilon$. Since Eq. (12.4) is stable too, so for $\frac{1}{2}\delta_1$ there exists a $\delta_2 > 0$ such that, for every function g, if $d[L(g), R(g)] \leq \delta_2$, then $d[g(x), f(x)] \leq \frac{1}{2}\delta_1$ for some solution f of (12.4). Assume that $d[g_1(x), g_2(x)] \leq \delta$ and $d[L(g_1), R(g_1)] \leq \delta$, where $\delta = \min(\frac{1}{2}\delta_1, \delta_2)$. Since $d[L(g_1), R(g_1)] \leq \delta \leq \delta_2$, so $d[g_1(x), f(x)] \leq \frac{1}{2}\delta_1$. Next

$$
d[g_2(x), f(x)] \leq d[g_2(x), g_1(x)] + d[g_1(x), f(x)] \leq \frac{1}{2}\delta_1 + \frac{1}{2}\delta_1 = \delta_1
$$

and consequently $d[L(g_2), R(g_2)] \leq \varepsilon$.

If the equation is app-stable, then it is inversely stable. Indeed, for $\varepsilon > 0$, let δ be such that in the definition of the approximate stability. Assume that, for the function g, there exists a solution f of (12.4) such that $d(g(x), f(x)) \leq \delta$. Since moreover $d[L(f), R(f)] = 0 \leq \delta$, so $d[L(g), R(g)] \leq \varepsilon$.

The approximate stability is not the consequence of the inverse stability. Indeed, the equation $F(f(x)) = 0$, where $F(x) = -x$ for $x < 0$ and $F(x) = 1$ for $x \geq 0$, is inversely stable, since it does not have any solution. We can prove it, by the indirect proof, using the functions $g_1(x) = \delta$ and $g_2(x) = 0$.

The stability is not the consequence of the approximate stability. E.g., Eq. (12.3) for $f : [0, +\infty) \rightarrow (0, +\infty)$ is app-stable since, if $|g_1(x) - g_2(x)| \leq \delta$ and $|g_1(x + y) - g_1(x) - g_1(y)| \leq \delta$, then

$$|g_2(x + y) - g_2(x) - g_2(y)|$$
$$\leq |g_2(x + y) - g_2(x) - g_2(y) - (g_1(x + y) - g_1(x) - g_1(y))|$$
$$+ |g_1(x + y) - g_1(x) - g_1(y)| \leq 3\delta + \delta = 4\delta.$$

Thus it is sufficient to put $\delta = \varepsilon/4$ in the definition of the approximate stability of (12.3). Equation (12.3) is not stable since it does not have any solution and the function $g(x) = x + \delta/2$ is a solution of the inequality $|g(x + y) - g(x) - g(y)| \leq \delta$.

12.14 Two Approximations

It is possible to consider the following question: are the two approximate solutions of a functional equation (12.4) near each other? More exactly: given $\varepsilon > 0$, does there exist a $\delta > 0$ such that, for every functions $g_1, g_2 : S_1 \rightarrow S_2$, if $d[L(g_1), R(g_1)] \leq \delta$ and $d[L(g_2), R(g_2)] \leq \delta$, then $d[g_1(x), g_2(x)] \leq \varepsilon$?

For Eq. (12.3) and $f : \mathbb{R} \rightarrow \mathbb{R}$ the answer is no. This is evident with the functions $g_1(x) = \delta$ and $g_2(x) = x$. For the stable equation, which has the only solution f, the answer is yes. Indeed, for $\varepsilon/2 > 0$ let δ be as in the definition of the stability. Thus, for the solutions g_1 and g_2 of the inequality (12.5), we have $d(g_1, f) \leq \varepsilon/2$ and $d(g_2, f) \leq \varepsilon/2$ and so $d(g_1, g_2) \leq \varepsilon$.

12.15 Stability in the Class

The functional equation is said to be stable (inversely stable, absolutely stable) in the class of functions K if it is stable (inversely stable, absolutely stable) according to the above definitions, in which the functions g and f belong to the class K.

The stability in the class K of the equation $L(f) = R(f)$ is in fact the stability of the conditional equation of the form: $f \in K \Rightarrow L(f) = R(f)$. The only difference is such that a function $f \notin K$ is a solution of this conditional equation, but it is not considered in the stability in that class. This difference has no effect on the stability.

We do not have any relation between the stability in the class K_1 and the stability in the class $K_2 \subset K_1$.

E.g., the equation $f(f(x)) = f(x)$ is stable in the class K_1 of functions from $\mathbb{R}$ to $\mathbb{R}$ (see Sect. 12.8) and it is unstable in the class K_2 of differentiable functions from $\mathbb{R}$ to $\mathbb{R}$. Indeed, assume that this equation is stable in the class K_2. Thus for $\varepsilon = 1$ there exists a $\delta > 0$ such that, for every differentiable function g, if $|g(g(x)) - g(x)| \le \delta$, then $|g(x) - x| \le 1$ (the function $f(x) = x$ is the only solution from the functions under the consideration in the class K_2). Let g be a differentiable function from $\mathbb{R}$ to the interval $[0, 2] \cap [0, \delta]$. Since $g(x) \in [0, \delta]$ and $g(g(x)) \in [0, \delta]$, so

$$|g(g(x)) - g(x)| \le \delta.$$

This yields that $|g(x) - x| \le 1$, but since $g(4) \le 2$, we have

$$|g(4) - 4| = 4 - g(4) \ge 4 - 2 = 2$$

and we obtain a contradiction.

Equation (12.3) for f from the free group G, generated by two elements, to $\mathbb{R}$ is unstable (see Sect. 12.3) and it is stable in the class of functions from G to $\mathbb{Z}$ (with arbitrary positive $\delta < 1$).

Let V be a normed non-complete space, B be a completion of V to the Banach space, and G an abelian group containing an element of infinite order. We put $K_1 = \{f : G \to B\}$, $K_2 = \{f : G \to V\}$ and let K_3 be the family of the functions from G to a finite-dimensional subspace of the space V. We have $K_3 \subset K_2 \subset K_1$. Equation (12.3) is stable in the classes K_1 and K_3 and it is unstable in the class K_2, since V is not complete [7].

Equation (12.3) is

a/ stable in the class $\{f : \mathbb{R} \to [0, +\infty)\}$—proof by the "direct method" as in [9];

b/ unstable in the class $\{f : \mathbb{R} \to (0, +\infty)\}$, since Eq. (12.3) does not have any solution in this class and the inequality (12.1) has a solution for every $\delta > 0$, e.g., $g(x) = \delta/2$;

c/ stable in the class $\{f : \mathbb{R} \to [1, +\infty)\}$, since the inequality does not have any solution for $\delta = 1$ (the inequality $|g(0)| = |g(x + 0) - g(x) - g(0)| < 1$ is impossible).

The translation equation $F(F(\alpha, x), y) = F(\alpha, x + y)$ for $F : I \times \mathbb{R} \to I$, where I is a non-degenerated interval in $\mathbb{R}$, is stable in the class of continuous functions [24]. It is unstable in the class of continuous functions for which the derivative of $F(., 0) : I \to I$ at the point α exists for every $\alpha \in I$ [24]. The problem of its stability in the class of all functions from $I \times \mathbb{R}$ to I is still open.

The equation $f(x)^2 = x^2$ is not inversely stable (see Sect. 12.10) and it is inversely stable in the class

$$K_2 = \{f : \mathbb{R} \to \mathbb{R} \setminus \{0\}\} \subset K_1 = \{f : \mathbb{R} \to \mathbb{R}\},$$

since this equation does not have the solution in this class.

If the equation is inversely stable in the class K_1, then is evidently inversely stable in the class $K_2 \subset K_1$.

The equations in the some above examples have not the solutions. These equations are not interesting. From here the following question.

Question Is it possible to replace these non-interesting equations in the above examples by the equations having the solutions?

Final conclusion *We have to be careful with the approximations, since*

a/ *the approximate solution of a functional equation does not have to be the approximation of a solution of this equation;*
b/ *the approximation of the solution does not have to be the approximate solution;*
c/ *the approximation of the approximation does not have to be the approximation.*

The situation is the same for the system and for the alternation of the functional equations and for the stabilities in the classes.

12.16   Stability of the Translation Equation

Theorem 12.1

(a) *Let the function $F : I \times \mathbb{R} \to I$, where I is an internal, be a solution of the translation equation*

$$F(F(x, t), s) = F(x, t + s), \tag{12.13}$$

and let S be the selector of the class of sets $F(x, \mathbb{R})$ for $x \in F(I, \mathbb{R}) =: I^*$ and for which card $F(x, \mathbb{R}) > 1$.

   If F is continuous with respect to the second variable for every $x \in S$ and for which at least one of the functions $F(., t) : I \to I$ is continuous, then it is of the form

$$F(x, t) = \begin{cases} h_n^{-1}[h_n(g(x)) + t] & \text{for } g(x) \in I_n, t \in \mathbb{R}, \\ g(x) & \text{for } g(x) \in g(I) \setminus \bigcup I_n, t \in \mathbb{R}, \end{cases} \tag{12.14}$$

where $g : I \to I$ is a continuous idempotent ($g \circ g = g$), $I_n \subset g(I)$ for $n \in N_1 \subset \mathbb{N}$ are open and disjoint non-empty intervals (named the non-degenerated orbits) and $h_n : I_n \to \mathbb{R}$ are the homeomorphisms.
(b) *The function F of the form (12.14) is a continuous solution of (12.13).*

This result has been proved in [26] in the case where $F(x, 0) = x$ (thus for the dynamical system), under the assumption that the function F is continuous.

Proof Part (a). We remark at the beginning that $F(I, t) = I^*$ for $t \in \mathbb{R}$. Evidently $F(I, t) \subset I^*$. If $x \in I^*$, then $x = F(x_1, t_1)$ for some $(x_1, t_1) \in I \times \mathbb{R}$ Thus $x = F(F(x_1, t_1 - t), t) \in F(I, t)$. From here $I^* \subset F(I, t)$. Moreover, if $x \in I^*$, then

$$F(x, 0) = F(F(x_1, t_1), 0) = F(x_1, t_1) = x.$$

This yields that, if $x_0 \in F(x, \mathbb{R})$ and $x \in I^*$, then $x \in F(x_0, \mathbb{R})$. Indeed, if $x_0 = F(x, t_1)$, then

$$F(x_0, -t_1) = F(F(x, t_1), -t_1) = F(x, 0) = x.$$

If $x_0 \in S \cap F(x, \mathbb{R})$, then the function $F(x_0, .) : \mathbb{R} \to I_0 := F(x, \mathbb{R})$ is an injection. Indeed, if $F(x_0, t_1) = F(x_0, t_2)$ for $t_1 < t_2$, then for every $\varepsilon > 0$ there exist τ_1 and τ_2 such that $0 < \tau_2 - \tau_1 \le \varepsilon$ and $F(x_0, \tau_1) = F(x_0, \tau_2)$. This yields that

$$x_0 = F(x_0, 0) = F(F(x_0, \tau_1), -\tau_1) = F(F(x_0, \tau_2), -\tau_1) = F(x_0, \tau_2 - \tau_1)$$

and thus $F(x_0, t + (\tau_2 - \tau_1)) = F(F(x_0, \tau_2 - \tau_1), t) = F(x_0, t)$. The function $F(x_0, t)$ as microperiodic and continuous is thus constant. So, we have a contradiction, since $\operatorname{card} F(x, \mathbb{R}) > 1$ and $x = F(x_0, t_1)$ for some $t_1 \in \mathbb{R}$, thus $F(x, t) = F(F(x_0, t_1), t) = F(x_0, t + t_1)$. From here the function $F(x_0, t)$ is a homeomorphism from $\mathbb{R}$ to I_0, which we denote by h^{-1}, and I_0 is an open interval. For $x \in I_0$ there exists a $t_0 \in \mathbb{R}$ such that $x = F(x_0, t_0) = h^{-1}(t_0)$. This yields that $t_0 = h(x)$. Consequently, we have

$$F(x, t) = F(F(x_0, t_0), t) = F(x_0, t_0 + t) = h^{-1}(t_0 + t) = h^{-1}(h(x) + t).$$

Every two sets $F(x_1, \mathbb{R})$ and $F(x_2, \mathbb{R})$ for $x_1, x_2 \in I$ are disjoint or identical. Indeed, assume that $x_0 \in F(x_1, \mathbb{R}) \cap F(x_2, \mathbb{R})$. We prove that $F(x_1, \mathbb{R}) \subset F(x_2, \mathbb{R})$. For $x \in F(x_1, \mathbb{R})$ we have $x = F(x_1, t_1)$ for some $t_1 \in \mathbb{R}$. Moreover $x_0 = F(x_1, t_2) = F(x_2, t_3)$ for some t_2, t_3 in $\mathbb{R}$, whence $x_1 = F(x_0, -t_2)$. From here

$$\begin{aligned} x &= F(F(x_0, -t_2), t_1) = F(x_0, t_1 - t_2) \\ &= F(F(x_2, t_3), t_1 - t_2) = F(x_0, t_3 + t_1 - t_2) \in F(x_2, \mathbb{R}). \end{aligned}$$

Analogously, we obtain that $F(x_2, \mathbb{R}) \subset F(x_1, \mathbb{R})$.

Let $s_0 \in \mathbb{R}$ be such that the function $F(., s_0) : I \to I$ is continuous. The set I^* is an interval since $I^* = F(I, s_0)$.

If $F(x_0, \mathbb{R})$ is a degenerated interval, e.g., $F(x_0, \mathbb{R}) =: \{x_1\}$, then

$$F(x_1, t) = F(F(x_0, 0), t) = F(x_0, t) = x_1$$

for $t \in \mathbb{R}$.

The function $F_1 = F|_{F(I,\mathbb{R}) \times \mathbb{R}}$ is thus of the form (12.12) with $g(x) = x$, where as the intervals I_n we take the different non-degenerate intervals $F(x, \mathbb{R})$ for $x \in F(I, \mathbb{R})$.

We prove that this function $F_1 : I^* \times \mathbb{R} \to I$ is continuous. At the beginning we prove that this function is right-hand continuous (i.e., for $x \to x_0+$ and $t \to t_0$).

1/ If $x_0 \in I_n$ for some $n \in N$, then for $x \in I$ we have

$$F_1(x, t) = h_n^{-1}[h_n(x) + t] \to F_1(x_0, t_0)$$

for $x \to x_0$ and $t \to t_0$.

2/ If $x_0 = \inf I_n$ for some $n \in N$, then since $x \to x_0+$ we can to admit that $x \in I_n$. From here $F_1(x, t) = h_n^{-1}[h_n(x)+t]$. The function h_n as a homeomorphism from I_n to $\mathbb{R}$ must be increasing or decreasing. If h_n is increasing, then $h_n(x) \to -\infty$ if $x \to x_0+$, and thus $h_n(x) + t \to -\infty$ if $x \to x_0+$ and $t \to t_0$. This implies that in this case

$$F_1(x, t) \to \inf I_n = x_0 = F_1(x_0, t_0).$$

If h_n is decreasing, the situation is analogous.

3/ Assume that, for every $n \in N$, the point x_0 is not in $[\inf I_n, \sup I_n)$ and that $x_n \to x_0$ and $t_n \to t_0$. Let

$$S_1 := \{x_n : x_n \text{ is not the fixed point of } F_1\}$$

and

$$S_2 := \{x_n : x_n \text{ is the fixed point of } F_1\}.$$

If the set S_1 is finite, then there exists an $n_0 \in \mathbb{N}$ such that, for $n > n_0$, we have

$$F_1(x_n, t_n) = x_n \to x_0 = F(x_0, t_0).$$

If the set S_2 is finite, then there exists an $n_1 \in \mathbb{N}$ such that, for $n > n_1$, there exists an interval $I_{k(n)}$ for which $x_n \in I_{k(n)}$ (the function $k(n)$ must not be injective). Since $x_n \to x_0$, so $\inf I_{k(n)} \to x_0$ and $\sup I_{k(n)} \to x_0$. We have

$$\inf I_{k(n)} < F_1(x_n, t_n) < \sup I_{k(n)},$$

thus

$$F_1(x_n, t_n) \to x_0 = F_1(x_0, t_0).$$

If the sets S_1 and S_2 are infinite, then the sequence x_n consists of two subsequences $x_{k(n)} \in S_1$ and $x_{l(n)} \in S_2$, whence by the above F_1 is continuous at the point (x_0, t_0).

The function F_1 is thus right-hand continuous. By the analogous reasoning we obtain that F_1 is left-hand continuous and thus it is continuous. Since

$$F(x, t) = F(F(x, s_0), t - s_0) = F_1(F(x, s_0), t - s_0),$$

the function $F(x, t)$ is continuous and consequently the idempotent function $F(x, 0)$ is continuous. We have $F(x, t) = F(F(x, 0), t) = F_1(F(x, 0), t)$ and this implies that F is of the form (12.12) with $g(x) = F(x, 0)$.

Part (b). It is easy to verify that the function of the form (12.14) is the solution of (12.13). The verification that this function is continuous is analogous as above. □

The set S in the Theorem 12.1 is countable, since it is a selector of the class of open disjoint intervals (e.g., the set of midpoints of these intervals). The supposition that at least one of the functions $F(., t)$ is continuous is essential in the above theorem. Indeed, let $h : I \to I$ be a discontinuous idempotent function. For the function $F(x, t) = h(x)$, the function $F(x, .)$ is continuous for every $x \in I$ and the function F is not of the form (12.14). If a solution F of (12.13) depends only on t, then it is the constant function, thus it has the form (12.14) (for $N_1 = \emptyset$).

Question Does there exist a solution F of (12.13), which depends on x and t, for which the function $F(x, .)$ is continuous for $x \in I$ and which is not of the form (12.14)?

Since the function of the form (12.14) is continuous (see [24] and above), so a solution F of (12.13) continuous with respect to second variable, for which at least one function $F(., t)$ is continuous, must be continuous.

If a solution F of (12.13) is Carathéodory, i.e., the function $F(x, .) : \mathbb{R} \to I$ is measurable for every $x \in I$ and $F(., t) : I \to I$ is continuous for every $t \in \mathbb{R}$, then the function $F|_{I \times (0, +\infty)}$ is continuous [2]. Let the sequence t_n be such that $t_n \to t_0 \leq 0$. We obtain

$$F(x, t_n) = F(F(x, -1 + t_0), t_n + 1 - t_0) \to F(F(x, -1 + t_0), 1) = F(x, t_0),$$

thus the function $F(x, .)$ is continuous at every point $t_0 \geq 0$. This yields that the function F is continuous, thus it has the form (12.14) by Theorem 12.1.

Unfortunately, a solution F of (12.13), which is only such that $F(x, 0)$ is continuous and $F(x, .)$ is measurable, may be discontinuous. E.g., for the function

$$F(x, t) = g^{-1}(g(x) + t) = g(g(x) + t),$$

where $g(x) = 1/x$ for $x > 0$ and $g(x) = x$ for $x \leq 0$, the function $F(x, .)$ is discontinuous only at the point $t = -g(x)$.

Let's notice yet that there exists the continuous solution F of (12.13) such that all functions $F(., t)$ are not Jordan-measurable [21].

The proof of Theorem 12.1 implies that the solution of (12.13) is of the form (12.14), where g is an idempotent (not necessarily continuous), I_n are the non-empty disjoint subsets of I and $h_n : I_n \to \mathbb{R}$ are the bijections, if and only if

$$\{t \in \mathbb{R} : F(x, t) = x\} \in \{\emptyset, \{0\}, \mathbb{R}\}$$

for every $x \in I$.

Note that the function $F(x, t) = x\varphi(t) : \mathbb{R} \times \mathbb{R} \to \mathbb{R}$, where $\varphi : \mathbb{R} \to \mathbb{R}$ is a discontinuous solution of the equation $\varphi(t + s) = \varphi(t)\varphi(s)$, is the discontinuous solution of Eq. (12.13), which is continuous with respect to the variable x for every t, and which is continuous with respect to variable t for $x = 0$.

Similarly, a solution $H : I \times \mathbb{R} \to I$ of the inequality (12.15) (below), which is continuous with respect to each variable, can be discontinuous, e.g., the function

$$H(x, t) = \frac{\varepsilon x t}{x^2 + t^2}, \qquad (x, t) \neq (0, 0),$$

and $H(0, 0) = 0$, which is discontinuous at the point $(0, 0)$.

The solution (12.14) is said to be simple if $\inf |I_n| > 0$, where $|I_n|$ is the length of the interval I_n. If $x_0 \in g(I)$, then $g(x_0) = x_0$, since g is an idempotent. From here, if $x_0 \in I_n \subset g(I)$, then $g(x_0) = x_0 \in I_n$ and $F(x_0, .) : \mathbb{R} \to I_n$ is a bijection. If $x_0 \in g(I) \setminus \bigcup I_n$, then $g(x_0) = x_0 \in I_n \subset g(I)$, thus $F(x_0, t) = g(x_0) = x_0$ for $t \in \mathbb{R}$, i.e., x_0 is a fixed point of F.

The following theorem has been proved in [24] as Theorem 1.1.

Theorem 12.2 *Let $I \subset \mathbb{R}$ be an interval. Suppose that $H : I \times \mathbb{R} \to I$ is continuous with respect to each variable and satisfies*

$$|H(H(x, t), s) - H(x, t + s)| \leq \varepsilon, \qquad x \in I, t, s \in \mathbb{R}. \tag{12.15}$$

Then there exists a continuous solution F of (12.13) such that

$$|F(x, t) - H(x, t)| \leq 10\varepsilon, \qquad x \in I, t \in \mathbb{R}. \tag{12.16}$$

Moreover in the proof in [24] of this theorem it is showed that this solution F is simple.

We suppose that all solutions of (12.13) considered still are continuous. This yields that every non-simple solution of (12.13) is approximated by the simple solutions of (12.13), i.e., for every $\varepsilon > 0$ and for every non-simple solution G of (12.13), there exists a simple solution F of (12.11) such that $|G(x, t) - F(x, t)| \leq \varepsilon$ for $(x, t) \in I \times \mathbb{R}$. The inverse is not true, i.e., it is not true that, for every $\varepsilon > 0$ and for every simple solution G of (12.13), there exists a non-simple solution F of (12.11) such that $|G(x, t) - F(x, t)| \leq \varepsilon$ for $(x, t) \in I \times \mathbb{R}$.

For the proof we need the following lemma.

Lemma 12.1 *If at least one fixed point x_0 of F is in a non-degenerated orbit J of G, then it is impossible that*

$$|G(x,t) - F(x,t)| \leq \varepsilon, \qquad (x,t) \in I \times \mathbb{R}$$

(a) for every $0 < \varepsilon < |J|/2$, if J is bounded;
(b) for every $\varepsilon > 0$, if J is unbounded.

Proof Part (a). In the opposite case we have

$$\frac{|J|}{2} > \varepsilon \geq \sup_{t \in \mathbb{R}} |G(x_0,t) - F(x_0,t)|$$

$$\geq \sup_{x \in J} |G(x_0,t(x)) - F(x_0,t(x))|$$

$$= \sup_{x \in J} |x - x_0| \geq \frac{|J|}{2},$$

where $t(x)$ is such that $G(x_0,t(x)) = x$ for $x \in J$ ($G(x_0,.) : \mathbb{R} \to J$ is a bijection), thus a contradiction.

Part (b). If J is unbounded, then the function $|G(x_0,t) - F(x_0,t)| = |G(x_0,t) - x_0|$ is unbounded, too, since $G(x_0,\mathbb{R}) = J$. $\qquad \square$

Put $S(F) := \{x \in I : F(x,t) = x \text{ for } t \in \mathbb{R}\}$ for $F : I \times \mathbb{R} \to I$.

Proposition 12.1 *Let $G : I \times \mathbb{R} \to I$ be a simple solution of (12.13) such that $G(x,0) = x$ for $x \in I$. This solution can be approximated by the non-simple solutions of (12.13) if and only if $\mathsf{Int}\, S(G) \neq \emptyset$.*

Proof "If" part. Let $a, b \in \mathbb{R}$ and $a < b$. The function $f(a,b,.,.) : (a,b) \times \mathbb{R} \to (a,b)$ of the form $f(a,b,x,t) = h^{-1}(h(a,b,x) + t)$, where

$$h(a,b,x) = \tan\left[\frac{\pi}{b-a}(x-a) - \frac{\pi}{2}\right]$$

for $x \in (a,b)$, is a continuous solution of (12.12). Assume that the non-degenerated interval $[c,d]$ is such that $[c,d] \subset S(G)$. Let $0 < \varepsilon < d - c$. The function

$$F(x,t) = \begin{cases} f\left(c + \frac{\varepsilon}{n+1}, c + \frac{\varepsilon}{n}, x, t\right) & \text{for } (x,t) \in \left(c + \frac{\varepsilon}{n+1}, c + \frac{\varepsilon}{n}\right) \times \mathbb{R} \text{ and } n \in \mathbb{N}, \\ G(x,t) & \text{for } x \in I \setminus \bigcup \left(c + \frac{\varepsilon}{n+1}, c + \frac{\varepsilon}{n}\right) \text{ and } t \in \mathbb{R}, \end{cases}$$

is the non-simple continuous solution of (12.13) for which

$$|G(x,t) - F(x,t)| \leq \varepsilon, \qquad (x,t) \in I \times \mathbb{R}.$$

It is evident if $(x,t) \in I \setminus \bigcup \left(c + \frac{\varepsilon}{n+1}, c + \frac{\varepsilon}{n}\right) \times \mathbb{R}$. If $(x,t) \in \left(c + \frac{\varepsilon}{n+1}, c + \frac{\varepsilon}{n}\right) \times \mathbb{R}$, then

$$|G(x,t) - F(x,t)| = \left|x - f\left(c + \frac{\varepsilon}{n+1}, c + \frac{\varepsilon}{n}, x, t\right)\right| \le \frac{\varepsilon}{n} - \frac{\varepsilon}{n+1} \le \varepsilon$$

since x and $f\left(c + \frac{\varepsilon}{n+1}, c + \frac{\varepsilon}{n}, x, t\right)$ are in the interval $\left(c + \frac{\varepsilon}{n+1}, c + \frac{\varepsilon}{n}\right)$.

"Only if" part. We remark as the beginning that, for the two open non-degenerated different intervals K_1 and K_2, if $K_1 \cap K_2 \ne \emptyset$ and K_2 is bounded, then a bound of K_2 is in K_1. Let $0 < \varepsilon < \frac{1}{2} \inf |J_n|$, where J_n for $n \in N_2$, are the non-degenerated orbits of G. Let F of the form (12.14) be a non-simple solution of (12.13), for which $|G(x,t) - F(x,t)| \le \varepsilon$ for $(x,t) \in I \times \mathbb{R}$. If $S(F) = I$, then of course there exists a fixed point of F in every non-degenerated orbit of G, thus we have a contradiction by Lemma 12.1. If F has the non-degenerated orbits, then there exists a non-degenerated orbit I^* of F for which $|I^*| < \inf |J_n|$. Thus $I^* \ne J_n$ for $n \in N_2$. If $I^* \cap J_n \ne \emptyset$ for some $n \in N_2$, then there exists a bound of I^*, which is in J_n. Since this bound is a fixed point of F, we obtain a contradiction by Lemma 12.1. From here $I^* \cap J_n = \emptyset$ for $n \in N_2$, hence

$$I^* \subset I \setminus \bigcup J_n = G(I,0) \setminus \bigcup J_n = S(G). \qquad \square$$

We remark that the supposition that $G(x,0) = x$ for $x \in I$ is used only in the end of the above proof for "only if" part.

Question Is this supposition essential in the part "only if" in the Proposition?

The Proposition yields that it is impossible to substitute the simple solution of (12.13) by the non-simple solution of (12.13) in the above mentioned Theorem 12.2. But there exist the solutions H of the inequality (12.15), which can be approximated by the non-simple solutions of (12.13). E.g., if for the function F in (12.16) there exists a non-simple solution F_1 of (12.13) such that

$$|F(x,t) - F_1(x.t)| \le \varepsilon, \qquad (x,t) \in I \times \mathbb{R}$$

(e.g., if $\text{Int } S(F) \ne \emptyset$), then

$$|H(x,t) - F_1(x,t)| \le |H(x,t) - F(x,t)| + |F(x,t) - F_1(x,t)| \le 10\varepsilon + \varepsilon = 11\varepsilon$$

for $(x,t) \in I \times \mathbb{R}$.

12.17 Stability of the Dynamical System

A continuous solution $F : I \times \mathbb{R} \to I$ of (12.13) (it is sufficient that F is continuous with respect to the second variable—see Sect. 12.15) for which

$$F(x,0) = x \qquad for \, x \in I \tag{12.17}$$

is called the dynamical system. This system is said to be stable if the system (12.13) and (12.17) is stable. It is proved in [24] that this system is stable in the class C of functions from $I \times \mathbb{R}$ to I continuous with respect to each variable if and only if $I = \mathbb{R}$. In this case $\delta = \varepsilon/9$ and the approximation F is the simple solution of (12.13) and (12.17). Here this approximation F may be replaced by the non-simple solution of (12.13) if and only if $\mathsf{Int}\, S(F) \neq \emptyset$ by the Proposition 12.1.

Equation (12.17) is stable in the class C. We present the proof given by B. Przebieracz (communicated by e-mail).

Proof Assume that the function $H : I \times \mathbb{R} \to I$ is such that $|H(x.0) - x| \leq \varepsilon$ for an $\varepsilon > 0$ and every $x \in I$. Let $a := \inf I$ and $b := \sup I$ (may be $a = -\infty$ and/or $b = +\infty$). Let $G(x, t) := H(x, t) - H(x, 0) + x$. Put

$$
K(x, t) = \begin{cases} a & \text{if } G(x, t) \leq a, \\ G(x, t) & \text{if } G(x, t) \in I, \\ b & \text{if } G(x, t) \geq b. \end{cases}
$$

We have $K(I, \mathbb{R}) \subset \mathsf{cl}\, I$. It cannot be that $K(I, \mathbb{R}) \subset I$. E.g., if $I = (0, +\infty)$, $H(x, t) = x + \varepsilon \exp t$ and $x_0 = \varepsilon(1 - \exp(-1))$, then $G(x_0, -1) = 0$ and thus $K(x_0, -1) = 0 \notin I$. Since $G(x, 0) = x \in I$, so $K(x, 0) = G(x, 0) = x$.
If $G(x, t) \in I$, then

$$
|K(x, t) - H(x, t)| = |H(x, t) - H(x, t) + H(x, 0) - x| \leq \varepsilon.
$$

If $G(x, t) \leq a$, then

$$
|K(x, t) - H(x, t)| = |a - H(x, t)| = H(x, t) - a \leq H(x, 0) - x \leq \varepsilon.
$$

If $G(x, t) \geq b$, then

$$
|K(x, t) - H(x, t)| = |b - H(x, t)| = b - H(x, t) = x - H(x, 0) \leq \varepsilon.
$$

From here $|K(x, t) - H(x, t)| \leq \varepsilon$ for $(x, t) \in I \times \mathbb{R}$. The above implies that the function K is good if I is closed.
Put

$$
F(x, t) = \begin{cases} (1 - |t|)K(x, t) + |t|H(x, t) & \text{if } |t| \leq 1 \text{ and } x \in I, \\ H(x, t) & \text{if } |t| > 1 \text{ and } x \in I. \end{cases}
$$

We remark that

a/ we have for $|t| \leq 1$ that

$$
|F(x, t) - H(x, t)| = (1 - |t|)|K(x, t) - H(x, t)| \leq \varepsilon
$$

and for $|t| > 1$ that

$$|F(x, t) - H(x, t)| = 0 \le \varepsilon,$$

b/ we obtain for $|t| > 1$ that $F(x, t) = H(x, t) \in I$ and for $0 < |t| \le 1$ that

$$F(x, t) = (1 - |t|)K(x, t) + |t|H(x, t) \in (K(x, t), H(x, t)) \cup (H(x, t), K(x, t)) \subset I$$

since $K(x, t) \in \text{cl } I$, $H(x, t) \in I$ and $F(x, 0) = K(x, 0) = x \in I$. This yields
that $K(x, t) \in I$ for $(x, t) \in I \times \mathbb{R}$. The proof is thus finished.

□

The translation equation (12.13) is stable in the class C for every interval I. Since
the system (12.13) and (12.17) is stable in the class C only for $I = \mathbb{R}$ [24], so for
$I \ne \mathbb{R}$ this system is not stable though the equations in this system are stable!

12.18   b-Stability and Inverse b-Stability

Hyers by his "direct method" has constructed in fact, for a solution g of the
inequality

$$|g(x + y) - g(x) - g(y)| \le \varepsilon$$

with a given $\varepsilon > 0$ the solution f of (12.3) $(f(x) = lim_{n \to +\infty}(g(2^n x)2^{-n})$ for
which the inequality

$$|g(x) - f(x)| \le \delta$$

is true with $\delta = \varepsilon$. This remark suggests the following definition: Eq. (12.4) is said
to be uniformly b-stable if, for every $\varepsilon > 0$, there exists a $\delta > 0$ such that for every
function g for which $d[L(g), R(g)] \le \varepsilon$ there exists a solution f of (12.4) for which
$d[g, f] \le \delta$. This definition may be formulated as follow: there exists a function
$\Psi : (0, +\infty) \to (0, +\infty)$ (the measure of the uniform b-stability) such that, for
every $\varepsilon > 0$ and every function g, if $d[L(g), R(g)] \le \varepsilon$, then there exists a solution
f of (12.4) such that $d[g, f] \le \Psi(\varepsilon)$.

If there exists a measure for which $\inf \Psi(0, +\infty) = 0$, then this uniform
b-stability is called normal. The Hyers definition of the stability quoted in the
introduction is different with the definition of this uniform b-stability. These
definitions are equivalent for the functions from a Banach space to a Banach space,
since $\delta = \varepsilon = \Psi(\varepsilon)$ in this case. They are not equivalent in general case. E.g.,
Eq. (12.3) for the functions from the free group generated by two elements to the
group of integers with normal metric is not uniformly b-stable [6] and it is stable
(every $\delta < 1$ is good here). Conversely, the equation

$$f(x + y) = f(x)f(y) \tag{12.18}$$

for $f : \mathbb{R} \to (0, +\infty)$ is universally b-stable and it is not stable. Indeed, it is proven in [1] that if $|g(x + y) - g(x)g(y)| \leq \varepsilon$, then g is the solution of the equation or $|g(x)| \leq \max(4, 4\varepsilon) =: \Psi(\varepsilon)$ for $x \in \mathbb{R}$. From here your equation is universally b-stable with $\Psi(\varepsilon)$ as the measure. This equation is not stable (see Sect. 12.7).

The above notion of the uniform b-stability is called uniform since δ does not depend on the function g. If δ depends on ε and g, then the equation is only b-stable. More precisely, Eq. (12.4) is said to be b-stable if, for every function g and for every $\varepsilon > 0$, there exists a $\delta > 0$ such that if $d[L(g), R(g)] \leq \varepsilon$, then $d[g, f] \leq \delta$ for some solution f of (12.4). We remark that the fact that δ depends on the function g is not sufficient to believe that the equation is not uniformly b-stable. E.g., we have for Eq. (12.16) that $|g(x)| \leq \max\{4, 4\varepsilon\} + |g(0)|$ if $|g(x + y) - g(x)g(y)| \leq \varepsilon$, but in spite of this Eq. (12.16) is uniformly b-stable.

The second example. The theorems in [10] imply that if

$$\left| 2g\left(\frac{x + y}{2}\right) - g(x) - g(y) \right| \leq \varepsilon \tag{12.19}$$

for a function g from a normed space to a Banach space, then there exists a Jensen function f, i.e., the solution of the Jensen equation

$$2f\left(\frac{x + y}{2}\right) = f(x) + f(y),$$

such that $|g(x) - f(x)| \leq \varepsilon + |g(0)|$. Here $\delta = \varepsilon + |g(0)|$ depends on ε and g and nevertheless this equation is uniformly b-stable. Indeed, if the function g satisfies (12.19), then the function $G(x) = g(x) - g(0)$ satisfies (12.19), too. Thus there exists a Jensen function F such that

$$|g(x) - [F(x) + g(0)]| = |G(x) - F(x)| \leq \varepsilon + G(0) = \varepsilon$$

and $F(x) + g(0)$ is the Jensen function.

It is proved in [4] that if

$$\left| g\left(\frac{x + y}{2}\right)^2 - g(x)g(y) \right| \leq \varepsilon$$

for g from an abelian 2-divisible group G to $\mathbb{C}$ and some ε, then

$$|g(x)| \leq \frac{1}{2}[A + \sqrt{A^2 + 4\delta}], \qquad x \in G,$$

where $A = \inf_{x \in G} |g(x)|$, or g is a solution of the Lobachevsky equation

$$f\left(\frac{x + y}{2}\right)^2 = f(x)f(y).$$

Thus this equation is superstable and b-stable. Hence the following question.

Question Is this equation uniformly b-stable?

The equation $\sin f(x) = 0$ for $f : \mathbb{R} \to \mathbb{R} \setminus \{k\pi\}_{k \in \mathbb{N}}$ is evidently b-stable in the class of bounded functions, since every equation, which has a bounded solution, is b-bounded in this class. Assume that it is uniformly b-stable. Thus, for $\varepsilon > 0$, there exists a $\delta(\varepsilon) > 0$ such that, for every function $g : \mathbb{R} \to \mathbb{R} \setminus \{k\pi\}_{k \in \mathbb{N}}$ for which $|\sin g(x)| \leq \varepsilon$, there exists a solution $f : \mathbb{R} \to \mathbb{R} \setminus \{k\pi\}_{k \in \mathbb{N}}$ of our equation such that $|g(x) - f(x)| \leq \delta(\varepsilon)$. We have thus that, for $g(x) = E(\delta(1)) + 1$ ($E(x)$ is the entire part of x) there exists a solution f of our equation such that $|g(x) - f(x)| \leq \delta(1)$. This solution f has the form $f(x) = -k(x)\pi$, where $k(x) \in \mathbb{N} \cup \{0\}$. From here

$$\delta(1) \geq |g(0) - f(0)| = E(\delta(1)) + 1 + k(0)\pi > \delta(1),$$

thus a contradiction. Therefore our equation is not uniformly b-stable.

This equation is not b-stable in the class of all functions $f : \mathbb{R} \to \mathbb{R} \setminus \{k\pi\}_{k \in \mathbb{N}}$. Indeed, for the function $g(x) = x$, the function $\sin g(x)$ is bounded and the function $x + k(x)\pi$ is unbounded if $k(x) \in \{k\pi\}_{k \in \mathbb{N}}$ for $x \in \mathbb{R}$. Our equation is uniformly b-bounded if $f : \mathbb{R} \to \mathbb{R}$, since for every function $g : \mathbb{R} \to \mathbb{R}$ we have

$$\left| g(x) - E\left(\frac{g(x)}{\pi}\right)\pi \right| = \left| \pi \left[\frac{g(x)}{\pi} - E\left(\frac{g(x)}{\pi}\right) \right] \right| \leq \pi$$

for $x \in \mathbb{R}$ and the function $E(g(x)/\pi)\pi$ is a solution of our equation.

However, it is easy to verify that, if Eq. (12.4) is normally stable (see Sect. 12.2(b)), then it is uniformly b-stable. We remark that, if for an equation there exists a normal measure Ψ, then there exists the measure, which is not normal, e.g., $\Psi_1(\varepsilon) = \max[a, \Psi(\varepsilon)]$ for $a > 0$, but non conversely.

Some of the theorems about the Ulam-Hyers stability are formulated in fact as the theorems about the uniform b-stability with the normal measure of uniform b-stability.

Equation (12.4) is called inversely b-stable if every approximation of a solution of (12.4) is an approximate solution of (12.4). It is possible to consider the uniform inverse b-stability. The b-stability and the inverse b-stability are not equivalent [17]. However, if the inverse stability of Eq. (12.4) is normal, then this equation is inversely b-stable. Indeed, assume that $d[g, f] \leq \delta$ for some $\delta > 0$ and some solution f of (12.12). Since the function Ψ is unbounded, there exists an $\varepsilon > 0$ such that $\Psi(\varepsilon) \geq \delta$. From here $d[g, f] \leq \Psi(\varepsilon)$, thus $d(L(g), R(f)) \leq \varepsilon$, since Eq. (12.4) is inversely stable. The supposition that the inverse stability is normal is here essential, since there exists an equation which is inversely stable, but not inversely b-stable. E.g., the equation $1/f(x) = 1$ for $f : \mathbb{R} \to \mathbb{R}$. Really, the inequality

$$|g(x) - 1| \leq \frac{\varepsilon}{1 + \varepsilon}$$

implies the inequality $|1/g(x) - 1| \leq \varepsilon$ and for the function $g(x) = \exp(-x^2)$ the function $g(x) - 1$ is bounded and the function $1/g(x) - 1$ is unbounded.

The b-stability is considered in many papers of stability. This b-stability and the stability are not equivalent [16]. Nevertheless, if the stability of Eq. (12.12) is normal, then this equation is b-stable (the proof as above).

12.19 Stability of Difference Equations

It is possible to consider the stability in a different sense than that of the Ulam-Hyers. E.g., we have for the difference equation the following definition. The solution a_n of a difference equation is said to be stable if, for every $\varepsilon > 0$, there exists a $\delta > 0$ such that, for every solution b_n of this equation, if $|a_1 - b_1| \leq \delta$, then $|a_n - b_n| \leq \varepsilon$ for every $n \in \mathbb{N}$. We remark that this stability is different from the Ulam-Hyers stability. Here stable is a solution and, in the case of Ulam-Hyers stability, stable is the equation.

The simple example. For the equation $a_{n+1} = \lambda a_n$ every solution is stable if $|\lambda| \leq 1$. Indeed, every solution of this equation is of the form $a_n = \lambda^{n-1} a_1$, thus for $\delta = \varepsilon$, if $|a_1 - b_1| \leq \delta = \varepsilon$ for a solution b_n of our equation, then

$$|a_n - b_n| = |\lambda^{n-1} a_1 - \lambda^{n-1} b_1| \leq |a_1 - b_1| \leq \varepsilon$$

for $n \in \mathbb{N}$.

If $|\lambda| > 1$, then every solution of this equation is not stable. Assume, for an indirect proof, that there exists a solution a_n, which is stable. Thus for $\varepsilon = 1$ there exists a $\delta > 0$ such that, for every solution b_n, if $|a_1 - b_1| \leq \delta$, then $|a_n - b_n| \leq 1$. We have for $b_n = \lambda^{n-1}(a_1 + \delta)$ that $|a_1 - (a_1 + \delta)| \leq \delta$, thus

$$1 \geq |\lambda^{n-1} a_1 - \lambda^{n-1}(a_1 + \delta)| = |\lambda|^{n-1} \delta \to +\infty$$

if $n \to +\infty$ and we obtain the contradiction.

This equation is Ulam-Hyers stable if $|\lambda| < 1$, i.e., for every $\varepsilon > 0$ there exists a $\delta > 0$ such that, for every sequence b_n for which $|b_{n+1} - \lambda b_n| \leq \delta$ for $n \in \mathbb{N}$, there exists a solution a_n of our equation for which $|b_n - a_n| \leq \varepsilon$ for $n \in \mathbb{N}$. Indeed, we have by the induction that

$$|b_n - \lambda^{n-1} b_1| \leq \delta(1 + |\lambda| + \ldots + |\lambda|^{n-2}) = \delta \frac{1 - |\lambda|^{n-1}}{1 - |\lambda|} \leq \delta \frac{1}{1 - |\lambda|} \leq \varepsilon$$

for $\delta = (1 - |\lambda|)\varepsilon$.

Our equation is not Ulam-Hyers stable for $|\lambda| \geq 1$, since we have for $b_n = n\delta$ that $|b_{n+1} - b_n| \leq \delta$ and the function $b_n - \lambda^{n-1} a_1$ is unbounded.

12.20 On the Stability of Idempotent Function with Constant Derivative

We consider the system

$$f(f(x)) = f(x) \quad \text{and} \quad f'(x) = a \tag{12.20}$$

of functional equations, where $f : I \to I$ is a differentiable function, I is non-degenerated interval in $\mathbb{R}$ and $a \in \mathbb{R}$.

We adjust the Ulam-Hyers stability of the Cauchy equation of the additive function [9] to this system. The system (12.20) is said to be stable if, for every $\varepsilon > 0$ there exists a $\delta > 0$ such that, for every differentiable function $h : I \to I$ such that

$$|h(h(x)) - h(x)| \le \delta \quad \text{and} \quad |h'(x) - a| \le \delta, \qquad x \in \mathbb{R}, \tag{12.21}$$

there exists a solution f of (12.20) for which

$$|h(x) - f(x)| \le \varepsilon, \qquad x \in \mathbb{R}. \tag{12.22}$$

Here in fact the stability of idempotent function with constant derivative is not defined; only the stability of the system (12.20) is defined, for which this function is the solution. Thus the title of the Sect. 12.20 is incorrect. The same function may be a solution of different stable or unstable systems [23]. E.g., the system (12.20) with $a = 1$ and the system: $f'(x) = 1$ and $f(0) = 0$ are equivalent, the system (12.20) is stable while the second system is not stable. Indeed, for $I = \mathbb{R}$ and the function $h(x) = (1 + \delta)x$ we have $|h'(x) - 1| \le \delta$ and $|h(0) - 0| \le \delta$ and the function $h(x) - x = \delta x$ is unbounded.

Every differentiable idempotent function $f : I \to I$ is the identity ($f(x) = x$) or it is constant ($f(x) = b \in I$). Indeed, we have $f(x) = x$ for $x \in f(I)$ and $f(I)$ is an interval. If $f(I) = I$, then $f(x) = x$ for $x \in I$. If $f(I)$ is degenerated, i.e., $I = b$, then $f(x) = b$ for $x \in I$. We prove that the case when $f(I)$ is non-degenerated and $f(I) \ne I$ is impossible. So, suppose that there exists a bound c of $f(I)$ for which $c \in \text{Int } I$. Assume that c is the left end of $f(I)$ (the reasoning is analogous if c is the right end of $f(I)$). We have thus

$$f'(x_0) = \lim_{x \to c+} \frac{f(x) - f(c)}{x - c} = \lim_{x \to c+} \frac{x - c}{x - c} = 1$$

and since $f(x) \ge f(c)$ for $x < c$, so

$$\frac{f(x) - f(c)}{x - c} \le 0.$$

From this we obtain a contradiction.

On the other hand, there exist the idempotent continuous functions, which are not of the above form, e.g., the function $f(x) = b$ for $x \in (-\infty, b] \cap I$, $f(x) = x$ for $x \in I \cap (b, c)$ and $f(x) = c$ for $x \in [c, +\infty) \cap I$, where $b, c \in I$, $b < c$. Moreover, we notice that, for every $\delta > 0$, there exist a differentiable function f with $|f(f(x)) - f(x)| \leq \delta$ and which is neither the constant function nor the identity function, e.g.,

$$f(x) = \frac{\delta}{2} \sin x.$$

Theorem 12.3 *The system* (12.20) *is stable for every* $a \in \mathbb{R}$ *(for* $a = 1$ *with* $\delta = \min(\varepsilon/3, 1/4)$, *for* $a = 0$ *with* $\delta = \min(\varepsilon/2, 1/2)$, *for* $0 \neq a \neq 1$ *with* δ *for which* (12.21) *does not have any solution).*

Proof We notice that if $f'(x) = a$, then $f(x) = ax + b$ for any b and, if f is idempotent, then $a(ax + b) + b = ax + b$, thus $a = 0$ or $a = 1$. If $a = 0$, then $f(x) = b \in I$ and, if a=1, then $f(x) = x$.

Let the function $h : I \to I$ be such that $|h(h(x)) - h(x)| \leq \delta$ and $|h'(x) - a| \leq \delta$ for every $x \in I$, with some $\delta > 0$ and $a \in \mathbb{R}$.

We prove that if $a > 0$ and $\delta \leq a/(2a + 2)$, then

$$|h(x) - x| \leq \frac{a + 2}{a} \delta, \qquad x \in I. \tag{12.23}$$

We have

$$|h(h(x)) - h(x)| \leq \delta \leq \frac{a + 2}{a} \delta, \qquad x \in I,$$

thus (12.23) is satisfied for $x \in h(I)$.

Since $|h'(x) - a| \leq \delta$, so

$$0 < a - \frac{a}{2a + 2} \leq a - \delta \leq h'(x) \leq a + \delta \leq a + \frac{a}{2a + 2}. \tag{12.24}$$

From here the function h is increasing.

Let $y_1 = \inf I$, $y_2 = \sup I$, $x_1 = \inf h(I)$, $x_2 = \sup h(I)$. Let us consider some cases.

1/ For $y_1 > -\infty$ and $y_2 = +\infty$ the function h is unbounded from above, since otherwise we would have for some $\theta(n)$

$$\frac{h(n) - h(y_1 + 1)}{n - (y_1 + 1)} = h'(\theta(n)) \to 0 \qquad \text{for } n \to +\infty,$$

which is a contradiction with (12.24).

a/ If $x_1 = y_1$, then $h(I) = I$ and the condition (12.23) is satisfied.

b/ If $x_1 > y_1$ and $y_1 \in I$, then we have $h(y_1) = x_1$ and $|h(x_1) - x_1| \leq \delta$. Since

$$h(x_1) - x_1 = h(x_1) - h(y_1) = h'(\theta)(x_1 - y_1)$$

for some θ, so

$$(a - \delta)(x_1 - y_1) \leq h'(\theta)(x_1 - y_1) = |h(x_1) - x_1| \leq \delta,$$

whence

$$x_1 - y_1 \leq \frac{\delta}{a - \delta}.$$

Since $\delta \leq a/(2a + 2)$, from (12.24) we obtain

$$|h(x) - x| \leq |h(x) - x_1| + |x_1 - x| \leq h'(\theta)(x - y_1) + (x_1 - x)$$

$$\leq \frac{\delta}{a - \delta}[h'(\theta) + 1] \leq \frac{\delta + a + 1}{a - \delta}\delta \leq \frac{a + 2}{a}\delta$$

for $x \in [y_1, x_1)$. Hence (12.23) is satisfied, because $h(I) = [x_1, +\infty)$.

2/ If $y_1 = -\infty$ and $y_2 = +\infty$, then, as above in 1/, we obtain by (12.24) that the function h is unbounded from above and from below. Thus $h(I) = \mathbb{R}$ and (12.23) is satisfied.

The proof of (12.23) is analogous in the other cases.

By (12.21) we have, for $a = 1$, that $|h(x) - x| \leq 3\delta$ if $0 < \delta \leq 1/4$. Since $f(x) = x$ is the solution of (12.20) in this case, so for every $\varepsilon > 0$ and $\delta = \min(\varepsilon/3, 1/4)$ we have (12.22). From here the system (12.20) is stable for $a = 1$.

For $a < 0$ and

$$\delta \leq \frac{a - a^2}{3a - 2}$$

we have

$$|h(x) - x| \leq 2\frac{a - 1}{a}\delta, \qquad x \in I. \tag{12.25}$$

The proof is analogous as above, only we take x_2 in place of x_1, since the function h is decreasing in this case.

By (12.23) and (12.25), for $a \neq 0$, there exist the positive constants $A(a)$ and $B(a)$ such that, for the solution h of (12.21), we have

$$|h(x) - x| \leq B(a)\delta \quad \text{for} \quad 0 < \delta \leq A(a)$$

and $x \in I$. Since also $|h(y) - y| \le B(a)\delta$, so for some θ we obtain

$$|[h'(\theta) - 1](x - y)| = |h'(\theta)(x - y) - (x - y)| = |h(x) - h(y) - (x - y)|$$
$$\le |h(x) - x| + |h(y) - y| \le 2B(a)\delta.$$

Moreover, since $|h'(x) - a| = |h'(x) - 1 - (a - 1)| \le \delta$, so $|a - 1| - \delta \le |h'(x) - 1|$. Let $0 \ne a \ne 1$. Assume that, for every positive $\delta < \min\{|a - 1|, A(a)\}$, there exists a solution of (12.21). This yields that $[|a - 1| - \delta]|x - y| \le 2B(a)\delta$ and

$$|x - y| \le 2B(a)\delta \, \frac{1}{|a - 1| - \delta}$$

for $x, y \in I$.

If the interval I is unbounded, then the last inequality is impossible. From here, for $\delta < \min\{|a - 1|, A(a)\}$, the system (12.21) does not have any solution.

If the interval I is bounded, then

$$0 < |I| \le 2B(a)\delta \, \frac{1}{|a - 1| - \delta},$$

where $|I|$ is the length of I. This implies that, if I is bounded and

$$\delta < \min\left\{|a - 1|, A(a), |I||a - 1|\frac{1}{2B(a) + |I|}\right\},$$

then the system (12.21) does not have any solution, too.

Thus, for $0 \ne a \ne 1$, there exists a $\delta > 0$ such that the system (12.21) does not have any solution. From here this δ is "good" for every $\varepsilon > 0$ in the definition of the stability of the system (12.20). Hence this system is stable for $0 \ne a \ne 1$.

We consider the case $a = 0$. We proceed with the following remark. Let α and β be real numbers for which $|\alpha - \beta| \le 2\delta$ and $|\alpha/\beta| \le \delta$ for some positive $\delta \le 1/2$. We have

$$\left|\alpha\left(\frac{\alpha}{\beta} - 1\right)\right| \le 2\delta^2$$

and

$$0 < 1 - \delta \le 1 - \frac{\alpha}{\beta},$$

thus

$$|\alpha| \le \frac{2\delta}{1 - \delta} \, \delta \le 2\delta.$$

We have $|h(x) - x| \leq \delta$ and $|h(y) - y| \leq \delta$ for $x, y \in h(I)$. Thus

$$|[h(x) - h(y)] - (x - y)| \leq 2\delta$$

and, moreover, for $x \neq y$,

$$\left| \frac{h(x) - h(y)}{x - y} \right| = |h'(\theta)| \leq \delta.$$

By the above remark, for $\alpha = h(x) - h(y)$ and $\beta = x - y$, we have $|h(x) - h(y)| \leq 2\delta$ for $x, y \in h(I)$, whence $\sup h(I) - \inf h(I) \leq 2\delta$. Let $c = \inf h(I)$ and $d = \sup h(I)$. Since $|h(h(x)) - h(x)| \leq \delta$, so

$$c - \delta \leq h(h(x)) - \delta \leq h(x) \leq h(h(x)) + \delta \leq d + \delta.$$

From here $h(x) \in [c - \delta, d + \delta]$ and, for the solution $f(x) = (c + d)/2$ of the system (12.20), we have

$$|h(x) - f(x)| \leq \frac{d + \delta - c + \delta}{2} \leq \frac{4\delta}{2} = 2\delta, \qquad x \in I.$$

Thus the system (12.20) is stable for $a = 0$ with $\delta = \min(\varepsilon/2, 1/2)$. $\qquad\square$

Remark Let $S(a, I)$, for $0 \neq a \neq 1$, be the set of all $\delta > 0$ for which the system (12.21) does not have any solution. We have, by the proof of Theorem 12.3, in the case $0 \neq a \neq 1$, that, if I is unbounded, then $\delta \in S(a, I)$ for

$$0 < \delta < \min\{|a - 1|, A(a)\}.$$

Moreover, if I is bounded, then $\delta \in S(a, I)$ for

$$0 < \delta < \min\left\{|a - 1|, A(a), \frac{|I||a - 1|}{2B(a) + |I|}\right\}.$$

Since $S(a, I) \neq \emptyset$ and, if $\delta \in S(a, I)$ and $0 < \delta_1 < \delta$, then $\delta_1 \in S(a, I)$, this set is an interval and $\inf S(a, I) = 0$. The function $h(x) = x$ is a solution of (12.21) for $\delta = |1 - a|$, thus $\sup S(a, I) \leq |1 - a|$. Moreover, the function $h(x) = b \in I$ is a solution of (12.21) for $\delta = |a|$, whence $\sup S(a, I) \leq \min\{|a|, |1 - a|\}$.

Problem To calculate $\sup S(a, I)$.

Remark Note that $\sup S(a, I)$ cannot be equal to $\min(|a|, |1 - a|)$. E.g., for $I = [0, 1]$, $a = 1/2$, $\delta = (\sqrt{2} - 1)/2$ the function

$$h(x) = \frac{\sqrt{2}x}{2}$$

is a solution of (12.21), whence

$$\sup S\left(\frac{1}{2}, [0, 1]\right) \le \frac{\sqrt{2} - 1}{2}$$

and $\min(|a|, |1 - a|) = 1/2$. According to the last remark, if $0 < \delta < 1/22$, then $\delta \in S(1/2, [0, 1])$, and consequently $1/24 \in S(1/2, [0, 1])$. For $a = 1/2$, $I = [0, 1/6]$, $\delta = 1/24$, the function $h(x) = x/2$ is a solution of (12.21), thus $1/24 \notin S(1/2, [0, 1/6])$. From here $S(1/2, [0, 1]) \ne S(1/2, [0, 1/6])$.

Let S be a metric space with the metric d. Recall that the equation $f(f(x)) = f(x)$ of the idempotent function $f : S \to S$ is said to be stable, if for every $\varepsilon > 0$, there exists a $\delta > 0$ such that, for every function $h : S \to S$ for which $d[h(h(x)), h(x)] \le \delta$ for $x \in S$, there exists a idempotent function f for which $d[h(x), f(x)] \le \varepsilon$ for $x \in S$.

Theorem 12.4

(a) *The equation of the idempotent function from S to S is stable (with $\delta = \varepsilon$).*

(b) *The equation of the idempotent function from the non-degenerated interval I to I with the usual metric, is stable in the class of continuous functions (with $\delta = \varepsilon/3$).*

(c) *The above equation is not stable in the class of differentiable functions.*

Proof Part (a). For the function $f(x) = x$ for $x \in h(S)$ and the function $f(x) = h(x)$ for $x \in S \setminus h(S)$, we have $d[h(x), f(x)] \le \delta$ for the function $h : S \to S$ such that $d[h(h(x)), h(x)] \le \delta$.

Note that unfortunately the above function f may be discontinuous, even if the function h is continuous. E.g., for the continuous function $h(x) = \delta \exp(x)$ for $x \le 0$ and $h(x) = x + \delta$ for $x > 0$, the function $f(x) = x$ for $x > 0$ and $f(x) = h(x)$ for $x \le 0$ is not continuous at the point 0.

But there exists here the continuous idempotent $f(x) = \max\{x, \delta\}$ for which $|h(x) - f(x)| \le \delta$. This idempotent is of the form (12.26) (below). There exists also continuous idempotent, which is not of the form (12.26), e.g., $f(x) = \max\{x, 0\}$.

Part (b). Assume that $h : I \to I$ is the continuous function such that $|h(h(x)) - h(x)| \le \delta$. The function $f : I \to I$ of the form

$$f(x) = \begin{cases} x, & \text{if } x \in [h(z_1), h(z_2)] \cap I, \\ h(z_1) & \text{if } x \in [z_1, h(z_1)) \cap I, \\ h(z_2) & \text{if } x \in (h(z_2), z_2] \cap I, \\ h(x) & \text{if } x \in I \setminus h(I) \text{ and } h(x) \in [h(z_1), h(z_2)], \\ h(z_1) & \text{if } x \in I \setminus h(I) \text{ and } h(x) \in [z_1, h(z_1)), \\ h(z_2) & \text{if } x \in I \setminus h(I) \text{ and } h(x) \in (h(z_2), z_2], \end{cases} \qquad (12.26)$$

where $z_1 = \inf h(I)$, $z_2 = \sup h(I)$, $h(z_1) := z_1$ if $z_1 \notin I$ and $h(z_2) := z_2$ if $z_2 \notin I$, is the continuous idempotent [19]. We prove that $|h(x) - f(x)| \leq 3\delta$ for $x \in I$ (by the considerations in [24], we have $|h(x) - f(x)| \leq 10\delta$). Indeed

a/ if $x \in [h(z_1), h(z_2)] \cap I$, then $x \in h(I)$ and thus $|h(x) - f(x)| = |h(x) - x| \leq \delta$,

b/ if $x \in [z_1, h(z_1)] \cap I$, then $x \in [z_1, z_2]$ and thus $|h(x) - x| \leq \delta$; moreover, we have $|h(z_1) - z_1| \leq \delta$ and since $x, z_1 \in [z_1, h(z_1)]$, so $|x - z_1| \leq |h(z_1) - z_1| \leq \delta$. From here

$$|h(x) - f(x)| = |h(x) - h(z_1)| \leq |h(x) - x| + |x - z_1| + |z_1 - h(z_1)| \leq 3\delta.$$

c/ for $x \in [h(z_2), z_2] \cap I$ the reasoning is analogous as above,

d/ for $x \in I \setminus h(I)$ and $h(x) \in [h(z_1), h(z_2)]$, we obtain $|h(x) - f(x)| = 0 \leq \delta$,

e/ if $x \in I \setminus h(I)$ and $h(x) \in [z_1, h(z_1)]$, then

$$|h(x) - f(x)| = |h(x) - h(z_1)| \leq |h(z_1) - z_1| \leq \delta,$$

f/ for $x \in I \setminus h(I)$ and $h(x) \in [h(z_2), z_2]$ the situation is analogous.

Part (c). For the function

$$h(x) = \begin{cases} 0 & \text{for } x \leq -\delta, \\ \frac{1}{4\delta}(x + \delta)^2 & \text{for } x \in (-\delta, \delta), \\ x & \text{if } x \geq \delta. \end{cases} \tag{12.27}$$

and $\delta > 0$ we have $|h(h(x)) - h(x)| \leq \delta$ for $x \in \mathbb{R}$ and the inequality

$$|h(x) - f(x)| \leq \varepsilon, \qquad x \in \mathbb{R}$$

is impossible for any differentiable idempotent function f and for any $\varepsilon > 0$. □

Remark In the paper [23, Theorem 3.1 and Corollary 3.8] it has been proved, in a quite long way, that, for every $\delta > 0$, $\delta \leq 2/5$, and every continuous function $H : I \times \mathbb{R} \to I$ with $|H(H(x, t), s) - H(x, t + s)| \leq \delta$ and $|H'(x, 0) - 1| \leq \delta$, there is a continuous function $F : I \times \mathbb{R} \to I$ for which $F(F(x, t), s) = F(x, t+s)$, $F'(x, 0) = 1$ and $|H(x, t) - F(x, t)| \leq 10\delta$. From here, if $H(x, t) = h(x)$, then $|h(h(x)) - h(x)| \leq \delta$ and $|h'(x) - 1| \leq \delta$. Thus there exists a continuous solution $F(x, t)$ of the translation equation such that $F'(x, 0) = 1$ and $|h(x) - F(x, t)| \leq 10\delta$, whence $|h(x) - F(x, 0)| \leq 10\delta$. For the differentiable idempotent function $f(x) = F(x, 0)$ we have thus $f'(x) = 1$ and $|h(x) - f(x)| \leq 10\delta$.

Remark The equation $f'(x) = 0$ for $f : \mathbb{R} \to \mathbb{R}$ is not stable, i.e., it is not true that, for every $\varepsilon > 0$, there exists a $\delta > 0$ such that, for every differentiable function $h : \mathbb{R} \to \mathbb{R}$ for which $|h'(x)| \leq \delta$, there exists a constant function $f(x)$ for which $|h(x) - f(x)| \leq \varepsilon$. The function $h(x) = \delta x$ is "good" here.

We have here a phenomenon: *the equations in the system* (12.20) *for a = 0 are unstable and the system is stable.*

In the theory of the stability of functional equations the other stabilities have been considered [19], e.g., the b-stability or the inverse stability.

We say that the system (12.20) is b-stable provided, for every differentiable function $h : I \rightarrow I$, if the condition (12.21) is true for a $\delta > 0$ (i.e., the function $|h(h(x)) - h(x)| + |h'(x) - a|$ is bounded), then (12.22) is satisfied for a solution f of (12.20) and an $\varepsilon > 0$.

The system (12.20) is said to be inversely stable provided, for every $\delta > 0$, there exists a $\varepsilon > 0$ such that, if for a differentiable function $h : I \rightarrow I$ there exists a solution f of (12.20) for which (12.22) is true, then (12.21) is satisfied.

Theorem 12.5 *The system* (12.20) *is b-stable if and only if the interval I is bounded and a = 0 or a = 1 (with $\varepsilon = |I|$) and it is inversely stable (with every $\varepsilon > 0$) if and only if $0 \neq a \neq 1$.*

Proof If I is bounded and $a = 0$ ($a = 1$), then $|h(x) - b| \leq |I|$ for every $b \in I$ and $x \in I$ ($|h(x) - x| \leq |I|$ for $x \in I$). Since $f(x) = b$ ($f(x) = x$) is the solution of (12.20), this system is b-stable in this case with $\varepsilon = |I|$.

If I is unbounded and $a = 0$ ($a = 1$), then (12.21) is satisfied by $h(x) = x$ ($h(x) = b \in I$) and $\delta = 1$. Since the constant function $f(x) = b \in I$ ($f(x) = x$) is the only solution of (12.20) and the function $|x - b|$ is unbounded thus the system (12.20) is not b-stable.

If $0 \neq a \neq 1$, then $h(x) = b \in I$ satisfies (12.21) with $\delta = |a|$. Since the system (12.20) does not have any solution in this case, so this system is not b-stable. By the same reasoning, this system in inversely stable (with arbitrary $\varepsilon > 0$), since (12.22) is not true for any function h.

In the case $a = 0$ the system (12.20) has only the constant solutions $f(x) = b \in I$. Assume that it is inversely stable. Thus there exists a $\varepsilon > 0$ such that, for every differentiable function $h : I \rightarrow I$, if there exists a $b \in I$ with $|h(x) - b| \leq \varepsilon$ for $x \in I$, then

$$|h(h(x)) - h(x)| \leq 1 \quad \text{and} \quad |h'(x)| \leq 1.$$

Let b and c be such that $b, c \in I$, $b < c$ and $c - b \leq \varepsilon$. Let $h : I \rightarrow [b, c]$ be a differentiable function for which there exists an x_0 such that $|h'(x_0)| > 1$. Then we have $|h(x) - b| \leq b - c \leq \varepsilon$, which is a contradiction.

In the case $a = 1$ the system (12.20) has the only solution $f(x) = x$. Assume that it is inversely stable. Thus there exists an $\varepsilon > 0$ such that, for every differentiable function $h : I \rightarrow I$, if $|h(x) - x| \leq \varepsilon$ for $x \in I$, then

$$|h(h(x)) - h(x)| \leq 1 \quad \text{and} \quad |h'(x)| \leq 1.$$

Let b and c be such that $b, c \in \text{Int } I$ and $b < c$. Put $\alpha = \min\{\varepsilon, c - b\}$ and let $g : \mathbb{R} \rightarrow [0, \alpha]$ be the differentiable function such that $g(x) = 0$ for $x \leq c$ and there exists an x_0 for which $g'(x_0) > 1$.

Put $h(x) = x - g(x)$ for $x \in I$. If $x \in (-\infty, c] \cap I$, then $h(x) = x \in I$.
If $x \in (c, +\infty) \cap I$, then

$$h(x) = x - g(x) \in [x - \alpha, x] \subset [c - \alpha, x] \subset [b, x] \subset I.$$

From here h is the function from I to I. Moreover $|h(x) - x| = g(x) \le \alpha \le \varepsilon$
and $|h'(x_0) - 1| = |-g'(x_0)| > 1$, thus a contradiction.   □

It is possible to consider the other stabilities of the system (12.20) [19].

Remark The results for the equation

$$|f(f(x)) - f(x)| + |f'(x) - a| = 0$$

are the same as above. Indeed, this equation and the system (12.20) are equivalent
and

$$|f(f(x)) - f(x)| + |f'(x) - a| \le \delta \Rightarrow |f(f(x)) - f(x)| \le \delta \text{ and } |f'(x) - a| \le \delta$$

and

$$|f(f(x)) - f(x)| \le \delta \text{ and } f'(x) - a| \le \delta \Rightarrow |f(f(x)) - f(x)| + |f'(x) - a| \le 2\delta.$$

Remark The equation of the idempotent function is b-stable and it is not inversely
stable.

References

1. Baker, J.A., Lawrence, J., Zorzitto, F.: The stability of the equation $f(x + y) = f(x)f(y)$. Proc. Amer. Math. Soc. **74**, 242–246 (1979)
2. Baron, K., Chojnacki W., Jarczyk W.: Continuity of solutions of the translation equation. Aequationes Math. **74**, 314–317 (2007)
3. Batko, B.: Superstability of the Cauchy equation with squares in finite-dimensional normed algebras. Aequationes Math. **89**, 785–789 (2015)
4. Chung, J., Lee, B., Ha, M.: On the superstability of Lobacevskii's functional equation with involution. J. Funct. Spaces **2016**, 7 pp. (2016). Article ID 1036094
5. Faiziev, V.A., Sahoo P.K.: On a problem of Z. Moszner. J. Inform. Math. Sci. **4**, 175–177 (2012)
6. Forti, G.L.: The stability of homomorphisms and amenability, with applications to functional equations. Abh. Math. Semin. Univ. Hamburg **57**, 215–226 (1987)
7. Forti, G.L., Schwaiger, J.: Stability of homomorphisms and completeness. C. R. Math. Rep. Acad. Sci. Can. **11**(6), 215–220 (1989)
8. Forti, G.L.: Sulla stabilità degli omomorfismi e sue applicazioni alle equazioni funzionali. Rend. Sem. Mat. Fis. Milano **58**(1988), 9–25 (1990)
9. Hyers, D.H.: On the stability of linear functional equation. Proc. Nat. Acad. Sci. U. S. A **27**, 221–224 (1941)
10. Jung, S.M.: Hyers-Ulam-Rassias stability of Jensen's equation and its application. Proc. Amer. Math. Soc. **126**, 3137–3143 (1998)

11. Jung, S.M., Min, S.: Stability of wave equation with a source. J. Funct. Spaces **2018**, 4 pp. (2018). Article ID 8274159
12. Kuczma, M.: An Introduction to the Theory of the Functional Equations and Inequalities. Cauchy's Equation and Jensen's Inequality. Birkhäuser, Basel (2009)
13. Łojasiewicz, S.: Sur le problème d'itération. Colloq. Math. **3**, 176–177 (1955)
14. Mach, A., Moszner, Z.: Unstable (stable) system of stable (unstable) functional equations. Ann. Univ. Paedagog. Crac. Studia Math. **9**, 43–47 (2010)
15. Moszner, Z.: Sur la définition de Hyers de la stabilité de l'équation fonctionnelle. Opuscula Math. **3**, 47–57 (1987)
16. Moszner, Z.: On the stability of functional equations. Aequationes Math. **77**, 33–88 (2009)
17. Moszner, Z.: On the inverse stability of functional equations. Banach Center Publ. **99**, 111–121 (2013)
18. Moszner, Z.: On the stability of the squares of some functional equations. Ann. Univ. Paedagog. Crac. Stud. Math. **14**, 81–104 (2015)
19. Moszner, Z.: Stability has many names. Aequationes Math. **90**, 983–999 (2016)
20. Moszner, Z.: On the normal stability of functional equations. Ann. Math. Sil. **30**, 111–128 (2016)
21. Moszner, Z.: Translation equation and the Jordan non-measurable continuous functions. Ann. Univ. Paedagog. Crac. Stud. Math. **16**, 117–120 (2017)
22. Moszner, Z.: On the geometric concomitants. Ann. Univ. Paedagog. Crac. Stud. Math. **18**, 53–58 (2019)
23. Moszner, Z., Przebieracz, B.: Is the dynamical system stable? Aequationes Math. **89**, 279–296 (2015)
24. Przebieracz, B.: On the stability of the translation equation and dynamical systems. Nonlinear Anal. **75**, 1980–1988 (2012)
25. Roberts, F.S.: On the indicator of plurality function. Math. Soc. Sci. **22**, 163–174 (1991)
26. Sibirsky, S.: Introduction to Topological Dynamics. Leiden, Noordhoff (1975)
27. Székelyhidi, L.: Note on a stability theorem. Can. Math. Bull. **25**, 500–501 (1982)
28. Ulam, S.M.: A Collection of Mathematical Problems. Interscience Publications, New York (1960)

Chapter 13
Subdominant Eigenvalue Location and the Robustness of Dividend Policy Irrelevance

Adam J. Ostaszewski

Abstract This paper, on subdominant eigenvalue location of a bordered diagonal matrix, is the mathematical sequel to an accounting paper by Gao et al. (J Bus Financ Acc 40:673–694, 2013). We explore the following characterization of dividend-policy irrelevance (DPI) to equity valuation in a multi-dimensional linear dynamics framework L: DPI occurs under L when discounting the expected dividend stream by a constant interest rate iff that rate is equal to the dominant eigenvalue of the canonical principal submatrix A of L. This is justifiably the 'latent' (or gross) rate of return, since the principal submatrix relates the state variables to each other but with dividend retention. We find that DPI reduces to the placement of the maximum eigenvalue of L between the dominant and subdominant eigenvalues of A. We identify a special role, and a lower bound, for the coefficient measuring the year-on-year dividend-on-dividend sensitivity in achieving robust equity valuation (independence of small variations in the dividend policy).

Keywords Dividend irrelevance · Dominant eigenvalue · Bordered diagonal matrix · Performance stability · Dividend-on-dividend sensitivity

Mathematics Subject Classification (2010) Primary 91B32, 91B38; Secondary 91G80, 49J55, 49K40

13.1 Introduction and Motivation

Accounting theory seeks to reconcile valuation of a firm based on *historically* observed variables ('primitives', that recognize value created *to date*) with its equity value, arrived at by markets in a *prospective* fashion: see [27]. The market's

A. J. Ostaszewski (✉)
Department of Mathematics, London School of Economics, London, UK
e-mail: a.j.ostaszewski@lse.ac.uk

© Springer Nature Switzerland AG 2019
J. Brzdęk et al. (eds.), *Ulam Type Stability*,
https://doi.org/10.1007/978-3-030-28972-0_13

valuation is theoretically modelled as the present value of future (expected) dividends and involves discounting by the (notional) *riskless* interest rate in force, say r per unit time. From the historic (accounting) side, various secondary composite variables have been derived from the primitives (with appropriate technical names such as 'residual income'—for a brief introduction see [8]), formalizing in one way or another a notion of *current* 'earnings'; the latter is then intended to identify equity value directly (as a dependent variable) and to provide empirically stable time series.

To arrive at such a composite accounting variable, assumptions are needed concerning the future evolution of the primitives—at least in a hypothetical 'steady state' context. (For a 'dynamic' alternative, drawing on the value of waiting, see [5] in this same volume.) The favourite mechanism for this context is a *linear state-space representation L*, thereby introducing subtle links—our main concern here—between accounting theory and mathematics.

It is noteworthy, though not of direct mathematical significance to this paper, that an encouraging feature for the use of a (linear) representation L is its flexibility in permitting inclusion, alongside state variables that recognize historic value creation (as above), additional 'information' state variables; these capture the (typical) dynamics of an embedded 'potential to create' value, an 'intangible' value, currently unrecognized in the accounts but feeding through to future recognized value (a matter central to the luckless 2014 attempt by Pfizer to bid for AstraZeneca—'the mega-merger that never was'). This partly bridges the historic-prospective divide. (The idea was introduced into the accounting literature of linear systems by Ohlson [21], and enabled him to include the accounting of 'goodwill' value—see [15]; for another example of an intangible, involving product 'image' and its valuation, see e.g. [13].)

Returning to mathematical concerns, we note that the eigenstructure of L (eigenvalue distribution) has to connect with economic consequences of an assumed 'steady state'—such as absence of *arbitrage* opportunities in equity valuation, and its relation to the notional riskless interest rate (above). A further fundamental insight, going back to Miller and Modigliani [18] in 1961, is that—under prescribed conditions (but see e.g. [4] for the effects of alternative informational assumptions)—the equity value should not depend on the distribution of value, be it impounded into the share price or placed in the share-holders' pockets (via dividend payouts); this is properly formalized below. (This is one aspect of capital structure irrelevance: equity value should not depend on debt versus equity issuance [17, 19].) The principle of *dividend policy irrelevancy* also carries implications for the linear dynamics. For a recent analysis of the connections see [7], where the basic result asserts that DPI occurs iff the riskless interest rate agrees with the dominant eigenvalue of the reduced linear system ('subsystem') obtained by the firm withholding (retaining) dividend payouts. One may call the latter the dominant 'latent' rate of the system L. Recall that it is the riskless rate r that is used in the present-value calculation above.

This is a *knife-edge* characterization in regard both to the riskless interest rate and the dominant eigenvalue, so it is natural to study *accounting robustness* in the DPI framework. That is the principal aim of this paper, achieved by studying an eigenvalue location problem, similar but distinct from one in control theory (reviewed shortly below). The delicacy of this matter is best seen in the light of Wilkinson's example in [28, §33] of a sparse 20×20 matrix with the integers $n = 1, 2, \ldots, 20$ on the diagonal (its eigenvalues) and all superdiagonal entries of 20; a small perturbation of ε in the bottom left-hand corner yields the characteristic polynomial to be $(\lambda - 20) \ldots (\lambda - 1) - 20^{19} \varepsilon$, so that for $\varepsilon := 10^{-10}$ the eigenvalues are these: 6 real ones which are to 1 decimal place 0.9, 2.1, 2.6, 18.4, 18.9, 20.0, and 7 conjugate complex pairs (all in mid-range values, in modulus) as follows:

$$4 \pm i1.1; \quad 5.9 \pm i1.9; \quad 8.1 \pm i2.5; \quad 10.5 \pm i2.7; \quad 12.9 \pm i2.5; \quad 15.1 \pm i1.9; \quad 17 \pm i1.1.$$

Turning now to the mathematical problem, consider, granted initial conditions, the performance of the following discrete-time system:

$$\left.\begin{aligned}
z_{t+1} &= Az_t &+ av_t &+ bd_t, \\
v_{t+1} &= &+ \alpha v_t, \\
d_{t+1} &= w^T z_t &+ \gamma v_t &+ \beta d_t.
\end{aligned}\right\} \tag{Ω}$$

Here β is the *dividend-on-dividend* 'year-on-year' growth; its effect is particularly significant—see below and Sect. 13.2 (Theorem 3). So the state variables at time t are $(z_t, v_t, d_t) \in \mathbb{R}^{n+2}$ – with d_t representing the time-t dividend, v_t an 'information' variable (as above), subjected to 'fading' over time by a factor α satisfying

$$0 \le \alpha < 1,$$

with A a real matrix (hereafter, the *reduced matrix* of the system, or the 'dividend-retention' matrix) that is constant over time. The performance of the system at time 0 is measured by the expression

$$P_0 = \sum_{t=1}^{\infty} R^{-t} d_t,$$

in which a discount factor R is applied to the sequence $d = \{d_t\}$ generated by (Ω). Here $R = 1 + r$ with r as above (the governing riskless interest rate per unit time), and so P_0 represents an initial *equity* valuation of the firm—in the sense motivated above. To guarantee convergence it is sufficient to assume that all components of any solution of (Ω) have growth below R; referring to the modulus of the dominant eigenvalue of the coefficient matrix in (Ω) by $\lambda_{\max}^{\Omega}$ this growth condition may be restated as

$$\lambda_{\max}^{\Omega} < R. \tag{13.1}$$

The bottom row vector in (Ω)

$$\omega_{\text{div}} = (w^T, \gamma, \beta)$$

is termed the *dividend policy*.

In this setting the full coefficient matrix is assumed constant, but not known to observers of the state variables (which are disclosed in the annual accounts). However, whereas A and the penultimate row involve value created over time through an initial (fixed) investment, it is the final row that generates the returns (over time) to the investors. Thus the equity P_0 should be regarded as a function of R and of the dividend-policy vector parameters set by the managers, that is

$$P_0(R; \omega_1, \ldots, \omega_n, \gamma, \beta) := \sum_{t=1}^{\infty} R^{-t} d_t. \tag{13.2}$$

One says that the valuation $P_0(.)$ exhibits *Dividend Policy Irrelevancy* (DPI) *at R* if the function $P_0(R; \ldots)$ is unchanged as the dividend-policy vector ω_{div} varies. A first problem is to determine circumstances under which the system exhibits dividend irrelevancy. Up to a technical side-condition (ensuring co-dependence of dividends and value creation) the short answer is that the *dominant eigenvalue* of A should agree with R—this was first proved by Ohlson in the special case $n = 1$, and then generalized in [7] (and also referred to in the earlier published monograph [23]).

Below we refine the notion of dividend irrelevancy in order to study the effects of a proximal *sub-dominant* eigenvalue. We first establish notation and some conventions. Begin by omitting hereafter explicit mention of the information variable v_t; we regard it as yet another state variable absorbed into A (with α then becoming an eigenvalue of the reduced matrix), and so we overlook its simple dynamics; we may now free up α and γ for other uses below. The eigenvalues $\lambda_1^A, \lambda_2^A, \ldots, \lambda_n^A$ of the reduced matrix A (*latent* relative to L, below) are listed in order of decreasing modulus. As these will be required to be real, positive and (generically) distinct, this is taken to mean

$$\lambda_1^A > \lambda_2^A > \ldots > \lambda_n^A > 0.$$

Whenever convenient (e.g. in proofs) we omit the superscript A. The system matrix L of (Ω) above, viewed as the *augmented matrix* of A, is now given by

$$L_A = A^{\lrcorner}(w, \beta) := \begin{bmatrix} A & b \\ w^T & \beta \end{bmatrix}$$

("A-bordered"), and is regarded as a function of the *real* vector (w^T, β). Its (possibly complex) eigenvalues will likewise be regarded as *functions* of (w^T, β) and denoted by λ_j^L, or more simply by κ_j, so that

$$\kappa_1 = \lambda_1^L, \kappa_2 = \lambda_2^L, \ldots, \kappa_{n+1} = \lambda_{n+1}^L,$$

which distinguish them more easily (from λ_j^A); each index here is identified through the functional conditions

$$\kappa_j(0, \ldots, 0, \beta) = \lambda_j^A \text{ for } j = 1, \ldots, n, \text{ and } \kappa_{n+1}(0, \ldots, 0, \beta) = \beta. \tag{13.3}$$

We write

$$\lambda_{\max}^L(w^T, \beta), \text{ or } \kappa_{\max}(w^T, \beta) := \max\{\kappa_j(w^T, \beta) : j = 1, \ldots, n+1\}.$$

Although $A^\lrcorner$ is not in general symmetric, we will contrive situations in which the eigenvalues of A interlace with those of $A^\lrcorner$: $\kappa_j \geq \lambda_j \geq \kappa_{j+1}$, just as in Cauchy's Interlace Theorem, cf. [11, Thm. 4.3.1], [3, Ch.7, §8 Th. 4], [12], at least for $j \geq 2$.

Since we are mostly concerned with the characteristic polynomial and eigenvalue location, we will be working in an equivalent *canonical setting* in which, firstly, A is diagonal and, secondly, as a further simplification, we suppose that for $j \leq n$ the *dividend significance coefficients* b_j are all non-zero. Rescaling by b_j the j-th equation of the diagonalized system gives what we term the equivalent *canonical system* in which the resulting *dividend significance coefficients are* $\delta_j = \pm 1$ (as, of course, we may also rescale by $-b_j$: see Remark 4 in Sect. 13.2.1). Thus L is replaced by

$$H(\omega) = \begin{bmatrix} \lambda_1^A & 0 & & 0 & \delta_1 \\ 0 & \lambda_2^A & & & \delta_2 \\ & & \cdots & & \\ 0 & 0 & & \lambda_n^A & \delta_n \\ \omega_1 & \omega_2 & & & \omega_{n+1} \end{bmatrix}, \tag{13.4}$$

where $\delta_j = \pm 1$ for each j. It is preferable to subsume β as ω_{n+1} (rather than as δ_{n+1}) into the 'canonical dividend-policy vector' corresponding to (w^T, β). Our first definitions all contain growth conditions analogous to (13.1) and are motivated by Proposition 1 below.

Definition 1 We say that the system (Ω) has dividend irrelevance at R if, for all $\omega = (\omega_1, \ldots, \omega_n, \beta)$ such that $|\lambda_{\max}^L(\omega, \beta)| < R$,

$$P_0(R; \omega, \beta) = P_0(R; o, \beta),$$

with o the zero vector.

Definition 2 We say that the system (Ω) has local dividend irrelevance at R for $\omega = (\omega_1, \ldots, \omega_n, \beta)$ if there is $\varepsilon > 0$ so that, for all $\omega' = (\omega_1', \ldots, \omega_n', \beta')$ such that $||\omega - \omega'|| < \varepsilon$ and $|\lambda_{\max}^L(\omega', \beta')| < R$,

$$P_0(R; \omega', \beta') = P_0(R; \omega, \beta).$$

Here the norm is Euclidean. The local definition is weaker in that it requires merely that the equity valuation be robust in respect of the accounting system (i.e. insensitive to minor accounting variations). However, in our model setting P_0, regarded as a function of $\omega_1, \ldots, \omega_{n+1}$, is a rational function in these variables (see Observation below in Sect. 13.2), so its local constancy for a given R is equivalent to global constancy for the same R. An intermediate definition permitting constant equity is the following

Definition 3 We say that the system (Ω) has bounded dividend irrelevance at R if, for some positive $\rho < R$ and all $\omega = (\omega_1, \ldots, \omega_n)$ such that $|\lambda_{max}^L(\omega, \beta)| < \rho$,

$$P_0(R; \omega, \beta) = P_0(R; o, \beta).$$

The example below identifies anomalous behaviour which these definitions offer as possible.

The requirement for dividend irrelevance amounts to discovering to what extent $P_0(R, d)$ depends only on the initial data: A, b, z_0, d_0.

In view of the role of the interest rate $R > 0$, it will be appropriate to make the following.

Blanket Assumption The eigenvalues $\lambda_1^A, \ldots, \lambda_n^A$ of A are all real and positive.

Notice that if $|\kappa_{max}(\omega, \beta)| < \lambda_2$, small enough variations in the dividend-policy vector will ensure that the inequality is preserved. This entails (see Proposition 1 below) that the system will have local dividend-policy irrelevance at more than one rate, namely at $R = \lambda_1$ and $R = \lambda_2$. Our contribution is to identify in Theorem 3 below a condition on β, namely that

$$\beta > 2\lambda_2 - \lambda_1,$$

requiring a lower bound on the dividend-on-dividend yearly growth, which ensures that $|\kappa_{max}(\omega, \beta)| > \lambda_2$ and thereby achieves uniqueness of the latent rate of return in this case: dividend-policy irrelevance occurs only at the one rate $R = \lambda_1$.

Example of Bounded-DPI at Both $R = \lambda_1$ and λ_2 We take $\lambda_1 = 2, \lambda_2 = 1.5$, $\beta = 0.5$, and $\delta_1 = -1, \delta_2 = +1, \omega_2 = 0.1$. Note that $2\lambda_2 - \lambda_1 = 3 - 2 = 1$, so $\beta < 2\lambda_2 - \lambda_1$.

Figure 13.1 shows the locus of the conjugate complex root pair for ω_1 running through the range 0.1–1.2 generated by Mathematica. The third root is below, but close to, 1.5. Note that the additional vertical line appears from the numerics (the computer routine switches the identity of the two conjugates). Variations in ω bounding the eigenvalues to $|\zeta| < \lambda_2$ keep both equity $P_0(\lambda_2)$ and $P_0(\lambda_1)$ constant.

Fig. 13.1 Eigenvalue locus as ω_1 varies. For a range of values of ω_1 all eigenvalues of $A^\lrcorner$ lie in the disc of radius λ_2^A (the second largest eigenvalue of A)

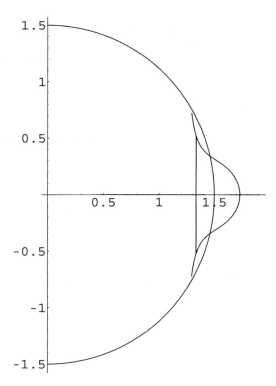

13.1.1 Continuous-Time Analogue

Letting D_t denote cumulative dividends, the analogous dynamic in continuous time may be formulated, indicating differentiation with a dot, as

$$\dot{z}_t = A z_t + b D_t,$$
$$\dot{D}_t = w^T z_t + w_{n+1} D_t.$$

Here $z_t = (x_t^1, \ldots, x_t^n)$ is now the state vector of flow variables and b and w are columns in R^n. Putting $Z_t = (x_t^1, \ldots, x_t^n, D_t)$, we arrive at the system

$$\dot{Z}_t = A^\lrcorner Z_t,$$

where as before $A^\lrcorner$ is A bordered by b and $\bar{w}^T = (w_1, \ldots, w_n, w_{n+1})$. Thus D_t has growth rate at most $\kappa_{\max}$, the largest eigenvalue of $A^\lrcorner$.

The equity is

$$P_0 = \int_0^\infty e^{-rt} dD_t = -r D_0 + r \int_0^\infty e^{-rt} D_t dt,$$

which converges provided that the dividend flow is of exponential growth less than r and in particular that

$$\lim_{t \to \infty} e^{-rt} D_t = 0. \tag{13.5}$$

This requires therefore that $\kappa_{\max} < r$.

The identical format allows interpretation of results in this paper leading to valid conclusions for the continuous-time framework. In fact set

$$\tilde{Z} = \int_0^\infty e^{-rt} Z_t dt$$

to obtain the discounted flow, i.e. the Laplace transform. We conclude that

$$-rZ_0 + r\tilde{Z} = A^\lrcorner \tilde{Z},$$

subject to $\kappa_{\max} < r$. Assuming $rI - A^\lrcorner$ is invertible we have

$$\tilde{Z} = (rI - A^\lrcorner)^{-1} Z_0,$$

whence

$$P_0 = P_0(\bar{w}) = (0, \ldots, 0, 1)^T (rI - A^\lrcorner(\bar{w}))^{-1} Z_0.$$

This formula was exploited in Ashton et al. in [1, 2] in a stochastic setting.

An identical formula can be developed for discrete time (with Laplace transform replaced by z-transform) and is the departure point for the purely algebraic argument developed in [7]. We note an important conclusion, which follows from the form of the adjugate matrix (or by invoking Cramer's Rule).

Observation In the framework above, the equity P_0 regarded as a function of $w_1, \ldots, w_{n+1}$ is a rational function in these variables.

13.1.2 Control-Theory Analogies

We remark on the contextual similarities between the accounting model and two standard control-theory settings and identify the differences. For background, see e.g. [25].

The first setting is the closest. Here we regard z_t as a state vector and d_t as a control variable, and write the system equation as

$$z_{t+1} = Az_t + d_t b,$$

with the control variable selected according to a non-standard (since it is in effect differential) feedback-law

$$d_{t+1} = w^T z_t + \beta d_t.$$

In these circumstances we interpret $P_0(R, d)$ defined by (13.2) as the system performance index (a discounted cost stream), and require that that it be either (1) independent of the feedback law parameters, or (2) independent of small variations in the parameters. This is akin to performance stability, or even a 'disturbance decoupling problem', except that there is no modelled disturbance here. Our analysis is similar to that of the standard pole-placement problem (requiring assignment of eigenvalues through feedback-law design). Where we differ is in the inevitable presence and effect of zeros as well as poles.

The alternative view is to regard z_t as the observation vector of the state $Z_t = (z_t, d_t)$, with state equation and observation vector defined respectively by

$$Z_{t+1} = H Z_t,$$
$$z_t = P Z_t.$$

Here P is the projection matrix from $\mathbb{R}^{n+1}$ to $\mathbb{R}^n$. In these circumstances the performance index of the system is given by $P_0(R, d)$, and we wish to ensure that the performance is dependent only on the evolution of observation and the initial state Z_0.

13.1.3 The Accounting Context

The dividend irrelevance question (in particular whether or not dividends should be irrelevant to stockholders) has been a live issue since the 1961 paper [18] of Modigliani and Miller. See, for example, [6]. The current quest for dividend irrelevance comes from the general possibility of restating equity in terms of an identically discounted alternative series based on accounting numbers, as first pointed out in 1936 by Preinreich [24]; see the discussion in the survey paper by Ohlson and Gao [23]. If (Ω) models the evolution of the firm and z_t models its observable accounting numbers, interest focuses on whether valuations are possible at time $t = 0$, based on the accounting numbers alone, that is to say in the absence of access to the currently unobservable information ω_{div}. See Proposition 2 below.

13.1.4   Organization of Material

The paper is organized as follows. In Sect. 13.2 we give our main theorems
(Theorems 1–4) and the auxiliary propositions on which they are based. Shorter
proofs are included here, but longer proofs are delayed till later. The results of
Sects. 13.2 and 13.3 are then used in Sect. 13.4 to perform a detailed study of
eigenvalue location of the augmented matrix. This is done by examining the two-
pole case first, and then estimating the distortion effects when other poles are
present. Some bifurcation analysis is conducted in Sect. 13.5 in circumstances
corresponding to Theorem 1. Section 13.6 and onwards contain longer proofs, or
such details as are not required for the analysis of Sect. 13.4.

13.2   Main Theorems and Auxiliary Propositions

In [7] it is shown that dividend irrelevance at R occurs iff R takes the value of
the dominant eigenvalue, here defined to be the largest in modulus (in the spirit
of the Perron-Frobenius context—see [11, Ch. 8], [26, Ch.1,2]), of the reduced
matrix A, which will forthwith be diagonal. Asymptotic considerations suggest this
result, since for generic initial conditions, and for large t the dominant eigenvalue
growth of A^J dwarfs into insignificance the other state components, both those
entering the accounting state vector and those entering the dividend (provided of
course that the dividend-policy vector gives the dominant growth component a non-
zero coefficient). Asymptotic considerations thus turn the multi-dimensional system
apparently into an essentially one-dimensional one, and it is to this that Ohlson's
Principle (initially proven in dimension one only) might apply—see Theorem 2
below. That is to say, assuming dividend and dominant state variable are inter-
linked, dividend irrelevance occurs if and only if R takes a unique value, that
value being the dominant eigenvalue of the dividend-retention matrix A (that of the
dominant state). (Of course, in the long run, observation of the dividend sequence
permits inference of the dividend-policy vector.)

   In this paper we offer an analysis of the quoted result based on algebraic
considerations, some complex analysis (including an inessential reference to Marden's 'Mean-Value Theorem for polynomials'), and graphical analysis. These
complement a standard textbook analysis based on Gerschgorin's circle theorem—
for which see e.g. [20, Th. 13.14] or [9, Th. 7.8d].

   Unsurprisingly, *the eigenvalues of A^J may be located arbitrarily,* but only
if no restrictions are placed on the dividend policy (w, β). Evidently, Dividend
Irrelevance must implicitly assume the convergence assumption (13.1) as a bound
on the eigenvalues of A^J. It transpires (see Proposition 3) that the dividend-policy
vector is restricted by this assumption to the interior of an appropriate polytope
in $\mathbb{R}^{n+1}$.

Conditions may be placed on the vector b such that, when $\omega = (w, \beta)$ lies in an open region of parameter space, it is the case that the dominant eigenvalue of the augmented matrix is real and lies *between the first largest and the second largest* eigenvalue of the reduced matrix. This is the substance of our first main result stated here and proved in Sect. 13.6.

Theorem 1 (An Eigenvalue Dominance Theorem) *Suppose that A has real positive distinct eigenvalues. In the canonical setting (13.4) we have as follows.*

(i) *If $sign[\delta_1] = -1$ and $sign[\delta_j] = +1$ for $j = 2, \ldots, n$, then the open set*

$$\{\omega : A^{\lrcorner} \text{ has real distinct roots and } \lambda_2 < \kappa_1(\omega) < \lambda_1\},$$

has non-empty intersection with the set

$$\{\omega : \omega_1 > 0, \ldots, \omega_{n+1} > 0\}.$$

Moreover, κ_2, the second largest eigenvalue of $A^{\lrcorner}$, is increasing in ω for small ω. Under these circumstances dividend irrelevance holds uniquely at $R = \lambda_1$.

(ii) *More generally, the open set*

$$\{\omega : A^{\lrcorner} \text{ has real distinct roots and } \lambda_2 < \kappa_1(\omega) < \lambda_1\},$$

has non-empty intersection with the set

$$\{\omega : \delta_1\omega_1 < 0, \delta_2\omega_2 > 0, \ldots, \delta_n\omega_n > 0\},$$

and again under these circumstances dividend irrelevance holds uniquely at $R = \lambda_1$.

(iii) *If $\delta_1\omega_1 < 0$ and $\delta_k\omega_k > 0$ for all $k = 2, .., n$, and Ω has all its eigenvalues in the disc $|\zeta| < \lambda_1$ of the complex ζ-plane, then $A^{\lrcorner}$ has an eigenvalue in the annulus*

$$\mathscr{A} := \{\zeta \in \mathbb{C} : \lambda_2 < |\zeta| < \lambda_1\}.$$

(iv) *If $\delta_1\omega_1 < 0$ and $\delta_k\omega_k < 0$ for all $k = 2, .., n$, then the system Ω has a real eigenvalue in the real interval (λ_2, λ_1).*

For a proof see Sect. 13.6.

Remark We see therefore that for an appropriate vector b there is a region of parameter space for which the eigenvalues of the *augmented* matrix $A^{\lrcorner}$ remain strictly bounded in modulus by λ_1, the dominant eigenvalue of A. Note the re-emergence of the side conditions $\delta_1\omega_1 < 0$ analogous to the condition $\omega_{12}\omega_{21} < 0$ in Ohlson's Theorem for $n = 1$ (see [23]).

We are able to provide some information about the extent of the subspace (see formula (13.14) of Sect. 13.3) where we obtain (when $\delta_1 < 0$) the upper bound on positive ω_1 of

$$\frac{1}{4}(\lambda_1 - \beta)^2 + \{\omega_2\delta_2 + \ldots\},$$

for the case $\delta_2\omega_2 > 0$. Moreover, Proposition 4 and calculations of Sect. 13.4 appear to imply that, even if ω_1 rises above this bound, the two particular roots of the characteristic polynomial of $A^\lrcorner$ which are forced into coincidence remain outside the disc $|\zeta| \leq \lambda_2$ in the complex ζ-plane (as they move asymptotically to a vertical towards $\Re(\zeta) = \lambda_2$), provided

$$\beta > 2\lambda_2 - \lambda_1.$$

By contrast, we find for $\omega_1\delta_1 < 0$ and $\omega_2\delta_2 < 0$ the top two roots of the augmented matrix $A^\lrcorner$ both approach λ_2 from opposite sides; this again is in keeping with the expectation that dividend irrelevance occurs only at the dominant root λ_1.

Our results link to work concerned with the real spectral radius of a matrix, see Hinrichsen and Kelb [10], which investigates by how much a matrix may be perturbed without moving its spectrum out of a given open set in the complex plane. In the cited work the open set of concern is usually either the unit disc or the open left half-plane, both in connection with stability issues. Our interest, however, focuses additionally on the open set described by the *annulus* $\mathscr{A}$ defined by the first and second largest eigenvalues of A (cf. Theorem 1). We note that there is a well-established Sturmian algorithm for counting the number of zeros of a polynomial in the unit disc in the complex plane (see Marden [16, §42, p. 148]), and so in principle the issue of Dividend Irrelevance is resolvable for a given policy vector ω_{div} by reference to the number of zeros in the unit circle of the two characteristic polynomials

$$\chi_{A^\lrcorner}(\kappa/\lambda_1^A), \qquad \chi_{A^\lrcorner}(\kappa/\lambda_2^A).$$

Specifically, the first should have $n + 1$ zeros and the second no more than n. The Schur-Cohn criterion [16, Th. 43.1], [9, §6.8] might perhaps also be invoked to count the number of roots in the unit disc.

13.2.1 Preliminaries

Our analysis is based on two results embodied in Proposition 1 and in the equivalences given in Proposition 2. The arbitrary placement of the zeros, the substance of Proposition 3, is also a consequence of Proposition 2.

Proposition 1 (Under the Assumption of Distinct Eigenvalues) *In the canonical setting (13.4) with*

$$\delta_j = \pm 1 :$$

for any $j \le n$ and fixed R, the equity $P_0(R, \omega)$ is locally or globally independent of ω iff $R = \lambda_j$ provided

$$R > \max\{|\kappa_k(\omega)| : k = 1, \ldots, n + 1\}, \tag{13.6}$$

in which case

$$d_0 + P_0(R; d) = -\frac{R Z_0^j}{\delta_j}.$$

The proof is in Sect. 13.7.

Remark 1 Apparently, if the eigenvalues of $A^\lrcorner$ all lie in the disc with radius any other eigenvalue of A, the Proposition permits *local* dividend irrelevance to occur at several rates of return. We will show below that subject to (13.6) such an anomalous behaviour is definitely excluded when $\omega_1 \ne 0$ and also $\omega_j \ne 0$ for some $1 < j \le n$.

Remark 2 In principle we might want to allow $\delta_j = -1$, to respect a restriction in the directional sense of a re-scaling of accounting variables (if appropriate); it transpires from the next Proposition that the sign of δ_j can be absorbed by ω_j and the choice of sign is only a matter of symbolic convenience, so that we can interpret $\delta_1 \omega_1 < 0$ as saying $\omega_1 > 0$. That said, it is important to realize that rescaling an accounting variable, say z_k by α, requires an inverse rescaling of the corresponding dividend-policy component, that is of ω_k by α^{-1} (in order to preserve the definition of dividend untouched). The right-hand side of the valuation equation perforce does not refer to the eigenvalues κ_j, despite the fact that these control the growth rates of the canonical accounting variables.

The following algebraic equivalences lie at the heart of all our arguments. Below we denote by $\chi_{A^\lrcorner}(\kappa, \omega_1, \ldots, \omega_{n+1})$ the characteristic polynomial of $A^\lrcorner(\omega_1, \ldots, \omega_{n+1})$.

Proposition 2 (Inverse Relations) *Put $\lambda_{n+1} = \beta = \omega_{n+1}$. The equations below are all equivalent.*

$$\chi_{A^\lrcorner}(\kappa, \omega_1, \ldots, \omega_{n+1}) = 0. \tag{13.7}$$

$$\prod_{j=1}^{n+1}(\kappa - \lambda_j) = \omega_1 \delta_1 \prod_{j=2}^{n}(\kappa - \lambda_j) + \omega_2 \delta_2 \prod_{j \ne 2}^{n}(\kappa - \lambda_j) \tag{13.8}$$

$$+ \ldots + \omega_n \delta_n \prod_{j \ne n}^{n}(\kappa - \lambda_j).$$

*Polar form for $j \leq n$ with **leading quadratic term**:*

$$\omega_1\delta_1 + \omega_2\delta_2 + \ldots + \omega_n\delta_n = (\kappa - \lambda_j)(\kappa - \beta) + \sum_{k \neq j}^{n} \frac{\omega_k\delta_k(\lambda_j - \lambda_k)}{\kappa - \lambda_k}. \tag{13.9}$$

Polar form for $j = n + 1$ and with $\kappa \neq \lambda_k$ for $k = 1, \ldots, n$:

$$\beta = \kappa - \frac{\omega_1\delta_1}{\kappa - \lambda_1} - \frac{\omega_2\delta_2}{\kappa - \lambda_2} - \ldots - \frac{\omega_n\delta_n}{\kappa - \lambda_n}. \tag{13.10}$$

In particular, with $j = 1$, putting

$$f(\kappa) := \{\omega_2\delta_2 + ..\} - (\kappa - \lambda_1)(\kappa - \beta) - \frac{\omega_2\delta_2(\lambda_1 - \lambda_2)}{\kappa - \lambda_2} - \ldots \tag{13.11}$$

we obtain the equivalent equation

$$f(\kappa) = -\omega_1\delta_1.$$

Proof of equivalence follows in Sect. 13.8. Each of the above identities enables a different analytic approach.

Our first conclusion regards the potentially arbitrary placement of the zeros of (13.7).

Proposition 3 (Zero Placement) *In the canonical setting of Proposition 1, for an appropriate choice of real vector ω the characteristic polynomial*

$$\chi_{A^{\dashv}}(\kappa, \omega) = |\kappa I - A^{\dashv}(\omega)|$$

may take the form

$$\kappa^{n+1} - p_0\kappa^n + p_2\kappa^{n-1} + \ldots + (-1)^{n+1}p_n, \tag{13.12}$$

for arbitrary choice of real coefficients $p_0, \ldots, p_n$. The transformation

$$(p_0, \ldots, p_n) \rightarrow (\omega_1, \ldots, \omega_{n+1})$$

is affine invertible. The roots $\kappa_1, \ldots, \kappa_{n+1}$ of the characteristic polynomial may therefore be located at will, subject only to the inclusion, for each selected complex root, of its conjugate.

This result is proved in Sect. 13.9.

Proposition 3 above indicates that in principle the region of parameter space in which the boundedness assumption (13.8) holds may be obtained as the transform under the above mentioned transformation of the set of vectors $(p_0, \ldots, p_n)$

satisfying a criterion derived from Cauchy's theorem on the Inclusion Radius [9, Th. 6.41], [16, Th. 27.1], namely

$$|p_n| + |p_{n-1}|\lambda_1 + \ldots + |p_0|\lambda_1^n < \lambda_1^{n+1}.$$

(Recall that the inclusion radius of the polynomial (13.12) is the unique positive root of the polynomial $|p_n| + |p_{n-1}|\kappa + \ldots + |p_0|\kappa^n - \kappa^{n+1}$.) Since the set of vectors $(p_0, \ldots, p_n)$ so described is the interior of a polytope, the corresponding region in parameter space is therefore likewise seen to be the interior of a polytope. Let us term this the **Cauchy polytope**.

Evidently $(0, \ldots, 0, \beta)$ is on the boundary of the Cauchy polytope, since then

$$\chi_{A^{\lrcorner}}(\kappa, \omega) = (\kappa - \lambda_1)..(\kappa - \lambda_n)(\kappa - \beta).$$

An immediate corollary is the following result, first announced for the case $n = 1$ by Ohlson at the 2003 International Conference on Advances in Accounting-based Valuation—see [23, Lemma 4.1; generalization of Lemma 4.1: Appendix 2].

Theorem 2 (Multivariate Ohlson Principle) *The system Ω has dividend irrelevance at R iff $R = \lambda_1$.*

Proof By varying ω we can place one eigenvalue of $A^{\lrcorner}$ in the interval (λ_2, λ_1), so by Proposition 1, there cannot be dividend irrelevance at λ_2 and below. Note that this means that $\delta_1\omega_1 \neq 0$ for the chosen ω. □

The situation with general placement of eigenvalues alters if β is a positive real, lies below the eigenvalues of A, and the dividend-policy vector ω of the canonical setting is non-negative in all its components. The formula (13.10) confines the non-real eigenvalues κ_j to an infinite strip, while the formula (13.8) allows us to confine all the eigenvalues still further when ω is itself bounded.

We refer to formula (13.9) as the **associated polar form**. This form offers a graphical approach to the analysis of the real root location, and some insight into complex root location; in particular, the **leading quadratic term** is responsible for unbounded root behaviour, as follows.

Proposition 4 (Unbounded Roots) *Fix ω_k for $k \neq j$ with*

$$A_j = \sum_{h \neq j} \omega_h \delta_h (\lambda_j - \lambda_h) \neq 0.$$

(i) *Subject to $\lambda_k < |\kappa|$ we have the asymptotic expansion*

$$\omega_1\delta_1 + \omega_2\delta_2 + \ldots + \omega_n\delta_n$$

$$= (\kappa - \lambda_j)(\kappa - \beta) + \sum_{s=1}^{\infty} \frac{1}{\kappa^s} \left(\sum_{h \neq j}^{n} \omega_h \delta_h \lambda_h^{s-1} (\lambda_j - \lambda_h) \right).$$

(ii) *For $j \leq n$ the unbounded roots as $\delta_j \omega_j \to -\infty$ behave asymptotically as follows:*

$$\kappa = \begin{cases} \pm\sqrt{|\omega_j|} + O(\omega_j^{-1/2}), & \text{for } \delta_j = 1, \\ \left[\frac{1}{2}(\lambda_j + \beta) + \frac{A_j}{2|\omega_j|} + +O(\omega_j^{-2})\right] \pm i\left[\sqrt{|\omega_j|} + O(\omega_j^{-1/2})\right], & \text{for } \delta_j = -1. \end{cases}$$

For the proof, see Sect. 13.10.

See Fig. 13.1 above and Fig. 13.5 in Sect. 13.4 for illustrative examples.

Remark 3 In the case $j = 1$ with $\delta_1 = -1$, we are of course assuming that $\omega_1 \to \infty$. If moreover $\delta_k \omega_k > 0$ for all $k = 2, \ldots, n$, and $\lambda_1 > \lambda_2 > \ldots > \lambda_n$, we have $A_1 = \sum_{h\neq 1} \omega_h \delta_h (\lambda_1 - \lambda_h) > 0$. Here the conjugate roots have real part approaching $\frac{1}{2}(\lambda_j + \beta)$ from the right. However, with other sign assumptions on $\delta_h \omega_h$, the sign of A_1 need not be positive, in particular if $\delta_h \omega_h < 0$ for all h.

Remark 4 By (13.8) we may rewrite the characteristic polynomial in the form

$$\frac{1}{\omega_j} \prod_{k=1}^{n+1}(\kappa - \lambda_k) - \frac{1}{\omega_j}\{\omega_1 \delta_1 \prod_{k=2}^{n}(\kappa - \lambda_k) + \omega_2 \delta_2 \prod_{k\neq 2}^{n}(\kappa - \lambda_k) + \ldots + \omega_n \delta_n \prod_{k\neq n}^{n}(\kappa - \lambda_k)\}.$$

For fixed ω_k, with $j \neq k$, pass to the limit as $|\omega_j| \to \infty$ to obtain the following equation of degree $n - 1$:

$$\prod_{k\neq j}^{n}(\kappa - \lambda_k) = 0.$$

Thus given the assumptions of the Proposition 4, **only two** complex roots can be unbounded.

Remark 5 Note that, by contrast, the unbounded roots for $\omega_1, \ldots, \omega_n$ fixed and β varying have the asymptotic behaviour $\kappa = \beta + O(\beta^{-1})$. Note also that, if $\sum_{k\neq j} \omega_k \delta_k (\lambda_j - \lambda_k) = 0$, then the error term O-behaviour alters.

Theorem 3 *With fixed ω_k for $k \neq 1$ such that $\sum_{k\neq 1} \omega_k \delta_k (\lambda_j - \lambda_k) \neq 0$, and with $\delta_1 = -1$ if*

$$\beta \geq 2\lambda_2 - \lambda_1,$$

the unbounded root locus does not enter the disc $|\zeta| \leq \lambda_2$ as $\omega_1 \to \infty$. So the system Ω has local dividend irrelevance at (ω, β) uniquely at $R = \lambda_1$.

Proof Under these circumstances the unbounded roots are outside the disc $|\zeta| \leq \lambda_2$, since they are confined to $\Re(\zeta) > \lambda_2$ by virtue of

$$\lambda_2 \le \frac{1}{2}(\lambda_1 + \beta).$$

By Proposition 1 the only remaining value available for R is thus λ_1.   □

Notation Below and throughout, $K(\varepsilon)$ denotes the real interval $[\beta - \eta(\varepsilon), \lambda_1 + \eta(\varepsilon)]$, where

$$\beta - \eta = \frac{(\beta + \lambda_1) - \sqrt{(\lambda_1 - \beta)^2 + 4\varepsilon}}{2}, \qquad \lambda_1 + \eta = \frac{(\beta + \lambda_1) + \sqrt{(\lambda_1 - \beta)^2 + 4\varepsilon}}{2};$$

$S(K, \pi/(n+1))$ comprises the two circles in the plane subtending angles of $\pi/(n+1)$ on $K(\varepsilon)$.

Proposition 5 (Strip-and-Two-Circles Theorem) *Suppose that* $\beta \le \lambda_n < \ldots < \lambda_1$, *that* $\omega_1 \ne 0$, *and that*

$$\delta_1 \omega_1, \ldots, \delta_n \omega_n \ge 0.$$

(i) *All the non-real roots of the characteristic equation (13.7) lie in the infinite strip of the complex ζ-plane given by*

$$\beta \le \Re(\zeta) \le \lambda_1.$$

(ii) *For $\varepsilon > 0$ arbitrary, if*

$$\omega_1 + \ldots + \omega_n \le \varepsilon, \qquad \delta_1 \omega_1, \ldots, \delta_n \omega_n \ge 0,$$

then all the roots of (13.7) lie in the star-shaped region $S(K, \pi/(n+1))$.

The proof is delayed to Sect. 13.11.

Remark 6 Taken together parts (i) and (ii) may operate simultaneously. These results should, however, be taken together with Gerschgorin's Circle Theorem, which implies immediately that the eigenvalues lie in the union of the discs in the complex ζ-plane given by $|\zeta - \lambda_j| \le |\omega_j|$ and by $|\zeta - \beta| \le |\omega_1| + \ldots |\omega_n|$. Thus the eigenvalues are bounded, not only to the above mentioned vertical strip but also to a horizontal strip of width $2\max\{|\omega_j| : j \le n\}$ around the real axis.

Remark 7 It is obvious that, for $\omega_2 = .. = \omega_n = 0$ and with $|\omega_1| \le \varepsilon$, the real roots of (13.7) lie in $K(\varepsilon)$ by continuity. Gerschgorin's Circle Theorem limits the real roots to the slightly larger interval $[\beta - \varepsilon, \lambda_1 + \varepsilon]$. Thus the two-circle result is merely a sharpening of the bounds.

Remark 8 If $\lambda_n < \beta$, less elegant improvements can be made so that K extends only as far as λ_1 on the left.

Fig. 13.2 Vertical strip and
two-circle bounds. Horizontal
bound implied by
Gerschgorin's Theorem

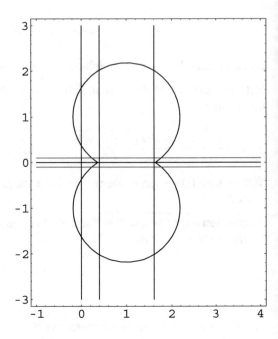

We can state, ahead of the proof of Proposition 5, our theorem on eigenvalue location.

Theorem 4 (Eigenvalue Bounds) *Suppose that $\beta \leq \lambda_n < \ldots < \lambda_1$ and that*

$$|\omega_1|, \ldots, |\omega_n| \leq \varepsilon.$$

Non-real eigenvalues lie in the rectangle bounded by $\gamma = \beta, \zeta = \lambda_1, \zeta = \pm\varepsilon$. Real eigenvalues lie in the interval $K(\varepsilon)$.

The theorem follows from Proposition 5—see Fig. 13.2. The two-circle result gives useful bounds only for the real roots.

Remark The above analysis does not yet exclude the possibility of all eigenvalues being located to the left of λ_2. We next offer a graphical analysis of the real-root locations in the following subsection, which shows that at least one root has to be to the right of λ_1 when $\omega_1 \neq 0$ and $\omega_j \neq 0$ for some $j > 1$.

13.3 Eigenvalue Location: General Analysis

Our ultimate purpose of establishing DPI (achieved in the next section) is to show that under suitable restrictions one can guarantee the existence of a real eigenvalue in the range (λ_2, λ_1). Specifically, we show that if $\delta_1 = -1$ and $\delta_2 = \ldots = +1$ there is a real eigenvalue in the range (λ_2, λ_1) for all small enough positive ω_1. The aim of this section is to identify (i) the real-root locus of the characteristic polynomial of $H(\omega)$ by examining the general features of the graph of the associated polar form of the characteristic equation given by (13.9), and (ii) the complex-root locus when the associated polar form has just two poles. The latter is a preliminary to our identification of (iii) an elementary estimate of the distortion effect of other poles.

In this section the real eigenvalues of $H(\omega)$ (see Eq. (13.4)) are studied as functions of ω_1 with the other components of ω fixed. Interest naturally focuses on ω_1 as the link coefficient with the dominant state vector. The two-pole case arises when $n = 3$ and is considered as a benchmark, with a view to understanding how the multi-pole situation deforms the benchmark case.

Treating ω_1 as a free variable, with the remaining dividend-policy coefficient fixed, we use (13.9) to study the map $\kappa \to \omega_1$ and its local inverses. We have, with f as in (13.11),

$$-\omega_1\delta_1 = f(\kappa) = \{\omega_2\delta_2 + ..\} - (\kappa - \lambda_1)(\kappa - \beta) - \frac{\omega_2\delta_2(\lambda_1 - \lambda_2)}{\kappa - \lambda_2} - ..., \qquad (13.13)$$

so that the graph of f, or of ω_1, against κ has $(n - 1)$ vertical asymptotes from right to left at $\kappa = \lambda_2, \lambda_3, \ldots \lambda_n$ all of which are manifestly simple poles. The asymptotes break up the concave leading quadratic term (if $\delta_1 < 0$) into n connected components corresponding to the intervals $(-\infty, \lambda_n), (\lambda_n, \lambda_{n-1}), \ldots, (\lambda_n, +\infty)$. The equation

$$\frac{\partial \omega_1}{\partial \kappa} = 0$$

is equivalent to an n-degree polynomial equation so its n roots contribute to at most n stationary points in the graph.

In the interval $(\lambda_{j+1}, \lambda_j)$ the component has an even, respectively an odd, number of stationary points depending on whether the sign of $\omega_{j+1}\delta_{j+1}\omega_j\delta_j$ is $+1$ or -1. In view of the behaviour of the leading quadratic term, not all the components can be monotone (possess a zero number of stationary points!). Thus at least one component is non-monotonic.

The components may be interpreted as graphs/loci of the eigenvalues $\kappa_j(\omega_1)$. More precisely, the differentiable local inverses of the mapping $\kappa \to \omega_1$ are the graphs of $\kappa_j(\omega_1)$. That is to say, each non-monotonic component must be first partitioned into monotone parts on either side of its stationary points. The labelling of these inverses from right to left respects the cyclic order on the set $\{1, \ldots, n\}$ together with one or other of the identifications

$$\lim_{\kappa \nearrow \lambda_j} \omega_1(\kappa) = \kappa_j, \qquad \lim_{\kappa \searrow \lambda_j} \omega_1(\kappa) = \kappa_j.$$

The latter may require the point at infinity on the asymptote $\kappa = \lambda_j$ to be considered as the intersection of consecutive loci.

Note that from (13.13) $f(\lambda_1) = 0$ and so $\kappa_1(0) = \lambda_1$. (This is consistent with the matrix $H(\omega) - \kappa I$ having a first column with zeros in all but the last row.)

We will see in Sect. 13.4.1 from these asymptotic features of the graph that, since $\kappa_1(0) = \lambda_1$, for all small enough positive ω_1 the eigenvalue $\kappa_1(\omega_1)$ is in the range (λ_2, λ_1), as we now demonstrate.

As a preview of the full argument of Sect. 13.4.2, with our assumption that $\delta_1 = -1$ and $\delta_2 = \delta_3 = \ldots = +1$, we note that we can arrange for $\kappa_1(\omega_1)$ to be large and positive in the vicinity to the right of λ_1 by taking $\omega_2 < 0$. With $\omega_2 < 0$ the domain of κ_1 is infinite, so that

$$\lim_{\omega_1 \to \infty} \kappa_1(\omega_1) = \lambda_2.$$

Thus the largest real eigenvalue κ_1 remains above λ_2. See Fig. 13.3 (in Sect. 13.4.1 below). Of course for small enough ω_1 the remaining roots $\kappa_j(\omega_1)$, even if complex, remain in an open vertical complex strip including the closed real interval $[\beta, \lambda_2]$.

We can similarly arrange for $\kappa_1(\omega_1)$ to be large and negative in the vicinity to the right of λ_1 by taking $\omega_2 > 0$. In view of the behaviour of the graph for large $\kappa > \lambda_1$, this implies the existence of two roots in (λ_1, λ_2) under these circumstances. With $\omega_2 > 0$ the domain of κ_1 is bounded, say by $\omega_1 \le \omega_1^* = \omega_1^*(\omega_2, \ldots, \omega_{n+1})$, and one has

$$\lim_{\omega_1 \searrow \omega_1^*} \kappa_1(\omega_1) = \lim_{\omega_1 \nearrow \omega_1^*} \kappa_2(\omega_1).$$

See Fig. 13.4 (below in Sect. 13.4.1). As the eigenvalue κ_1 remains above λ_2, dividend irrelevance can occur only at λ_1. An upper bound for ω_1^* is provided by the maximum value of the leading quadratic term

$$-(\kappa - \lambda_1)(\kappa - \beta) + \{\omega_2 \delta_2 + ..\}$$

obtained by evaluation at $\kappa = \frac{1}{2}(\lambda_1 + \beta)$, namely

$$\frac{1}{4}(\lambda_1 - \beta)^2 + \{\omega_2 \delta_2 + \ldots\}. \tag{13.14}$$

This gives β, the coefficient at the previous date's dividend, a significant bounding role.

In order to understand the qualitative behaviour of the complex eigenvalues of the characteristic equation, we study the associated polar form on an open interval between the two adjacent poles at λ_{j+1} and λ_j for $j \ge 2$. As a first step, we study the contribution to $f(\kappa)$ arising in (13.13) only from the two terms corresponding to the two adjacent poles λ_{j+1} and λ_j. In the subsequent section we identify how the presence of the other terms in (13.13) perturbs this simple analysis.

13.3.1 Root Locus for Two-Pole Case

For present purposes, by scaling and a shift of origin, as $\lambda_{j+1} < \lambda_j$ we may take $\lambda_{j+1} = 0$ and $\lambda_j = \tau > 0$. It is convenient to study the corresponding terms of $f(\kappa)$ by introducing $T > 0$ and looking at the two functions

$$f_1(x) = \frac{1}{x} + \frac{T}{x-\tau} = \frac{(1+T)x - \tau}{x(x-\tau)},$$

$$f_2(x) = \frac{1}{x} - \frac{T^2}{x-\tau}.$$

The first of these has its zero at $x = \tau/(1+T)$ in $(0, \tau)$ and maps $(0, \tau)$ bijectively onto the reals. The second function is more awkward; provided $T \neq 1$, it has a zero outside $(0, \tau)$ at $\tau/(1 - T^2)$. More information is provided in the Circle Lemma below.

Circle Lemma *For $0 < T \neq 1$, the function $f_2(x)$ has a positive local minimum value K_+ at x^+, and positive local maximum value K_- at x_-, where*

$$x^{\pm} = \frac{\tau}{1 \pm T}, \qquad K_{\pm} = \frac{(1 \pm T)^2}{\tau}.$$

The range of $f_2(x)$ on $(0, \tau)$ omits an interval of positive values (K_-, K_+).
  For $K_- < K < K_+$ the equation

$$f_2(z) = K, \text{ equivalently } z^2 - (\tau + (1 - T^2)K^{-1})z + \tau K^{-1} = 0, \qquad (13.15)$$

has conjugate complex roots in the complex ζ-plane describing a circle centred at the real number $(x^+ + x^-)/2$ with radius $|x^+ - x^-|/2$.
  The real part moves from x^+ towards x^-. Hence for $0 < T < 1$, as K decreases from K_+ the real part increases and for $T > 1$ it decreases.

Note See below for a more general analysis of the behaviour of the real part near a local minimum. For $T = 1$ the function f_2 is symmetric about $x = \tau/2$ and is asymptotic to zero at infinity; a limiting version of the lemma is thus still valid, but the conjugate roots lie on the vertical line $\Re(\zeta) = \tau/2$ for $0 < K < K_+ = 4/\tau$.

Proof The conjugate roots $z, \bar{z} = x \pm iy$ satisfy

$$x^2 + y^2 = z\bar{z} = \tau K^{-1}.$$

In view of the dependence of the real part on K as given by $x = (\tau + (1 - T^2)K^{-1})/2$, we see that the term τK^{-1} can be absorbed by a shift of origin on the x axis. Hence the locus as K varies is a circle. Reference to the extreme locations $x^{\pm}$ identifies the shifted centre, and the radius.   □

13.3.2 Multi-Pole Case: A Distortion Estimate

The presence of additional poles outside $(\lambda_{j+1}, \lambda_j)$ distorts the result obtained in the Circle Lemma above for the two-pole contribution. Provided all the eigenvalues λ_j are well-separated, i.e. the ratio of adjacent intervals does not vary greatly (see the calculation of τ/c below), the distortion is controlled by the value of $\sum_{k \notin \{j, j+1\}} \omega_k \delta_k (\lambda_j - \lambda_k)$. In the next subsection we take note of a third-derivative test (based on Taylor's Theorem) which identifies bifurcation behaviour of coincident real roots. For an application see Sect. 13.4.5.1.

We again work in the standardized co-ordinate system with the adjacent poles at $x = 0$ and $x = \tau$. Put $\xi := \tau/2$; then the decomposition, valid for $x \in (0, \tau)$ and $c > 0$,

$$\frac{1}{x+c} = \frac{A_0(x)}{x} 1_{[\xi, a]}(x) - \frac{A_\tau(x)}{x - \tau} 1_{[0, \xi]}(x)$$

yields the following bounds:

$$0 < \frac{\tau}{2c + \tau} < A_0(x) < \frac{c}{c + \tau} < 1, \text{ for } \xi < x < a,$$

$$0 < \frac{\tau}{2c + \tau} < A_\tau(x) < \frac{\tau}{c}, \text{ for } 0 < x < \xi.$$

These and an amendment of the parameter T in $f_2(x)$ to a variable coefficient $T = T(x)$ for x in the closed interval $[0, \tau]$ enable us to account for the presence of terms other than $f_2(x)$ in $f(x)$ (as defined in Proposition 2), by absorbing their contributions into $T(x)$. On $[0, \tau]$, the 'adjusted' T inherits from f boundedness, continuity and indeed two-fold differentiability. Thus we have

$$f(x) = \frac{1}{x} - \frac{T}{x - \tau}, \qquad f' = -\frac{1}{x^2} + \frac{S}{(x - \tau)^2},$$

where $S = S(x) = T(x) - (x - \tau)T'(x) \sim T(\tau)$, for x close to τ. Expansion round x yields

$$T(\tau) = T(x) + (\tau - x)T'(\tau) + \frac{1}{2}(\tau - x)^2 T''(x) + o((\tau - x)^3), \qquad \text{as } x \to \tau,$$

so

$$S = T(\tau) - \frac{1}{2}(x - \tau)^2 T''(x) + o((\tau - x)^3).$$

The main point of this is to observe how S perturbs the complex roots (cf. Fig. 13.7). The Eq. (13.15) of the Circle Lemma now gives us the following.

Proposition 6 (Distortion Estimate) *The equation*

$$K = \frac{1}{x} - \frac{T(x)}{x - \tau}$$

is equivalent to $f_2(x) = K + \bar{T}$ *for*

$$\bar{T} = T'(x) + \frac{1}{2}(\tau - x)T''(x) + o((\tau - x)^2).$$

The equation $f(x) = K$ *is equivalent to*

$$x = \frac{\tau}{2} + \frac{(1 - S^2)}{2K} \pm \frac{i}{2}\sqrt{\delta(K^{-1})},$$

with $\delta(k) = \tau^2 K_+ K_- (k - k_+)(k_- - k)$, *where*

$$0 < k_-^{-1} = K_- = \frac{(1 - S(x))^2}{\tau} < K < \frac{(1 + S(x))^2}{\tau} = K_+ = k_+^{-1},$$

$$S = T(\tau) - \frac{1}{2}(x - \tau)^2 T''(x) + o((\tau - x)^3).$$

We note that $\frac{1}{2}T''(x^+) = -f'(x^+) + o((\tau - x^+))$.

13.3.3 Analysis of the Real Part via Taylor's Theorem

Proposition 7 (Third-Derivative Test: Real Part Follows f''/f''') *Suppose that* $f(\kappa)$ *has a local minimum/maximum at* $\kappa = \kappa^*$. *Let* $\kappa(\omega)$ *denote the local solution for* κ *over the complex domain of the equation* $g(\kappa) = 0$ *for*

$$g(\kappa) = \begin{cases} f(\kappa) - \omega^2, & \text{for } \kappa^* \text{ a local minimum,} \\ f(\kappa) + \omega^2, & \text{for } \kappa^* \text{ a local maximum,} \end{cases}$$

with $\omega > 0$ *small and with* $\kappa(0) = \kappa^*$. *If* $f'''(\kappa^*) \neq 0$, *then the locus* $\kappa(\omega)$ *satisfies near 0:*

$\Re(\kappa(\omega))$ *is increasing if* $f''(\kappa^*)/f'''(\kappa^*) > 0$, *and is otherwise decreasing.*

For a proof see Sect. 13.12. Proposition 7 implies that the quadratic terms of the associated polar form $f(\kappa)$ have no effect on the local behaviour of the real part at a bifurcation.

13.4   Eigenvalue Location: Some Cases

In this section we consider the case $n = 2$, and the two cases with n general when $\delta_k \omega_k$ are of constant sign for $k = 2, \ldots, n$, as referred to in Theorem 1.

13.4.1   The Case n = 2

This case is in fact typical, despite having the simplifying structure that one root of the cubic characteristic polynomial χ_H is always real. There may thus be two more real roots, or two conjugate complex roots.

In view of earlier comments, we need to consider only the case $\delta_1 \omega_1 < 0$. Interpret this as saying $\delta_1 = -1$ and $\omega_1 > 0$.

(a) For now assume $\delta_2 \neq 0$.(For $\delta_2 = 0$ see (b) below.) Taken together Figs. 13.3 and 13.4 tell it all. They graph the implicit relation between ω_1 and the eigenvalues κ as given by the equation $\chi_H(\kappa; \omega_1) = 0$, treating κ as the independent variable and ω_1 as dependent. To derive the root locus for real roots κ rotate the graphs, so that ω_1 becomes the independent variable. Then each branch of the graph yields the κ_j as the dependent eigenvalues in decreasing magnitude.

If $\delta_2 \omega_2 < 0$, then for increasing ω_1, as in Fig. 13.3, the dominant root of $H(\omega)$ decreases down to λ_2 (in the limit).

If $\delta_2 \omega_2 > 0$, then for increasing ω_1, as in Fig. 13.4, the first/second root of $H(\omega)$, respectively, decrease/increase into coincidence in the interval (λ_2, λ_1). Thereafter, the root locus of the conjugate pair behaves as illustrated below in Fig. 13.5. That is, the real part decreases towards $\frac{1}{2}(\lambda_1 + \beta)$ and the imaginary parts tend to infinity.

Fig. 13.3 Graph of $\omega_1(\kappa)$ with $\delta_1 = -1$, $\delta_2 = +1$, $\omega_2 < 0$. Leading quadratic in green; κ-axis horizontal

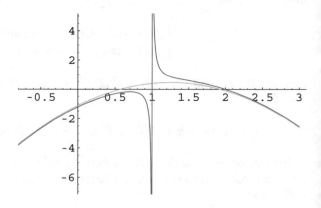

Fig. 13.4 Graph of
$\omega_1(\kappa)$ with $\delta_1 = -1$,
$\delta_2 = +1$, $\omega_2 > 0$. Leading
quadratic in green; κ-axis
horizontal

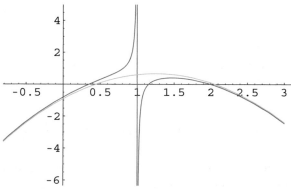

Fig. 13.5 Locus of the
conjugate roots in the
complex ζ-plane for $\delta_1 < 0$
as ω_1 increases with ω_2 fixed.
Vertical asymptote is
$\Re(\zeta) = \frac{1}{2}(\lambda_1 + \beta)$

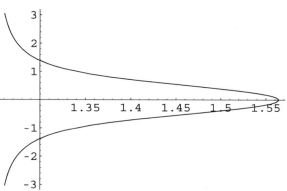

(b) We consider now the limiting case $\delta_2 = 0$. As the characteristic polynomial is

$$\chi_H(\kappa) = (\kappa - \lambda_1)(\kappa - \lambda_2)(\kappa - \beta) - \omega_1\delta_1(\kappa - \lambda_2)$$
$$= (\kappa - \lambda_2)[(\kappa - \lambda_1)(\kappa - \beta) - \omega_1\delta_1],$$

this has one root at $\kappa = \lambda_2$ which, by Proposition 1, precludes DPI occurring
at $R = \lambda_2$. Note that the expression in square brackets corresponds to an
analogous problem with reduced order (effectively the $n = 1$ case), and this
feature, of a reduction of order, occurs for general n.

Remark Of course, for $\delta_1\omega_1 < 0$, the characteristic polynomial of $H(\omega)$ for $\delta_2 = 0$
has two additional real roots in the interval (β, λ_1) for small $|\omega_1|$. For larger $|\omega_1|$
the conjugate roots lie in the complex ζ-plane on the vertical $\Re(\zeta) = \frac{1}{2}(\lambda_1 + \beta)$.
This is because

$$\kappa = \frac{(\beta + \lambda_1) \pm \sqrt{(\lambda_1 - \beta)^2 + 4\omega_1\delta_1}}{2} = \frac{1}{2}(\lambda_1 + \beta) \pm i\{\sqrt{|\omega_1|} + O(|\omega_1|^{-1/2})\}.$$

As before, if $\beta \geq 2\lambda_2 - \lambda_1$, then $\lambda_2 \leq \frac{1}{2}(\lambda_1 + \beta)$ and the conjugate roots do not enter the open disc $|\zeta| < \lambda_2$; however, if $\beta < 2\lambda_2 - \lambda_1$, then $\frac{1}{2}(\lambda_1 + \beta) < \lambda_2$, which means that two roots will be in the open disc for a range of values of $|\omega_1|$ with the third on the boundary. Eventually the conjugate pair exits the annulus $\mathscr{A}$ (see Fig. 13.1).

13.4.2 General Case $\delta_1\omega_1 < 0$ with $\delta_j\omega_j > 0$ for $j = 2, \ldots, n$

We fix arbitrarily $\delta_j\omega_j > 0$ for $j = 2, \ldots, n$. The assumption $\omega_1\delta_1 < 0$ is without loss of generality interpreted as $\delta_1 = -1$ with variable $\omega_1 > 0$. The analysis now proceeds similarly. Here we have the identity connecting eigenvalue κ and parameter ω_1 in the shape of the associated polar form (13.9):

$$\omega_1 = f(\kappa) := -(\kappa - \lambda_1)(\kappa - \beta) + \{\omega_2\delta_2 + ..\} - \frac{\omega_2\delta_2(\lambda_1 - \lambda_2)}{\kappa - \lambda_2} - \ldots.$$

We note that the quadratic term reflects the behaviour of the two equations which remain when all the variables other than the dividend and the dominant accounting variable are ignored (again equivalent to taking $\omega_2 = .. = \omega_n = 0$). Our analysis identifies the locations of all the $n + 1$ real roots of $f(\kappa) = 0$ in order to discuss the equation $f(\kappa) = \omega_1$.

Noting that $f(\lambda_j+) = -\infty$ and $f(\lambda_j-) = +\infty$, by the Intermediate Value Theorem, there exist roots $\kappa_j(0)$ for $j = 3, \ldots, n$ of the equation $f(\kappa) = 0$ which are real and satisfy

$$\lambda_j < \kappa_j(0) < \lambda_{j-1}.$$

Finally, since $f(-\infty) = -\infty$ and $f(\lambda_n-) = +\infty$, there exists a root $\kappa_{n+1}(0) < \lambda_n$.

Evidently the equation $f(\kappa) = \omega_1$ similarly has roots $\kappa_j(\omega_1)$ for any ω_1 in the respective intervals for $j = 3, \ldots, n + 1$. Thus we have 'interlacing' for $j \geq 2$:

$$\kappa_{n+1}(\omega_1) < \lambda_n < \kappa_n(\omega_1) < \ldots < \lambda_j < \kappa_j(\omega_1) < \lambda_{j-1} < \ldots < \kappa_3(\omega_1) < \lambda_2.$$

Now $f(\lambda_2+) = -\infty$ and $f(\lambda_1) = 0$ with $f(+\infty) = -\infty$. There are thus two generic possibilities.

1. The equation $f(\kappa) = 0$ has a root in (λ_2, λ_1). In this case, as ω_1 increase from zero there are initially two real roots $\kappa_1(\omega_1)$ and $\kappa_2(\omega_2)$ of the equation $f(\kappa) = \omega_1$ which lie in (λ_2, λ_1) and which move into coincidence. Thereafter they become complex conjugates which behave qualitatively as in the case $n = 2$. In particular, $\kappa_3, \ldots, \kappa_{n+1}$ are increasing in ω_1 (under the assumptions of this case), and since

$$\kappa_1 + .. + \kappa_{n+1} = \lambda_1 + \lambda_2 + \ldots \lambda_n + \beta,$$

with the right-hand side constant, we see that $\Re(\kappa_1) = \frac{1}{2}(\kappa_1 + \kappa_2)$ is decreasing in ω_1. As the lower bound is $\frac{1}{2}(\lambda_1 + \beta)$, the complex roots certainly do not enter the disc $|\zeta| \leq \lambda_2$ provided $\frac{1}{2}(\lambda_1 + \beta) \geq \lambda_2$ i.e. $\beta \geq 2\lambda_2 - \lambda_1$. (Note that $2\lambda_2 - \lambda_1 < \lambda_1$.)

2. The function $f(\kappa)$ is negative in (λ_2, λ_1) and the equation $f(\kappa) = 0$ has its remaining root in $(\lambda_1, +\infty)$. In this case the convergence assumption is violated for small ω_1 in that there are eigenvalues of $H(\omega)$ outside the disc $|\zeta| < \lambda_1$.

Finally, the special case arises when $f(\kappa) = 0$ has a double root at $\kappa = \lambda_1$. In this case as ω_1 increases from zero the remaining two roots are conjugate complex and again behave as in Figs. 13.3 and 13.4.

13.4.3 General n : Case $\delta_1\omega_1 < 0$ with $\delta_j\omega_j < 0$ for $j = 2, \ldots, n$

This proceeds similarly. The root κ_1 decreases towards λ_2 as ω_1 increases. Likewise the roots κ_j for $j = 2, \ldots, n - 1$ decrease towards λ_{j+1}. As these latter roots remain bounded, the remaining two roots κ_n and κ_{n+1} may be real, but will be the unbounded complex conjugates for large enough ω_1 (indeed the remaining component of the graph is n-shaped). This time the real part increases towards $\frac{1}{2}(\beta + \lambda_1)$ as ω_1 increases.

13.4.4 General n : Case when $\delta_k\omega_k = 0$ for some $k = 2, \ldots, n$ with $\delta_j\omega_j > 0$ for Remaining Indices j

We find that in all these cases DPI cannot hold at $R = \lambda_2$, by Proposition 1.

Clearly, if $\delta_2 = 0$, then $\kappa = \lambda_2$ is an eigenvalue of the system, i.e. there is an eigenvalue in the annulus $\mathscr{A}$ and so DPI cannot hold at $R = \lambda_2$.

In general, if $\delta_k\omega_k = 0$ for just one of $k = 3, \ldots, n$, then $\chi_H(\lambda_k) = 0$, and in fact

$$\chi_H(\kappa) = \prod_{j=1}^{n+1}(\kappa - \lambda_j) - \sum_{j=1}^{n}\omega_j\delta_j\prod_{h\neq j}^{n}(\kappa - \lambda_h)$$

$$= (\kappa - \lambda_k)\Big[\prod_{h=1,h\neq k}^{n+1}(\kappa - \lambda_h) - \sum_{j=1,j\neq k}^{n}\omega_j\delta_j\prod_{h\neq j}^{n}(\kappa - \lambda_h)\Big]$$

$$= (\kappa - \lambda_k)\cdot\chi_K(\kappa),$$

with K the matrix obtained from H by omitting the k-th row and column, i.e. a reduction of order occurs. But now the assumptions for K agree with those made in the case of the previous subsection. It now follows that $\chi_K(\kappa)$ has an eigenvalue in the annulus $\mathscr{A}$, and so again DPI cannot hold at $R = \lambda_2$.

Finally, if $\delta_k \omega_k = 0$ for several among $k = 3, \ldots, n$ and $\delta_j \omega_j > 0$ for the remaining indices j, then a further reduction of order occurs, with the same conclusion that an eigenvalue exists in the annulus $\mathscr{A}$. So here too, DPI cannot hold at $R = \lambda_2$.

Note that in both scenarios the locus of κ_1 decreases as ω_1 increases from zero.

13.4.5 Effect of the Dividend-on-Dividend Multiplier: A Distortion Example

We conclude this section by illustrating the effect of the policy parameter $\beta = \omega_3$ on the three eigenvalues in the case $\omega_1 = \omega_2 = 0.1$. In the range $\beta < \lambda_2$ we see in the illustrative example of Fig. 13.6 that the root κ_1 decreases whilst the root κ_2 increases as β increases; κ_3 increases for all β, as might be expected, with λ_2 as supremum. Intuitively speaking, the push away from the origin created by the two increasing roots κ_2 and κ_3 causes the location of the coincident root $\kappa_1 = \kappa_2$ to execute a jump up to a new coincidence location above λ_1, by way of a continuous root locus in the complex ζ-plane (see the Remark on bifurcation in the next section). The push can in fact be physically interpreted. The partial-fraction-expansion terms in (13.10) may be regarded as modelling electric charges placed at the pole locations λ_j and acting according to an inverse distance law (see Marden [16, §1.3 p. 7]). Thus for β large enough to ensure both κ_2 and κ_1 have been re-located above λ_1, we see the locus of κ_1 resume its downward path towards the origin (but tending in the limit only as far as the barrier λ_1), while κ_2 resumes its upward path away from the origin. The locus dynamics are investigated more properly in the next section.

Fig. 13.6 Graph of $\omega_3(\kappa)$ with $\delta_1 = -1$, $\delta_2 = +1$, ω_1, $\omega_2 > 0$; κ-axis horizontal

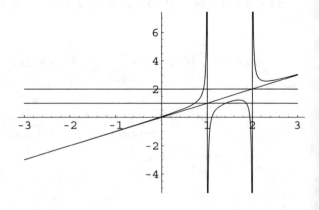

Fig. 13.7 Root locus of
conjugate roots as β varies
for $\omega_1, \omega_2 > 0$ fixed
$(\delta_1 = -1,\ \delta_2 = +1)$

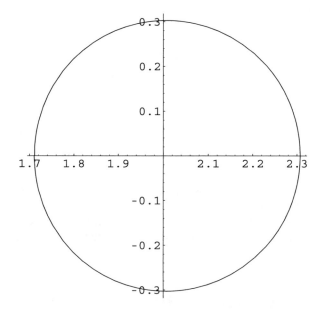

The conjugate root locus for this case is shown in Fig. 13.7. For β varying over
the range corresponding to the presence of complex roots (with ω_1 and ω_2 fixed),
the root locus of the conjugate pair appears close to being a circle $|\kappa - \lambda_2|^2 = const.$
as a consequence of the Circle Lemma and Proposition 6.

13.4.5.1 A Case Study: Distorted Circle

Since $\omega_1 = \omega_2 = 0.1$ we have here

$$f(\kappa) = 10\kappa + \frac{1}{\kappa - 2} - \frac{1}{\kappa - 1},$$

so the equation $f(x) = K$ is equivalent to $f_2(\kappa) = K - 10\kappa$. The essential reason
for the closeness observed here is the slow variation of κ_3 in the vicinity of λ_2.
Indeed, by (13.8) we have

$$\chi(\kappa) = (\kappa - \lambda_1)(\kappa - \lambda_2)(\kappa - \beta) - \{\omega_1\delta_1(\kappa - \lambda_2) + \omega_2\delta_2(\kappa - \lambda_1)\},$$

from which we compute that

$$\kappa_1 + \kappa_2 + \kappa_3 = \lambda_1 + \lambda_2 + \beta,$$
$$\kappa_1\kappa_2\kappa_3 = \lambda_1\lambda_2\beta - \omega_1\delta_1\lambda_2 - \omega_2\delta_2\lambda_1,$$

and so

$$
\begin{aligned}
(\kappa_1 - \lambda_1)(\kappa_2 - \lambda_1) &= \kappa_1\kappa_2 - \lambda_1(\kappa_1 + \kappa_2) + \lambda_1^2 \\
&= (\lambda_1\lambda_2\beta - \omega_1\delta_1\lambda_2 - \omega_2\delta_2\lambda_1)\kappa_3^{-1} - \lambda_1(\lambda_1 + \beta + (\lambda_2 - \kappa_3)) \\
&\quad + \lambda_1^2 \\
&= \beta\lambda_1(\lambda_2 - \kappa_3)\kappa_3^{-1} - \lambda_1(\lambda_2 - \kappa_3) - (\omega_1\delta_1\lambda_2 + \omega_2\delta_2\lambda_1\lambda_2^{-1}) \\
&= \lambda_1(\lambda_2 - \kappa_3)[\beta\kappa_3^{-1} - 1] - (\omega_1\delta_1\lambda_2 + \omega_2\delta_2\lambda_1\lambda_2^{-1}).
\end{aligned}
$$

So the positive sensitivity to variation in β is slight, as asserted, though in fact it is increasing.

Of course, writing λ_2 in place of κ_3, all the dependence on β is formally lost; however, note that the remaining constant is inadequate as an explanation of the actual radius.

13.4.6 Qualitative Behaviour in the Most General Case

The picture emerging from our analysis for $\delta_1\omega_1 < 0$ is that the dominant and subdominant eigenvalues remain in the annulus (of Theorem 1) provided

$$
\beta > 2\lambda_1 - \lambda_2.
$$

Other eigenvalues, when real, interlace between the eigenvalues of A, with κ_j remaining associated with λ_h for $h = j$ or $h = j + 1$. One pair of roots become unbounded as a complex conjugate pair and are asymptotic on one or other side of $\Re(\zeta) = \frac{1}{2}(\lambda_1 + \beta)$ as $\omega_1 \to +\infty$, depending on the remaining parameters of ω. Evidently a regime change occurs with the unbounded root locus being vertical. Other conjugate pairs execute deformed circles as described by the Circle Lemma and Proposition 6 of Sect. 13.3.1.

13.5 Differential Properties of Eigenvalues: Some Bifurcation Analysis

The purpose of this section is to analyse briefly the root locus in the ζ-plane. We conduct a partial analysis mostly concentrated on the dynamics of the dominant eigenvalue as ω_1 changes (with the remaining policy parameters fixed) with a view to completing the proof of the Dominance Theorem in the next section. Our starting point is the following proposition, which follows from (13.8) by implicit differentiation. It is suggested by our earlier Circle Lemma of Sect. 13.3.1 and Proposition 6 concerning the third derivative.

Proposition 8 (Under the Assumption of Distinct Eigenvalues) *In the canonical setting of Proposition 1, let the dividend-policy vector be represented by $\omega = (\omega_1, \ldots, \omega_{n+1})$ with $\omega_{n+1} = \beta$. The eigenvalues $\kappa_j = \lambda_j^L$ of the augmented matrix $A^{\lrcorner}$, viewed as functions of $\omega = (\omega_1, \ldots, \omega_{n+1})$, satisfy the following differential properties expressed in terms of the constants $\lambda_j = \lambda_j^A$:*

(i) for $1 \le h \le n+1$, $1 \le k \le n$

$$\frac{\partial \kappa_h}{\partial \omega_k} = -\frac{\delta_j (\lambda_1 - \kappa_h)(\lambda_2 - \kappa_h) \ldots (\lambda_{k-1} - \kappa_h)(\lambda_{k+1} - \kappa_h) \ldots (\lambda_n - \kappa_h)}{(\kappa_1 - \kappa_h)(\kappa_2 - \kappa_h) \ldots (\kappa_{h-1} - \kappa_h)(\kappa_{h+1} - \kappa_h) \ldots (\kappa_{n+1} - \kappa_h)};$$

$$(13.16)$$

(ii) for $1 \le h \le n+1$

$$\frac{\partial \kappa_h}{\partial \omega_{n+1}} = \frac{(\lambda_1 - \kappa_h)(\lambda_2 - \kappa_h) \ldots (\lambda_n - \kappa_h)}{(\kappa_1 - \kappa_h)(\kappa_2 - \kappa_h) \ldots (\kappa_{h-1} - \kappa_h)(\kappa_{h+1} - \kappa_h) \ldots (\kappa_{n+1} - \kappa_h)}.$$

$$(13.17)$$

Proof From the identity

$$\chi_H(\kappa) = (\kappa - \kappa_1)(\kappa - \kappa_2) \ldots (\kappa - \kappa_{n+1}),$$

we have

$$\chi_H'(\kappa) = \frac{d}{d\kappa} \chi_H(\kappa) = (\kappa - \kappa_2) \ldots (\kappa - \kappa_{n+1}) + (\kappa - \kappa_1)(\kappa - \kappa_3) \ldots (\kappa_{n+1} - \kappa) + \ldots,$$

so that

$$\chi_H'(\kappa_j) = \prod_{h \ne j} (\kappa_j - \kappa_h).$$

So, for $\omega = \omega_0 := (0, \ldots, 0, \beta)$, as then $\kappa_1 = \lambda_1$,

$$\chi_H'(\kappa_1) = (\lambda_1 - \lambda_2) \ldots (\lambda_1 - \lambda_n)(\lambda_1 - \omega_{n+1}).$$

From the identity

$$\chi_H(\kappa_j(\omega), \omega) = 0,$$

by implicit differentiation,

$$\frac{\partial \kappa_1}{\partial \omega_j} = -\frac{\partial \chi}{\partial \omega_j} \div \left(\frac{\partial \chi}{\partial \kappa} \right)_{\kappa = \kappa_1} = \frac{\partial \chi}{\partial \omega_j} \div \prod_{h>1} (\kappa_h - \kappa_1),$$

except at the critical points ω defined by

$$\prod_{j>1}(\kappa_j(\omega) - \kappa_1(\omega)) = 0,$$

e.g. where the locus of $\kappa_1(\omega)$ crosses $\kappa_2(\omega)$. The result of the Proposition now follows directly from (13.8). $\qquad\Box$

Remark For the assumption of distinct roots to hold we must manifestly disregard the non-generic critical points, which are those points ω where any two of the functions κ_j agree in value; of particular importance to us are points ω where $\kappa_1(\omega)$ may cease to be the largest eigenvalue (in modulus), as for instance when it agrees in value with $\kappa_2(\omega)$. The first formula when $j = 1$ is to be read as

$$\frac{\partial \kappa_1}{\partial \omega_1} = -\frac{\delta_1(\lambda_2 - \kappa_1)\ldots(\lambda_n - \kappa_1)}{(\kappa_2 - \kappa_1)\ldots(\kappa_n - \kappa_1)(\kappa_{n+1} - \kappa_1)},$$

and note that, at $\omega = \omega_0 := (0,\ldots,0,\beta)$, we have, by (13.3), $\kappa_{n+1} = \beta$ and for $j = 1,\ldots,n$:

$$\kappa_j = \lambda_j,$$

$$\frac{\partial \kappa_1}{\partial \omega_1} = \frac{\delta_1}{(\lambda_1 - \beta)}, \qquad (13.18)$$

and

$$\frac{\partial \kappa_1}{\partial \omega_j} = 0, \text{ for } j > 1.$$

Equation (13.18) implies that the choice of a β value close to λ_1 will accelerate the growth rate κ_1 of the leading canonical variable (see Sect. 13.7) Z^1 relative to the first dividend-policy coefficient.

Technical Point In the arguments that follow, it is important to realize that when the roots κ_j and κ_{j+1} are complex conjugates, then for real κ_1 the following signature property is satisfied:

$$\text{sign}[(\kappa_j - \kappa_1)(\kappa_{j+1} - \kappa_1)] = +1,$$

just as when κ_j and κ_{j+1} were real and both below κ_1. (Since the quadratic has no real roots, it is positive definite here.)

Corollary 1 (Bifurcation Behaviour Near $\kappa_1 = \kappa_2$) *Assume the eigenvalues of A are real and distinct and that for $j = 3,\ldots,n$, κ_j is real and satisfies $\lambda_{j+1} < \kappa_j < \lambda_j$. At any bifurcation point, for small enough positive increments in ω_1, the conjugate complex roots κ_1 and κ_2 move away from the origin if $\delta_1 > 0$, and towards the origin if $\delta_1 < 0$.*

Remark The corollary is thus in keeping with the intuition expressed in connection with Figs. 13.6 and 13.7 where we alluded to the push away from the origin for the root κ_2 regarded as a function of $\beta = \omega_3$.

Proof Suppose that $\delta_1 = -1$. Suppose that $\kappa_1 = \kappa_2$ occurs at some point $\omega_1 = \omega_1^*$. If now $\omega_1 = \omega_1^* + \Delta\omega$ with $\Delta\omega > 0$, put $\kappa = \Re(\kappa_1)$ and $\beta = \Im(\kappa_1)$, so that

$$\kappa_1 = \kappa + i\varepsilon, \qquad \kappa_2 = \kappa - i\varepsilon,$$

$$(\kappa_1 - \lambda_j) = \rho_j e^{i\theta_j}, \qquad (\kappa_1 - \kappa_{j+1}) = \rho_j' e^{i\phi_j}.$$

We need to be sure which of $\kappa \pm i\varepsilon$ is to be interpreted as κ_1 and κ_2 and whether the display above is correct. In fact, either interpretation is valid, and leads to the same conclusion; we return to this issue in a moment. Thus, since $\rho_j < \rho_j'$, for $\Delta\omega$ small enough we shall have

$$\theta_j > \phi_j,$$

so that

$$\frac{\rho_j e^{i\theta_j}}{\rho_j' e^{i\phi_j}} = \frac{\rho_j}{\rho_j'} \exp[i(\theta_j - \phi_j)],$$

and hence that

$$\frac{\Delta\kappa_1}{\Delta\omega} = -\frac{1}{2\varepsilon i} \frac{\rho_2 e^{i\theta_2}}{\rho_2' e^{i\phi_2}} \cdot \ldots \cdot \frac{\rho_n e^{i\theta_n}}{\rho_n' e^{i\phi_n}} = \frac{1}{2\varepsilon} \frac{\rho_2}{\rho_2'} \cdot \ldots \cdot \frac{\rho_n}{\rho_n'} \exp[i(\frac{\pi}{2} + \psi)],$$

where ψ is small and *positive*. That is, the remaining ratios pull $\Delta\kappa_1$ in the same direction, towards the origin. Note that if we switch the interpretation of $\kappa \pm i\varepsilon$ around, then the angles θ_j, ϕ_j change sign, making ψ small and *negative*. However, the sign of $(\kappa_2 - \kappa_1)$ also switches.

In conclusion, the conjugate complex roots κ_1 and κ_2 initially move closer to the origin if $\delta_1 < 0$. See Figs. 13.7 and 13.8 above for an illustration in the case $n = 2$, where the third root of $\chi_{A^{\lrcorner}}$ is evidently real. Note the vertical asymptote in the complex ζ-plane for $\Re(\zeta) = \frac{1}{2}(\lambda_2 + \beta)$ identified by Proposition 4. □

Remark (Bifurcation Behaviour Elsewhere) Assuming that the first repeated root is not the dominant root, one may attempt to repeat the argument at the other locations to observe a tug of war between those ratios below the coincidence location pulling one way and those above it pulling the other way. (We have noted in Sect. 13.4.5 the electric force field interpretation.) Who wins this tug of war is determined by the geometric considerations, and so we discover that there will be a critical point λ, a watershed, such that to the right of λ the complex roots move towards the origin, whereas to the left they move away from it.

Corollary 2 *Assume all the eigenvalues λ_j are real and positive, at some ω κ_1 is the maximal eigenvalue (in modulus) and is real, and further*

$$\lambda_3 < \kappa_1 < \lambda_1.$$

Then

(i) *for $j = 1, 2$*

$$sign[\partial \kappa_1/\partial \omega_1] = sign[\delta_1]sign[(\kappa_1 - \lambda_2)],$$
$$sign[\partial \kappa_1/\partial \omega_2] = -sign[\delta_2],$$

(ii) *for $j = 3, \ldots, n$*

$$sign[\partial \kappa_1/\partial \omega_j] = -sign[\delta_j]sign[(\kappa_1 - \lambda_2)],$$

and finally for $j = n + 1$

$$sign[\partial \kappa_1/\partial \omega_{n+1}] = -1.$$

Proof Counting the signs gives

$$sign[(\kappa_2 - \kappa_1)(\kappa_3 - \kappa_1)\ldots(\kappa_{n+1} - \kappa_1)] = (-1)^n$$
$$sign[(\lambda_1 - \kappa_1)(\lambda_3 - \kappa_1)\ldots(\lambda_n - \kappa_1)] = (-1)^{n-2}.$$

For $j \geq 3$:

$$sign\,[(\lambda_1 - \kappa_1)(\lambda_2 - \kappa_1)\ldots(\lambda_{j-1} - \kappa_1)(\lambda_{j+1} - \kappa_1)\ldots(\lambda_n - \kappa_1)]$$
$$= (-1)^{(n-3)}sign[(\lambda_2 - \kappa_1)] = (-1)^{(n-4)}sign[(\kappa_1 - \lambda_2)].$$

$\square$

Corollary 3 *Suppose all the eigenvalues of A are real and positive, and ω is small enough so that*

$$|\kappa_{n+1}| < \ldots < |\kappa_2| < \kappa_1,$$

and κ_j is real with

$$\lambda_{j+1} < \kappa_j < \lambda_{j-1}.$$

Then, for each h with κ_h real,

$$sign[\partial \kappa_h / \partial \omega_h] = sign[\delta_h],$$

$$sign[\partial \kappa_h / \partial \omega_j] = sign[\delta_h]sign[(\kappa_h - \lambda_h)]sign[(\kappa_h - \lambda_j)], \quad for \, h \neq j \leq n,$$

$$sign[\partial \kappa_h / \partial \omega_{n+1}] = sign[(\kappa_h - \lambda_h)],$$

where $h > 1$. In particular, provided $\kappa_2 < \kappa_1$,

$$sign[\partial \kappa_2 / \partial \omega_1] = -sign[\delta_1]sign[(\kappa_2 - \lambda_2)], \qquad sign[\partial \kappa_2 / \partial \omega_2] = sign[\delta_2],$$

$$sign[\partial \kappa_2 / \partial \omega_j] = sign[\delta_j]sign[(\kappa_2 - \lambda_2)] \, for \, 3 \leq j \leq n,$$

$$sign[\partial \kappa_2 / \partial \omega_{n+1}] = sign[(\kappa_2 - \lambda_2)].$$

Proof Recalling for $h \leq n$ that

$$\frac{\partial \kappa_h}{\partial \omega_h} = -\frac{\delta_h(\lambda_1 - \kappa_h)(\lambda_2 - \kappa_h) \ldots (\lambda_{h-1} - \kappa_h)(\lambda_{h+1} - \kappa_h) \ldots (\lambda_n - \kappa_h)}{(\kappa_1 - \kappa_h)(\kappa_2 - \kappa_h) \ldots (\kappa_{h-1} - \kappa_h)(\kappa_{h+1} - \kappa_h) \ldots (\kappa_{n+1} - \kappa_h)},$$

we compute that

$$sign[(\kappa_1 - \kappa_h)(\kappa_2 - \kappa_h) \ldots (\kappa_{h-1} - \kappa_h)(\kappa_{h+1} - \kappa_h) \ldots (\kappa_{n+1} - \kappa_h)] = (-1)^{(n+1-h)},$$

$$sign[(\lambda_1 - \kappa_h)(\lambda_2 - \kappa_h) \ldots (\lambda_{h-1} - \kappa_h)(\lambda_{h+1} - \kappa_h) \ldots (\lambda_n - \kappa_h)] = (-1)^{(n-h)},$$

and so

$$= (-1)^{(n-h)}sign[(\lambda_h - \kappa_h)]$$

$$= sign[(\lambda_1 - \kappa_h)(\lambda_2 - \kappa_h) \ldots (\lambda_{h-1} - \kappa_h)(\lambda_h - \kappa_h)(\lambda_{h+1} - \kappa_h) \ldots (\lambda_n - \kappa_h)]$$

$$= sign[(\lambda_j - \kappa_h)]sign[(\lambda_1 - \kappa_h)(\lambda_2 - \kappa_h) \ldots (\lambda_{j-1} - \kappa_h)(\lambda_{j+1} - \kappa_h) \ldots (\lambda_n - \kappa_h)].$$

□

13.6 Proof of the Dominance Theorem

We may now put together the analysis of the last sections to deduce our main result concerning the location of the dominant eigenvalue of $H(\omega)$.

Proof of Eigenvalue Dominance Theorem We consider the first part of the theorem only, as the more general result follows by a restatement of the same argument. By selecting the sign of δ_1 as (-1) and of δ_h for $h > 1$ as $(+1)$, we can arrange, given (13.18), for the eigenvalue function $\kappa_1(\omega)$ identified by the condition $\kappa_1(\omega_0) = \lambda_1$ to be decreasing in ω in the region

$$\{\omega : \omega_1 > 0, \ldots, \omega_{n+1} > 0\},$$

and so to remain below λ_1. We have, however, to ensure that $\kappa_1(\omega)$ remains the maximal root. Recall that $\omega_0 = (0, \ldots, 0, \beta)$. Since the remaining eigenvalues $\lambda_j = \kappa_j(\omega_0)$ are below λ_1 we may, by continuity, ensure that the eigenvalues functions $\kappa_2(\omega), \ldots, \kappa_{n+1}(\omega)$ of $H(\omega)$ also lie strictly below λ_1 and that moreover $\kappa_2(\omega) < \kappa_1(\omega)$. $\square$

Remark By Corollary 3 it is possible that, following a path in parameter space, the locus of κ_2 intersects that of κ_1. Note, however, that if upon intersection at ω^* we were thereafter to have $\kappa_1(\omega') < \kappa_2(\omega')$ for ω' close to ω^*, then provided the remaining eigenvalues remain below $\kappa_1(\omega^*)$, the signs of all the derivatives $\partial \kappa_1 / \partial \omega_j$ and those of $\partial \kappa_2 / \partial \omega_j$ would switch, i.e. both loci would turn around, a contradiction. Thus, subject to the assumption about the remaining eigenvalues, this implies that in fact ω^* is at the boundary of that region in policy-parameter space where κ_1 and κ_2 are both real. Moreover, according to (13.16) the graph has infinite slope at ω^*. We illustrate this point in the following simple example with $n = 1$, $\lambda_1 = 1$ and $|\beta| < 1$.

Example For $\delta_1 = -1$, let

$$A^{\lrcorner} = \begin{bmatrix} 1 & -1 \\ \omega_1 & \beta \end{bmatrix}.$$

The characteristic polynomial is $\kappa^2 - (1 + \beta)\kappa + (\beta + \omega_1)$. Here

$$\frac{d\kappa_1}{d\omega} = \frac{1}{(1 + \beta - 2\kappa_1)} = \frac{-b_1}{(\kappa_2 - \kappa_1)},$$

since $\kappa_1 + \kappa_2 = 1 + \beta$. The roots are real for

$$\omega_1 \le \frac{1}{4}[(1 + \beta)^2 - 4\beta] = \frac{1}{4}(1 - \beta)^2 = \omega_1^*(\beta),$$

and we have

$$\kappa_1 = \frac{1}{2}\left(1 + \beta + \sqrt{(1 - \beta)^2 - 4\omega_1}\right)$$

decreasing down to $\kappa_1 = \frac{1}{2}(1 + \beta)$ as ω_1 increases, and analogously

$$\kappa_2 = \frac{1}{2}\left(1 + \beta - \sqrt{(1 - \beta)^2 - 4\omega_1}\right)$$

increasing up to $\kappa_2 = \frac{1}{2}(1 + \beta)$. See Fig. 13.8.

We note that as the roots become complex the real part stays constant at $(1+\beta)/2$, i.e. the root locus bifurcates and the conjugate roots move orthogonally to the real axis; there being no poles in this simple case, there is no 'push' on the real part,

Fig. 13.8 For fixed β the root locus of $\kappa_1(\omega_1)$ (in red) and $\kappa_2(\omega_1)$ (in green) with vertical κ-axis

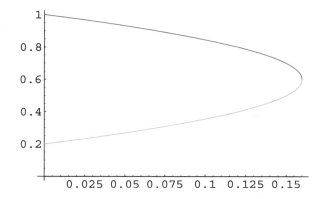

neither away nor towards the origin. In particular, provided the roots are in the unit circle, they remain in the annulus $\beta = \lambda_2 < |\zeta| < 1 = \lambda_1$.

13.7 Obtaining Dividend Irrelevance

In this section we prove the results in Proposition 1.

After a change of accounting state variables from z_t to, say Z_t, the system becomes

$$\begin{bmatrix} Z_{t+1} \\ d_{t+1} \end{bmatrix} = H \begin{bmatrix} Z_t \\ d_t \end{bmatrix},$$

where

$$H = H(\omega) = \begin{bmatrix} \lambda_1 & 0 & & 0 & b_1 \\ 0 & \lambda_2 & & & b_2 \\ & & \ddots & & \\ 0 & 0 & & \lambda_n & b_n \\ \omega_1 & \omega_2 & & & \omega_{n+1} \end{bmatrix},$$

with $\omega_{n+1} = \beta$, the new augmented matrix (with the same eigenvalues as the original augmented matrix $A^{\lrcorner}$) and where $\lambda_1, \ldots, \lambda_n$ are the eigenvectors of A assumed presented in decreasing modulus size (with λ_1 largest).

Evidently the characteristic polynomial

$$\chi_H(\kappa) = \chi_H(\kappa, \omega_1, \ldots, \omega_n, \omega_{n+1}) = |\kappa I - H|$$

is the same as $\chi_{A^{\lrcorner}}(\kappa)$. We assume its eigenvalues $\kappa_1, \ldots, \kappa_{n+1}$ have distinct modulus.

The force of our implicit assumptions on the **dividend significance coefficients** is that they are all non-zero: $b_j \neq 0$ for all j. Otherwise, the eigenvalues of the augmented matrix would include all the eigenvalues of the reduced matrix. (To see this expand the characteristic determinant by the j-th row.)

Henceforth we assume the canonical variables have been re-scaled by b_j and we may therefore take for the **canonical dividend significance coefficients** the symbol δ_j with the additional stipulation that

$$|\delta_j| = 1 \text{ for } j = 1, .., n.$$

As a first step we note the consequence for dividend irrelevance of the non-zero dividend significance coefficients. Writing $Z = (\ldots, Z^j, \ldots)$ and fixing j, there are coefficients $l_1, .., l_{n+1}$ such that

$$d_t = \sum_h l_h \kappa_h^t.$$

Now the equation

$$Z^j(t+1) = \lambda_j Z^j(t) + \delta_j \sum_h l_h \kappa_h^t,$$

with the solution also given by the eigenvalues of the augmented matrix

$$Z^j(t) = \sum_h L_h \kappa_h^t,$$

must satisfy

$$\sum_h L_h \kappa_h^t (\kappa_h - \lambda_j) = \delta_j \sum_h l_h \kappa_h^t$$

for all t. Hence

$$L_h = \frac{\delta_j l_h}{\kappa_h - \lambda_j}.$$

We thus have, assuming $R > |\kappa_j|$ for all j, that the dividend series converges, and

$$P_0(R; d) = \sum_{t=1}^{\infty} R^{-t} d_t = \sum_h \frac{1}{R - \kappa_h} l_h \kappa_h = \frac{1}{\delta_j} \sum_h \frac{\kappa_h - \lambda_j}{R - \kappa_h} L_h \kappa_h.$$

Consequently, if $R = \lambda_j$ is permitted, then with subscript indicating time

$$P_0(R; d) = -\frac{1}{\delta_j} \sum_h L_h \kappa_h = -\frac{Z_1^j}{\delta_1} = -\frac{\lambda_j Z_0^j + \delta_j d_0}{\delta_1} = -\frac{\lambda_j Z_0^j}{\delta_1} + d_0.$$

This indeed depends only on the initial data. See Ashton [1] for a discussion of this formula. The earliest form of this equation is due to Ohlson in 1989, though published later in [22].

We recall that the basis of this calculation is the identity

$$P_0(R; d) = \sum_{t=1}^{\infty} R^{-t} d_t = \sum_{t=1}^{\infty} R^{-t} \sum_h l_h \kappa_h^t = \sum_h l_h \sum_{t=1}^{\infty} R^{-t} \kappa_h^t$$

$$= \sum_h l_h \frac{\kappa_h/R}{1 - \kappa_h/R} = \sum_h \frac{1}{R - \kappa_h} l_h \kappa_h.$$

We will thus obtain dividend irrelevance at $R = \lambda_j$ provided all the eigenvalues κ_j are in modulus less than R.

13.8 Derivation of Equivalences

We begin by expanding by the bottom row

$$|H - \kappa I| = \begin{vmatrix} \lambda_1 - \kappa & 0 & & 0 & \delta_1 \\ 0 & \lambda_2 - \kappa & & & \delta_2 \\ & & \cdots & & \\ 0 & 0 & & \lambda_n - \kappa & \delta_n \\ \omega_1 & \omega_2 & & & \lambda_{n+1} - \kappa \end{vmatrix}_{n+1} = 0$$

to obtain

$$(-1)^n \omega_1 D_1(\kappa) - \ldots - \omega_n D_n(\kappa) + \prod_{j=1}^{n+1} (\lambda_j - \kappa) = 0,$$

or

$$(-1)^n \omega_1 D_1(\kappa) + (-1)^{n-1} \omega_2 D_2(\kappa) \ldots - \omega_n D_n(\kappa) = (-1)(-1)^{n+1} \prod_{j=1}^{n+1} (\kappa - \lambda_j),$$

where

$$D_1(\kappa) = \begin{vmatrix} 0 & 0 & 0 & \delta_1 \\ \lambda_2 - \kappa & & 0 & \delta_2 \\ & \cdots & & \\ 0 & & \lambda_n - \kappa & \delta_n \end{vmatrix}_n$$

$$= (-1)^{n-1} \delta_1 \prod_{j=2}^{n} (\lambda_j - \kappa) = \delta_1 \prod_{j=2}^{n} (\kappa - \lambda_j).$$

Similarly,

$$
D_2(\kappa) = \begin{vmatrix}
\lambda_1 - \kappa & 0 & 0 & \delta_1 \\
0 & 0 & 0 & \delta_2 \\
& & \lambda_3 - \kappa & \\
0 & 0 & \lambda_n - \kappa & \delta_n
\end{vmatrix}_n
$$

$$
= (-1)^{n-2}\delta_2 \prod_{\substack{j=2}}^{n}(\lambda_j - \kappa) = -\delta_2 \prod_{\substack{j \neq 2}}^{n}(\kappa - \lambda_j).
$$

This yields the equation

$$
\omega_1\delta_1 \prod_{j=2}^{n}(\kappa - \lambda_j) + \omega_2\delta_2 \prod_{\substack{j \neq 2}}^{n}(\kappa - \lambda_j) + \ldots + \omega_n\delta_n \prod_{\substack{j \neq n}}^{n}(\kappa - \lambda_j) = \prod_{j=1}^{n+1}(\kappa - \lambda_j).
$$

Dividing by $\prod_{\substack{h \neq j}}^{n}(\kappa - \lambda_h)$,

$$
(\kappa - \lambda_j)(\kappa - \beta) = \omega_j\delta_j + \sum_{h \neq j}\omega_h\delta_h\frac{\kappa - \lambda_j}{\kappa - \lambda_h} = \omega_j\delta_j + \sum_{h \neq j}\omega_h\delta_h\left(1 - \frac{\lambda_j - \lambda_h}{\kappa - \lambda_h}\right),
$$

or

$$
\omega_1\delta_1 + \omega_2\delta_2 + \ldots = (\kappa - \lambda_j)(\kappa - \beta) + \sum_{h \neq j}^{n}\frac{\omega_h\delta_h(\lambda_j - \lambda_h)}{\kappa - \lambda_h},
$$

as required.

Dividing by $\prod_{j=1}^{n}(\kappa - \lambda_j)$, we obtain

$$
\kappa - \beta = \frac{\omega_1\delta_1}{\kappa - \lambda_1} + \frac{\omega_2\delta_2}{\kappa - \lambda_2} + \ldots + \frac{\omega_n\delta_n}{\kappa - \lambda_n}.
$$

13.9 Invertible Parametrization and Zero Placement

This section is devoted to a proof of Proposition 3. Let us write

$$
\chi_A(\kappa) = |\kappa I - A| = \sum_{s=0}^{n}(-1)^s a_s \kappa^{n-s} = \prod_{j=1}^{n}(\kappa - \lambda_j),
$$

so that

$$a_s = \sum_{j_1 < \ldots < j_s} \lambda_{j_1} \ldots \lambda_{j_s}$$

denotes the elementary symmetric function summing the zeros of $\chi_A(\kappa)$ taken s at a time. Thus

$$a_0 = 1, \qquad a_1 = \lambda_1 + \ldots + \lambda_n, \qquad \ldots \qquad a_n = \lambda_1 \ldots \lambda_n.$$

Hence

$$\prod_{j=1}^{n+1} (\kappa - \lambda_j) = (\kappa - \beta) \left[\sum_{s=0}^{n} (-1)^s a_s \kappa^{n-s} \right]$$

$$= \kappa^{n+1} - (a_1 + \beta a_0) \kappa^n + \ldots + (-1)^s [a_{s+1} + \beta a_s] \kappa^{n-s}$$

$$+ \ldots + (-1)^{n+1} \beta a_n.$$

As a first step we compute that

$$\lambda_2 + \ldots + \lambda_n = a_1 - \lambda_1,$$

and that

$$\sum_{1 < u < v} \lambda_u \lambda_v = \sum_{u < v} \lambda_u \lambda_v - \lambda_1 \sum_{1 < v} \lambda_v = \sum_{u < v} \lambda_u \lambda_v - \lambda_1 \left(\sum_v \lambda_v - \lambda_1 \right)$$

$$= a_1 - \lambda_1 (a_1 - \lambda) = a_2 - a_1 \lambda_1 + \lambda_1^2.$$

Similarly,

$$\sum_{1 < u < v < w} \lambda_u \lambda_v \lambda_w = \sum_{u < v < w} \lambda_u \lambda_v \lambda_w - \lambda_1 \sum_{1 < v < w} \lambda_v \lambda_w$$

$$= a_3 - \lambda_1 [a_2 - a_1 \lambda_1 + \lambda_1^2]$$

$$= a_3 - a_2 \lambda_1 + a_1 \lambda_1^2 - \lambda_1^3.$$

The pattern is now clear, and we shall show by induction that

$$\sum_{1 < j_1 < \ldots < j_s} \lambda_{j_1} \ldots \lambda_{j_s} = a_s - a_{s-1} \lambda_1 + \lambda_1^2 a_{s-2} + \ldots + (-1)^s a_0 \lambda_1^s.$$

Indeed,

$$\sum_{j_1 < \ldots < j_s} \lambda_{j_1} \ldots \lambda_{j_s} = \sum_{j_1 < \ldots < j_s} \lambda_{j_1} \ldots \lambda_{j_s} - \lambda_1 \sum_{1 < j_2 < \ldots < j_s} \lambda_{j_2} \ldots \lambda_{j_s}$$

$$= a_s - \lambda_1 (a_{s-1} + \ldots + (-1)^{s-1} \lambda_1^{s-1})$$

$$= a_s - a_{s-1} \lambda_1 + \lambda_1^2 a_{s-2} + \ldots + (-1)^s a_0 \lambda_1^s.$$

Note that

$$a_n - a_{n-1} \lambda_1 + \lambda_1^2 a_{n-2} + \ldots + (-1)^n a_0 \lambda_1^n = 0,$$

so

$$\lambda_2 \ldots \lambda_n = \lambda_1^{-1} a_n = a_{n-1} - \lambda_1 a_{n-2} + \ldots + (-1)^{n-1} \lambda_1^{n-1}.$$

Our next step is to observe that the coefficients in the polynomial on the right-hand side of identity (13.8) may be expanded as follows:

$$D(\kappa) = \omega_1 \delta_1 \prod_{j \neq 1}^n (\kappa - \lambda_j) + \omega_2 \delta_2 \prod_{j \neq 2}^n (\kappa - \lambda_j) + \ldots + \omega_n \delta_n \prod_{j \neq n}^n (\kappa - \lambda_j)$$

$$= (\omega_1 \delta_1 + \omega_2 \delta_2 + \ldots + \omega_n \delta_n) \kappa^{n-1} - (\omega_1 \delta_1 [\lambda_2 + \ldots] + \ldots) \kappa^{n-2}$$

$$+ (\omega_1 \delta_1 [\lambda_2 \lambda_3 + \ldots] + \ldots) \kappa^{n-3} +$$

$$\ldots + (-1)^s (\omega_1 \delta_1 \bar{a}_s(1) + \ldots) \kappa^{n-s} + \ldots + (-1)^{n-1} [\sum_{j=1}^n \omega_j \delta_j \prod_{h \neq j}^n \lambda_h],$$

where

$$\bar{a}_s(h) = \sum_{\substack{j_1 < \ldots < j_s \\ j_k \neq h}} \lambda_{j_1} \ldots \lambda_{j_s}$$

(i.e. the summation refers to the omission of h from any of the components $j_1 \ldots j_s$). Note also that

$$\prod_{h \neq j}^n \lambda_h = \frac{a_n}{\lambda_j}.$$

We now consider for constants p_s the identity

$$\prod_{j=1}^{n+1}(\kappa - \lambda_j) - \left\{\omega_1\delta_1\prod_{j\neq1}^{n}(\kappa - \lambda_j) + \omega_2\delta_2\prod_{j\neq2}^{n}(\kappa - \lambda_j) + \ldots + \omega_n\delta_n\prod_{j\neq n}^{n}(\kappa - \lambda_j)\right\}$$

$$= \kappa^{n+1} - p_0\kappa^n + p_2\kappa^{n-1} + \ldots + p_n.$$

Comparing sides, we obtain

$$p_0 = (a_1 + \beta a_0),$$

$$p_1 = a_2 + \beta a_1 - (\omega_1\delta_1 + \omega_2\delta_2 + \ldots + \omega_n\delta_n),$$

$$p_2 = a_3 + \beta a_2 - (\omega_1\delta_1\lambda_1 + \ldots),$$

$$\ldots$$

$$p_n = a_{n+1} + \beta a_n - \left(\omega_1\delta_1\lambda_1^{-1} + \ldots\right).$$

Now given any p_0, we select β, so that

$$\beta = p_0 - a_1.$$

For the remaining equations, we have

$$p_1 - a_2 - \beta a_1 = \omega_1\delta_1 + \omega_2\delta_2 + \ldots + \omega_n\delta_n,$$

$$p_2 - a_3 - \beta a_2 = \omega_1\delta_1(a_1 - \lambda_1) + \ldots,$$

$$p_3 - a_4 - \beta a_3 = \omega_1\delta_1(a_2 - s_1\lambda_1 + \lambda_1^2) + \ldots,$$

$$\ldots$$

$$p_n - a_{n+1} - \beta a_n = \omega_1\delta_1(a_{n-1} - a_{n-2}\lambda_1 + (-1)^{n-1}a_0\lambda_1^{n-1}) + \ldots,$$

where $a_{n+1} = 0$, i.e.

$$N\begin{bmatrix} \delta_1\omega_1 \\ \delta_2\omega_2 \\ \ldots \\ \delta_n\omega_n \end{bmatrix} = \begin{bmatrix} p_1 - a_2 - \beta a_1 \\ p_2 - a_3 - \beta a_2 \\ \ldots \\ p_n - a_{n+1} - \beta a_n \end{bmatrix}.$$

Here the coefficient matrix N is given as follows:

$$
N = \begin{bmatrix}
1 & & 1 \\
a_1 - \lambda_1 & & a_1 - \lambda_n \\
a_2 - a_1\lambda_1 + \lambda_1^2 & \cdots & a_2 - a_1\lambda_n + \lambda_n^2 \\
\cdots & & \cdots \\
a_{n-1} - \lambda_1 a_{n-2} + \ldots + (-1)^{n-1}\lambda_1^{n-1} & & a_{n-1} - \lambda_n a_{n-2} + \ldots + (-1)^{n-1}\lambda_n^{n-1}
\end{bmatrix}.
$$

Its determinant is equal, up to a possible sign change, to the van der Monde determinant $V(\lambda_1, \ldots, \lambda_n)$. Hence N is non-singular and the equation may be solved for any given vector $(p_1, \ldots, p_n)$. To see this, note that N may be reduced to the alternant matrix $A(0, \ldots, n - 1)$ in the variables $(-\lambda_1), \ldots (-\lambda_n)$:

$$
\begin{vmatrix}
1 & 1 & \cdots & 1 \\
a_1 - \lambda_1 & a_1 - \lambda_2 & \cdots & a_1 - \lambda_n \\
a_2 - a_1\lambda_1 + \lambda_1^2 & & & \\
\cdots & & & \\
\sum_{i=0}^{n-1}(-1)^i a_{n-1-i} \lambda_1^i & & & \sum_{i=0}^{n-1}(-1)^i a_{n-1-i} \lambda_n^i
\end{vmatrix}
$$

$$
= \begin{vmatrix}
1 & 1 & \cdots & 1 \\
-\lambda_1 & -\lambda_2 & \cdots & -\lambda_n \\
\lambda_1^2 & & & \lambda_n^2 \\
\cdots & & & \\
(-\lambda_1)^{n-1} & & & (-\lambda_n)^{n-1}
\end{vmatrix}
$$

$$
= V(-\lambda_1, \ldots, -\lambda_n)
$$

(taking a_1 times the first row, a_2 times the second row and so on).

It is now easy to find the inverse transformation by applying the elementary row operations just used to the original matrix equation. This leads to the following result. Putting $g_j = p_j - a_{j+1} - \beta a_j$, the original equations

$$
N \begin{bmatrix} \delta_1\omega_1 \\ \delta_2\omega_2 \\ \cdots \\ \delta_n\omega_n \end{bmatrix} = \begin{bmatrix} g_1 \\ g_2 \\ \cdots \\ g_n \end{bmatrix}
$$

now transform to

$$
V \begin{bmatrix} \delta_1\omega_1 \\ \delta_2\omega_2 \\ \cdots \\ \delta_n\omega_n \end{bmatrix} = \begin{bmatrix} h_1 \\ h_2 \\ \\ h_n \end{bmatrix},
$$

where

$$h_1 = g_1,$$
$$h_2 = g_2 - a_1 h_1,$$
$$h_3 = g_3 - a_2 h_1 + a_1 h_2,$$
$$\dots$$
$$h_n = g_n - a_{n-1} h_1 + a_{n-2} h_2 - \dots \pm a_1 h_{n-1}.$$

Note that with the sign adjustment $h_n' = (-1)^n h_n$ the equations specify g_j as a convolution (taking $a_0 := 1$). Now we have

$$\begin{bmatrix} \delta_1 \omega_1 \\ \delta_2 \omega_2 \\ \dots \\ \delta_n \omega_n \end{bmatrix} = V^{-1} \begin{bmatrix} h_1 \\ h_2 \\ \\ h_n \end{bmatrix},$$

where the inverse V^{-1} is given by (see Klinger [14]) the matrix with jk entry

$$(-1)^{j+k} \frac{\bar{a}_{n-j}(k)}{\prod_{l=1}^{k-1}(\lambda_k - \lambda_l) \prod_{h=k+1}^{n}(\lambda_h - \lambda_k)};$$

here $\bar{a}_s(j)$ is the elementary symmetric function as above (sum over the s-fold product omitting the variable λ_j).

13.10 Asymptotics of the Unbounded Roots

Proof of Proposition 4 Assume that $\kappa \neq \lambda_j$ for $j = 1, \dots, n$. We rewrite (13.9) as

$$\omega_1 \delta_1 + \omega_2 \delta_2 + \dots + \omega_n \delta_n = (\kappa - \lambda_j)(\kappa - \beta) + \sum_{k \neq j}^{n} \frac{\omega_k \delta_k (\lambda_j - \lambda_k)}{\kappa}$$

$$+ \sum_{k \neq j}^{n} \omega_k \delta_k (\lambda_j - \lambda_k) \left[\frac{1}{\kappa - \lambda_k} - \frac{1}{\kappa} \right],$$

so that

$$\omega_1\delta_1 + \omega_2\delta_2 + \ldots + \omega_n\delta_n = (\kappa - \lambda_j)(\kappa - \beta) + \frac{1}{\kappa}\left[\sum_{k\neq j}^{n} \omega_k\delta_k(\lambda_j - \lambda_k)\right]$$

$$+ \sum_{k\neq j}^{n} \omega_k\delta_k(\lambda_j - \lambda_k)\left[\frac{\lambda_\kappa}{\kappa(\kappa - \lambda_k)}\right].$$

Iterating, we obtain

$$\omega_1\delta_1 + \omega_2\delta_2 + \ldots + \omega_n\delta_n = (\kappa - \lambda_j)(\kappa - \beta) \qquad (13.19)$$

$$+ \sum_{s=1}^{N} \frac{1}{\kappa^s}\left[\sum_{k\neq j}^{n} \omega_k\delta_k\lambda_\kappa^s(\lambda_j - \lambda_k)\right]$$

$$+ \sum_{k\neq j}^{n} \omega_k\delta_k\lambda_\kappa^{N-1}(\lambda_j - \lambda_k)\frac{1}{\kappa^N(\kappa - \lambda_k)}.$$

and hence the assertion of the Proposition follows provided $\kappa > |\lambda_k|$. The alternative direct derivation by expanding $\kappa^{-1}(1 - \lambda_k/\kappa)^{-1}$ as a geometric series is less informative about the convergence of the series.

We may in principle use the identity (13.19) (valid for all large enough κ) recursively to obtain an asymptotic expansion (in ω_j) for the unbounded roots.

To obtain the first term of the expansion, let $|\kappa| \to \infty$ and consider the quadratic approximation

$$\omega_1\delta_1 + \omega_2\delta_2 + \ldots + \omega_n\delta_n \sim \kappa^2 - (\lambda_j + \beta)\kappa + \lambda_j\beta$$

$$= [\kappa - \frac{1}{2}(\lambda_j + \beta)]^2 - \frac{1}{4}(\lambda_j - \beta)^2,$$

so that $\omega_j\delta_j$ is large and positive and

$$\kappa = \frac{1}{2}(\lambda_j + \beta) \pm \sqrt{\frac{1}{4}(\lambda_j - \beta)^2 + \sum_k \omega_k\delta_k}.$$

Hence for $\delta_j = -1$ we have the first term of the expansion to be

$$\kappa = \frac{1}{2}(\lambda_j + \beta) \pm i\sqrt{|\omega_j|}.$$

Now write

$$\kappa = \alpha + i\sqrt{|\omega_j|}, \qquad \alpha = \hat{\kappa}_j + \varepsilon, \qquad \hat{\kappa}_j = \frac{1}{2}(\lambda_j + \beta), \qquad A = \sum_{k\neq j}^{n} \omega_k\delta_k(\lambda_j - \lambda_k).$$

With this notation, taking only one term in (13.19) we have with ω_j large and negative that

$$-\omega_j + \sum_{k \neq j}^{n} \omega_k \delta_k = [\kappa - \hat{\kappa}_j]^2 - \frac{1}{4}(\lambda_j - \beta)^2 + \frac{A}{\kappa}$$

$$= [i\sqrt{\omega_j} + \varepsilon]^2 - \frac{1}{4}(\lambda_j - \beta)^2 + \frac{A}{\alpha + i\sqrt{\omega_j}}$$

$$= -\omega_j + 2i\varepsilon\sqrt{\omega_j} + \varepsilon^2 - \frac{1}{4}(\lambda_j - \beta)^2 - A\frac{\alpha - i\sqrt{\omega_j}}{\omega_j - \alpha^2},$$

so to first order in ε

$$\sum_{k \neq j}^{n} \omega_k \delta_k = 2i\varepsilon\sqrt{\omega_j} - \frac{1}{4}(\lambda_j - \beta)^2 - A\frac{\hat{\kappa}_j + \varepsilon - i\sqrt{\omega_j}}{\omega_j - \hat{\kappa}_j^2}.$$

Writing $\varepsilon = u + iv$ and taking real and imaginary parts gives the two equations

$$\frac{-A}{\omega_j - \hat{\kappa}_j^2}u - 2v\sqrt{\omega_j} = \gamma_j = \sum_{k \neq j}^{n} \omega_k \delta_k + \frac{1}{4}(\lambda_j - \beta)^2 + A\frac{\hat{\kappa}_j}{\omega_j - \hat{\kappa}_j^2},$$

$$2u\sqrt{\omega_j} - \frac{A}{\omega_j - \hat{\kappa}_j^2}v = -\frac{A\sqrt{\omega_j}}{\omega_j - \hat{\kappa}_j^2}.$$

Note that for $|\omega_j| \to \infty$

$$\gamma_j \to \sum_{k \neq j}^{n} \omega_k \delta_k + \frac{1}{4}(\lambda_j - \beta)^2.$$

Now the determinant of the two equations in u and v above is positive and equal to

$$\Delta = \frac{A^2}{(\omega_j - \hat{\kappa}_j^2)^2} + 4\omega_j.$$

Solving for u and v, we have

$$u\Delta = -\frac{A\gamma_j}{\omega_j - \hat{\kappa}_j^2} - \frac{2A\omega_j}{\omega_j - \hat{\kappa}_j^2} = -2A - \frac{2\hat{\kappa}_j^2 + \gamma_j}{\omega_j - \hat{\kappa}_j^2}A,$$

$$v\Delta = \frac{-A\sqrt{\omega_j}}{(\omega_j - \hat{\kappa}_j^2)^2} - 2\gamma_j\sqrt{\omega_j},$$

so that

$$u = -\frac{A}{2\omega_j} + O(\omega_j^{-2}), \qquad v = -\frac{\gamma_j}{4\sqrt{\omega_j}} + O(\omega^{-5/2}).$$

The result for u assumes that $A \neq 0$ and further that

$$2\hat{\kappa}_j^2 + \gamma_j \neq 0,$$

i.e.

$$\sum_{k \neq j}^{n} \omega_k \delta_k + \frac{3}{4}(\lambda_j + \beta)^2 \neq 0.$$

13.11   Strip-and-Two-Circles Theorem

We prove Proposition 5. For part (i) we argue as follows. Suppose, for all j, that $\omega_j \delta_j \geq 0$ and $\beta \leq \lambda_n$. Suppose z satisfies

$$z - \beta = \frac{\omega_1 \delta_1}{z - \lambda_1} + \frac{\omega_2 \delta_2}{z - \lambda_2} + \ldots + \frac{\omega_n \delta_n}{z - \lambda_n}.$$

If z is strictly to the right of λ_1, then we may also assume that z has positive imaginary part (otherwise switch to the conjugate root $\bar{z}$). The argument of $z - \lambda_j$ is thus positive for each j, and that of $1/(z - \lambda_j)$ negative, i.e. has negative imaginary part. The right-hand side therefore sums to a complex number with negative imaginary part. However, $z - \beta$ has positive imaginary part.

If z is to the left of β, then we may suppose it has negative imaginary part. The argument of $\lambda_j - z$ is for each j thus positive, as also for $\beta - z$. Now apply the previous reasoning to the identity

$$\beta - z = \frac{\omega_1 \delta_1}{\lambda_1 - z} + \frac{\omega_2 \delta_2}{\lambda_2 - z} + \ldots + \frac{\omega_n \delta_n}{\lambda_n - z}.$$

For part (ii), let ε_j be arbitrary real for $j = 1, \ldots n$. We will apply Marden's 'Mean-Value Theorem for polynomials' (Marden [16, §2.8 p.23]) to the polynomials h_j for $j = 1, \ldots, n$ and the polynomial $f(z)$ as defined by

$$f(z) = \prod_{h=1}^{n+1}(z - \lambda_h), \qquad h_j(z) = \varepsilon_j \prod_{h \neq j}^{n}(z - \lambda_h).$$

We must, however, first find for each j the location of the roots of the equation $f(z) = h_j(z)$. The roots are of course $z = \lambda_k$ for $k \neq j$ taken together with the two real roots of

$$(z - \beta)(z - \lambda_j) = \varepsilon_j,$$

which are to the left of β and the right of λ_j. The exact and approximate formulas are

$$u_j^\pm = \frac{(\beta + \lambda_j) \pm \sqrt{(\lambda_j - \beta)^2 + 4\varepsilon_j}}{2} \sim \beta - \frac{\varepsilon_j}{4(\lambda_j - \beta)}, \qquad \lambda_j + \frac{\varepsilon_j}{4(\lambda_j - \beta)},$$

and require that

$$-\frac{1}{4}(\lambda_j - \beta)^2 \leq \varepsilon_j.$$

Thus the roots of all the equations lie in the interval $K = (u_1^-, u_1^+)$. By Marden's Theorem in the special case of real positive scalars m_j summing to unity, the roots of

$$f(z) = \sum m_j h_j(z)$$

lie in the star-shaped region $S(K, \pi/(n+1))$ (cf. Proposition 5). Thus if we take $\varepsilon_j = \varepsilon$ small and $m_j \varepsilon = \delta_j \omega_j$ so that

$$\delta_1 \omega_1 + \ldots + \delta_n \omega_n = \varepsilon, \quad \text{with } \delta_1 \omega_1, \ldots, \delta_n \omega_n \geq 0,$$

then indeed $\sum m_j = 1$ and all the roots of (13.7) lie in the said star-shaped region.

In fact, one may take $\varepsilon_1 = \varepsilon$ small and $m_1 \varepsilon_1 = \delta_1 \omega_1$, and for $j > 1$, $\varepsilon_j = \mu :=$ $\min\{\lambda_j - \lambda_{j+1} : \text{for } j > 1\}$ and $m_j \mu = \delta_j \omega_j > 0$, leading to the restriction

$$1 = m_1 + \ldots + m_n = \frac{\delta_1 \omega_1}{\varepsilon} + \frac{1}{\mu}(\delta_2 \omega_2 + \ldots),$$

i.e.

$$\delta_1 \omega_1 + \frac{\varepsilon}{\mu}(\delta_2 \omega_2 + \ldots + \delta_n \omega_n) = \varepsilon \qquad (\delta_1 \omega_1, \ldots, \delta_n \omega_n \geq 0).$$

13.12 The Third-Derivative Test

The result in Sect. 13.3.3 is a consequence of the following by specializing $g(\kappa) := f(\kappa^* + \kappa) - f(\kappa^*)$ when $f'(\kappa^*) = 0$.

Proposition 9 *For g with $g(0) = 0$, $g'(0) = 0$, $g''(0) > 0$, $g'''(0) \neq 0$, the solution $\kappa = \kappa(\omega)$ over the complex domain of the equation*

$$g(\kappa) = -\omega^2$$

with $\omega > 0$ small and subject to $\kappa(0) = 0$, satisfies

$$\Re(\kappa(\omega)) \text{ initially increasing if } g'''(0) > 0, \text{ and}$$
$$\text{initially decreasing if } g'''(0) < 0.$$

The imaginary part is initially increasing and satisfies $|\Im(\kappa(\omega))| > \omega$.

Proof Below we work to order $o(\omega)$. Without loss of generality we assume that $g''(0) = +2$. (Otherwise rescale g and ω^2 by $2/g''(0)$.) Set $G := g'''(0)/6 \neq 0$; then, by Taylor's Theorem, the first approximation to the equation is

$$g(\kappa) = \kappa^2 + o(\kappa^2) = -\omega^2,$$

with solution $\kappa = \pm i\omega$. So we introduce correction terms by putting

$$\kappa = \alpha + i(\omega + \beta)$$

and solve a second approximation

$$g(\kappa) = \kappa^2 + G\kappa^3 + o(\kappa^3) = -\omega^2.$$

Substitution for κ gives after cancellation of the term $-\omega^2$

$$0 = [\alpha^2 - (2\omega\beta + \beta^2) + 2i\alpha(\omega + \beta)] + G[\alpha^3 + 3i\alpha^2(\omega + \beta) \qquad (13.20)$$
$$-3\alpha(\omega + \beta)^2 - i(\omega + \beta)^3].$$

Equate real and imaginary parts to 0; cancelling the second by $(\omega + \beta)$ (non-zero, w.l.o.g.) gives

$$3G\alpha^2 + 2\alpha - G(\omega + \beta)^2 = 0.$$

The roots corresponding to the imaginary part, for ω and β sufficiently small, are

$$\frac{-1 \pm \sqrt{1 + 3G^2(\omega + \beta)^2}}{3G} = \frac{1}{3G}\{\frac{1}{2}3G^2(\omega + \beta)^2 - \frac{1}{8}9G^4(\omega + \beta)^4 + ..\}$$

$$= \frac{1}{2}G(\omega + \beta)^2 + \dots,$$

So we may neglect α^2 and β^2 in what follows, which leads to the approximation

$$\alpha = G\omega\beta. \qquad \square$$

Claim For $\omega > 0$ small enough, $\beta > 0$. In particular, α is the same sign as G. (So, also, conversely, if α is the same sign as G, then $\beta > 0$.)

Proof The equation for the real part of (13.20), ignoring $(2\omega\beta + \beta^2)$ and cancelling by $\alpha \neq 0$, gives

$$G\alpha^2 + \alpha - 3G(\omega + \beta)^2 = 0.$$

Its two roots $\alpha_{\pm}$ have negative product $-3(\omega + \beta)^2$, and

$$\alpha_{\pm} = -\frac{1}{2G}\left[1 \pm \sqrt{1 + 12G^2(\omega + \beta)^2}\right]$$

$$= -\frac{1}{2G}\left[1 \pm \{1 + 6G^2(\omega + \beta)^2\}\right] + o(\omega)$$

$$= -G^{-1}, \qquad 3G(\omega + \beta)^2 \text{ to order } o(\omega).$$

The root near $-1/G$ is ruled out by the continuity of the root locus, which tends to 0 as $\omega \to 0$. For $G > 0$ the positive root is near $3G(\omega + \beta)^2 \sim 6G\omega\beta$, and so $\beta > 0$, as $\omega > 0$. For $G < 0$, the negative root is $3G(\omega + \beta)^2 \sim 6G\omega\beta$, and again $\beta > 0$, as $\omega > 0$. $\qquad \square$

References

1. Ashton, D.: The cost of equity capital and a generalization of the dividend growth model. Account. Bus. Res. **26**, 34–18 (1995)
2. Ashton, D., Cooke, T., Tippett, M., Wang, P.: Linear information dynamics, aggregation, dividends and "dirty surplus". Account. Bus. Res. **34**, 277–299 (2004)
3. Bellman, R.: Introduction to matrix analysis. Reprint of the second (1970) edition. In: Classics in Applied Mathematics, vol. 19 (SIAM, Philadelphia, 1997)
4. Bhattacharya, S.: Imperfect information, dividend policy, and "the bird in the hand" fallacy. Bell J. Econ. **10**, 259–270 (1979)

5. Davies, R.O., Ostaszewski, A.J.: Optimal forward contract design for inventory: a value-of-waiting analysis. In: Brzdek, J., Popa, D., Rassias, T.M. (eds.) Ulam Type Stability, pp. 73–96. Springer, Cham (2019)
6. Dybvig, P.H., Zender, J.F.: Capital structure and dividend irrelevance with asymmetric information. Rev. Financ. Stud. **4**, 201–219 (1991)
7. Gao, Z., Ohlson, J.A., Ostaszewski, A.J.: Dividend policy irrelevancy and the construct of earnings. J. Bus. Financ. Acc. **40**, 673–694 (2013)
8. Gietzmann, M.B., Ostaszewski, A.J.: Predicting firm value: the superiority of q-theory over residual income. Account. Bus. Res. **34**, 349–377 (2004)
9. Henrici, P.: Applied and Computational Complex Analysis. Power Series-Integration-Conformal Mapping-Location of Zeros, vol. I, Reprinted 1988. (Wiley, Hoboken, 1974)
10. Hinrichsen, D., Kelb, B.: Stability radii and spectral value sets for real matrix perturbations. In: Systems and Networks: Mathematical Theory and Applications, vol. II, pp. 217–220. Invited and Contributed Papers (Akademie-Verlag, Berlin, 1994)
11. Horn, R.A., Johnson, C.R.: Matrix Analysis (Cambridge University Press, Cambridge, 1985)
12. Hwang, S.-K.: Cauchy's interlace theorem for eigenvalues of Hermitian matrices. Am. Math. Mon. **111**, 157–159 (2004)
13. Jack, A., Johnson, T., Zervos, M.: A singular control model with application to the goodwill problem. Stoch. Process. Appl. **118**, 2098–2124 (2008)
14. Klinger, A.: The Vandermonde matrix. Am. Math. Mon. **74**, 571–574 (1967)
15. Lo, K., Lys, T.: The Ohlson model: contribution to valuation theory, limitations, and empirical applications. J. Acc. Audit. Financ. **15**, 337–367 (2000)
16. Marden, M.: The Geometry of the Zeros of a Polynomial in a Complex Variable (American Mathematical Society, New York, 1949)
17. Miller, M.H., Modigliani, F.: The cost of capital, corporation finance and the theory of investment. Am. Econ. Rev. **48**, 261–297 (1958)
18. Miller, M.H., Modigliani, F.: Dividend policy, growth, and the valuation of shares. J. Bus. **34**, 411–433 (1961)
19. Miller, M.H., Modigliani, F.: Corporate income taxes and the cost of capital: a correction. Am. Econ. Rev. **53**, 433–443 (1963)
20. Noble, B.: Applied Linear Algebra (Prentice-Hall, Upper Saddle River, 1969)
21. Ohlson, J.: Earnings, book values, and dividends in equity valuation. Contemp. Account. Res. **11**, 661–687 (1995)
22. Ohlson, J.: Accounting earnings, book value, and dividends: the theory of the clean surplus equation (part I). In: Brief, R.P., Peasnell, K.V. (eds.) Clean Surplus: A Link Between Accounting and Finance, pp. 165–230 (Garland Publishing, Princeton, 1996)
23. Ohlson, J.A., Gao, Z.: Earnings growth and equity value. Found. Trends Acc. **1**, 1–70 (1981)
24. Preinreich, G.: The fair value and yield of common stock. Acc. Rev. **11**, 130–140 (1936)
25. Rugh, W.J.: Linear System Theory (Prentice-Hall, Upper Saddle River, 1996)
26. Seneta, E.: Non-negative Matrices and Markov Chains, 2nd edn. Revised reprint (1st ed. 1973) (Springer, New York, 1981)
27. Tippett, M., Warnock, T.: The Garman-Ohlson structural system. J. Bus. Financ. Acc. **24**, 1075–1099 (1997)
28. Wilkinson, J.H.: The Algebraic Eigenvalue Problem (Oxford University Press, Oxford, 1965)

Chapter 14
A Fixed Point Theorem in Uniformizable Spaces

Lahbib Oubbi

Abstract We provide a fixed point theorem in uniformizable spaces, extending former results of G. L. Forti, and of J. Brzdęk.

Keywords Fixed point theorem · Uniformizable space

Mathematics Subject Classification (2010) Primary 37C25; Secondary 54E15

14.1 Introduction

In [7], Forti noticed that most of the results concerning the Ulam-Hyers stability of functional equations are shown using, roughly speaking, the same argument. The proofs consist of modifying the given equation, generally depending on several variables, in order to get an operator T acting on functions from a domain X into a metric space (Y, d) whose fixed point, if any, is a solution of the given functional equation, approximating the given function f. He then gave a fixed point theorem showing that, under appropriate conditions, the operator T admits a fixed point so that the equation is stable. Forti's theorem has been applied by Brzdęk in an easier fashion to show the stability of the Cauchy equation, the Jensen equation, and the quadratic one, see [2]. A variant of the same theorem from [4] (see also [3]) has been used by Zhang [8] and later by Bahyrycz and Olko [1] to show the (hyper-) stability of a very general linear equation. For more information on connections between the fixed point theory and Ulam stability we refer to [5].

In this note, we provide a theorem similar to Forti's one, but in a more general setting; for a survey of related results see [6].

L. Oubbi (✉)
Department of Mathematics, Team GrAAF, Laboratory LMSA, Center CeReMar, Ecole Normale Supérieure, Mohammed V University in Rabat, Takaddoum, Rabat, Morocco
e-mail: oubbi@daad-alumni.de

© Springer Nature Switzerland AG 2019
J. Brzdęk et al. (eds.), *Ulam Type Stability*,
https://doi.org/10.1007/978-3-030-28972-0_14

14.2 Preliminaries

A pseudo-metric on a non-empty set Z is any mapping $d : Z \times Z \to [0, +\infty[$ such that:

(P-1) $\forall x \in Z, d(x, x) = 0$,
(P-2) $\forall (x, y) \in Z^2, d(x, y) = d(y, x)$,
(P-3) $\forall (x, y, z) \in Z^3, d(x, y) \le d(x, z) + d(z, y)$.

A uniformizable space is any non-empty set Z endowed with a family $(d_i)_{i \in I}$ of semi-metrics such that $d_i(x, y) = 0$ for every $i \in I$ if and only if $x = y$. Such a uniformizable space will be denoted by $(Z, (d_i)_{i \in I})$.

A sequence $(x_n)_n \subset Z$ is said to be Cauchy if, for every $i \in I$, $d_i(x_n, x_m)$ converges to zero as n, m tend to $+\infty$. It is said to converge to some $x \in Z$, if, for all $i \in I$, $d_i(x_n, x)$ tends to zero as n tends to $+\infty$. If every Cauchy sequence in Z converges to some point of Z, we say that Z is sequentially complete.

If $T_n : S \to S$ is a self mapping of some non-empty set S, $n \in \mathbb{N}$, we will denote by

$$\overset{m}{\underset{n=k}{\circ}} T_n$$

the composition mapping $T_k \circ T_{k+1} \circ \cdots \circ T_m$, with $k \le m$.

14.3 The Results

The following theorem generalizes a former result of the author, presented in the international conference on functional equations and inequalities (ICFEI 2017) held in Będlewo (Poland) in July 2017.

Theorem 14.1 *Let S be a non-empty set and $(Z, (d_i)_{i \in I})$ a sequentially complete uniformizable space. Given mappings $f : S \to Z$, $G : S \to S$, $H : Z \to Z$ and, for every $i \in I$, $\delta_i : S \to \mathbb{R}^+$, and $\beta_i : \mathbb{R}^+ \to \mathbb{R}^+$. Assume that, for every $i \in I$, β_i is non-decreasing and the conditions (14.1), (14.2) and (14.3) below hold:*

$$d_i(H \circ f \circ G(s), f(s)) \le \beta_i \circ \delta_i(s), \quad \forall s \in S, \tag{14.1}$$

there exists a sequence $(i_j)_{j \in \mathbb{N}_0} \subset I$ such that $i_0 = i$ and

$$d_{i_j}(H(x), H(y)) \le \beta_{i_j}(d_{i_{j+1}}(x, y)), \quad x, y \in Z, j \ge 0, \tag{14.2}$$

$$\sum_{n=1}^{\infty} \left(\overset{n}{\underset{j=0}{\circ}} \beta_{i_j} \right) \circ \delta_{i_n} \circ G^n(s) < +\infty, \quad \forall s \in S. \tag{14.3}$$

Then there exists a mapping $F : S \to Z$ *such that:*

$$F(s) = \lim_{n \to +\infty} H^n \circ f \circ G^n(s), \tag{14.4}$$

$$\forall i \in I, \ d_i(H^n \circ f \circ G^n(s), F(s)) \leq \sum_{k=n}^{+\infty} \left(\mathop{\circ}_{j=0}^{k} \beta_{i_j} \right) \circ \delta_{i_k} \circ G^k(s), \ s \in S. \tag{14.5}$$

In particular

$$d_i(F(s), f(s)) \leq \sum_{n=0}^{\infty} \left(\mathop{\circ}_{j=0}^{n} \beta_{i_j} \right) \circ \delta_{i_n} \circ G^n(s), \ \forall s \in S. \tag{14.6}$$

If in addition H is continuous, then F is a fixed point of the operator $\Lambda : Z^S \to Z^S$, $g \mapsto H \circ g \circ G$. Finally, if for every $i \in I$, β_i is subadditive, then F is the unique fixed point of Λ which satisfies (14.5).

Proof If we set $Q_n := H^n \circ f \circ G^n$, then using (14.2), we get for all $s \in S$ and all $i \in I$:

$$d_i(Q_{n+1}(s), Q_n(s)) \leq \beta_{i_0} \left(d_{i_1} (Q_n(G(s)), Q_{n-1}(G(s))) \right).$$

Using (14.2) and the fact that β_i is non-decreasing, we obtain

$$d_i(Q_{n+1}(s), Q_n(s)) \leq \beta_{i_0} \circ \beta_{i_1} \left(d_{i_2} \left(Q_{n-1}(G^2(s)), Q_{n-2}(G^2(s)) \right) \right).$$

Then, step by step, we get:

$$d_i(Q_{n+1}(s), Q_n(s)) \leq \left(\mathop{\circ}_{k=0}^{n-1} \beta_{i_k} \right) (d_{i_n} (Q_1(G^n(s)), f(G^n(s)))).$$

Using (14.1), we come to

$$d_i(Q_{n+1}(s), Q_n(s)) \leq \left(\mathop{\circ}_{k=0}^{n} \beta_{i_k} \right) \circ \delta_{i_n} \circ G^n(s).$$

Therefore, for $n < m$ and $i \in I$, one has:

$$d_i(Q_n(s), Q_m(s)) \leq \sum_{k=n}^{m-1} d_i(Q_{k+1}(s), Q_k(s))$$

$$\leq \sum_{k=n}^{m-1} \left(\mathop{\circ}_{j=0}^{k} \beta_{i_j} \right) \circ \delta_{i_k} \circ G^k(s). \tag{14.7}$$

Since, by assumption (14.3), the series

$$\sum_{k=0}^{\infty} \left(\underset{j=0}{\overset{k}{\circ}} \beta_{ij} \right) \circ \delta_{i_k} \circ G^k(s)$$

converges for every $s \in S$, the sequence $(Q_n(s))_n$ is Cauchy in Z. The latter being sequentially complete, $(Q_n(s))_n$ converges to some

$$F(s) := \lim_{n \to \infty} H^n \circ f \circ G^n(s) \in Z.$$

If in (14.7) we let m tend to infinity, the so-defined function F satisfies (14.5) as required. Taking $n = 0$, we get (14.6).

Now, if H is continuous, for every $s \in S$, one has:

$$(H \circ F \circ G)(s) = H\{F[G(s)]\}$$
$$= H\{\lim_n (H^n \circ f \circ G^n)(G(s))\}$$

H being continuous $\qquad = \lim_n H^{n+1} \circ f \circ G^{n+1}(s)$

$$= F(s).$$

Therefore F is a fixed point of Λ. For the unicity, assume that $R : S \to X$ satisfies $H \circ R \circ G = R$ together with (14.5). Then for every $i \in I$, we have:

$$d_i(R(s), F(s)) \le d_i(R(s), Q_n(s)) + d_i(Q_n(s), F(s)).$$

Due to (14.5)), we get

$$d_i(R(s), F(s)) \le 2 \sum_{k=n}^{+\infty} \left(\underset{j=0}{\overset{k}{\circ}} \beta_{ij} \right) \circ \delta_{i_k} \circ G^k(s).$$

Since

$$\sum_{k=n}^{+\infty} \left(\underset{j=0}{\overset{k}{\circ}} \beta_{ij} \right) \circ \delta_{i_k} \circ G^k(s)$$

tends to 0 as n tends to infinity, $d_i(R(s), F(s)) = 0$. As i was arbitrary, $R(s) = F(s)$.

Denote by υX (resp. βX) the realcompactification (Stone-Čech compactification) of a Hausdorff Tychonoff space X. This is the smallest realcompact (resp. compact) space containing X as a dense topological subspace such that every continuous function f from X into $\mathbb{R}$ (resp. into a compact space) extends continuously to υX (resp. βX). Denote by f^υ (resp. f^β) such an extension. By $C_b(X)$ we will mean the algebra of bounded continuous functions from X into $\mathbb{K} \in \{\mathbb{R}, \mathbb{C}\}$. Notice that X is realcompact (resp. compact) if and only if $X = \upsilon X$ (resp. $X = \beta X$).

Corollary 14.1 *Assume that (X, τ) is a Tychonoff space, that $f : S \to X$, $G : S \to S$, and $H : X \to X$ are given mappings, and that, for every $g \in C(X)$, there exist mappings $\delta_g : S \to \mathbb{R}^+$ and $\beta_g : \mathbb{R}^+ \to \mathbb{R}^+$ so that β_g is non-decreasing and*

$$|g(H \circ f \circ G(s)) - g(f(s))| \le \beta_g \circ \delta_g(s), \quad \forall s \in S. \tag{14.8}$$

Assume that, for every $g \in C(X)$, there is a sequence $(g_n)_{n \in \mathbb{N}_0} \subset C(X)$ such that $g_0 = g$ and

$$|g_n(H(x)) - g_n(H(y))| \le \beta_g(|g_{n+1}(x) - g_{n+1}(y)|), \quad \forall x, y \in X, \ n \in \mathbb{N}_0 \tag{14.9}$$

$$\sum_{n=0}^{\infty} \left(\underset{k=0}{\overset{n}{\circ}} \beta_{g_k} \right) \circ \delta_{g_n} \circ G^n(s) < +\infty, \quad \forall s \in S. \tag{14.10}$$

Then there exists $F : S \to \upsilon X$ s.t., for every $g \in C(X)$,

$$|g^\upsilon(F(s)) - g(f(s))| \le \sum_{n=0}^{\infty} \left(\underset{k=0}{\overset{n}{\circ}} \beta_{g_k} \right) \circ \delta_{g_n} \circ G^n(s), \quad \forall s \in S. \tag{14.11}$$

If in addition H is continuous and, for every $g \in C(X)$, β_g is subadditive, then F is the unique mapping from S into υX which satisfies both $H^\upsilon \circ F \circ G = F$ and, for every $g \in C(X)$ and every $n \in \mathbb{N}$,

$$|g(H^n \circ f \circ G^n(s)) - g^\upsilon(F(s))| \le \sum_{k=n}^{\infty} \left(\underset{k=0}{\overset{n}{\circ}} \beta_{g_k} \right) \circ \delta_{g_n} \circ G^n(s), \quad \forall s \in S. \tag{14.12}$$

Similarly,

Corollary 14.2 *Assume that (X, τ) is a Tychonoff space, that $f : S \to X$, $G : S \to S$, and $H : X \to X$ are given mappings, and that, for every $g \in C_b(X)$, there exist mappings $\delta_g : S \to \mathbb{R}^+$ and $\beta_g : \mathbb{R}^+ \to \mathbb{R}^+$ so that β_g is non-decreasing and*

$$|g(H \circ f \circ G(s)) - g(f(s))| \le \beta_g \circ \delta_g(s), \quad \forall s \in S. \tag{14.13}$$

Assume that, for every $g \in C_b(X)$, there is a sequence $(g_n)_{n \in \mathbb{N}_0} \subset C_b(X)$ such that $g_0 = g$ and

$$|g_n(H(x)) - g_n(H(y))| \le \beta_g(|g_{n+1}(x) - g_{n+1}(y)|), \quad \forall x, y \in X, \ n \in \mathbb{N}_0 \tag{14.14}$$

$$\sum_{n=0}^{\infty} \left(\underset{k=0}{\overset{n}{\circ}} \beta_{g_k} \right) \circ \delta_{g_n} \circ G^n(s) < +\infty, \quad \forall s \in S. \tag{14.15}$$

Then there exists $F : S \to \beta X$ such that, for every $g \in C_b(X)$,

$$|g^\beta(F(s)) - g(f(s))| \le \sum_{n=0}^{\infty} \left(\underset{k=0}{\overset{n}{\circ}} \beta_{g_k} \right) \circ \delta_{g_n} \circ G^n(s), \ \forall s \in S. \tag{14.16}$$

If in addition H is continuous and, for every $g \in C_b(X)$, β_g is subadditive, then F is the unique mapping from S into βX which satisfies both $H^\beta \circ F \circ G = F$ and, for every $g \in C_b(X)$ and every $n \in \mathbb{N}$,

$$|g(H^n \circ f \circ G^n(s)) - g^\beta(F(s))| \le \sum_{k=n}^{\infty} \left(\underset{k=0}{\overset{n}{\circ}} \beta_{g_k} \right) \circ \delta_{g_n} \circ G^n(s), \ \forall s \in S. \tag{14.17}$$

Notice that if we take in Theorem 14.1, in Corollary 14.1 and in Corollary 14.2, $d_{i_k} = d_i$ for every $i \in I$, we get the main fixed point theorems previously obtained by the author and presented in the International Conference on Functional Equations and Inequalities in Będlewo, July 2017.

Remark 14.1 The subadditivity of β_i in Theorem 14.1 and of β_g in Corollary 14.1 and Corollary 14.2 can be released. The proofs remain unchanged.

References

1. Bahyrycz, A., Olko, J.: On stability of the general linear equation. Aequationes Math. **89**, 1461–1474 (2015)
2. Brzdęk, J.: On a method of proving the Hyers-Ulam stability of functional equations on restricted domains. Aust. J. Math. Anal. Appl. **6**(1), 1–10 (2009). Article 4
3. Brzdęk, J., Ciepliński, K.: A fixed point approach to the stability of functional equations in non-Archimedean metric spaces. Nonlinear Anal. **74**, 6861–6867 (2011)
4. Brzdęk, J., Chudziak, J., Páles, Z.: A fixed point approach to stability of functional equations. Nonlinear Anal. **74**, 6728–6732 (2011)
5. Brzdęk, J., Cădariu, L., Ciepliński, K.: Fixed point theory and the Ulam stability. J. Funct. Spaces **2014**, 16 (2014). Art. ID 829419
6. Brzdęk, J., Ciepliński, K., Leśniak, Z.: On Ulam's type stability of the linear equation and related issues. Discret. Dyn. Nat. Soc. **2014**, 14 (2014). Art. ID 536791
7. Forti, G.L.: Comments on the core of the direct method for proving Hyers-Ulam stability of functional equations. J. Math. Anal. Appl. **295**, 127–133 (2004)
8. Zhang, D.: On Hyperstability of generalized linear equations in several variables. Bull. Aust. Math. Soc. **92**, 259–267 (2015)

Chapter 15
Symmetry of Birkhoff-James Orthogonality of Bounded Linear Operators

Kallol Paul, Debmalya Sain, and Puja Ghosh

Abstract We survey the recent developments in the study of symmetry of Birkhoff-James orthogonality of bounded linear operators between Banach spaces and Hilbert spaces. We also present some new results, along with the corresponding proofs, that have not been published before. In the last section we suggest some future directions for research, in particular connected to the notion of Ulam stability.

Keywords Birkhoff-James orthogonality · Symmetry of orthogonality · Bounded linear operators · Ulam stability

Mathematics Subject Classification (2010) Primary 47L05; Secondary 46B20, 39B82

15.1 Introduction

The purpose of this short chapter is to survey the recent developments in the study of symmetry of Birkhoff-James orthogonality of bounded linear operators between Banach spaces and Hilbert spaces. We also present some new results, along with the corresponding proofs, that have not been published before. Before establishing the relevant notations and terminologies to be used throughout this article, we would like to draw the attention of the reader to the importance of different notions of orthogonality in studying the geometry of Banach spaces. We believe that our

K. Paul (✉)
Department of Mathematics, Jadavpur University, Kolkata, India

D. Sain
Department of Mathematics, Indian Institute of Science, Bangalore, India

P. Ghosh
Department of Mathematics, Sabang Sajanikanta Mahavidyalaya, Paschim Medinipur, West Bengal, India

© Springer Nature Switzerland AG 2019
J. Brzdęk et al. (eds.), *Ulam Type Stability*,
https://doi.org/10.1007/978-3-030-28972-0_15

comments in the next paragraph would serve the purpose of motivating the reader towards further exploration of orthogonality and related topics in order to gain a better understanding of the difficulties arising in the study of geometry of Banach spaces, as opposed to the Hilbert spaces.

The usual notion of orthogonality in an inner product space has no unique standard counterpart in case of Banach spaces. However, there are several notions of orthogonality in a Banach space, each of which generalizes some particular aspect of the usual inner product orthogonality. It is perhaps already clear from our previous comments that the various orthogonality types are not equivalent in general. Indeed, the study of interconnections between the different orthogonality types in the general setting of Banach spaces is an active and fruitful area of research in the study of geometry of Banach spaces. We refer the reader to [1, 3, 18, 20] for a detailed study of different orthogonality types and their interrelations in the setting of Banach spaces. We would like to emphasize that the differences between the orthogonality types in Banach spaces highlight the ideal behaviour of inner product orthogonality. Furthermore, these differences essentially result in a more complicated Banach space geometry in contrast to its Hilbert space counterpart. In essence, in order to describe the geometry of a Banach space, it is often helpful to consider the various notions of orthogonality types and their respective properties.

Letters $\mathbb{X}$, $\mathbb{Y}$ denote Banach spaces. We reserve the symbol $\mathbb{H}$ for a Hilbert space. Let θ denote the zero element of the concerned space. Throughout this article, we assume the underlying field to be $\mathbb{R}$, i.e., we consider only real Banach spaces and real Hilbert spaces. Let $B_{\mathbb{X}} = \{x \in \mathbb{X} : \|x\| \leq 1\}$ and $S_{\mathbb{X}} = \{x \in \mathbb{X} : \|x\| = 1\}$ be the unit ball and the unit sphere of the Banach space $\mathbb{X}$ respectively. Let $B(\mathbb{X}, \mathbb{Y})$, $K(\mathbb{X}, \mathbb{Y})$ respectively denote the Banach space of all bounded linear operators and compact linear operators from $\mathbb{X}$ to $\mathbb{Y}$, endowed with the usual operator norm. We write $B(\mathbb{X}, \mathbb{Y}) = B(\mathbb{X})$ and $K(\mathbb{X}, \mathbb{Y}) = K(\mathbb{X})$ if $\mathbb{X} = \mathbb{Y}$. It is immediate that in case of finite-dimensional spaces $\mathbb{X}$ and $\mathbb{Y}$, we have, $B(\mathbb{X}, \mathbb{Y}) = K(\mathbb{X}, \mathbb{Y})$. Let $\mathbb{X}^*$ denote the dual space of $\mathbb{X}$. Although we would be considering only one orthogonality type, namely Birkhoff-James orthogonality, let us also mention the notion of isosceles orthogonality to put things in perspective.

Definition 15.1 ([2, 16]) For any two elements $x, y \in \mathbb{X}$, x is said to be orthogonal to y in the sense of Birkhoff-James, written as $x \perp_B y$, if $\|x\| \leq \|x + \lambda y\| \, \forall \lambda \in \mathbb{R}$.

Definition 15.2 ([3]) For any two elements $x, y \in \mathbb{X}$, x is said to be isosceles orthogonal to y, written as $x \perp_I y$, if $\|x + y\| = \|x - y\|$.

A quick glance at the definitions of the above mentioned orthogonality types reveals the following properties: Birkhoff-James orthogonality is homogeneous, i.e, $x \perp_B y$ implies $\alpha x \perp_B \beta y$ for every $x, y \in \mathbb{X}$ and for every scalars $\alpha, \beta \in \mathbb{R}$. However, Birkhoff-James orthogonality is not symmetric, i.e., $x \perp_B y$ does not necessarily imply that $y \perp_B x$. On the other hand, isosceles orthogonality is symmetric but not homogeneous. This observation leads us to the question of symmetry of Birkhoff-James orthogonality. This question has been explored by many mathematicians, including Radon [25], James [15] and Day [10]. Indeed, for spaces having dimension at least 3, we have the following useful characterization of inner product spaces:

Theorem 15.1 ([15, Th.1]) *Let* $\mathbb{X}$ *be a Banach space of dimension at least* 3. *Then* $\mathbb{X}$ *is a Hilbert space (i.e., the norm on* $\mathbb{X}$ *comes from an inner product) if and only if Birkhoff-James orthogonality is symmetric in* $\mathbb{X}$.

In view of the above result, it is perhaps natural to ask the following local question:

Question In the setting of a Banach space, what can be said about the symmetry of Birkhoff-James orthogonality *at a particular point of the space.*

In order to address the above question, the following two notions were introduced in [28]:

Definition 15.3 An element $x \in \mathbb{X}$ is said to be left symmetric if $x \perp_B y \Rightarrow y \perp_B x$ for all $y \in \mathbb{X}$.

Definition 15.4 An element $x \in \mathbb{X}$ is said to be right symmetric if $y \perp_B x \Rightarrow x \perp_B y$ for all $y \in \mathbb{X}$.

It is interesting to investigate and characterize the left symmetric and the right symmetric points in a Banach space. However, in this article our focus is on exploring the left symmetric and the right symmetric points in the space of all bounded linear operators between Banach spaces and Hilbert spaces, endowed with the usual operator norm. It is easy to see that the concept of Birkhoff-James orthogonality remains meaningful in the space of bounded (compact) linear operators between Banach spaces, with the usual operator norm. Indeed, for any two elements $T, A \in B(\mathbb{X})$, T is orthogonal to A in the sense of Birkhoff-James, written as $T \perp_B A$, if

$$\|T\| \leq \|T + \lambda A\| \ \forall \lambda \in \mathbb{R}.$$

Since $B(\mathbb{X}, \mathbb{Y})$ is not an inner product space, the question of symmetry of Birkhoff-James orthogonality in $B(\mathbb{X}, \mathbb{Y})$ is a valid one. In case of a general Banach space $\mathbb{X}$, it is easy to see that for $T, A \in B(\mathbb{X})$, $T \perp_B A$ may not imply $A \perp_B T$, or conversely. Indeed, on $(\mathbb{R}^3, \|.\|_2)$, let us consider the following two linear operators with respect to the usual basis of the space: Let $T = \begin{pmatrix} 1 & 0 & 0 \\ 0 & 1/2 & 0 \\ 0 & 0 & 1/2 \end{pmatrix}$ and $A = \begin{pmatrix} 0 & 0 & 0 \\ 0 & 1 & 0 \\ 0 & 0 & 0 \end{pmatrix}$. Then it can be easily verified by applying elementary arguments that $T \perp_B A$ but $A \not\perp_B T$. This evidently illustrates that the left symmetric and the right symmetric linear operators between Banach (Hilbert) spaces are special elements in the corresponding operator space.

The notions of strict convexity and smoothness in a Banach space play important roles in our discussion of symmetric points in operator spaces. Although these concepts are well-known and are almost standard, let us mention the relevant definitions for the sake of completeness.

Definition 15.5 A Banach space $\mathbb{X}$ is said to be strictly convex if every element of the unit sphere is an extreme point of the unit ball. Equivalently, $\mathbb{X}$ is strictly convex if and only if given any two elements $x, y \in \mathbb{X} \setminus \{\theta\}$, with $\|x + y\| = \|x\| + \|y\|$, there exists $\lambda > 0$ such that $y = \lambda x$.

Definition 15.6 Let $\mathbb{X}$ be a Banach space and $x \in \mathbb{X} \setminus \{\theta\}$. $\mathbb{X}$ is said to be smooth at x if there exists a unique norm one functional $f \in S_{\mathbb{X}^*}$ such that $f(x) = \|f\| \|x\| = \|x\|$. Equivalently, $\mathbb{X}$ is smooth at a non-zero point x in $\mathbb{X}$ if and only if there exists a unique supporting hyperplane to $B(\theta, \|x\|) = \{y \in \mathbb{X} : \|y\| \leq \|x\|\}$ at the point x.

In order to study the Birkhoff-James orthogonality of bounded linear operators between Banach spaces, it is often helpful to consider the set of points on the unit sphere of the domain space at which the concerned operators attain their norm. In view of this, the following definition was introduced in [29]:

Definition 15.7 Let $\mathbb{X}, \mathbb{Y}$ be Banach spaces and let $T \in B(\mathbb{X}, \mathbb{Y})$. The norm attainment set of M_T of T is defined as $M_T = \{x \in S_{\mathbb{X}} : \|Tx\| = \|T\|\}$.

The norm attainment set of a bounded linear operator T defined between Hilbert spaces is always a unit sphere of some subspace of the domain space, if $M_T \neq \emptyset$. However, the structure of M_T for operators defined between Banach spaces is yet to be known. The idea of studying the norm attainment set of a bounded linear operator is relatively new and it can effectively applied in exploring many areas of geometry of Banach spaces, including the study of isometries, smooth operators and extreme contraction between Banach spaces. We refer the readers to [24, 33, 34] for more information on this topic. Indeed, it is easy to observe that T is a scalar multiple of an isometry if and only if $M_T = S_{\mathbb{X}}$. On the other hand, it follows from Theorems 4.1 and 4.2 of [24] that if $\mathbb{X}$ is a finite-dimensional smooth Banach space then $T \in B(\mathbb{X})$ is a smooth point if and only if M_T is a doubleton, i.e., T attains norm at only one pair of points. The relation between the norm attainment set M_T and orthogonality of operators in $B(\mathbb{X})$ has been studied by Sain and Paul [30] and Sain et al. [31]. In the context of a finite-dimensional Hilbert space $\mathbb{H}$, Bhatia and Šemrl [5] and Paul et al. [21, 22] independently proved that for every $T, A \in B(\mathbb{H})$, $T \perp_B A$ if and only if there exists $x \in M_T$ such that $Tx \perp_B Ax$. On the other hand, Benítez et al. proved in [4] that a finite-dimensional real Banach space $\mathbb{X}$ is an inner product space if and only if for every $T, A \in B(\mathbb{X})$, $T \perp_B A$ implies that there exists $x \in M_T$ such that $Tx \perp_B Ax$. We would like to remark that both these results give a nice relation between the geometry of the underlying space $\mathbb{X}$ and orthogonality of operators in $B(\mathbb{X})$. We would like to note that this particular connection has been utilized to a great extent in order to study the symmetry of Birkhoff-James orthogonality of bounded linear operators. It should be noted that if $\mathbb{H}$ is a complex

Hilbert space then Theorem 2.5 of [35] gives a complete characterization of right symmetric bounded linear operators in $B(\mathbb{H})$, in terms of isometry and coisometry. However, the arguments used in [35] are no longer applicable in general, if we allow the operators to be defined on a Banach space instead of a Hilbert space. This observation serves as another motivation for studying the symmetry of Birkhoff-James orthogonality of bounded linear operators between Banach spaces.

15.2 Brief Outline of the Article

We begin with an exploration of right symmetric and left symmetric bounded linear operators in $B(\ell_\infty^n)$ and $B(\ell_1^n)$. We next study the same problem, in the context of a Hilbert space, for both finite-dimensional and infinite-dimensional spaces. Our next objective is to study the left symmetry and the right symmetry of bounded linear operators on a two-dimensional strictly convex Banach space. As we will see, for the corresponding result in higher dimensions, we require the additional assumption of smoothness. We also mention the disjointness of the class of smooth operators and the class of right symmetric operators on a finite-dimensional strictly convex and smooth Banach space. We further make note of the fact that it is possible to explore the connection between the rank of a bounded linear operator and the operator being a right symmetric point in the corresponding operator space. Finally, we furnish some new results on symmetry of Birkhoff-James orthogonality in the operator space, along with the detailed proofs, in the setting of infinite-dimensional Banach spaces. Let us mention here that the new results presented here actually improve some of the earlier known results in this context.

15.3 Symmetry of Birkhoff-James Orthogonality in Operator Spaces

We begin with a study of symmetric points in the operator spaces $B(\ell_1^n)$ and $B(\ell_\infty^n)$. In this context, whenever we consider the matrix representation of a bounded linear operator, it is assumed that the matrix representation is with respect to the standard ordered basis of the space $\mathbb{R}^n$. In order to study the symmetric properties of orthogonality of linear operators on $(\mathbb{R}^n, \|.\|_1)$ in the sense of Birkhoff-James, it is useful to make note of the following easy lemma [13].

Lemma 15.1 ([13, Lemma 2.1]) *Extreme points (or, their scalar multiples) of the closed unit ball are the only right symmetric points of ℓ_1^n.*

Using the above lemma, it is possible to completely characterize the right symmetric operators in $B(\ell_1^n)$. Indeed, the following theorem was proved in [13]:

Theorem 15.2 ([13, Th.2.2]) *Let $T = (t_{ij})$ be a bounded linear operator in $B(\ell_1^n)$. Then T is right symmetric if and only if T attains norm at all extreme points and images of the extreme points under T are scalar multiples of extreme points of the unit ball.*

We next state a result [13] that characterizes left symmetric operators in $B(\ell_1^n)$.

Theorem 15.3 ([13, Th.2.3]) *Let $T = (t_{ij})$ be a bounded linear operator in $B(\ell_1^n)$. Then T is left symmetric if and only if T attains norm at only one extreme point, the image of which is a left symmetric point in ℓ_1^n and the images of other extreme points are zero.*

We next state some analogous results from [12], regarding the symmetry of Birkhoff-James orthogonality in $B(\ell_\infty^n)$. The following result characterizes the left symmetric points in $B(\ell_\infty^n)$. We would like to note that the unit ball of ℓ_∞^2 has only two pair of extreme points $\pm(1, 1)$ and $\pm(1, -1)$.

Theorem 15.4 ([12, Th.2.3]) *Let $T = (t_{ij})$ be a linear operator on ℓ_∞^2. Then T is left symmetric if and only if T attains norm at only one extreme point, the image of which is a left symmetric point of ℓ_∞^2 and image of the other extreme point is zero.*

When $n \geq 3$, we have the following theorem [12] regarding left symmetric operators in $B(\ell_\infty^n)$.

Theorem 15.5 ([12, Th.2.5]) *Let T be a linear operator on ℓ_∞^n, where $n \geq 3$. Then T is left symmetric if and only if T is the zero operator.*

The next theorem [12] characterises right symmetric points in $B(\ell_\infty^n)$.

Theorem 15.6 ([12, Th.2.1]) *Let $T = (t_{ij})$ be a bounded linear operator on ℓ_∞^n. Then T is right symmetric if and only if for each $i \in \{1, 2, \ldots, n\}$ exactly one term of $t_{i1}, t_{i2}, \ldots, t_{in}$ is non-zero and all non-zero terms are of the same magnitude.*

Our next objective is to state some results from [11] on symmetric points in $B(\mathbb{H})$, where $\mathbb{H}$ is a finite-dimensional Hilbert space.

Theorem 15.7 ([11, Th.2.7]) *Let $\mathbb{H}$ be a finite-dimensional Hilbert space and let $T \in B(\mathbb{H})$. Then T is right symmetric if and only if $M_T = S_{\mathbb{H}}$, i.e., T is a scalar multiple of an isometry.*

If the dimension of the underlying Hilbert space is infinite then we have the following result for compact linear operators [11]:

Theorem 15.8 ([11, Th.2.8]) *Let $\mathbb{H}$ be an infinite-dimensional Hilbert space and let $T \in K(\mathbb{H})$. Then T is right symmetric if and only if T is the zero operator.*

Remark 15.1 In case of a complex Hilbert space $\mathbb{H}$, symmetry of Birkhoff-James orthogonality in the corresponding operator space was also studied by Turnšek in [35]. However, the technique used in [11], for the finite-dimensional case and for compact operator on an infinite-dimensional Hilbert space are completely different from the one used by Turnšek.

In the next theorem [11] we consider the converse question of characterizing the left symmetric operators in $K(\mathbb{H})$.

Theorem 15.9 ([11, Th.2.10]) *Let $\mathbb{H}$ be a Hilbert space and $T \in K(\mathbb{H})$. Then T is left symmetric if and only if T is the zero operator.*

The following corollary is immediate from the above theorem.

Corollary 15.1 *Let $\mathbb{H}$ be a finite-dimensional Hilbert space and let $T \in B(\mathbb{H})$. Then T is left symmetric if and only if T is the zero operator.*

Remark 15.2 If T is a bounded linear operator in $B(\mathbb{H})$ then it is not necessarily true that $M_T \neq \emptyset$. However, if $T \in B(\mathbb{H})$ attains norm then following the same method as used in the proof of Theorem 2.10 in [11], it can be proved that T is left symmetric if and only if T is the zero operator.

In view of the results on symmetry of Birkhoff-James orthogonality of operators in $B(\ell_1^n)$, $B(\ell_\infty^n)$ and $B(\mathbb{H})$ stated above, it is apparent that Banach spaces and Hilbert spaces behave very differently in this context. Motivated by this observation, it is natural to investigate the same question of symmetry of Birkhoff-James orthogonality in the framework of general Banach spaces. Our next results from [32] characterizes the left symmetric operator(s) defined on a two-dimensional strictly convex Banach space.

Theorem 15.10 ([32, Th.2.1]) *Let $\mathbb{X}$ be a two-dimensional strictly convex Banach space. Then $T \in B(\mathbb{X})$ is left symmetric if and only if T is the zero operator.*

For the corresponding result in case of higher dimensional Banach spaces, the following lemma turns out to be useful. Moreover, we would like to remark that the proof of the next lemma in [32] also gives an alternative approach towards probing the norm attainment set of a bounded linear between Banach spaces, in comparison to the last part of Theorem 2.3 in [29].

Lemma 15.2 ([32, Lemma.2.1]) *Let $\mathbb{X}$ be a Banach space. Let $T \in B(\mathbb{X})$ and $x \in M_T$. If in addition, both x and Tx are smooth points in $\mathbb{X}$ then for any $y \in \mathbb{X}$, we have, $x \perp_B y \Rightarrow Tx \perp_B Ty$.*

When the dimension of $\mathbb{X}$ is strictly greater than 2, the above lemma can be effectively applied to obtain the following theorem [32] regarding left symmetric bounded linear operator(s) in $B(\mathbb{X})$.

Theorem 15.11 ([32, Th.2.2]) *Let $\mathbb{X}$ be an n-dimensional strictly convex and smooth Banach space. $T \in B(\mathbb{X})$ is left symmetric if and only if T is the zero operator.*

The next result [32] gives a rather surprising connection between right symmetric bounded linear operators and the smoothness of the concerned operator, with the additional assumption of the underlying Banach space being finite-dimensional, strictly convex and smooth.

Theorem 15.12 ([32, Th.2.3]) *Let* $\mathbb{X}$ *be a finite-dimensional strictly convex and smooth Banach space. Let* $T \in B(\mathbb{X})$ *be smooth. Then* T *is not right symmetric.*

When $\mathbb{X}$ is not necessarily strictly convex or smooth, we have the following two theorems [32] regarding right symmetric operators.

Theorem 15.13 ([32, Th.2.4]) *Let* $\mathbb{X}$ *be an* $n-$*dimensional Banach space. Let* $x_0 \in S_{\mathbb{X}}$ *be a left symmetric point. Let* $T \in B(\mathbb{X})$ *be such that* $M_T = \{\pm x_0\}$ *and* x_0 *is an eigenvector of* T. *Then either of the following is true:*

(i) rank $T \geq n - 1$.
(ii) T *is not a right symmetric point in* $B(\mathbb{X})$.

Theorem 15.14 ([32, Th.2.5]) *Let* $\mathbb{X}$ *be an* $n-$*dimensional Banach space. Let* $T \in B(\mathbb{X})$ *be such that* $M_T = \{\pm x_0\}$ *and* $\ker T$ *contains a non-zero left symmetric point. Then either of the following is true:*

(i) $I \perp_B T$ *and* $T \perp_B I$, *where* $I \in B(\mathbb{X})$ *is the identity operator on* $\mathbb{X}$.
(ii) T *is not a right symmetric point in* $B(\mathbb{X})$.

Till this point, we have only stated some of the known results regarding the symmetry of Birkhoff-James orthogonality in operator spaces, without discussing the proofs. Now we are going to present some new results, along with their corresponding proofs, in this context. First, we need the following lemma, the proof of which can be found in [29] and [32].

Lemma 15.3 *Let* $\mathbb{X}$ *be a Banach space,* $T \in B(\mathbb{X})$ *and* $x \in M_T$. *Then*

(i) If $Tx \perp_B Ty$ *then* $x \perp_B y$.
(ii) If in addition, both x *and* Tx *are smooth points in* $\mathbb{X}$ *then for any* $y \in \mathbb{X}$, *we have,* $x \perp_B y \Rightarrow Tx \perp_B Ty$.

Using the above lemma, we now obtain a complete characterization of left symmetric operators in $K(\mathbb{X})$, under the additional assumptions that $\mathbb{X}$ is reflexive, strictly convex and smooth. Let us observe that the following theorem improves Theorem 15.11.

Theorem 15.15 *Let* $\mathbb{X}$ *be a reflexive, strictly convex and smooth Banach space and dimension of* $\mathbb{X}$ *is at least two. Then* $T \in K(\mathbb{X})$ *is left symmetric if and only if* T *is the zero operator.*

Proof Since $\mathbb{X}$ is reflexive and T is compact so $M_T \neq \emptyset$ i.e., there exists $x \in S_X$ such that $\|Tx\| = \|T\|$. We complete the proof in the following five steps:

Step 1 We claim that x is right symmetric.

If possible suppose that x is not right symmetric. Then there exists $y \in S_{\mathbb{X}}$ such that $y \perp_B x$, but $x \not\perp_B y$. Since $\mathbb{X}$ is strictly convex so y is an exposed point and there exists a subspace H_y of codimension one such that $y \perp_{SB} H_y$, see [23]. Consider a linear operator $A : \mathbb{X} \longrightarrow \mathbb{X}$ defined as $A(z = ay + h) = aTx$, where a is a scalar and $h \in H_y$. Clearly A is a compact operator. Then $1 = \|z\| = \|ay + h\| > |a|$, if $a \neq 0, h \neq 0$ and $\|Az\| = \|aTx\| \leq |a| \|T\|$. Also $\|Az\| = \|T\|$, if $|a| = 1$. So $M_A = \{\pm y\}$. Since $\mathbb{X}$ is smooth and $y \perp_B x$, we have $x \in H_y$, so $Ax = 0$, from which it follows that $Tx \perp_B Ax$. As $x \in M_T$, $T \perp_B A$. Now $x \not\perp_B y$, and $x \in M_T$, so by Lemma 15.3, we get $Tx \not\perp_B Ty$, i.e., $Ay \not\perp_B Ty$. Since $M_A = \pm\{y\}$, by Theorem 2.1 of Sain [30], it follows that $A \not\perp_B T$, which contradicts that T is left symmetric. Hence x must be right symmetric.

Step 2 We claim that $Ty = 0$ if $y \in S_{\mathbb{X}}$ and $y \perp_B x$.

Since $\mathbb{X}$ is strictly convex so y is an exposed point and there exists a subspace H_y of codimension one such that $y \perp_{SB} H_y$. Consider a linear operator $A : \mathbb{X} \longrightarrow \mathbb{X}$ defined as $A(z = ay + h) = aTy$, where a is a scalar and $h \in H_y$. Clearly A is a compact operator. Then $1 = \|z\| = \|ay + h\| > |a|$, if $a \neq 0, h \neq 0$ and $\|Az\| = \|aTy\| \leq |a| \|T\|$. Also $\|Az\| = \|T\|$, if $|a| = 1$. So $M_A = \{\pm y\}$. Since $\mathbb{X}$ is smooth and $y \perp_B x$, we have $x \in H_y$, so $Ax = 0$, from which it follows that $Tx \perp_B Ax$. As $x \in M_T$, $T \perp_B A$. But $Ay = Ty$ and $M_A = \{\pm y\}$ implies that $A \not\perp_B T$. Thus T is not left symmetric, a contradiction. Hence, $y \perp_B x \Rightarrow Ty = 0$.

Step 3 We claim that x is left symmetric.

If possible suppose there exists $y \in S_{\mathbb{X}}$ such that $x \perp_B y$, but $y \not\perp_B x$. We claim $Ty = 0$. If possible let $Ty \neq 0$. As before since $\mathbb{X}$ is strictly convex so y is an exposed point and there exists a subspace H_y of codimension one such that $y \perp_{SB} H_y$. Define a linear operator A on $\mathbb{X}$ such that $A(z = ay + h) = aTy$, where a is a scalar and $h \in H_y$. Clearly, $M_A = \pm\{y\}$. As $Ay \not\perp_B Ty$, applying Theorem 2.1 of [30], we conclude that $A \not\perp_B T$. Since $\mathbb{X}$ is smooth and $x \perp_B y$, applying Lemma 15.3 we get, $Tx \perp_B Ty$. It is easy to check that $Ax = bTy$ for some scalar b. So $Tx \perp_B Ax$. Since $x \in M_T$, we have, $T \perp_B A$. Thus we have, $T \perp_B A$ but $A \not\perp_B T$, which contradicts our assumption that T is left symmetric. This completes the proof of our claim that $Ty = 0$. Thus we have $x \perp y$, $y \not\perp_B x$ and $Ty = 0$. Now, from Theorem 2.3 of James [16], it follows that there exists a scalar k such that $kx + y \perp_B x$. As T is left symmetric, by Step 2, we get $T(kx + y) = 0$. Since $Ty = 0$ and $Tx \neq 0$, it now follows that $k = 0$. So $y \perp_B x$, a contradiction to our choice of y. Therefore x is left symmetric. Hence x is a symmetric point in $\mathbb{X}$.

Step 4 We show that Tx is left symmetric.

If possible let Tx be not left symmetric. Then there exists $y \in S_{\mathbb{X}}$ such that $Tx \perp_B y$ but $y \not\perp_B Tx$. Since $\mathbb{X}$ is strictly convex so x is an exposed point and there exists a subspace H_x of codimension one such that $x \perp_{SB} H_x$. Consider a linear operator $A : \mathbb{X} \longrightarrow \mathbb{X}$ defined as $A(z = ax + h) = ay$, where a is a scalar and $h \in H_x$. Clearly A is a compact operator. Then $1 = \|z\| = \|ax + h\| > |a|$, if $a \neq 0, h \neq 0$ and $\|Az\| = \|ay\| \leq |a|$. Also $\|Az\| = 1$, if $|a| = 1$. So $M_A = \{\pm x\}$. Since A is compact and $Ax \not\perp_B Tx$ so $A \not\perp_B T$ but $tx \perp_B Ax$ implies that $T \perp A$. This is a contradiction to the fact that T is left symmetric. Hence for each $x \in M_T$, Tx is left symmetric.

Step 5 We construct an operator A such that $A \perp_B T$ but $T \not\perp_B A$.

Consider a unit vector $y \in H$. Then $y \perp_B H_y$ where H_y is a subspace of H of codimension one in H. Let $v \in S_{\mathbb{X}}$ such that $Tx \perp_B v$. Then since T is left symmetric by Step 3, Tx is left symmetric and so $v \perp_B Tx$. Let $\|x + y\| = r$, by orthogonality and strict convexity $1 < r < 2$. Choose $w = (1 - t)Tx + tv$ for some $t \in (0, 1)$ so that $w \in B(v, \epsilon)$ with $0 < \epsilon < \frac{2-r}{1+r}$. This choice is possible, for, $\|w - v\| = (1 - t)\|Tx - v\| \leq (1 - t)(1 + \|T\|)$ and one can consider t so that $(1-t)(1+\|T\|) < \frac{2-r}{1+r}$. Now, any element $z \in \mathbb{X}$ can be written as $z = ax + by + h$ where a, b are scalars and $h \in H_y$. Consider a linear operator $A : \mathbb{X} \longrightarrow \mathbb{X}$ defined as $Az = av + bw$. Then clearly A is compact. Clearly $T \perp_B A$ since $x \in M_T$ and $Tx \perp_B Ax$. We next show that $A \not\perp_B T$. For this we first claim that $z \notin M_A$ if $ab < 0$. Clearly $\|A\left(\frac{x+y}{\|x+y\|}\right)\| = \frac{\|w+v\|}{r} > 1+\epsilon$ and so $\|A\| > 1$. Clearly $x, y \notin M_A$. Let $z = -ax + by + h \in S_{\mathbb{X}}$ where $a > 0, b > 0$. Then by using orthogonality we have, $1 = \|z\| = \| - ax + by + h\| > | a |$. Similarly $| b |< 1$. Now

$$
\begin{aligned}
\|Az\| &= \| - av + bw\| \\
&= \|(b - a)v + b(w - v)\| \\
&< | b - a | +\|w - v\| \\
&< 1 + \epsilon \\
&< \|A(\frac{x + y}{\|x\| + \|y\|})\|
\end{aligned}
$$

This shows that $z \notin M_A$. Similarly considering $z = ax - by + h \in S_{\mathbb{X}}$ where $a > 0, b > 0$ we can show that $z \notin M_A$. So if $z = ax + by + h \in M_A$ then we must have $ab > 0$. Next our claim is that $Tz \notin (Az)^-$ for all $z \in M_A$. Let $z = ax + by + h \in M_A$, then $ab > 0$ and $Az = av + bw$, $Tz = aTx$. Consider $a > 0$. Then $b > 0$. Then $av+bw = av+b(1-t)x+btv = (a+bt)v+b(1-t)Tx \Rightarrow \|av+bw-b(1-t)Tx\| = \|(a+bt)v\| < \|(a+bt)v+b(1-t)Tx\| = \|av+bw\|$, since $v \perp_B Tx$. Thus, $Tx \notin (av + bw)^-$. Hence, by Sain [28, Prop. 2.1], $aTx \notin (av + bw)^- \Rightarrow Tz \notin (Az)^-$. Similarly, $a < 0, b < 0$ implies that $Tz \notin (Az)^-$. Using Theorem 2.1 of [33] we get $T \not\perp_B A$. This shows that T is not left symmetric. This completes the proof of the theorem.

We next show that smooth compact operators on a reflexive, strictly convex, smooth Banach space are not right symmetric. We would further like to note that the following result, which happens to be the final one in this article, improves Theorem 15.12.

Theorem 15.16 *Let T be a compact linear operator on a reflexive, strictly convex, smooth Banach space $\mathbb{X}$. Let us further assume that T is smooth. Then T is not right symmetric.*

Proof If possible let T be smooth and right symmetric. Since T is a compact operator on a reflexive space $\mathbb{X}$ so T must attain its norm. The smoothness of T ensures that $M_T = \{\pm x\}$ for some $x \in S_{\mathbb{X}}$. We claim that x is left symmetric. If not, then there exists $y \in S_{\mathbb{X}}$ such that $x \perp_B y$ but $y \not\perp_B x$. Since $\mathbb{X}$ is strictly convex so y is an exposed point and there exists a subspace H_y of codimension one such that $y \perp_{SB} H_y$. Consider a linear operator $A : \mathbb{X} \longrightarrow \mathbb{X}$ defined as

$A(z = ay + h) = aTx$, where a is a scalar and $h \in H_y$. Clearly A is a compact operator. Then $1 = \|z\| = \|ay + h\| > | a |$, if $a \neq 0, h \neq 0$ and $\|Az\| = \|aTx\| \leq | a | \|T\|$. Also $\|Az\| = \|T\|$, if $| a | = 1$. So $M_A = \{\pm y\}$. Since $x \in M_T$ and $x \perp_B y$ so we get $Tx \perp_B Ty$ i.e., $Ay \perp_B Ty$. So $A \perp_B T$. Again by Lemma 15.3 $Ay \not\perp_B Ax$, since $M_A = \{\pm y\}$ and $y \not\perp x$. Thus $Tx \not\perp_B Ax$. As T is compact and $M_T = \{\pm x\}$ so $T \not\perp_B A$. This shows that T is not right symmetric and so our claim is established. Since $x \in M_T$ so there exists a hyperspace H_x such that $x \perp_B H_x$ and $Tx \perp_B T(H_x)$. Since x is left symmetric so $H_x \perp_B x$. Consider $z_0 = x + h_0$ where $h_0 \in H_x \|h_0\| \neq 0$ and $\|Th_0\| > \|T\|$. As $\mathbb{X}$ is strictly convex so orthogonality is left unique and hence $h_0 \perp_B x$ implies that $z_0 \not\perp_B x$. Let $z = \frac{z_0}{\|z_0\|}$. Considering the element Tz, Th_0 we get a scalar d such that $(dTz + Th_0) \perp_B Tz$. We claim that $d \neq 0$. If $d = 0$ then $Th_0 \perp_B Tz$ and so $Th_0 \perp_B Tx + Th_0$, which is not possible as $\|Th_0 - (Tx + Th_0)\| = \|Tx\| = \|T\| < \|Th_0\|$. Thus $d \neq 0$. As before we get a hyperspace H_z such that $z \perp_{SB} H_z$ and we can define a compact linear operator A on $\mathbb{X}$ such that $A(az + h) = a(dTz + Th_0)$ where a is a scalar and $h \in H_z$. Then $A \perp_B T$. We claim that $T \not\perp_B A$. If not, then by Theorem 2.1 of [24] we get $Tx \perp_B Ax$. Let $x = bz + h$ for some scalar b and $h \in H_z$. Clearly $b \neq 0$, since $z \not\perp_B x$. Now $Ax = b(dTz + Th_0)$, $Tx \perp_B Ax$, $Tx \perp_B Th_0$ and so by smoothness of Tx we get $Tx \perp_B Tz$. Then by Lemma 15.3 we get $x \perp_B z$. Since x is left symmetric so we get $z \perp_B x$, a contradiction. Thus T is not right symmetric. This completes the proof of the theorem.

15.4   Future Directions for Research and Ulam Stability

In this short article we have surveyed some of the known results and have also presented some new results, in the context of symmetry of Birkhoff-James orthogonality in operator spaces. However, there are some important questions in this area that remain unanswered. One of the major unsolved problems in this particular area of study is to characterize right symmetric points in the space of bounded linear operators between general Banach spaces, even with the assumption of smoothness and strict convexity. Theorems 15.12 and 15.16 give a useful necessary condition for a bounded linear operator to be right symmetric, namely, the concerned operator can not be smooth. It will be interesting to obtain a tractable sufficient condition for a bounded linear operator to be right symmetric. On the other hand, regarding left symmetric bounded linear operators, it will be interesting to study the problem without the assumptions of smoothness and strict convexity.

Some other possibilities for further research could be connected with the notion of Ulam stability (for more details we refer to [6–8, 14, 17, 19, 26, 27]). One of more abstract versions of that notion concerns the stability of mathematical theorems (motivated by a more general notion of stability arising naturally in problems of mechanics). It reads as follows (see [36], Page 63, chapter VI *Some Questions in Analysis*, section 1 *Stability*):

When is it true that by changing "a little" the hypotheses of a theorem one can still assert that the thesis of the theorem remains true or "approximately" true?

As J. Brzdęk has suggested to us, we could consider such modifications for instance for Theorem 15.1. Following the notion of hyperstability (see [9]), we may ask: under what conditions on a function $\phi : X^2 \to \mathbb{R}$ we can replace in the theorem the Birkhoff-James orthogonality by the approximate orthogonality $\perp_B^\phi$ understood in the following way (cf., e.g., [9] for somewhat similar ideas).

Definition 15.8 Let $\phi : X^2 \to \mathbb{R}$. For any two elements $x, y \in \mathbb{X}$, x is said to be ϕ-orthogonal to y in the sense of Birkhoff-James, written as $x \perp_B^\phi y$, if

$$\|x\| \le \|x + \lambda y\| + \phi(x, y) \ \forall \lambda \in \mathbb{R}. \tag{15.1}$$

A similar issue also can be considered with the last inequality replaced by the condition

$$\|x\| \le \phi(x, y)\|x + \lambda y\| \ \forall \lambda \in \mathbb{R};$$

or even with the condition

$$\|x\| \le \phi(\|x + \lambda y\|) \ \forall \lambda \in \mathbb{R},$$

but this time with $\phi : \mathbb{R} \to \mathbb{R}$.

We could go one step further and ask about a possibility to replace, in Theorem 15.1, the symmetry by a ϕ-symmetry understood as follows: if $x, y \in X$ and $x \perp_B y$, then $x \perp_B^\phi y$.

Similar problems could be raised in connection with many other results presented in this paper.

Acknowledgements The authors gratefully acknowledge the contribution of Prof. J. Brzdęk, specially in connection with the interrelation between Ulam stability and our research on symmetry of Birkhoff-James orthogonality which might open up the possibility of further research in this direction.

References

1. Alonso, J., Martini, H., Wu, S.: On Birkhoff orthogonality and isosceles orthogonality in normed linear spaces. Aequationes Math. **83**, 153–189 (2012)
2. Birkhoff, G.: Orthogonality in linear metric spaces. Duke Math. J. **1**, 169–172 (1935)
3. Benítez, C.: Orthogonality in normed linear spaces: a classification of the different concepts and some open problems. Rev. Mat. Univ. Complut. Madrid **2**, 53–57 (1989)
4. Benítez, C., Fernández, M., Soriano, M.L.: Orthogonality of matrices. Linear Algebra Appl. **422**, 155–163 (2007)
5. Bhatia, R., Šemrl, P.: Orthogonality of matrices and distance problem. Linear Algebra Appl. **287**, 77–85 (1999)

6. Brzdęk, J., Ciepliński, K.: Hyperstability and superstability. Abstr. Appl. Anal. **2013**, 13 (2013). Article ID 401756
7. Brzdęk, J., Ciepliński, K., Leśniak, Z.: On Ulam's type stability of the linear equation and related issues. Discret. Dyn. Nat. Soc. **2014**, 14 (2014). Art. ID 536791
8. Brzdęk, J., Popa, D., Raşa, I., Xu, B.: Ulam Stability of Operators. Elsevier, Oxford (2018)
9. Chmieliński, J.: Orthogonality preserving property and its Ulam stability. In: Rassias, Th.M., Brzdęk, J. (eds.) Functional Equations in Mathematical Analysis, pp. 33–58. Springer Optimization and Its Applications, vol. 52. Springer, Berlin (2011)
10. Day, M.M.: Some characterization of inner product spaces. Trans. Am. Math. Soc. **62**, 320–337 (1947)
11. Ghosh, P., Sain, D., Paul, K.: Orthogonality of bounded linear operators. Linear Algebra Appl. **500**, 43–51 (2016)
12. Ghosh, P., Sain, D., Paul, K.: On symmetry of Birkhoff-James orthogonality of linear operators. Adv. Oper. Theory **2**, 428–434 (2017)
13. Ghosh, P., Sain, D., Paul, K.: Symmetric properties of orthogonality of linear operators on $(\mathbb{R}^n, \|.\|_1)$. Novi Sad J. Math. **47**, 41–46 (2017)
14. Hyers, D.H., Isac, G., Rassias, Th.M.: Stability of Functional Equations in Several Variables. Progress in Nonlinear Differential Equations and Their Applications, vol. 34. Birkhäuser, Boston (1998)
15. James, R.C.: Inner product in normed linear spaces. Bull. Am. Math. Soc. **53**, 559–566 (1947)
16. James, R.C.: Orthogonality and linear functionals in normed linear spaces. Trans. Am. Math. Soc. **61**, 265–292 (1947); **69**, 90–104 (1958)
17. Jung, S.-M.: Hyers-Ulam-Rassias Stability of Functional Equations in Nonlinear Analysis. Springer, New York (2011)
18. Kapoor, O.P., Prasad, J.: Orthogonality and characterizations of inner product spaces. Bull. Aust. Math. Soc. **19**, 403–416 (1978)
19. Moszner, Z.: Stability has many names. Aequationes Math. **90**, 983–999 (2016)
20. Partington, J.R.: Orthogonality in normed spaces. Bull. Aust. Math. Soc. **33**, 449–455 (1986)
21. Paul, K.: Translatable radii of an operator in the direction of another operator. Sci. Math. **2**, 119–122 (1999)
22. Paul, K., Hossein, Sk.M., Das, K.C.: Orthogonality on $B(H, H)$ and minimal-norm operator. J. Anal. Appl. **6**, 169–178 (2008)
23. Paul, K., Sain, D., Jha, K.: On strong orthogonality and strictly convex normed linear spaces. J. Inequal. Appl. **2013:242**, 7 (2013)
24. Paul, K., Sain, D., Ghosh, P.: Birkhoff-James orthogonality and smoothness of bounded linear operators. Linear Algebra Appl. **506**, 551–563 (2016)
25. Radon, J.: *Über eine Besondere Art Ebener Konvexer Kurven*. Leipziger Berichre. Math. Phys. Klasse. **68**, 23–28 (1916)
26. Rassias, Th.M.: On the stability of the linear mapping in Banach spaces. Proc. Am. Math. Soc. **72**, 297–300 (1978)
27. Rassias, Th.M.: On the stability of functional equations and a problem of Ulam. Acta Appl. Math. **62**, 23–130 (2000)
28. Sain, D.: Birkhoff-James orthogonality of linear operators on finite dimensional Banach spaces. J. Math. Anal. Appl. **447**, 860–866 (2017)
29. Sain, D.: On the norm attainment set of a bounded linear operator. J. Math. Anal. Appl. **457**, 67–76 (2018)
30. Sain, D., Paul, K.: Operator norm attainment and inner product spaces. Linear Algebra Appl. **439**, 2448–2452 (2013)
31. Sain, D., Paul, K., Hait, S.: Operator norm attainment and Birkhoff-James orthogonality. Linear Algebra Appl. **476**, 85–97 (2015)
32. Sain, D., Ghosh, P., Paul, K.: On symmetry of Birkhoff-James orthogonality of linear operators on finite-dimensional real Banach spaces. Oper. Matrices **11**, 1087–1095 (2017)

33. Sain, D., Paul, K., Mal, A.: A complete characterization of Birkhoff-James orthogonality in infinite dimensional normed space. J. Oper. Theory **80**(2), 399–413 (2018)
34. Sain, D., Ray, A., Paul, K.: Extreme contractions on finite-dimensional polygonal Banach spaces. J. Convex Anal., **26**(3), 877–885 (2019)
35. Turnšek, A.: On operators preserving James' orthogonality. Linear Algebra Appl. **407**, 189–195 (2005)
36. Ulam, S.M.: Problems in Modern Mathematics. Dover Phoenix Editions, New York (2004)

Chapter 16
Ulam Stability of Zero Point Equations

Adrian Petruşel and Ioan A. Rus

Abstract In this paper, we will study different kind of Ulam stability concepts for the zero point equation. Our approach is based on weakly Picard operator theory related to fixed point and coincidence point equations.

Keywords Coincidence point equation · Zero point equation · Root equation · Fixed point equation · Functional equation · Difference equation · Differential equation · Integral equation · Convergent iterative algorithm · Retraction on the solution set · Retraction-displacement condition · Ulam inequation corresponding to an equation · Ulam stability · Ulam-Hyers stability · Ulam-Hyers-Rassias stability

Mathematics Subject Classification (2000) Primary 47H10, 54H25, 34D10, 36B20, 45M10, 39A30, 39B12, 65J15

16.1 Introduction

There is a large number of papers on Ulam stability for:

- functional equations [5, 8, 10, 21, 22, 28, 32–37, 64–66, 85, 86];
- difference equations [16, 17, 46, 62, 69, 72, 83, 90];
- differential equations [1, 3, 4, 26, 38–41, 47, 54, 63, 73–76, 88, 89, 91];

A. Petruşel (✉)
Babeş-Bolyai University, Cluj-Napoca, Romania

Academy of Romanian Scientists, Bucharest, Romania
e-mail: petrusel@math.ubbcluj.ro

I. A. Rus
Babeş-Bolyai University, Cluj-Napoca, Romania
e-mail: iarus@math.ubbcluj.ro

© Springer Nature Switzerland AG 2019
J. Brzdęk et al. (eds.), *Ulam Type Stability*,
https://doi.org/10.1007/978-3-030-28972-0_16

- integral equations [4, 28, 71, 72, 75, 76];
- fixed point equations [12, 13, 23, 25, 44, 56, 59, 68, 72, 75, 76, 78];
- coincidence point equations [27, 59, 60, 72, 75, 76].

In this paper we will study different kind of Ulam stability concepts for zero point equations. Our approach is based on weakly Picard operator theory related to fixed point and coincidence point equations.

The structure of our paper is as follows:

1. Introduction
2. Preliminaries

   (a) Retraction-displacement condition in the metric fixed point theory
   (b) Operatorial equations and corresponding Ulam inequations
   (c) Basic notions in Ulam stability

3. Ulam stability of the zero point equations
4. Abstract models of Ulam stability. Examples
5. Some research directions.

16.2 Preliminaries

Let X be a nonempty set and $T : X \to X$ be an operator. Any solution of the fixed point equation

$$x = T(x), \ x \in X, \tag{16.1}$$

is called a fixed point of T.

We denote by F_T the fixed point set of T, i.e., $F_T := \{x \in X \mid T(x) = x\}$ and by $Graph(T) := \{(x, T(x)) \in X \times X \mid x \in X\}$ the graphic of T.

If (X, d) is a metric space, then, by definition, $T : X \to X$ is said to be a weakly Picard operator if

$$T^n(x) \to x^*(x) \in F_T \text{ as } n \to \infty, \text{ for all } x \in X. \tag{16.2}$$

If T is a weakly Picard operator, then we can define the following set retraction $T^\infty : X \to F_T$ given by $T^\infty(x) = \lim_{n \to \infty} T^n(x)$.

A weakly Picard operator $T : X \to X$ for which there exists a function $\psi : \mathbb{R}_+ \to \mathbb{R}_+$ increasing, continuous in 0 and satisfying $\psi(0) = 0$, such that

$$d(x, T^\infty(x)) \le \psi(d(x, T(x))), \text{ for all } x \in X, \tag{16.3}$$

is called a ψ-weakly Picard operator. In particular, if $\psi(t) = ct$ (for some $c > 0$), then T is said to be a c-weakly Picard operator.

By definition, a weakly Picard operator T with a unique fixed point is called a Picard operator. In this case, we write $F_T = \{x^*\}$. As a consequence $T^\infty(x) = x^*$, for every $x \in X$.

Moreover, a Picard operator T for which the condition (16.3) is satisfied is called a ψ-Picard operator.

For more considerations on weakly Picard operator theory see [11, 12, 57, 77, 79].

Definition 16.1 Let (X, d) be a metric space and $T : X \to X$ be an operator. Then, T is called:

(i) an l-contraction if $l \in {]}0, 1{[}$ and

$$d(T(x), T(y)) \le l d(x, y), \text{ for every } x, y \in X.$$

(ii) a graphic l-contraction if $l \in {]}0, 1{[}$ and

$$d(T(x), T^2(x)) \le l d(x, T(x)), \text{ for every } x \in X.$$

Any contraction is a graphic contraction but not reversely.

Example 16.1 Let $X := [0, 1] \cup [2, 3]$ and $T : X \to X$ be defined by

$$T(x) := \begin{cases} \frac{1}{2}x, & x \in [0, 1] \\ \frac{1}{2}x + \frac{3}{2}, & x \in [2, 3]. \end{cases}$$

Then T is a continuous graphic $\frac{1}{2}$-contraction, but it isn' t a contraction.

The following result was proved in [77]. It is also known as Graphic Contraction Principle.

Theorem 16.1 ([77]) *Let (X, d) be a complete metric space and $T : X \to X$ be a graphic l-contraction. Then:*

(1) the sequence $(T^n(x))_{n \in \mathbb{N}}$ converges in (X, d) and $\sum_{n \in \mathbb{N}} d(T^n(x), T^{n+1}(x)) <$

∞*, for each $x \in X$;*
If, in addition to our initial assumptions, we suppose that, for each $x \subset X$, we have

$$\lim_{n \to \infty} T(T^n(x)) = T(\lim_{n \to \infty} T^n(x)),$$

then we have the following conclusions:
(2) $F_T = F_{T^n} \ne \emptyset$, for all $n \in \mathbb{N}^*$;
(3) T is a weakly Picard operator;
(4) $d(x, T^\infty(x)) \le \frac{1}{1-l} d(x, T(x))$, for every $x \in X$, i.e., T is a ψ-weakly Picard operator with $\psi(t) = \frac{t}{1-l}$, for $t \in \mathbb{R}_+$.

Remark 16.1 There are many relevant examples of graphic contractions. For example, generalized contractions such as Kannan mappings, Ćirić-Reich-Rus mappings, Chatterjea mappings, Zamfirescu mappings, Hardy-Rogers mappings, Berinde mappings, Suzuki mappings are graphic contractions. For other examples and related considerations see [57].

16.2.1 Retraction-Displacement Condition in the Metric Fixed Point Theory

We start our considerations with some examples.

Example 16.2 Let (X, d) be a complete metric space and $T : X \to X$ be a graphic l-contraction with closed graphic. Then, by the Graphic Contraction Principle, we have that $F_T \neq \emptyset$, and $T^n(x) \to x^*(x) \in F_T$ as $n \to \infty$, for all $x \in X$. The operator $T^\infty : X \to F_T$ defined by $x \mapsto x^*(x)$, is a retraction of X onto the fixed point set of T, F_T. Moreover,

$$d(x, T^\infty(x)) \le \frac{1}{1-l} d(x, T(x)), \ \forall \, x \in X.$$

Remark 16.2 In the conditions of the above example, let $Y \subset X$ be such that $F_T \subset Y$. Then, for each $y \in Y$, there exists a fixed point $x \in F_T$ such that

$$d(y, x) \le \frac{1}{1-l} d(y, T(y)).$$

Indeed, for a given $y \in Y$, the corresponding element with the above property is $x := T^\infty(y) \in F_T$.

Remark 16.3 If (X, d) is a complete metric space and $T : X \to X$ is an l-contraction, then $F_T = \{x^*\}$, $T^\infty(x) = x^*$ for all $x \in X$ and

$$d(x, x^*) \le \frac{1}{1-l} d(x, T(x)), \ \text{for every } x \in X.$$

Example 16.3 Let (X, d) be a generalized complete metric space (with a generalized metric $d : X \times X \to \mathbb{R}_+ \cup \{+\infty\}$). Let $X = \bigcup_{i \in I} X_i$ be the canonical decomposition of X into metric spaces, with respect to the generalized metric d (see [58, 79]).

Let $T : X \to X$ be an l-contraction and denote

$$E_T := \{x \in X \mid d(x, T(x)) < +\infty\}.$$

We suppose that $E_T \neq \emptyset$. Let $J := \{i \in I \mid X_i \cap E_T \neq \emptyset\}$.

Then $T(X_i) \subset X_i$, for all $i \in J$ and $E_T = \bigcup_{i \in J} X_i$. It is clear that $T(E_T) \subset E_T$ and $F_T = F_T \cap E_T$. Each pair $(X_i, d), i \in I$ is a complete metric space and $T\big|_{X_i} : X_i \to X_i$ is an l-contraction.

By the above example we immediately obtain the following result.

Theorem 16.2 *Let (X, d) be a generalized complete metric space (with a generalized metric $d : X \times X \to \mathbb{R}_+ \cup \{+\infty\}$), $T : X \to X$ be an l-contraction such that $E_T \neq \emptyset$. Then:*

(i) $F_T \neq \emptyset$, and $F_T \subset E_T$;

(ii) $T^n(x) \to T^\infty(x) \in F_T$ as $n \to \infty$, for all $x \in E_T$, i.e., $T\big|_{E_T} : E_T \to E_T$ is a weakly Picard operator;

(iii) $T^\infty : E_T \to F_T$ is a retraction and

$$d(x, T^\infty(x)) \le \frac{1}{1-l} d(x, T(x)), \text{ for all } x \in E_T,$$

i.e., the operator $T\big|_{E_T} : E_T \to E_T$ satisfies a retraction-displacement condition (see[12, 77, 79]).

Remark 16.4 In the conditions of the above theorem, let $Y \subset E_T$ be such that $F_T \subset Y$. Then, for each $y \in Y$, there exists $x \in F_T$ such that

$$d(y, x) \le \frac{1}{1-l} d(y, T(y)). \tag{16.4}$$

We notice again that $x := T^\infty(y)$ (i.e., the value on y of the retraction T^∞) is an element x with the property (16.4).

Remark 16.5 For an adequate understanding of the above theorem, it is very useful to compare it with other Contraction Principles in generalized metric spaces, see [21, 22, 58, 64, 79].

Recall that $\varphi : \mathbb{R}_+ \to \mathbb{R}_+$ is said to be a comparison function (see [11, 70]) if it is increasing and $\varphi^k(t) \to 0$ as $k \to +\infty$, for each $t > 0$ As a consequence, we also have $\varphi(t) < t$, for each $t > 0$, $\varphi(0) = 0$ and φ is continuous in 0. Moreover, $\varphi : \mathbb{R}_+ \to \mathbb{R}_+$ is said to be a strict comparison function (see [70]) if it is a comparison function and $\lim_{t \to \infty} (t - \varphi(t)) = \infty$. For example, $\varphi(t) = \frac{t}{t+1}, t \in \mathbb{R}_+$ is a strict comparison function.

Example 16.4 Let (X, d) be a generalized complete metric space (with a generalized metric $d : X \times X \to \mathbb{R}_+ \cup \{+\infty\}$) and $T : X \to X$ be a graphic φ-contraction (where the mapping $\varphi : \mathbb{R}_+ \to \mathbb{R}_+$ is a strict comparison function), such that $F_T \neq \emptyset$. Denote

$$E_T := \{x \in X \mid d(x, T(x)) < +\infty\}.$$

Obviously, $E_T \neq \emptyset$. If we define

$$\psi_\varphi(t) := \sup\{s \in \mathbb{R}_+ | s - \varphi(s) \leq t\},$$

then $T : E_T \to E_T$ is a ψ_φ-WPO and $T^\infty : E_T \to F_T$ satisfies the relation

$$d(x, T^\infty(x)) \leq \psi_\varphi(d(x, T(x))), \quad \text{for all } x \in E_T.$$

For other examples in which a retraction r from the space to the fixed point set of an operator T appear as well as some estimates of $d(x, r(x))$ in terms of $d(x, T(x))$ see: [12, 77, 79].

The above considerations give rise to the following notions.

Definition 16.2 Let (X, d) be a metric space and $T : X \to X$ be an operator with $F_T \neq \emptyset$. Let $r : X \to F_T$ be a retraction and $\psi : \mathbb{R}_+ \to \mathbb{R}_+$. By definition the operator T satisfies the (ψ, r) retraction-displacement condition if:

(i) ψ is increasing, continuous at 0 and $\psi(0) = 0$;
(ii) $d(x, r(x)) \leq \psi(d(x, T(x)))$, for all $x \in X$.

Definition 16.3 Let (X, d) be a generalized complete metric space (with a generalized metric $d : X \times X \to \mathbb{R}_+ \cup \{+\infty\}$), $T : X \to X$ an operator and $E_T := \{x \in X \mid d(x, T(x)) < +\infty\}$. We suppose that $F_T \neq \emptyset$ and $T(E_T) \subset E_T$. Let $r : E_T \to F_T$ a retraction and $\psi : \mathbb{R}_+ \to \mathbb{R}_+$. By definition the operator T satisfies the (ψ, r) retraction-displacement condition if:

(i) ψ is increasing, continuous at 0 and $\psi(0) = 0$;
(ii) $d(x, r(x)) \leq \psi(d(x, T(x))$, for all $x \in E_T$.

16.2.2 Operatorial Equations and Corresponding Ulam Inequations

The Case of Coincidence Equations Let (X, d) and (Y, ρ) be two metric spaces, and $T, R : X \to Y$ two operators. We consider the coincidence point equation

$$T(x) = R(x) \tag{16.5}$$

with the solution set

$$C(T, R) := \{x \in X \mid T(x) = R(x)\} \neq \emptyset,$$

and for each $\varepsilon > 0$ the inequations

$$\rho(T(x), R(x)) \leq \varepsilon$$

with the solution set S_ε.

The Ulam Problem (See [76]) Is the Following In which conditions there exist a retraction $r_\varepsilon : S_\varepsilon \to C(T, R)$ and a function $\theta : \mathbb{R}_+ \to \mathbb{R}_+$ such that

$$d(x, r_\varepsilon(x)) \le \theta(\varepsilon), \text{ for all } x \in S_\varepsilon, \text{ and for every } \varepsilon > 0.$$

If, in the above problem, there exists a function $\theta : \mathbb{R}_+ \to \mathbb{R}_+$ such that $\theta(0) = 0$ and is continuous in 0, then the coincidence point equation is said to be Ulam stable.

Consider in the above coincidence equation that $(Y, \rho) = (X, d)$ and $R = 1_X$. Thus, we have:

The Case of Fixed Point Equations Let (X, d) be a metric space, $T : X \to X$ an operator. We consider the fixed point equation

$$x = T(x)$$

with the solution set $F_T := C(T, 1_X) \ne \emptyset$, and for each $\varepsilon > 0$ the inequations

$$d(x, T(x)) \le \varepsilon,$$

with the solution set S_ε.

The Ulam problem consists in finding sufficient conditions under which there exist a retraction $r_\varepsilon : S_\varepsilon \to F_T$ and a function $\theta : \mathbb{R}_+ \to \mathbb{R}_+$ such that

$$d(x, r_\varepsilon(x)) \le \theta(\varepsilon), \text{ for all } x \in S_\varepsilon \text{ and for every } \varepsilon > 0.$$

If such a mapping θ exists such that $\theta(0) = 0$ and θ is continuous in 0, then we call the fixed point equation to be Ulam stable.

If, in the coincidence point equation (16.5), we consider that $(Y, \|\cdot\|)$ is a normed space (with O the zero (null) element of the space) and $R : X \to Y$ is defined by $R(x) = O$, then we have:

The Case of Zero Point Equation Let (X, d) be a metric space, $(Y, \|\cdot\|)$ a normed space (with O its zero (null) element) and $T : X \to Y$ an operator. We consider the zero point equation

$$T(x) = O$$

with the solution set, $Z_T \ne \emptyset$, and for each $\varepsilon > 0$ the inequations

$$\|T(x)\| \le \varepsilon,$$

with the solution set, S_ε.

The **Ulam problem** consists in finding conditions under which there exist a retraction $r_\varepsilon : S_\varepsilon \to F_T$ and a function $\theta : \mathbb{R}_+ \to \mathbb{R}_+$ such that

$$\|x - r_\varepsilon(x)\| \leq \theta(\varepsilon), \text{ for all } x \in S_\varepsilon \text{ and for every } \varepsilon > 0.$$

If a such θ exists such that $\theta(0) = 0$ and θ is continuous in 0, then the zero point equation is called Ulam stable.

Remark 16.6 If, in the above considerations, we work with generalized metric spaces, then (see [75]):

(1) if $d(x, y) \in \mathbb{R}_+^m$, then $\varepsilon = (\varepsilon_1, \ldots, \varepsilon_m)$, with $\varepsilon_i > 0, i = \overline{1, m}$;
(2) if $d(x, y) \in s(\mathbb{R}_+)$, then $\varepsilon = (\varepsilon_1, \ldots, \varepsilon_n, \ldots)$ with $\varepsilon_i > 0, i \in \mathbb{N}^*$;
(3) if $d(x, y) \in C([a, b], \mathbb{R}_+)$, then $\varepsilon \in C([a, b], \mathbb{R}_+^*)$;
(4) if $d(x, y)$ is an element in a cone K of an ordered Banach space, then ε is an element in the interior of K, i.e. $\varepsilon \in \overset{\circ}{K}$;
(5) if $d(x, y) \in K$, then $\theta : K \to K$.

16.2.3 Basic Notions in Ulam Stability

Since fixed point equations and zero point equations are particular cases of coincidence point equations we consider in what follow the coincidence point equations. The problem is what conditions should we ask for the function (operator !) θ which appear in Ulam stability. See for example, [16, 17, 32, 35, 36, 38–41, 66, 68, 72, 75, 76, 85, 86]. In what follow we consider the point of view presented in [72, 75] and [76].

Let (X, d) and (Y, ρ) be two metric spaces, and $T, R : X \to Y$ two operators. We consider the coincidence point equation

$$T(x) = R(x) \tag{16.6}$$

with the solution set, $C(T, R) \neq \emptyset$, and for each $\varepsilon > 0$ the inequation

$$\rho(T(x), R(x)) \leq \varepsilon$$

with the solution set S_ε.

Definition 16.4 By definition, Eq. (16.6) is Ulam-Hyers stable if there exist a retraction $r_\varepsilon : S_\varepsilon \to C(T, R)$ and $c > 0$ such that

$$d(x, r_\varepsilon(x)) \leq c\varepsilon, \ \forall \, x \in S_\varepsilon, \ \forall \, \varepsilon > 0.$$

In other words, Eq. (16.6) is Ulam-Hyers stable if it is Ulam stable with a function $\theta(t) = ct$, for all $t \in \mathbb{R}_+$.

Definition 16.5 By definition, Eq. (16.6) is generalized Ulam-Hyers stable if there exist a retraction $r_\varepsilon : S_\varepsilon \to C(T, R)$ and an increasing function $\theta : \mathbb{R}_+ \to \mathbb{R}_+$, continuous in 0 with $\theta(0) = 0$ such that

$$d(x, r_\varepsilon(x)) \leq \theta(\varepsilon), \text{ for all } x \in S_\varepsilon \text{ and for every } \varepsilon > 0.$$

So, Eq. (16.6) is generalized Ulam-Hyers stable if it is Ulam stable and if, in addition, the function θ is increasing.

Let us consider now the set $\mathbb{M}(X, Y)$ of operators from a nonempty set X to a metric space (Y, ρ). If $T, R : \mathbb{M}(X, Y) \to \mathbb{M}(X, Y)$ are two given operators, then the coincidence problem for T and R is the following

$$T(f) = R(f). \tag{16.7}$$

There are two possibilities to define Ulam stability.

The first one is to endow $\mathbb{M}(X, Y)$ with a (generalized or not) metric and to study Ulam stability with respect to this metric. Thus, in this case, we have the following three notions: Ulam stability, Ulam-Hyers stability and generalized Ulam-Hyers stability.

The second possibility is to consider Ulam stability in terms of the metric ρ on Y. In this case, the Ulam inequations are the following

$$\rho(T(f)(x), R(f)(x)) \leq \varepsilon, \text{ for each } x \in X \text{ (where } \varepsilon > 0).$$

Denote by S_ε the solution set for these inequations, for each $\varepsilon > 0$.

In this case, we have the following concepts.

Definition 16.6 By definition, Eq. (16.7) is Ulam stable with respect to the metric ρ of Y, if there exist, for each $\varepsilon > 0$, a retraction $r_\varepsilon : S_\varepsilon \to C(T, R)$ and a function $\theta : \mathbb{R}_+ \to \mathbb{R}_+$, continuous in 0 with $\theta(0) = 0$ such that

$$\rho(f(x), r_\varepsilon(f)(x)) \leq \theta(\varepsilon), \text{ for all } f \in S_\varepsilon, \ x \in X \text{ and } \varepsilon > 0.$$

Definition 16.7 By definition, Eq. (16.7) is Ulam-Hyers stable with respect to the metric ρ of Y, if there exist $c > 0$ and, for each $\varepsilon > 0$, a retraction $r_\varepsilon : S_\varepsilon \to C(T, R)$ such that

$$\rho(f(x), r_\varepsilon(f)(x)) \leq c\varepsilon, \text{ for all } f \in S_\varepsilon, \ x \in X \text{ and for } \varepsilon > 0.$$

Definition 16.8 By definition, Eq. (16.7) is generalized Ulam-Hyers stable with respect to the metric ρ of Y, if there exist, for each $\varepsilon > 0$, a retraction $r_\varepsilon : S_\varepsilon \to C(T, R)$ and an increasing function $\theta : \mathbb{R}_+ \to \mathbb{R}_+$, continuous in 0 with $\theta(0) = 0$ such that

$$\rho(f(x), r_\varepsilon(f)(x)) \leq \theta(\varepsilon), \text{ for all } f \in S_\varepsilon, \ x \in X \text{ and for } \varepsilon > 0.$$

The papers of Aoki [5] and Rassias [65] have generated other types of Ulam stability on the function spaces changing the standard Ulam inequations (see [21, 22, 28, 35, 36, 66, 72, 75, 76]).

Let $\chi : X \to \mathbb{R}_+$ be a given functional. Related to Eq. (16.7), we consider the following inequations

$$\rho(T(f)(x), R(f)(x)) \leq \varepsilon \chi(x), \text{ for each } x \in X; \text{ (where } \varepsilon > 0).$$

We denote the solution set of these inequations by $S_{\varepsilon,\chi}$.

In this context, we have the following concepts.

Definition 16.9 By definition, Eq. (16.7) is Ulam-Hyers-Aoki-Rassias stable with respect to the metric ρ of Y and to the function $\chi : X \to \mathbb{R}_+$, if there exist $c > 0$ and, for each $\varepsilon > 0$, a retraction $r_\varepsilon : S_{\varepsilon,\chi} \to C(T, R)$ such that

$$\rho(f(x), r_\varepsilon(f)(x)) \leq c\varepsilon \chi(x), \text{ for all } f \in S_{\varepsilon,\chi}, \ x \in X \text{ and for } \varepsilon > 0.$$

Definition 16.10 By definition, Eq. (16.7) is generalized Ulam-Hyers-Aoki-Rassias stable with respect to the metric ρ of Y and to the function $\chi : X \to \mathbb{R}_+$, if there exist $c > 0$, an increasing function $\theta : \mathbb{R}_+ \to \mathbb{R}_+$, continuous in 0 with $\theta(0) = 0$ and, for each $\varepsilon > 0$, a retraction $r_\varepsilon : S_{\varepsilon,\chi} \to C(T, R)$ such that

$$\rho(f(x), r_\varepsilon(f)(x)) \leq \theta(\varepsilon)\chi(x), \text{ for all } f \in S_{\varepsilon,\chi}, \ x \in X \text{ and for } \varepsilon > 0.$$

Remark 16.7 For a better understanding of the above definitions, see the concepts given in [16, 26, 32, 33, 36, 38, 39, 41, 54, 62, 63, 68, 72–76, 85, 88, 89] and the results presented in [17, 21, 22, 28, 35, 40, 86, 90].

16.3   Ulam Stability of Zero Point Equations

Let (X, d) be a metric space, $(Y, +, \rho)$ be a metric abelian group and $T : X \to Y$ be an operator. We denote by O the zero (null) element of $(Y, +)$. We consider the zero point equation

$$T(x) = O. \tag{16.8}$$

We denote by Z_T the solution set of (16.8) and, for each $\varepsilon > 0$, by S_ε the solution set of the the corresponding Ulam inequations

$$\rho(T(x), O) \leq \varepsilon. \tag{16.9}$$

Following the approaches given in Sect. 16.2.3 of this paper, the basic notions in Ulam stability theory for (16.8) are given as follows.

Definition 16.11 By definition, the Ulam problem for the zero point equation (16.8) is the following:

In which conditions, for $\varepsilon > 0$, there exist a retraction $r_\varepsilon : S_\varepsilon \to Z_T$ and a function $\theta : \mathbb{R}_+ \to \mathbb{R}_+$, such that

$$\rho(x, r_\varepsilon(x)) \leq \theta(\varepsilon), \text{ for all } x \in S_\varepsilon, \text{ and for } \varepsilon > 0 \,?$$

We have the following concepts.

Definition 16.12 By definition, the zero point equation (16.8) is Ulam stable if, for $\varepsilon > 0$, there exist a retraction $r_\varepsilon : S_\varepsilon \to Z_T$ and a function $\theta : \mathbb{R}_+ \to \mathbb{R}_+$ continuous in 0 with $\theta(0) = 0$ such that

$$\rho(x, r_\varepsilon(x)) \leq \theta(\varepsilon), \text{ for all } x \in S_\varepsilon, \text{ and for } \varepsilon > 0.$$

Definition 16.13 By definition, the zero point equation (16.8) is Ulam-Hyers stable if, for $\varepsilon > 0$, there exist a retraction $r_\varepsilon : S_\varepsilon \to Z_T$ and $c > 0$ such that

$$\rho(x, r_\varepsilon(x)) \leq c\varepsilon, \text{ for all } x \in S_\varepsilon, \text{ and for } \varepsilon > 0.$$

Definition 16.14 By definition, the zero point equation (16.8) is generalized Ulam-Hyers stable if, for $\varepsilon > 0$, there exist a retraction $r_\varepsilon : S_\varepsilon \to Z_T$ and an increasing function $\theta : \mathbb{R}_+ \to \mathbb{R}_+$ continuous in 0 with $\theta(0) = 0$ such that

$$\rho(x, r_\varepsilon(x)) \leq \theta(\varepsilon), \text{ for all } x \in S_\varepsilon, \text{ and for } \varepsilon > 0.$$

Let us consider now the set $\mathbb{M}(X, Y)$ of operators from a nonempty set X to a metric abelian group $(Y, +, \rho)$ (with O its zero (null) element). Let $T : \mathbb{M}(X, Y) \to \mathbb{M}(X, Y)$ be an operator and consider the zero point equation for T

$$T(f) = o, \tag{16.10}$$

where $o : X \to Y$ is given by $o(x) = O$.

For this problem we have two possibilities to define Ulam stability. The first possibility is to endow $\mathbb{M}(X, Y)$ with a (generalized or not) metric and to study the Ulam stability of the zero point equation (16.10) with respect to this metric. As a consequence, in this case, we have the following notions: Ulam problem, Ulam stability, Ulam-Hyers stability and generalized Ulam-Hyers stability.

The second possibility is to consider Ulam stability of the zero point equation (16.10) in terms of the metric ρ on Y. In this case, the Ulam inequations are the following

$$\rho(T(f)(x), O) \leq \varepsilon, \text{ for each } x \in X \,(\text{where } \varepsilon > 0). \tag{16.11}$$

As a consequence, in this case, we have, with respect to the metric ρ, the following stabilities: Ulam stability, Ulam-Hyers stability and generalized Ulam-Hyers stability (see Definitions 16.6–16.8).

Let $\chi : X \to \mathbb{R}_+$ be a given functional. Related to Eq. (16.11), we consider the following inequations

$$\rho(T(f)(x), O) \leq \varepsilon \chi(x), \text{ for each } x \in X; \text{ (where } \varepsilon > 0). \tag{16.12}$$

We denote the solution set of these inequations by $S_{\varepsilon,\chi}$.

Definition 16.15 By definition, Eq. (16.10) is Ulam-Hyers-Aoki-Rassias stable with respect to the metric ρ of Y and to the function $\chi : X \to \mathbb{R}_+$, if there exist $c > 0$ and, for each $\varepsilon > 0$, a retraction $r_\varepsilon : S_{\varepsilon,\chi} \to Z_T$ such that

$$\rho(f(x), r_\varepsilon(f)(x)) \leq c\varepsilon \chi(x), \text{ for all } f \in S_{\varepsilon,\chi}, \ x \in X \text{ and for } \varepsilon > 0.$$

Remark 16.8 There are various zero point (null point, roots) theorems in the literature. One of the most interesting zero point theorem was proved by Miranda [49]. Actually, the result was presented, without proof, by H. Poincaré in 1883 (see [61]) and then rediscovered by Miranda in 1940. Some extensions of Miranda-Poincaré-Bolzano theorem, including the case of infinite dimensional spaces, were given in [6, 7, 30, 45, 48, 50, 55, 87].

16.4   Abstract Models of Ulam Stability: Examples

The first abstract model of Ulam stability for the zero point equation is the following.

I. Let $(X, \| \cdot \|)$ be a Banach space and $T : X \to X$ be an operator. We consider the zero point equation

$$T(x) = O \tag{16.13}$$

and, for each $\varepsilon > 0$, the corresponding Ulam inequation

$$\|T(x)\| \leq \varepsilon. \tag{16.14}$$

We denote by S_ε the solution set of (16.14).

Let us suppose that there exists an operator $P : X \to X$ such that:

(i) $Z_T = F_P$, i.e., Eq. (16.13) is equivalent to the fixed point equation

$$x = P(x); \tag{16.15}$$

(ii) P is a c-weakly Picard operator, i.e., $c > 0$ and

$$\|x - P^\infty(x)\| \le c\|x - P(x)\|, \text{ for each } x \in X;$$

(iii) there exists $k > 0$ such that

$$\|x - P(x)\| \le k\|T(x)\|, \text{ for each } x \in X.$$

Under the above assumptions the zero point equation (16.13) is Ulam-Hyers stable.

Indeed, by (i), we notice that P^∞ is a retraction of X onto Z_T. By (ii) and (iii), if $x \in S_\varepsilon$, then $P^\infty(x) \in Z_T$ and

$$\|x - P^\infty(x)\| \le c\|x - P(x)\| \le ck\|T(x)\| \le ck\varepsilon.$$

Example 16.5 Let $(X, \langle\rangle)$ be a real Hilbert space and $T : X \to X$ be an operator. We suppose:

(1) there exists $m > 0$ such that

$$\langle T(x) - T(y), x - y \rangle \ge m\|x - y\|^2, \text{ for each } x, y \in X,$$

i.e., T is strictly monotone;

(2) T is L-Lipschitz, i.e., $L > 0$ and

$$\|T(x) - T(y)\| \le L\|x - y\|, \text{ for every } x, y \in X.$$

It is well-known (see, for example, [19, 20, 80, 84]) that, in the above conditions, there exists $\gamma > 0$ such that the operator $P : X \to X$ defined by

$$P(x) = x - \gamma T(x)$$

is an l-contraction, with l depending on m, L and γ.

Since the assumptions (i)–(iii) of our abstract model are satisfied (with $c := \frac{1}{1-l}$ and $k := \gamma$), we have the following Ulam stability theorem.

Theorem 16.3 *Let $(X, \|\cdot\|)$ be a Banach space and $T : X \to X$ be an operator. If the above assumptions (1) and (2) are satisfied, then the zero point equation (16.13) is Ulam-Hyers stable.*

Remark 16.9 Similar examples and results can be given in the case of global Newton-Kantorovich algorithms [11, 20, 24, 51, 52, 82].

II. Our second abstract model for Ulam-Hyers stability of zero point equation is defined in the following setting.

Let $(X, \|\cdot\|_X)$ be a normed space and $(Y, \|\cdot\|_Y)$ be a Banach space. Let $\mathbb{M}(X, Y)$ be the set of all mappings from X to Y. We consider on $\mathbb{M}(X, Y)$ the complete generalized metric d given by

$$d(f, g) := \sup_{x \in X} \| f(x) - g(x) \|_Y .$$

Let $T : \mathbb{M}(X, Y) \to \mathbb{M}(X^m, Y)$ be an operator, where $m \in \mathbb{N}, m \geq 1$. The problem we consider now is to study the zero point equation

$$T(x) = o, \tag{16.16}$$

where $o : X \to Y$ is given by $o(x) = O$. In this respect, we consider the operator $T_\Delta : \mathbb{M}(X, Y) \to \mathbb{M}(X, Y)$ defined by

$$f \mapsto T_\Delta(f), \text{ where } T_\Delta(f)(x) := Tf(x, \cdots, x).$$

Let us suppose that $Z_T = Z_{T_\Delta}$ and there exists an operator $P : \mathbb{M}(X, Y) \to \mathbb{M}(X, Y)$ having the following properties:

(a) $F_P = Z_{T_\Delta}$;
(b) $P(E_P) \subset E_P$, where $E_P := \{f \in \mathbb{M}(X, Y) | d(f, P(f)) < \infty\}$;
(c) there exists $k > 0$ such that, for each $\varepsilon > 0$, the following implication holds

$$\|T_\Delta(f)(x)\|_Y \leq \varepsilon, \text{ for every } x \in X \Rightarrow \|f(x) - P(f)(x)\|_Y \leq k\varepsilon, \text{ for every } x \in X;$$

(d) the operator $P : E_P \to E_P$ is c-weakly Picard.

Then, the zero point equation (16.16) is Ulam-Hyers stable.
Indeed, let us consider, for each $\varepsilon > 0$, the Ulam inequation

$$\|T(f)(x_1, \cdots, x_m)\|_Y \leq \varepsilon, \text{ for every } x_1, \cdots, x_m \in X. \tag{16.17}$$

We denote by S_ε the solution set of (16.17). Let $f \in S_\varepsilon$. Then, we get that $f \in E_P$ and $P^\infty(f) \in Z_T$. Moreover, we have

$$d(f, P^\infty(f)) \leq cd(f, P(f)) \leq ck\varepsilon, \text{ for every } f \in S_\varepsilon.$$

We will illustrate the above abstract model by the following example.

Example 16.6 Let $(X, \|\cdot\|_X)$ be a normed space and $(Y, \|\cdot\|_Y)$ be a Banach space. We consider the following Cauchy type functional equation

$$f(x_1 + x_2) - f(x_1) - f(x_2) = O, \text{ for every } x_1, x_2 \in X, \tag{16.18}$$

where $f : X \to Y$.

If we define the operator $T : \mathbb{M}(X, Y) \to \mathbb{M}(X^2, Y)$ by

$$T(f)(x_1, x_2) := f(x_1 + x_2) - f(x_1) - f(x_2),$$

then the Cauchy functional equation (16.18) is equivalent to the zero point equation

$$T(f) = o, \tag{16.19}$$

where $o : X \times X \to Y$ is given by $o(x_1, x_2) = O$.

In this case, T_Δ has the following form

$$T_\Delta(f)(x) = f(2x) - 2f(x).$$

Now, we consider the operator $P : \mathbb{M}(X, Y) \to \mathbb{M}(X, Y)$ defined by

$$P(f)(x) := \frac{1}{2}f(2x), \ x \in X.$$

It is obvious that $P : E_P \to E_P$ is a 2-weakly Picard operator (by Theorem 16.2.) and P satisfies the above assumptions (a) and (b), with $c = 2$ and $k = \frac{1}{2}$. Thus, if $f \in S_\varepsilon$, then $f \in E_P$ and $P^\infty(f) \in Z_T$. Moreover, we have that

$$d(f, P^\infty(f)) \le \varepsilon.$$

Thus, we have proved the well-known Hyers stability theorem for the Cauchy functional equation (16.18).

Remark 16.10 In a similar manner we can study Ulam-Hyers stability for other functional equations, see [2, 29, 35, 36]. Moreover, we can consider a suitable metric on $\mathbb{M}(X, Y)$ for the study of Ulam-Hyers-Aoki-Rassias stability.

16.5 Some Research Directions

We will present now some research directions related to the concepts and the results of this paper.

16.5.1 The Root Equation

Let $(X, \|\cdot\|)$ be a Banach space and $T : X \to X$ be an operator. Let y be an element of Y. We consider the root equation

$$T(x) = y. \tag{16.20}$$

The problem is to use the first abstract model for the zero point equation

$$T(x) - y = O, \qquad (16.21)$$

in order to study the Ulam-Hyers stability of the root equation (16.20).

References: [14, 15, 19, 42, 51, 53, 67, 78, 81].

16.5.2 Volterra Type Equation

Let $K \in C([a, b] \times [a, b] \times \mathbb{R})$. We consider the following Volterra integral equation

$$\int_a^t K(t, s, x(s))ds = 0, \text{ for } t \in [a, b]. \qquad (16.22)$$

By a solution of (16.22) we understand a continuous function $x : [a, b] \to \mathbb{R}$ which satisfies (16.22), for every $t \in [a, b]$.

The problem is to utilize the technique proposed in Sect. 16.4 for the study of the Ulam-Hyers stability of the Volterra equation (16.22).

16.5.3 Kuratowski's Equation

Let (X, d) be a metric spaces and $\alpha_K : P_b(X) \to \mathbb{R}_+$ be Kuratowski's measure of non-compactness, where $P_b(X)$ denotes the family of all nonempty and bounded subsets of X. Let us consider the zero point equation

$$\alpha_K(A) = 0. \qquad (16.23)$$

The problem is to study the Ulam-Hyers stability of Eq. (16.23).

References: [9, 24, 70, 79].

16.5.4 More General Spaces

Another research direction is related to the study of Ulam-Hyers stability for the zero point equation in more general settings: generalized metric spaces (in various senses), quasi-metric spaces, pseudo-metric spaces, b-metric spaces, gauge spaces and other distance spaces.

References: [18, 31, 43, 75, 79].

References

1. Abbas, S., Benchohra, M., Petruşel, A.: Ulam stability for Hilfer type fractional differential inclusions via the weakly Picard operator theory. Fract. Calc. Applied Anal. **20**, 384–398 (2017)
2. Aczél, J.: Lectures on Functional Equations and Their Applications. Academic, New York (1966)
3. András, Sz., Kolumban, J.J.: On the Ulam-Hyers stability of first order differential systems with nonlocal initial conditions. Nonlinear Anal. **82**, 1–11 (2013)
4. András, Sz., Mészáros, A.R.: Ulam-Hyers stability of dynamic equation on time scales via Picard operators. Appl. Math. Comput. **219**, 4853–4864 (2013)
5. Aoki, T.: On the stability of the linear transformation in Banach spaces. J. Math. Soc. Japan **2**, 64–66 (1950)
6. Avramescu, C.: A generalization of Mirands's theorem. In: Seminar on Fixed Point Theory Cluj-Napoca, vol. 3, pp. 121–128 (2002)
7. Azagra, D., Gómez, J., Jaramillo J.A.: Rolle's theorem and negligibility of points in infinite dimensional Banach spaces. J. Math. Anal. Appl. **213**, 487–495 (1997)
8. Baker, J.A.: The stability of certain functional equations. Proc. Am. Math. Soc. **112**, 729–732 (1991)
9. Ban, A.I., Gal, S.G.: Defects of Properties in Mathematics. World Scientific, Singapore (2002)
10. Belitskii, G., Tkachenko, V.: One-Dimensional Functional Equations. Birkhäuser, Basel (2003)
11. Berinde, V.: Iterative Approximations of Fixed Points. Springer, Berlin (2007)
12. Berinde, V., Petruşel, A., Rus, I.A, Şerban, M.A.: The retraction-displacement condition in the theory of fixed point equation with a convergent iterative algorithm. In: Rassias, Th.M., Gupta, V. (eds.) Mathematical Analysis, Approximation Theory and Their Applications, pp. 75–106. Springer Berlin (2016)
13. Bota-Boriceanu, M., Petruşel, A.: Ulam-Hyers stability for operatorial equations. Analele Univ. Al.I. Cuza Iaşi **57**, 65–74 (2011)
14. Brown R.F., Hales, A.W.: Primitive roots of unity in H-manifolds. Am. J. Math. **92**, 612–618 (1970)
15. Brown R.F., Jiang, B., Schirmer, H.: Roots of iterates of maps. Topol. Appl. **66**, 129–157 (1995)
16. Brzdęk, J., Popa, D., Xu, B.: Note on nonstability of the linear recurrence. Abh. Math. Sem. Univ. Hamburg **76**, 183–189 (2006)
17. Brzdęk, J., Popa, D., Xu, B.: The Hyers-Ulam stability of nonlinear recurrences. J. Math. Anal. Appl. **335**, 443–449 (2007)
18. Brzdęk, J., Karapınar, E., Petruşel, A.: A fixed point theorem and the Ulam stability in generalized dq-metric spaces. J. Math. Anal. Appl. **467**, 501–520 (2018)
19. Buică, A.: Coincidence Principles ands Applications. Presa Universitară Clujeană, Cluj-Napoca (2001, in Romanian)
20. Buică, A., Rus, I.A., Şerban, M.A.: Zero point principle of ball-near identity operators and applications to implicit operator problem. Fixed Point Theory. **21**(1), (2020), to appear
21. Cădariu, L., Radu, V.: A general fixed point method for the stability of Jensen functional equation. Bul. Şt. Univ. Politehnica din Timişoara **51**, 63–72 (2006)
22. Cădariu, L., Radu, V.: The alternative of fixed point and stability results for functional equations. Int. J. Appl. Math. Stat. **7**, 40–58 (2007)
23. Chidume, C.E., Măruşter, Şt.: Iterative methods for the computation of fixed points of demicontractive mappings. J. Comput. Appl. Math. **234**, 861–882 (2010)
24. Deimling, K.: Nonlinear Funct. Anal. Springer, Berlin (1995)
25. Dontchev, A.L., Rockafellar, R.T.: Implicit Functions and Solution Mappings. Springer Mathematics Monographs. Springer, Dordrecht (2009)
26. Egri, E.: Ulam stabilities of a first order iterative functional-differential equation. Fixed Point Theory **12**, 321–328 (2011)

27. Fierro, R., Martinez, C., Morales, C.H.: The aftermath of the intermediate value theorem. Fixed Point Theory Appl. **2004**(3), 243–250 (2004)
28. Găvruţă, P., Găvruţă, L.: A new method for the generalized Hyers-Ulam-Rassias stability. Int. J. Nonlinear Anal. Appl. **1**, 11–18 (2010)
29. Ghermănescu, M.: Functional Equations. Ed. Acad. R.P.R., Bucureşti (1960, in Romanian)
30. Gilbert, W.J.: Generalizations of Newton's method. Fractals **9**, 251–262 (2001)
31. Granas, A., Dugundji, J.: Fixed Point Theory. Springer, Berlin (2003)
32. Gruber, P.M.: Stability of isometries. Trans. Am. Math. Soc. **245**, 263–277 (1978)
33. Hyers, D.H.: The stability of homomorphism and related topics. In: Rassias Th.M. (ed.) Global Analysis – Analysis on Manifolds, pp. 140–153. Teubner, Leipzig (1983)
34. Hyers, D.H., Ulam, S.M.: On approximate isometries. Bull. Am. Math. Soc. **51**, 288–292 (1945)
35. Hyers, D.H., Isac, G., Rassias, Th.M.: Stability of Functional Equations in Several Variables. Birkhäuser, Basel (1998)
36. Jung, S.-M.: Hyers-Ulam-Rassias Stability of Functional Equations in Mathematical Analysis. Hadronic Press, Florida (2001)
37. Jung, S.-M.: A fixed point to the stability of isometries. J. Math. Anal. Appl. **329**, 879–890 (2007)
38. Jung, S.-M.: Hyers-Ulam stability of linear partial differential equations of first order. Appl. Math. Lett. **22**, 70–74 (2009)
39. Jung, S.-M., Lee, K.-S.: Hyers-Ulam stability of first order linear partial differential equations with constant coefficients. Math. Ineq. Appl. **10**, 261–266 (2007)
40. Jung, S.-M., Rassias, Th. M.: Generalized Hyers-Ulam stability of Riccati differential equation. Math. Ineq. Appl. **11**, 777–782 (2008)
41. Jung, S.-M., Rezaei, H.: A fixed point approach to the stability of linear differential equations. Bull. Malays. Math. Sci. Soc. **38**, 855–865 (2015)
42. Kantorovich, L.V., Akilov, G.P.: Analyse fonctionelle. MIR, Moscou (1981)
43. Kirk, W., Shahzad, N.: Fixed Point Theory in Distance Spaces. Springer, Cham (2014)
44. Krantz, S.G., Parks, H.R.: The Implicit Function Theorem: History, Theory, and Applications. Birkhäuser, Basel (2013)
45. Kulpa, W.: The Poincaré-Miranda theorem. Am. Math. Mon. **104**, 545–550 (1997)
46. La Salle, J.P.: The Stability of Dynamical Systems. SIAM, Philadelphia (1976)
47. Lungu, N., Rus, I.A.: Ulam stability of nonlinear hyperbolic partial differential equations. Carpath. J. Math. **24**, 403–408 (2008)
48. Mawhin, J.: Variations on Poincaré-Miranda's theorem. Adv. Nonlinear Stud. **13**, 209–217 (2013)
49. Miranda, C.: Un'osservazione su un teorema di Brouwer. Boll. Un. Mat. Ital. **3**, 5–7 (1940)
50. Morales, C.H.: A Bolzano's theorem in the new millenium. Nonlinear Anal. **51**, 679–691 (2002)
51. Nirenberg, L.: Variational and topological methods in nonlinear problems. Bull. Am. Math. Soc. **4**, 267–302 (1981)
52. Ortega, J.M.: Numerical Analysis. Academic, New York (1972)
53. Ortega, J.M., Rheionboldt, W.C.: On a class of approximate iterative processes. Arch. Ration. Mech. Anal. **23**, 352–365 (1967)
54. Otrocol, D., Ilea V.: Ulam stability for a delay differential equation. Cen. Eur. J. Math. **11**, 1296–1303 (2013)
55. Pavel, N.H.: Zeros of Bouligand-Nagumo fields. Libertas Math. **9**, 13–36 (1989)
56. Petruşel, A., Petruşel, G.: A study of a general system of operator equations in b-metric spaces via the vector approach in fixed point theory. J. Fixed Point Theory Appl. **19**, 1793–1814 (2017)
57. Petruşel, A., Rus, I.A.: Graphic contraction principle and applications. Mathematical Analysis and Apoplications (Th. M. Rassias, P. Pardalos - Eds.), Springer Berlin (2019), to appear
58. Petruşel, A., Rus, I.A., Şerban, M.A.: Fixed point for operators on generalized metric spaces. CUBO A Math. J. **10**, 45–66 (2008)

59. Petruşel, A., Petruşel, G., Yao, J.-C.: Fixed point and coincidence point theorems in b-metric spaces with applications. Appl. Anal. Discret. Math. **11**, 199–215 (2017)
60. Petruşel, A., Petruşel, G., Yao, J.-C.: Contributions to the coupled coincidence point problem in b-metric spaces with applications. Filomat **31**, 3173–3180 (2017)
61. Poincaré, H., Sur certaines solutions paiticulieres du probleme des trois corps, C.R. Acad. Sci. Paris **91**, 251–252 (1883)
62. Popa, D.: Hyers-Ulam stability of the linear recurrence with constant coefficients. Adv. Differ. Equ. **2**, 101–107 (2005)
63. Popa, D., Raşa, I.: Hyers-Ulam stability of the Laplace operator. Fixed Point Theory **19**, 379–382 (2018)
64. Radu, V.: The fixed point alternative and the stability of functional equations. Fixed Point Theory **4**, 91–96 (2003)
65. Rassias, Th.M.: On the stability of the linear mapping in Banach spaces. Proc. Am. Math. Soc. **72**, 297–300 (1978)
66. Rassias, Th.M., Brzdek J. (eds.): Functional Equations in Mathematical Analysis. Springer, Berlin (2012)
67. Reem, D.: The open mapping theorem and the fundamental theorem of algebra. Fixed Point Theory **9**, 249–266 (2008)
68. Reich, S., Zaslawski A.J.: A stability result in fixed point theory. Fixed Point Theory **6**, 113–118 (2005)
69. Rus, I.A.: An abstract point of view in the nonlinear difference equations. In: Itinerant Seminar on Functional Equations, Approximation and Convexity, pp. 272–276. Ed Carpatica, Cluj-Napoca (1999)
70. Rus, I.A.: Fixed Point Structure Theory. Cluj University Press, Cluj-Napoca (2006)
71. Rus, I.A.: Gronwall lemma approach to the Hyers-Ulam-Rassias stability of an integral equation. In: Pardalos, P., Rassias, Th.M., Khan, A.A. (eds.) Nonlinear Analysis and Variational Problems, pp. 147–152. Springer, New York (2009)
72. Rus, I.A.: Remarks on Ulam stability of the operatorial equations. Fixed Point Theory **10**, 305–320 (2009)
73. Rus, I.A.: Ulam stability of ordinary differential equations. Studia Univ. Babeş-Bolyai Math. **54**, 125–133 (2009)
74. Rus, I.A.: Ulam stability of ordinary differential equations in a Banach space. Carpath. J. Math. **26**, 103–107 (2010)
75. Rus, I.A.: Ulam stability of operatorial equations. In: Rassias, Th.M,, Brzdek, J. (eds.) Functional Equations in Mathematical Analysis, pp. 287–305. Springer, New York (2012)
76. Rus, I.A.: Results and problems in Ulam stability of operatorial equations and inclusions. In: Rassias, Th.M. (ed.) Handbook of Functional Equations: Stability Theory, pp. 323–352. Springer, New York (2014)
77. Rus, I.A.: Relevant classes of weakly Picard operators. An. Univ. Vest Timişoara, Mat.-Inf. **54**, 3–19 (2016)
78. Rus, I.A., Aldea, F.: Fixed points, zeros and surjectivity Studia Univ. Babeş Bolyai Math. **45**, 109–116 (2000)
79. Rus, I.A., Petruşel, A., Petruşel, G.: Fixed Point Theory. Cluj University Press, Cluj-Napoca (2008)
80. Sburlan, S.: Monotone semilinear equations in Hilbert spaces and applications. Creat. Math. Inf. **17**, 32–37 (2008)
81. Seda, V.: Surjectivity of an operator. Czechoslov. Math. J. **40**, 46–63 (1990)
82. Smale, S.: The fundamental theorem of algebra and complexity theory. Bull. Am. Math. Soc. **4**, 1–36 (1981)
83. Száz, A.: Generalizations of an asymptotic stability theorem of Bahyrycz, Páles and Piszczek on Cauchy differences to generalized cocycles. Studia Univ. Babeş-Bolyai Math. **63**, 109–124 (2018)
84. Tarsia, A.: Differential equations and implicit functions: a generalization of the near operators theorem. Topol. Math. Nonlinear Anal. **11**, 115–133 (1998)

85. Trif, T.: On the stability of a general gamma-type functional equation. Publ. Math. **60**, 47–61 (2002)
86. Trif, T.: Hyers-Ulam-Rassias stability of a linear functional equation with constant coefficients. Nonlinear Funct. Anal. Appl. **11**, 881–889 (2006)
87. Vrahatis, M.N.: A short proof and a generalization of Miranda's existence theorem. Proc. Am. Math. Soc. **107**, 701–703 (1989)
88. Wang, J., Lv, L., Zhou, Y.: New concepts and results in stability of fractional differential equations. Commun. Nonlinear Sci. Numer. Simul. **17**, 2530–2538 (2012)
89. Wang, J., Zhou, Y., Fečkan, M.: Nonlinear impulsive problems for fractional differential equations and Ulam stability. Comput. Math. Appl. **64**, 3389–3405 (2012)
90. Xu, M.: Hyers-Ulam-Rassias stability of a system of first order linear recurences. Bull. Korean Math. Soc. **44**, 841–849 (2007)
91. Zhou, D.Y.: Basic Theory of Fractional Differential Equations. World Scientific Publishing, Singapore (2014)

Chapter 17
Cauchy Difference Operator in Some Orlicz Spaces

Stanisław Siudut

Abstract Let $(G, \cdot, \lambda)$ be a measurable group with a complete, left-invariant and finite measure λ. If φ is a convex φ-function satisfying conditions $\varphi(u)/u \to 0$ as $u \to 0$, $\varphi(u)/u \to \infty$ as $u \to \infty$, $f : G \to \mathbb{R}$ and the Cauchy difference $\mathscr{C}f(x, y) = f(x \cdot y) - f(x) - f(y)$ of f belongs to $\mathscr{L}^{\varphi}_{\lambda \times \lambda}(G \times G, \mathbb{R})$, then there exists unique additive $A : G \to \mathbb{R}$ such that $f - A \in \mathscr{L}^{\varphi}_{\lambda}(G, \mathbb{R})$. Moreover, $\|f - A\|_{\varphi} \leq K\|\mathscr{C}f\|_{\varphi}$, where $K = 1$ if $\lambda(G) \geq 1$, $K = 1 + (\lambda(G))^{-1}$ if $\lambda(G) < 1$. Similar result we also obtain without associativity of $\cdot$ but with $f \in \mathscr{L}^{\varphi}_{\lambda}(G, \mathbb{R})$ and with measurability of $\mathscr{C}f$. In this case $A = 0$ and the Cauchy difference $\mathscr{C} : L^{\varphi}(G, \mathbb{R}) \to L^{\varphi}(G \times G, \mathbb{R})$ is linear continuous and continuously invertible on its image, where $L^{\varphi}(G, \mathbb{R})$ denotes the space of equivalence classes of functions in $\mathscr{L}^{\varphi}(G, \mathbb{R})$. Moreover, $\mathscr{C}$ is compact iff $L^{\varphi}(G, \mathbb{R})$ has a finite dimension.

Let $(G, \cdot, \lambda)$ be a measurable group with a complete, left-invariant and σ-finite measure λ such that $\lambda(G) = \infty$. If φ is a φ-function, $f : G \to \mathbb{R}$ and $\mathscr{C}f \in \mathscr{L}^{\varphi}_{\lambda \times \lambda}(G \times G, \mathbb{R})$, then there exists a unique additive $A : G \to \mathbb{R}$, which is equal to f λ a.e.

Keywords Cauchy difference operator · Orlicz spaces · Ulam stability

Mathematics Subject Classification (2010) Primary 39B82; Secondary 46E30

17.1 Introduction

Let $(G, \cdot)$ be a semigroup. If $f : G \to \mathbb{R}$, then we call the expression

$$\mathscr{C}f(x, y) = f(x \cdot y) - f(x) - f(y)$$

S. Siudut (✉)
Institute of Mathematics, Pedagogical University, Kraków, Poland
e-mail: stanislaw.siudut@up.krakow.pl

© Springer Nature Switzerland AG 2019
J. Brzdęk et al. (eds.), *Ulam Type Stability*,
https://doi.org/10.1007/978-3-030-28972-0_17

the Cauchy difference of f on $G \times G$. The Cauchy equation

$$\mathscr{C} f(x, y) = 0$$

is Ulam stable when any function approximately satisfying it is near to a solution of this equation (such a solution will be called an additive function). Various definitions and examples of such stability are presented, e.g., in [2, 3, 6], [8, Chapter XVII], [11, 13–17].

Let λ be a complete measure defined on a σ-algebra Σ of subsets of G. By $\lambda \times \lambda$ we understand the completion of the product of the measure λ times itself. In this note φ-function is called a function $\varphi : [0, \infty) \rightarrow [0, \infty)$, continuous, non-decreasing and such that $\varphi(t) = 0$ iff $t = 0$ and

$$\lim_{t \to \infty} \varphi(t) = \infty.$$

Let φ be a φ-function. We consider two examples of Orlicz classes $\mathscr{L}_{\lambda}^{\varphi}(G, \mathbb{R})$ and $\mathscr{L}_{\lambda \times \lambda}^{\varphi}(G \times G, \mathbb{R})$ with Luxemburg's F-norms (quasi-norms in another terminology), both F-norms are denoted by $\| \|$. For the notions of qusi-norm and quasi-normed spaces see [19].

Then the stability question can be stated as follows: *Does there exist a $K \in \mathbb{R}_+$ such that, for every $f : G \to \mathbb{R}$ with*

$$\mathscr{C} f \in \mathscr{L}_{\lambda \times \lambda}^{\varphi}(G \times G, \mathbb{R}),$$

there exists an additive function $A : G \to \mathbb{R}$ such that

$$\| f - A \| \le K \| \mathscr{C} f \|?$$

Here, we only introduce some basic notations and prove some introductory lemmas. For the definition and further properties of Orlicz spaces we refer the reader to [10, 12] and [7].

Let (X, Σ, ν) be a measure space. Let φ be a φ-function. By $\mathscr{L}_{\nu}^{\varphi}(X, \mathbb{R})$ ($\mathscr{L}_{\nu}^{\varphi}$ for short) we denote the set of all φ-integrable functions on this measure space, i.e.,

$$\mathscr{L}_{\nu}^{\varphi} = \mathscr{L}_{\nu}^{\varphi}(X, \mathbb{R})$$

$$= \{ f \in \mathscr{M}_{\Sigma} \mid \int_{X} \varphi(k|f(t)|) d\nu(t) < \infty \text{ for some } k > 0 \},$$

where

$$\mathscr{M}_{\Sigma} := \{ f : X \to \mathbb{R} \mid f \text{ is } \Sigma\text{-measurable} \}.$$

We shall consider two elements of $\mathcal{L}_v^\varphi$ as equivalent if they are equal Σ-a.e. The space L_v^φ of equivalence classes of elements $f \in \mathcal{L}_v^\varphi$ becomes a Fréchet space with the Luxemburg's F-norm defined by

$$\|f\| = \inf\{u > 0 \mid \int_X \varphi(\frac{1}{u}|f(t)|)dv(t) \le u\}.$$

If φ is a convex φ-function then L_v^φ becomes a Banach space with the norm

$$\|f\| = \inf\{u > 0 \mid \int_X \varphi(\frac{1}{u}|f(t)|)dv(t) \le 1\}.$$

If we assume additionally that the measure v is σ-finite and

$$(0_1) \ \frac{\varphi(u)}{u} \to 0 \text{ as } u \to 0, \qquad (\infty_1) \ \frac{\varphi(u)}{u} \to \infty \text{ as } u \to \infty,$$

then one can define Orlicz's norm in L_v^φ by

$$\|f\|^O = \sup\{\int_X f(x)z(x)dv(x) \mid z \in \mathcal{M}_\Sigma \text{ and } \int_X \varphi^*(|z(x)|)dv(x) \le 1\},$$

where φ^* is the complementary (in the Young sense) function for φ, that is

$$\varphi^*(v) = \sup_{u \ge 0}(uv - \varphi(u))$$

for $v \ge 0$. The inequalities

$$\|f\| \le \|f\|^O \le 2\|f\|$$

hold and

$$\|f\|^O = \inf_{t>0}\{t^{-1} + t^{-1}\int_X \varphi(t|f(x)|)dv(x)\}.$$

Moreover, the following two Hölder inequalities

$$|\int_X f(t)w(t)dv(t)| \le \|f\|^O \|w\|_{L_v^{\varphi^*}},$$

$$|\int_X f(t)w(t)dv(t)| \le \|f\|\|w\|_{L_v^{\varphi^*}}^O$$

are valid for functions $f \in L_v^\varphi$, $w \in L_v^{\varphi^*}$.

Example 17.1 The functions

$$\varphi(u) = (u+1)ln(1+u) - u,$$

$$\varphi^*(v) = e^v - v - 1$$

are mutually complementary and satisfy (0_1) and (∞_1).

Let v be a nonatomic measure. Since

$$\frac{\varphi(u)}{u^p} \to 0 \text{ as } u \to \infty$$

for $p > 1$, from [10, Theorems 3.3 and 3.4] we obtain that $L^p \neq L_v^\varphi$ and the norms $\| \|_p$ and $\| \|$ are not equivalent for all $1 \leq p \leq \infty$.

If the mappings f and g are equal almost everywhere with respect to the measure v we write

$$f \overset{v}{=} g .$$

Let G be a nonempty set endowed with the binary operation written multiplicatively. Let (G, Σ, λ) be a complete measure space, where λ is not identically 0. Given such a measure space (G, Σ, λ) and $A \in \Sigma$, $y \in G$ we define

$$A_y = \{t \in G \mid ty \in A\}, \qquad _yA = \{t \in G \mid yt \in A\}.$$

The following two conditions will be assumed further on

(r) $A_y \in \Sigma$ and $\lambda(A_y) = \lambda(A)$ for all $A \in \Sigma$, $y \in G$,
(l) $_yA \in \Sigma$ and $\lambda(_yA) = \lambda(A)$ for all $A \in \Sigma$, $y \in G$.

Note that condition (r) ((l), respectively) is automatically satisfied if G is a group and the σ-algebra Σ and the measure λ are invariant with respect to right (left, resp.) translations.

Let $f : G \to \mathbb{R}$, $y \in G$. Then we define $_yf$, $f_y : G \to \mathbb{R}$ by

$$_yf(x) := f(y \cdot x), \quad f_y(x) := f(x \cdot y).$$

Lemma 17.1 ([13, p. 203]) *Let (G, Σ, λ) be a complete measure space satisfying* (r) *and let $y \in G$, $f \in \mathscr{L}_\lambda^1(G, \mathbb{R})$. Then*

$$f_y \in \mathscr{L}_\lambda^1(G, \mathbb{R}) \text{ and } \int_G f_y(t)d\lambda(t) = \int_G f(t)d\lambda(t). \tag{17.1}$$

If we replace the binary operation $(x, y) \mapsto x \cdot y$ by $(x, y) \mapsto y \cdot x$, then we get

Lemma 17.2 *If (G, Σ, λ) satisfies* (l), *then for all $y \in G$ and $f \in \mathscr{L}_\lambda^1(G, \mathbb{R})$, the function $_y f : G \to \mathbb{R}$ belongs to $\mathscr{L}_\lambda^1(G, \mathbb{R})$ and*

$$\int_G {}_y f(t) d\lambda(t) = \int_G f(t) d\lambda(t). \qquad (17.2)$$

Lemma 17.3 *Assume that $\lambda(G) < \infty$. If φ is a convex φ-function satisfying conditions (0_1), (∞_1) and $f \in \mathscr{L}_\lambda^\varphi(G, \mathbb{R})$, then the functions $m_2, m_3 : G \times G \to \mathbb{R}$, defined by*

$$m_2(x, y) = f(x), \qquad m_3(x, y) = f(y),$$

belong to $\mathscr{L}_{\lambda \times \lambda}^\varphi(G \times G, \mathbb{R})$ and

$$\|m_2\| = \|m_3\| \leq K \|f\|,$$

where $K = 1$ if $\lambda(G) \leq 1$ and $K = 2\lambda(G)$ if $\lambda(G) > 1$. Moreover, if (G, Σ, λ) satisfies (l) *or* (r) *and the function $m_1(x, y) = f(x \cdot y)$ is measurable in the product space, then*

$$m_1 \in \mathscr{L}_{\lambda \times \lambda}^\varphi(G \times G, \mathbb{R})$$

and $\|m_1\| = \|m_2\|$.

Proof Suppose that the condition (r) is satisfied. Applying the definition of $\mathscr{L}^\varphi$ and Lemma 17.1 we have, for a certain $k > 0$,

$$
\begin{aligned}
\infty &> \int_G \varphi(k|f(x)|) d\lambda(x) \\
&= \int_G \varphi(k|f_y(x)|) d\lambda(x) \\
&= \frac{1}{\lambda(G)} \int_G d\lambda(y) \int_G \varphi(k|f|_y(x)) d\lambda(x),
\end{aligned}
$$

and, by the Fubini-Tonelli theorem, the mapping $m_1 : (x, y) \mapsto f(x \cdot y)$ belongs to $\mathscr{L}_\mu^\varphi(G \times G, \mathbb{R})$.

Similarly, $m_2, m_3 \in \mathscr{L}_\mu^\varphi(G \times G, \mathbb{R})$. It is clear that $\|m_2\| = \|m_3\|$ and, from Lemma 17.1 and Fubini theorem, $\|m_2\| = \|m_1\|$.

Now we prove that $\|m_2\| \leq \|f\|$ when $\lambda(G) \leq 1$:

$$
\begin{aligned}
\|m_2\| &= \inf\{u > 0 \mid \int_{G \times G} \varphi(u^{-1}|f(x)|) d\mu(x, y) \leq 1\} \\
&= \inf\{u > 0 \mid \int_G \varphi(u^{-1}|f(x)|) d\lambda(x) \leq \frac{1}{\lambda(G)}\} \\
&\leq \inf\{u > 0 \mid \int_G \varphi(u^{-1}|f(x)|) d\lambda(x) \leq 1\} = \|f\|.
\end{aligned}
$$

Finally, if $\lambda(G) > 1$ then

$$
\|m_2\| \leq \|m_2\|^O = \inf_{t>0}\{t^{-1}(1 + \int_{G \times G} \varphi(t|f(x)|)d\mu(x, y))\}
$$

$$
= \lambda(G) \inf_{t>0}\{t^{-1}(\frac{1}{\lambda(G)} + \int_G \varphi(t|f(x)|)d\lambda(x))\}
$$

$$
\leq \lambda(G) \inf_{t>0}\{t^{-1}(1 + \int_G \varphi(t|f(x)|)d\lambda(x))\} = \lambda(G)\|f\|^O
$$

$$
\leq 2\lambda(G)\|f\|.
$$

The main results in Sect. 17.2 are Theorems 17.1 and 17.2. These theorems give an affirmative answer to the question of stability in the case of Orlicz's classes $\mathscr{L}_\lambda^\varphi(G, \mathbb{R})$ and $\mathscr{L}_{\lambda \times \lambda}^\varphi(G \times G)$ provided that $(G, \cdot, \lambda)$ is complete measurable group (see Definition 17.1). In the case

$$
\lambda \times \lambda(G) = \infty
$$

there occurred even the superstability, that is

$$
\mathscr{C}f \in \mathscr{L}_{\lambda \times \lambda}^\varphi(G \times G)
$$

entails $f = A$ a.e. for an additive A.

In Sect. 17.3 we investigate some properties of the Cauchy difference treated as a linear operator which maps the Orlicz space L_λ^φ into the Orlicz space $L_{\lambda \times \lambda}^\varphi$. For example, a necessary and sufficient condition under which $\mathscr{C}$ is compact is presented.

17.2   Stability of the Cauchy Equation

In the sequel μ denotes the complete product of the measure λ times itself.

Proposition 17.1 *Let G be a semigroup and let (G, Σ, λ) be a complete measure space satisfying at least one of the conditions* (r) ,(l) . *We assume additionally that the mapping*

$$
S : G \times G \ni (x, y) \mapsto (x, x \cdot y) \in G \times G
$$

is measurable, i.e. $S^{-1}(U)$ is μ-measurable for every μ-measurable set $U \subset G \times G$. Let $f : G \to \mathbb{R}$ be such a function that

$$
\mathscr{C}f \in \mathscr{L}_\mu^\varphi(G \times G, \mathbb{R})
$$

for a certain convex φ-function φ satisfying conditions (0_1) and (∞_1). If

$$0 < \lambda(G) < \infty,$$

then there exists a function $g : G \to \mathbb{R}$ such that:

(i) $\mathscr{C}g \overset{\mu}{=} 0$,
(ii) $f - g \in \mathscr{L}_\lambda^\varphi(G, \mathbb{R})$,
(iii) $\|f - g\| \le K\|\mathscr{C}f\|$, *where $K = 1$ if $\lambda(G) \ge 1$, $K = \frac{2}{\lambda(G)}$ if $\lambda(G) < 1$,*
(iv) $g_1 \overset{\lambda}{=} g$ *for each function $g_1 : G \to \mathbb{R}$ satisfying the conditions $\mathscr{C}g_1 \overset{\mu}{=} 0$,*
$f - g_1 \in \mathscr{L}_\lambda^\varphi(G, \mathbb{R})$.

Proof Suppose that condition (r) is satisfied and write

$$\lambda_n = \frac{\lambda}{\lambda(G)}.$$

Applying Hölder's inequality we get $\mathscr{L}_\mu^\varphi \subset \mathscr{L}_\mu^1$. Therefore $\mathscr{C}f \in \mathscr{L}_\mu^1$ and by the Fubini theorem the function

$$\phi(y) := \int_G (f(x \cdot y) - f(x) - f(y))d\lambda_n(x)$$

$$= \int_G (f(x \cdot y) - f(x))d\lambda_n(x) - f(y)$$

is defined for λ-almost all $y \in G$, so there exists $M \in \Sigma$, $\lambda(M) = 0$ such that $\phi(y)$ is defined for $y \in G \backslash M$. Furthermore, ϕ is measurable.
Define the function $g : G \to \mathbb{R}$ by

$$g(y) = \begin{cases} \int_G (f(x \cdot y) - f(x))d\lambda_n(x) & \text{for } y \in G \backslash M, \\ 0 & \text{for } y \in M. \end{cases}$$

Clearly $g - f \overset{\lambda}{=} \phi$, thus $g - f$ is measurable. Using Lemma 17.1, for $y, z, y \cdot z \in G \backslash M$, we have

$$g(y) + g(z) = \int_G (f(x \cdot y) - f(x))d\lambda_n(x) + \int_G (f(x \cdot z) - f(x))d\lambda_n(x)$$

$$= \int_G (f((x \cdot y) \cdot z) - f(x \cdot z) + f(x \cdot z) - f(x))d\lambda_n(x)$$

$$= \int_G (f(x \cdot (y \cdot z)) - f(x))d\lambda_n(x) = g(y \cdot z).$$

Thus

$$g(y) + g(z) = g(y \cdot z)$$

for $(y, z) \in G^2 \backslash (M \times G \cup G \times M \cup S^{-1}(G \times M))$. To prove (i) we only have to show that $M \times G \cup G \times M \cup S^{-1}(G \times M)$ is μ-measurable and that its measure is zero.

Clearly $\mu(M \times G) = \mu(G \times M) = 0$, therefore we have to prove that

$$S^{-1}(G \times M) = \{(y, z) \in G \times G \mid y \in M_z\}$$

is measurable and that its measure is zero. But this is a simple implication of the fact that S is measurable and of the Fubini Theorem (as by the condition (r), $M_z \in \Sigma$ and $\lambda(M_z) = 0$ for every $z \in G$). This completes the proof of (i).

We have

$$|g(y) - f(y)| = |\int_G \mathscr{C}f(x, y) d\lambda_n(x)| \tag{17.3}$$

$$\leq \int_G |\mathscr{C}f(x, y)| d\lambda_n(x), \qquad \text{for } \lambda \text{ almost all } y \in G.$$

Applying Jensen's inequality we obtain for $u > 0$ and almost all $y \in G$

$$\varphi(\frac{1}{u}|g(y) - f(y)|) \leq \int_G \varphi(\frac{1}{u}|\mathscr{C}f(x, y)|) d\lambda_n(x),$$

which implies

$$\lambda(G) \int_G \varphi(\frac{1}{u}|g(y) - f(y)|) d\lambda(y)$$

$$\leq \int_G d\lambda(y) \int_G \varphi(\frac{1}{u}|\mathscr{C}f(x, y)|) d\lambda(x)$$

$$= \int_{G \times G} \varphi(\frac{1}{u}|\mathscr{C}f(x, y)|) d\mu(x, y) \leq 1, \qquad \text{for } u > \|\mathscr{C}f\|.$$

Taking $u = \eta + \|\mathscr{C}f\|, \eta > 0$ we obtain

$$\lambda(G) \int_G \varphi(\frac{1}{u}|g(y) - f(y)|) d\lambda(y) \leq 1,$$

and if $\lambda(G) \geq 1$, then

$$\int_G \varphi(\frac{1}{u}|g(y) - f(y)|) d\lambda(y) \leq 1,$$

which implies that $\|g - f\| \leq u$ for all $\eta > 0$. It proves (ii) and (iii) if $\lambda(G) \geq 1$.

The case $\lambda(G) < 1$. Define the sets N, $\tilde{N}$, M by

$$N = \{z \mid z : G \to \mathbb{R} \text{ is measurable and } \int_G \varphi^*(|z(y)|)d\lambda(y) \leq 1\},$$

$$\tilde{N} = \{\tilde{z} \mid \tilde{z} : G \times G \to \mathbb{R} \text{ and } \tilde{z}(x, y) = z(y) \text{ for some } z \in N\},$$

$$M = \{w \mid w : G \times G \to \mathbb{R} \text{ is measurable and } \int_{G \times G} \varphi^*(|w(x, y)|)d\mu(x, y) \leq 1\}.$$

In view of the inclusion $\tilde{N} \subset M$ and (17.3) we obtain the following inequalities

$$\sup_{z \in N} \int_G \lambda(G)|g(y) - f(y)|z(y)d\lambda(y)$$

$$\leq \sup_{z \in N} \int_G \int_G |\mathscr{C}f|(x, y)z(y)d\lambda(x)d\lambda(y)$$

$$\leq \sup_{\tilde{z} \in \tilde{N}} \int_G \int_G |\mathscr{C}f|(x, y)\tilde{z}(x, y)d\lambda(x)d\lambda(y)$$

$$\leq \sup_{w \in M} \int_{G \times G} |\mathscr{C}f|(x, y)w(x, y)d\mu(x, y).$$

Consequently,

$$\lambda(G)\|g - f\|^O \leq \|\mathscr{C}f\|^O.$$

From this we get

$$\|g - f\| \leq \frac{2}{\lambda(G)}\|\mathscr{C}f\|,$$

which proves (ii) and (iii) if $\lambda(G) < 1$.

Uniqueness of g. If $\mathscr{C}g_1 \overset{\mu}{=} 0$ and $f - g_1 \in \mathscr{L}_\lambda^\varphi$, then

$$h := g - g_1 = (f - g_1) - (f - g) \in \mathscr{L}_\lambda^\varphi \subset \mathscr{L}_\lambda^1$$

and $\mathscr{C}h \overset{\mu}{=} 0$, so $\mathscr{C}h \in \mathscr{L}_\mu^1$. Therefore, by the Fubini theorem

$$0 = \int_G \mathscr{C}h(x, y)d\lambda(x)$$

$$= \int_G (h_y(x) - h(x))d\lambda(x) - h(y)\lambda(G)$$

$$= -h(y)\lambda(G)$$

for λ-almost all $y \in G$. Thus $\|h\|_1 = 0$, which proves (iv). The proof for case (l) is analogous.

We can now sharpen the statement of Proposition 17.1, (iii).

Proposition 17.2 *If $\lambda(G) \geq 1$, then the constant $K = 1$ in the statement of Proposition 17.1 (iii) is best possible. It means that for every $0 < \delta < 1$ there is a convex φ-function satisfying the conditions (0_1), (∞_1) such that the estimate*

$$\|f - g\| \leq (1 - \delta)\|\mathscr{C}f\|,$$

is false for some f with $\mathscr{C}f \in \mathscr{L}_\mu^\varphi$.

If $0 < \lambda(G) < 1$, then the best possible constant K_b belongs to the interval

$$\left[\frac{1}{\lambda(G)}, 1 + \frac{1}{\lambda(G)}\right].$$

Moreover, replacing the Luxemburg's norm by the Orlicz's one we obtain the following inequality

$$\|f - g\|^O \leq (\lambda(G))^{-1}\|\mathscr{C}f\|^O, \tag{17.4}$$

where $(\lambda(G))^{-1}$ cannot be decreased.

Proof First we calculate $\|f - 0\|/\|\mathscr{C}f\|$ and $\|f - 0\|^O/\|\mathscr{C}f\|^O$ for $f = 1$ and $\varphi(t) = t^p/p$, where $p > 1$ (see Proposition 17.1 (iv)). After easy calculations (cf. [7, Section 9]) we obtain

$$\frac{\|1\|}{\|\mathscr{C}1\|} = \frac{(\varphi^{-1}(\frac{1}{\lambda(G)}))^{-1}}{(\varphi^{-1}(\frac{1}{(\lambda(G))^2}))^{-1}} = \frac{1}{\sqrt[p]{\lambda(G)}} \to 1 \text{ as } p \to \infty.$$

Thus $K_b = 1$ in the case $\lambda(G) \geq 1$.

Now we assume $0 < \lambda(G) < 1$. The estimate (17.4) result from the proof of Proposition 17.1 (ii), (iii) in this case. For $\varphi(t) = t^p/p$ we have $\varphi^*(t) = t^q/q$, where $1/p + 1/q = 1$, and therefore

$$\frac{\|1\|^O}{\|\mathscr{C}1\|^O} = \frac{\lambda(G)(\varphi^{*-1}(\frac{1}{\lambda(G)}))}{\lambda(G)^2(\varphi^{*-1}(\frac{1}{(\lambda(G))^2}))} = \frac{1}{\lambda(G)}\sqrt[q]{\lambda(G)} \to \frac{1}{\lambda(G)} \text{ as } p \to 1+.$$

Thus (in Orlicz's norm)

(I) $\frac{1}{\lambda(G)}$ is best possible in the case $0 < \lambda(G) < 1$.

Moreover,

$$\frac{\|1\|}{\|\mathscr{C}1\|} = \frac{1}{\sqrt[p]{\lambda(G)}} \to \frac{1}{\lambda(G)} \text{ as } p \to 1+.$$

This proves that $K_b \in [1/\lambda(G), \infty)$.

The above localization of K_b can be improved in the following way: Let $r = (\lambda(G))^{-1}$ and Φ be a set of all convex φ-functions satisfying conditions $(0)_1, (\infty)_1$. Let us observe that

$$\{r^2\varphi \mid \varphi \in \Phi\} = \Phi \quad \text{and} \quad \mathscr{L}^\varphi = \mathscr{L}^{r^2\varphi}.$$

This yields

$$K_b = \sup \frac{\|f - g\|_\varphi}{\|\mathscr{C}f\|_\varphi} = \sup \frac{\|f - g\|_{r^2\varphi}}{\|\mathscr{C}f\|_{r^2\varphi}}, \tag{17.5}$$

where the supremum is taken over the set Γ defined by

$$\Gamma = \{(\varphi, f) \mid \varphi \in \Phi, \ f \in \mathbb{R}^G \text{ is such that } \mathscr{C}f \in \mathscr{L}_\mu^\varphi \text{ and } \|\mathscr{C}f\| \neq 0\}.$$

Clearly,

$$\|\mathscr{C}f\|_{r^2\varphi} = \|\mathscr{C}f\|_{\varphi,n}, \tag{17.6}$$

where the subscript n indicates that $\|\mathscr{C}f\|_{\varphi,n}$ is calculated with respect to the normalized measure.

Let

$$B = \|f - g\|_{\varphi,n}.$$

Dividing both $f - g$ and $\mathscr{C}f$ by B we see, that one can assume without loss of generality that $\|f - g\|_{\varphi,n} = 1$. Let us observe that inequality of the type (20) in [7, p. 251] is true also for Luxemburg's norm. Furthermore, formula (22) holds true for Luxemburg's norm, provided that the Luxemburg's norm of u is equal to 1 (cf. [7, p. 251]).

From these observations we obtain

$$\|f - g\|_{r^2\varphi} \leq 1 + \int_G r^2\varphi(|f(t) - g(t)|)d\lambda(t) \leq 1 + r\|f - g\|_{\varphi,n}. \tag{17.7}$$

Since $\|f - g\|_{\varphi,n} = 1$, from (17.5), (17.6) and (17.7) we deduce that

$$K_b \leq \sup \left(\frac{\|f - g\|_{\varphi,n}}{\|\mathscr{C}f\|_{\varphi,n}} + r\frac{\|f - g\|_{\varphi,n}}{\|\mathscr{C}f\|_{\varphi,n}} \right) \leq 1 + \frac{1}{\lambda(G)},$$

where the supremum is taken over the set Γ (the last inequality is a consequence of the first part of the proof, because $\lambda_n(G) = 1$). The proof is complete.

We propose the following two open problems.

(Q1) Find K_b as a function of $t = \lambda(G)$ when $\lambda(G)$ is less than 1.
(Q2) Examine whether the limit of $K_b(t)$ as $t \to 1-$ is equal to 1.

To formulate the main stability theorem we need the following definition.

Definition 17.1 ([5, §59 and p. 164]) We say that $(G, \cdot, \sum, \lambda)$ is a complete measurable group, iff

(a) $(G, \cdot)$ is a group,
(b) (G, Σ, λ) is a σ-finite measure space, λ is not identically zero and is complete,
(c) the σ-algebra Σ and the measure λ are invariant with respect to left translations,
(d) $\lambda \times \lambda$ is the completion of the product measure,
(e) the transformation $S : G \times G \ni (x, y) \mapsto (x, x \cdot y) \in G \times G$ is measurability preserving, i.e. S and S^{-1} are measurable.

It is worth mentioning that under the assumptions (a)–(e) and $\lambda(G) < \infty$ the measure λ is invariant under translations and under symmetry with respect to zero, where zero means the unit of the group G, (cf. [18, p. 115$_{5-2}$]). This implies that the measure $\lambda \times \lambda$ on $G \times G$ has the same properties. Therefore the families I_1, I_2 of all subsets of G and $G \times G$, respectively, of measure zero are p.l.i. ideals in G and $G \times G$, respectively. These ideals are conjugate (by the Fubini Theorem). This fact is still true if $\lambda(G) = \infty$, because the measure λ restricted to I_1 is invariant under translations and under symmetry with respect to zero (a simple consequence of [5, Theorem D, p. 259], cf. also [15]). Definitions of p.l.i. ideals and conjugate p.l.i. ideals one can found in [8, pp. 437–439].

Theorem 17.1 *Let (G, Σ, λ) be a complete measurable group. Let $f : G \to \mathbb{R}$ be such a function that $\mathscr{C}f \in \mathscr{L}_\mu^\varphi(G \times G, \mathbb{R})$ for a certain convex φ-function φ satisfying conditions (0_1), (∞_1). If $\lambda(G) < \infty$, then there is a unique additive map $A : G \to \mathbb{R}$ such that*

$$f - A \in \mathscr{L}_\lambda^\varphi(G, \mathbb{R}).$$

Moreover

$$\|f - A\| \le K\|\mathscr{C}f\|,$$

where $K = 1$ if $\lambda(G) \ge 1$ and $K = 1 + \frac{1}{\lambda(G)}$ if $\lambda(G) < 1$.

Proof In view of Proposition 17.1 there exists the function $g : G \to \mathbb{R}$ satisfying the conditions (i)–(iv) of Proposition 17.1. Due to (i) and [8, Theorem 17.6.1 on p. 444], there exists a unique additive function $A : G \to \mathbb{R}$ such that

$$A \stackrel{\lambda}{=} g.$$

This and conditions (ii), (iii) and (iv) from Proposition 17.1 together with Proposition 17.2 makes the proof complete.

Example 17.2 Let $B : \mathbb{R} \to \mathbb{R}$ be a discontinuous solution of the Cauchy equation

$$B(x + y) = B(x) + B(y)$$

such that $B_{|\mathbb{Q}} = 0$. Since every measurable solution of the Cauchy equation is continuous (cf. [8, p. 218] and [1]), B is not measurable. Consider the set $G = [0, 1)$ with the binary operation $+_1$ defined for all pairs (x, y) of elements of G by

$$x +_1 y = x + y \text{ for } x + y < 1, \qquad x +_1 y = x + y - 1 \text{ otherwise.}$$

Let λ be the Lebesgue measure restricted to the σ-algebra Σ of all Lebesgue-measurable subsets of G. Observe that (G, Σ, λ) is a complete measurable group with $\lambda(G) = 1$. Since $B(1) = 0$, $B_{|G}$ is a homomorphism of G into $\mathbb{R}$. Clearly, this homomorphism is not measurable. Therefore $B_{|G}$ is not λ-equal to 0. Denoting by f the function $1 + B_{|G}$ we have

$$\mathscr{C} f \in L^\varphi_\mu(G \times G, \mathbb{R}) \text{ and } \|\mathscr{C} f\| > 0$$

although f does not belong to $\mathscr{L}^\varphi_\lambda$. By Theorem 17.1 there exists a unique additive map $A : G \to \mathbb{R}$ such that

$$f - A \in \mathscr{L}^\varphi_\lambda(G, \mathbb{R}).$$

This fact and additivity of $B_{|G}$ entail

$$A(x) = B(x) \text{ for all } x \in G.$$

The next theorem is a counterpart of [15, Theorem 3.1].

Theorem 17.2 *Let $(G, \cdot, \lambda)$ be a measurable group with a complete, left-invariant and σ-finite measure λ such that $\lambda(G) = \infty$. If φ is a φ-function, $f : G \to \mathbb{R}$ and*

$$\mathscr{C} f \in \mathscr{L}^\varphi_\mu(G \times G, \mathbb{R}),$$

then there exists a unique additive $A : G \to \mathbb{R}$ which is equal to f λ a.e. The Cauchy equation

$$\mathscr{C} f(x, y) = 0$$

is satisfied for almost all $(x, y) \in G \times G$.

Proof First observe that the class $\mathscr{L}^{\varphi}$ is closed under addition and left translations and that from the Fubini Theorem

$$f \in \mathscr{L}^{\varphi}_{\mu} \text{ implies } f(\cdot, y) \in \mathscr{L}^{\varphi}_{\lambda} \text{ for almost all } y \in G.$$

Making use of the just stated facts and arguing analogously as in the proof of [15, Theorem 3.1] (with X replaced by G, $\mathscr{L}^+_p(X, \mathbb{R})$ replaced by $\mathscr{L}^{\varphi}_{\lambda}$), we can easily obtain the assertion.

As a trivial consequence of Theorems 17.1 and 17.2 we obtain the following result.

Remark 17.1 Let G be a complete measurable group with measure λ and let φ be a φ-function. If $\lambda(G) < \infty$ we additionally assume that φ is convex and that it satisfies conditions (0_1), (∞_1). Then the pair

$$(\mathscr{L}^{\varphi}_{\lambda}(G, \mathbb{R}), \mathscr{L}^{\varphi}_{\mu}(G \times G, \mathbb{R}))$$

has the double difference property, i.e. for every $f : G \to \mathbb{R}$ such that

$$\mathscr{C} f \in \mathscr{L}^{\varphi}_{\mu}(G \times G, \mathbb{R})$$

there exists an additive $A : G \to \mathbb{R}$ such that $f - A \in \mathscr{L}^{\varphi}_{\lambda}(G, \mathbb{R})$ (cf. [9]).

17.3   Cauchy Difference as a Linear Operator

We begin with

Theorem 17.3 *Let G be a non-empty set with a binary operation $\cdot$. Let (G, Σ, λ) be a measure space satisfying at least one of the conditions* (r) *,* (l) *. If*

$$0 < \lambda(G) < \infty,$$

$f \in \mathscr{L}^{\varphi}_{\lambda}(G, \mathbb{R})$ *for a certain convex φ-function φ satisfying conditions (0_1), (∞_1) and the mapping*

$$G^2 \ni (x, y) \mapsto f(x \cdot y) \in \mathbb{R}$$

is measurable in the product space, then

(j) $\mathscr{C} f \in \mathscr{L}^{\varphi}_{\mu}(G \times G, \mathbb{R})$;
(jj) $\|f\| \leq K \|\mathscr{C} f\|$, *where K is the constant defined in the preceding theorem;*
(jjj) $\|\mathscr{C} f\| \leq K_1 \|f\|$, *where $K_1 = 3$ if $\lambda(G) \leq 1$, $K_1 = 6\lambda(G)$ if $\lambda(G) > 1$;*
(jv) $\mathscr{C} f \overset{\mu}{=} 0 \Leftrightarrow f \overset{\lambda}{=} 0$.

Proof Since the functions m_1, m_2, m_3 defined in Lemma 17.3 belong to $\mathscr{L}_\mu^\varphi$, we get $\mathscr{C}f \in \mathscr{L}_\mu^\varphi$, which proves (j).

From Lemma 17.3

$$\|\mathscr{C}f\| = \|m_1 - m_2 - m_3\| \leq \|m_1\| + \|m_2\| + \|m_3\| = K_1 \|f\|,$$

which gives (jjj).

Observe that for every $y \in G$

$$\mathscr{C}f(\cdot, y) = f_y - f - f(y) \in \mathscr{L}_\lambda^\varphi \subset \mathscr{L}_\lambda^1,$$

and therefore by Lemma 17.1

$$\int_G \mathscr{C}f(x, y)d\lambda_n(x) = -f(y). \tag{17.8}$$

Hence, for every $y \in G$,

$$|f(y)| \leq \int_G |\mathscr{C}f(x, y)|d\lambda_n(x).$$

We obtain the proof of (jj) by repeating the method which is used in the proof of part (iii) of Propositions 17.1 and 17.2. Notice that the associativity of $\cdot$ and the property of mapping S are used in the proof of Proposition 17.1 only to obtain

$$g(y) + g(z) = g(y \cdot z)$$

for μ-almost all $(y, z) \in G^2$. Because now $g \overset{\lambda}{=} 0$, these assumptions are superfluous. Since (jj) implies (jv), the proof is completed.

The following equality

$$\{f \in L_\lambda^\varphi(G, \mathbb{R}) \mid \mathscr{C}f \in L_\mu^\varphi(G \times G, \mathbb{R})\}$$

$$= \{f \in L_\lambda^\varphi(G, \mathbb{R}) \mid (x, y) \mapsto f(x \cdot y) \text{ is measurable in the product space}\}$$

results from properties of measurable functions and Theorem 17.3. Clearly, the set

$$D_\mathscr{C} = \{f \in L_\lambda^\varphi(G, \mathbb{R}) \mid \mathscr{C}f \in L_\mu^\varphi(G \times G, \mathbb{R})\}$$

is a linear subspace of $L_\lambda^\varphi(G, \mathbb{R})$.

Theorem 17.4 *Let G be a non-empty set with a binary operation $\cdot$. Let (G, Σ, λ) be a measure space satisfying at least one of the conditions* (r) , (l) . *If*

$$0 < \lambda(G) < \infty,$$

then the linear operator

$$\mathscr{C} : D_{\mathscr{C}} \ni f \mapsto \mathscr{C}f \in L^{\varphi}_{\mu}(G \times G, \mathbb{R})$$

is continuous and invertible. Moreover, the inverse operator (defined on $\mathscr{C}(D_{\mathscr{C}})$) is continuous and $\mathscr{C}^{-1}$ has the form

$$\mathscr{C}^{-1}h(z) = \begin{cases} -(\lambda(G))^{-1} \int_G h(x, z)d\lambda(x) & \text{if (r) occurs,} \\ -(\lambda(G))^{-1} \int_G h(z, y)d\lambda(y) & \text{if (l) occurs,} \end{cases}$$

for almost all $z \in G$.

If $D_{\mathscr{C}} = L^{\varphi}_{\lambda}$ then the following statements are equivalent:

(p) *$\mathscr{C}$ is compact;*

(q) *there is a finite σ-algebra Σ_0 on G such that $\Sigma_0 \subset \Sigma$ and such that each set in Σ differs from a set in Σ_0 by a λ-null set.*

Proof The continuity of $\mathscr{C}$ results from Theorem 17.3 (jjj). The existence and continuity of $\mathscr{C}^{-1}$ results from Theorem 17.3 (jv), (jj), respectively.

The form of $\mathscr{C}^{-1}$ is a consequence of the proof of Theorem 17.3, (17.8).

(q) $\Rightarrow$ (p): From [4, Exercise 3, p. 141] both the spaces L^{φ}_{λ}, L^{φ}_{μ} have finite dimensions. This proves the compactness of $\mathscr{C}$.

(p) $\Rightarrow$ (q): Assume that $\mathscr{C}$ is compact and (q) does not hold. Thus, the space L^{φ}_{λ} is infinite-dimensional. The operator $H : L^{\varphi}_{\mu} \to L^{\varphi}_{\lambda}$ defined by

$$(Hz)(y) = -\int_G z(x, y)d\lambda_n(x)$$

is continuous (cf. the proof of Proposition 17.1 (iii)). By the formula (17.8), $H\mathscr{C}f = f$ for all $f \in L^{\varphi}_{\lambda}$. Since the product of a continuous linear operator with a compact operator is compact, $H\mathscr{C} = I$ is compact on the infinite-dimensional space L^{φ}_{λ}, a contradiction.

References

1. Alexiewicz, A., Orlicz, W.: Remarque sur l'équation fonctionnelle $f(x + y) = f(x) + f(y)$. Fund. Math. **33**, 314–315 (1945)
2. Brzdęk, J., Popa, D., Raşa, I.: Hyers-Ulam stability with respect to gauges. J. Math. Anal. Appl. **453**, 620–628 (2017)
3. Brzdęk, J., Popa, D., Raşa, I., Xu, B.: Ulam Stability of Operators. Mathematical Analysis and Its Applications, vol. 1. Academic, London (2018)
4. Cohn, D.L.: Measure Theory, 2nd edn. Springer, New York (2013)
5. Halmos, P.R.: Measure Theory. Springer, New York (1974)
6. Hyers, D.H., Rassias, T.M.: Approximate homomorphisms. Aequationes Math. **44**, 125–153 (1992)

7. Krasnosel'skiǐ, M.A., Rutickiǐ, Ja.B.: Convex functions and Orlicz spaces. Problems of Contemporary Mathematics Gosudarstv. Izdat. Fiz.-Mat. Lit., Moscow (1958)
8. Kuczma, M.: An Introduction to the Theory of Functional Equations and Inequalities. Polish Scientific Publishers (PWN) and Silesian University, Warszawa (1985)
9. Laczkovich, M.: Functions with measurable differences. Acta Math. Acad. Sci. Hungar. **35**(1-2), 217–235 (1980)
10. Maligranda, L.: Orlicz spaces and interpolation. Campinas (1989) https://www.researchgate.net/publication/44477854_Orlicz_spaces_and_interpolation_by_Lech_Maligranda
11. Moszner, Z.: Stability has many names. Aequationes Math. **90**, 983–999 (2016)
12. Musielak, J.: Orlicz spaces and modular spaces. Lecture Notes in Mathematics, vol. 1034. Springer, Berlin (1983)
13. Siudut, S.: Cauchy difference operator in $\mathscr{L}^p$ spaces. Funct. Approx. Coomment. Math. **XXV**, 201–211 (1997)
14. Shulman, E.V.: Group representations and stability of functional equations. J. Lond. Math Soc. (2) **54**(1), 111–120 (1996)
15. Tabor, J.: Stability of the Cauchy type equations in $\mathscr{L}_p$ norms. Results Math. **32**(1-2), 145–158 (1997)
16. Tabor, J.: Lipschitz stability of the Cauchy and Jensen equations. Results Math. **32**(1-2), 133–144 (1997)
17. Tabor, J. Tabor, J.: Stability of the Cauchy type equations in the class of differentiable functions. J. Approx. Theory **98**(1), 167–182 (1999)
18. Yamasaki, Y.: Measures on Infinite Dimensional Spaces. World Scientific, Singapore (1985)
19. Yosida, K.: Functional Analysis, 3rd edn. Springer, Berlin (1971)

Chapter 18
Semi-Inner Products and Parapreseminorms on Groups and a Generalization of a Theorem of Maksa and Volkmann on Additive Functions

Árpád Száz

Abstract By using inner products and paraprenorms on groups, we prove a natural generalization of a basic theorem of Gyula Maksa and Peter Volkmann on additive functions.

Keywords Groups · Semi-inner products · Parapreseminorms · Additive functions

Mathematics Subject Classification (2010) Primary 39B52, 39B62; Secondary 20A99, 46C50

18.1 Introduction

In this paper, by using inner products and paraprenorms on groups, we shall prove a natural generalization of the following basic theorem of Maksa and Volkmann [82].

Theorem 18.1 *For functions $f : G \to E$ from a group G to a real or complex inner product space E, the inequality*

$$\| f(xy) \| \geq \| f(x) + f(y) \| \qquad (x, y \in G)$$

implies

$$f(xy) = f(x) + f(y) \qquad (x, y \in G).$$

Á. Száz (✉)
Department of Mathematics, University of Debrecen, Debrecen, Hungary
e-mail: szaz@science.unideb.hu

© Springer Nature Switzerland AG 2019
J. Brzdęk et al. (eds.), *Ulam Type Stability*,
https://doi.org/10.1007/978-3-030-28972-0_18

Remark 18.1 For the origins of this striking theorem, see Volkmann [134], Maksa [81] and Kurepa [74]. The latter author also studied the converse inequality and provided two illustrating examples.

The $G(\cdot) = \mathbb{R}(+)$ and $E(+) = \mathbb{R}(+)$ particular case of Theorem 18.1 was later also proved, in a completely different way, by Kwon et al. [75] without citing the works of the above mentioned authors.

Remark 18.2 Before the inequalities

$$| f(x) + f(y) | \leq | f(x + y) | \qquad \text{and} \qquad \| f(x) + f(y) \| \leq \| f(x + y) \|,$$

the squared and normed Cauchy equations

$$f(x + y)^2 = \big(f(x) + f(y) \big)^2 \qquad \text{and} \qquad \| f(x + y) \| = \| f(x) + f(y) \|$$

were also intensively investigated by a great number of mathematicians.

See, for instance, Robinson [101], Hosszú [56], Vincze [131–133], Fischer and Muszély [35, 36], Haruki [54, 55], Dhombres and Aczél [1, 23], Swiatak and Hosszú [115, 116], Kuczma [68], Skof [110–112], Ger [40–42, 44], Schöpf [105], Piejko [97], Batko and Tabor [8, 9], Ger and Koclega [45], Tabor and Tabor [125, 126], Fochi [37], Oikhberg and Rosenthal [92], Kannappan [63] and Dong and Chen [25].

Remark 18.3 Later, some similar results have also been proved for some analogous inequalities derived from the quadratic equation and its generalizations by Gillányi [47], Rätz [100], Fechner [31], Eqorachi et al. [30] and Manar and Elqorachi [83],

Moreover, several other functional inequalities, implying additivity or quadraticity, have also been intensively investigated by Park et al. [94], Roh and Chang [102], Lee et al. [77], Chung et al. [22], Dong and Zheng [26] and Park et al. [95].

18.2 Semi-inner Products

The following definition of semi-inner products has been taken from [121] and [12]. (For some immediate generalizations, see [13] and [14].)

It is in accordance with that of [119], but differs from the one introduced by Lumer [78]. (For some further developments, see [51] and [27].)

Our former definitions of semi-inner products may also be modified according to the ideas of Bognár [10], Antoine and Grossmann [4] and Drygas [28].

However, in the present paper, it seems convenient to adhere to the following less general, original notion.

Definition 18.1 Let X be an additively written group. Then, a function P of X^2 to $\mathbb{C}$ is called a *semi-inner product* on X if for any $x, y, z \in X$ we have

(a) $P(x, x) \geq 0,$
(b) $P(y, x) = \overline{P(x, y)},$
(c) $P(x + y, z) = P(x, z) + P(y, z).$

Remark 18.4 The above semi-inner product P will be called an *inner product* if

(d) $P(x, x) = 0$ implies $x = 0$ for all $x \in X.$

To illustrate the appropriateness of Definition 18.1, the following example has been suggested to me by Zoltán Boros.

Example 18.1 If a is an additive function of X to an inner product space H and

$$P(x, y) = \langle a(x), a(y) \rangle$$

for all $x, y \in X$, then P is a semi-inner product on X. Moreover, P is an inner product if and only if a is injective. That is, $a^{-1}(0) = \{x : a(x) = 0\} = \{0\}.$

Note that, despite this, P may be a rather curious function even if $X = \mathbb{R}^n$ and $H = \mathbb{R}$. Namely, by Kuczma [69, p. 292], there exists a discontinuous, injective additive function of $\mathbb{R}^n$ to $\mathbb{R}$. In the case $n = 1$, by Makai [79], Kuczma [69, p. 293] and Baron [6, 7], we can state even more.

The most basic consequences of Definition 18.1 can be listed in the following

Theorem 18.2 *If P is a semi-inner product on X, then for any $x, y, z \in X$ and $k \in \mathbb{Z}$ we have*

(1) $P(x + y, z) = P(y + x, z),$
(2) $P(x, z + y) = P(x, y + z),$
(3) $P(x, y + z) = P(x, y) + P(x, z),$
(4) $P(k x, y) = k P(x, y) = P(x, k y),$
(5) $z = x + y - (y + x)$ *implies* $P(z, z) = 0.$

Proof By using (b) and (c), and the additivity of complex conjugation, we can easily see that (3) is true.

Hence, by the $\mathbb{Z}$-homogeneity of additive functions of one group to another [13, 120], it is clear that (4) is also true.

Moreover, from (c) and (3), by using the commutativity of the addition in $\mathbb{C}$, we can see that (1) and (2) are also true.

Now, if z is as in (5), then by using (c), (4) and (1) we can also easily see that

$$P(z, z) = P(x + y - (y + x), z) = P(x + y, z) - P(y + x, z) = 0.$$

Therefore, (5) is also true. □

Remark 18.5 Note that, in particular, (4) yields

$$P(0, y) = 0 = P(x, 0) \qquad \text{and} \qquad P(-x, y) = -P(x, y) = P(x, -y)$$

for all $x, y \in X.$

Moreover, from (5), we can see that $x + y = y + x$ for all $x, y \in X$ if in particular P is an inner product. Therefore, only commutative groups can have inner products. This simple, but striking fact was first observed by Zoltán Boros.

Remark 18.6 In addition to Definition 18.1 and Theorem 18.2, we can also easily note that the real and imaginary parts (first and second coordinate functions) P_1 and P_2 of P, defined by

$$P_1(x, y) = 2^{-1} \left(P(x, y) + \overline{P(x, y)} \right) = 2^{-1} \left(P(x, y) + P(y, x) \right)$$

and

$$P_2(x, y) = (i\,2)^{-1} \left(P(x, y) - \overline{P(x, y)} \right) = i^{-1} 2^{-1} \left(P(x, y) - P(y, x) \right)$$

for all $x, y \in X$, also have the same commutativity and bilinearity properties as P. Furthermore, by properties (a) and (b), for any $x, y \in X$ we have

(1) $P_1(x, x) = P(x, x)$ and $P_2(x, x) = 0$,
(2) $P_1(y, x) = P_1(x, y)$ and $P_2(y, x) = -P_2(x, y)$.

Thus, in particular P_1 is also a semi-inner product on X. However, because of its skew-symmetry, P_2 cannot be a semi-inner product on X whenever $P_2 \neq 0$.

Remark 18.7 Conversely, one can easily see that if P_1 is a real-valued semi-inner product on X and P_2 is a nonzero, skew-symmetric, biadditive function of X^2 to $\mathbb{R}$, then

$$P = (P_1, P_2) = P_1 + i\,P_2$$

is a complex-valued semi-inner product on X.

18.3   The Induced Seminorms

Because of (a) in Definition 18.1, we may naturally introduce the following

Definition 18.2 If P is a semi-inner product on the group X, then for any $x \in X$ we define

$$p(x) = \sqrt{P(x, x)}.$$

Example 18.2 If in particular P is as in Example 18.1, then

$$p(x) = \sqrt{\langle a(x), a(x) \rangle} = \| a(x) \|$$

for all $x \in X$. Thus, for any $x \in X$, we have $p(x) = 0$ if and only if $a(x) = 0$.

Concerning the function p, introduced in Definition 18.2, we can easily prove

Theorem 18.3 *If P is a semi-inner product on X, then for any $x, y \in X$ and $k \in \mathbb{Z}$ we have*

(1) $p(x) \geq 0$,
(2) $p(kx) = |k|\, p(x)$,
(3) $p(x+y) = p(y+x)$,
(4) $p(k(x+y)) = p(kx + ky)$,
(5) $z = x + y - (y+x)$ *implies* $p(z) = 0$,
(6) $p(x+y)^2 = P_1(x+y, x) + P_1(x+y, y)$,
(7) $p(x+y)^2 = p(x)^2 + p(y)^2 + 2\, P_1(x, y)$.

Proof By Definition 18.2 and Theorem 18.2, it is clear that (1), (2) and (5) are true. Moreover, by using Remark 18.6, we can see that

$$p(x) = \sqrt{P_1(x, x)}$$

and

$$
\begin{aligned}
p(x+y)^2 &= P_1(x+y,\, x+y) = P_1(x+y, x) + P_1(x+y, y) \\
&= P_1(x, x) + P_1(y, x) + P_1(x, y) + P_1(y, y) \\
&= p(x)^2 + 2\, P_1(x, y) + p(y)^2 .
\end{aligned}
$$

Therefore, (6) and (7) are also true.

Hence, by the symmetry of P_1, the commutativity of the addition in $\mathbb{R}$ and the nonnegativity of p, it is clear that (3) is also true.

Moreover, by using (2), (7) and Theorem 18.2, we can see that

$$p(k(x+y))^2 = k^2\, p(x+y)^2 = k^2\, p(x)^2 + k^2\, p(y)^2 + 2\, k^2\, P_1(x, y)$$

and

$$
\begin{aligned}
p(kx + ky)^2 &= p(kx)^2 + p(ky)^2 + 2\, P_1(kx, ky) \\
&= k^2\, p(x)^2 + k^2\, p(y)^2 + 2\, k^2\, P_1(x, y) .
\end{aligned}
$$

Therefore, $p(k(x+y))^2 = p(kx + ky)^2$, and thus by the nonnegativity of p assertion (4) also holds. □

Remark 18.8 Note that, in particular, (2) yields

$$p(0) = 0 \qquad \text{and} \qquad p(-x) = p(x)$$

for all $x \in X$. Thus, p is an even function.

Remark 18.9 Moreover, to feel the importance of (4), one can note that for any $x, y \in X$, we have $2(x + y) = 2x + 2y$ if and only if $y + x = x + y$.

Therefore, if x and y do not commute, then $2(x + y) \neq 2x + 2y$. However, by (4), we still have $p\big(2(x + y)\big) = p(2x + 2y)$.

Remark 18.10 In addition to Theorem 18.3, we can also note that P is an inner product on X if and only if $p(x) = 0$ implies $x = 0$ for all $x \in X$.

Now, by using Theorems 18.3 and 18.2, one can also easily establish the following

Theorem 18.4 *If P is a semi-inner product on X, then for any $x, y \in X$ we have*

(1) $p(x - y)^2 = p(x + y)^2 - 4P_1(x, y)$,
(2) $p(x - y)^2 = 2p(x)^2 + 2p(y)^2 - p(x + y)^2$.

Moreover, as an immediate consequence of Theorems 18.4 and 18.3, one can state.

Theorem 18.5 *If P is a semi-inner product on X, then for any $x, y \in X$ we have*

(1) $P_1(x, y) = 4^{-1}\big(p(x + y)^2 - p(x - y)^2\big)$,
(2) $P_1(x, y) = 2^{-1}\big(p(x + y)^2 - p(x)^2 - p(y)^2\big)$.

Remark 18.11 Now, similar polar formulas for $P_2(x, y)$ cannot be proved. Therefore, in accordance with Remark 18.7, P can be recovered from p only in the real-valued case. (See [13, Example 5.6].)

Moreover, in the present generality, the usual Schwarz's inequality cannot also be proved. In [12], by improving the argument of [11], we could only prove the following weakened form of it. (See also [13] and [14].)

Lemma 18.1 *If P is a semi-inner product on X, then for any $x, y \in X$ we have*

$$| P_1(x, y) | \leq p(x) p(y).$$

Proof By using Theorems 18.3 and 18.2, we can see that

$$p(nx + my)^2 = p(nx)^2 + p(my)^2 + 2P_1(nx, my)$$
$$= n^2 p(x)^2 + m^2 p(y)^2 + 2nm P_1(x, y),$$

and thus

$$-2P_1(x, y) \leq (n/m) p(x)^2 + (m/n) p(y)^2$$

for all $n, m \in \mathbb{N}$.

Therefore, by the definition of rational numbers, we actually have

$$-2P_1(x, y) \leq r p(x)^2 + r^{-1} p(y)^2$$

for all $r \in \mathbb{Q}$ with $r > 0$.

Hence, by using that each real number is a limit of a sequence of rational numbers and the operations in $\mathbb{R}$ are sequentially continuous, we can already infer that

$$-2\, P_1(x, y) \leq \lambda\, p\, (x)^2 + \lambda^{-1}\, p\, (y)^2$$

for all $\lambda \in \mathbb{R}$ with $\lambda > 0$.

Now, if $p\,(x) \neq 0$ and $p\,(y) \neq 0$, then by taking $\lambda = p\,(y)/p\,(x)$ we can see that

$$-2\, P_1(x, y) \leq 2\, p\,(x)\, p\,(y),$$

and thus $-P_1(x, y) \leq p\,(x)\, p\,(y)$ also holds.

While, if for instance $p\,(x) = 0$, then we can see that

$$-2\, P_1(x, y) \leq \lambda^{-1}\, p\,(y)^2$$

for all $\lambda > 0$, and thus $-2\, P_1(x, y) \leq 0$. Therefore, $-P_1(x, y) \leq 0$, and thus $-P_1(x, y) \leq p\,(x)\, p\,(y)$ trivially holds.

Consequently, the inequality

$$-P_1(x, y) \leq p\,(x)\, p\,(y)$$

is always true. Hence, by using Remarks 18.5 and 18.8, we can also see that

$$P_1(x, y) = -P_1(-x, y) \leq p\,(-x)\, p\,(y) = p\,(x)\, p\,(y).$$

Therefore, by the definition of the absolute value, the required inequality is also true. $\qquad \square$

The above weak Schwarz inequality allows us to prove the subadditivity of p. Thus, some of the results on semi-inner product spaces can be extended to semi-inner product groups.

Theorem 18.6 *If P is a semi-inner product on X, then for any x, $y \in X$ we have*

(1) $p\,(x + y) \leq p\,(x) + p\,(y)$,   *(2)* $|\, p\,(x) - p\,(y)\,| \leq p\,(x - y)$.

Proof By using Theorem 18.3 and Lemma 18.1, we can see that

$$p\,(x + y)^2 = P_1(x + y, x) + P_1(x + y, y) \leq p\,(x + y)\, p\,(x) + p\,(x + y)\, p\,(y).$$

Therefore, by the nonnegativity of p, inequality (1) is also true.

Now, to complete the proof, it remains to note only that (2) can be derived from (1) on the usual way. $\qquad \square$

Remark 18.12 Theorems 18.3 and 18.6 show that the function p is a *seminorm* on X. Moreover, from Remark 18.10, we can see that p is a *norm* on X if and only if P is an inner product on X.

Remark 18.13 Instead of a norm p on a group X, several authors (see [124, p. 111]) prefer to use a metric d on X which is translation-invariant in the sense that

$$d(x+z, \ y+z) = d(x,y) \qquad \text{and} \qquad d(z+x, \ z+y) = d(x,y)$$

for all $x, y, z \in X$.

Note that these conditions can be reformulated in several different forms. For instance, the second condition is equivalent to the requirement that

$$d(0, y) = d(x, x+y) \qquad \left(d(x,y) = d(0, -x+y) \right)$$

for all $x, y \in X$. Thus, d may, for instance, be naturally called upper left-translation-invariant if $d(0, y) \le d(x, x+y)$ for all $x, y \in X$.

Moreover, if p is as in Definition 18.2, then we can easily see that the function d, defined by

$$d(x, y) = p(-x+y)$$

for all $x, y \in X$, is a translation-invariant semimetric on X such that

$$d(kx, ky) = |k| \, d(x,y)$$

for all $k \in \mathbb{Z}$ and $x, y \in X$.

Remark 18.14 Metrics derivable from norms have formerly been explicitly studied by Oikhberg and Rosenthal [92], Šemrl [106, 107] and Chmieliński [20, 21].

Moreover, the equivalence of norms to norms derived from inner products have been studied by Joichi [59] and Chmieliński [19]. However, equivalences of metrics to metrics derived from norms seem not to be investigated.

18.4 Some Further Properties of the Induced Seminorm

Assertions (3)–(4) and (6)–(7) of Theorem 18.3 can be extended to all finite and certain infinite families of elements of X. However, in the sequel, we shall only need the last statement of the following

Theorem 18.7 *If P is a semi-inner product on X, then for any $x, y, z \in X$ and $k \in \mathbb{Z}$ we have*

(1) $p(x+y+z) = p(x+z+y)$,
(2) $p\big(k(x+y+z)\big) = p(kx+ky+kz)$,
(3) $p(x+y+z)^2 = P_1(x+y+z, x) + P_1(x+y+z, y) + P_1(x+y+z, z)$,

(4) $p(x+y+z)^2 = p(x)^2 + p(y)^2 + p(z)^2 + 2P_1(x,y) + 2P_1(x,z) + 2P_1(y,z)$,

(5) $p(x+y+z)^2 = p(x+y)^2 + p(x+z)^2 + p(y+z)^2 - p(x)^2 - p(y)^2 - p(z)^2$.

Proof Assertion (1) is immediate from Definition 18.2 by Remark 18.6. To prove (2) and (3), note that by (7) in Theorem 18.3 we have

$$p(x+y+z)^2 = p(x+y)^2 + p(z)^2 + 2P_1(x+y,\, z)$$

$$= p(x)^2 + p(y)^2 + 2P_1(x,y) + p(z)^2 + 2P_1(x,z) + 2P_1(y,z)$$

$$= p(x)^2 + p(y)^2 + p(z)^2 + 2P_1(x,y) + 2P_1(x,z) + 2P_1(y,z)$$

$$= p(x)^2 + p(y)^2 + p(z)^2 + p(x+y)^2 - p(x)^2 - p(y)^2$$

$$+ p(x+z)^2 - p(x)^2 - p(z)^2 + p(y+z)^2 - p(y)^2 - p(z)^2$$

$$= p(x+y)^2 + p(x+z)^2 + p(y+z)^2 - p(x)^2 - p(y)^2 - p(z)^2.$$

$\square$

Remark 18.15 The above *parallelepiped law* (5) plays a similar role in characterization of inner product spaces as the *parallelogram identity* established in assertion (2) of Theorem 18.4.

Their importance, in this context, was first recognized by Fréchet [38] and Jordan and von Neumann [60]. (See also Amir [3], Istrătescu [58] and Alsina, Sikorska and Tomás [2], for instance.)

Whenever p is even, the parallelogram identity can be derived from the parallelepiped law by taking $z = -y$. Moreover, if X is commutative and p is even, then they are actually equivalent. (For a proof, see [84] and [122].)

Remark 18.16 In this respect, it is also worth noticing that if p is a function of the group X to $\mathbb{R}$ such that for any $x,y,z \in Y$ we have

(1) $p(x+y)^2 = p(y+x)^2$,

(2) $p(x+y+z)^2 = p(x+y)^2 + p(x+z)^2 + p(y+z)^2 - p(x)^2 - p(y)^2 - p(z)^2$,

then by [114, Proposition 13.25] of Stetkaer there exist a unique additive function a of X to $\mathbb{R}$ and a unique symmetric biadditive function A of X^2 to $\mathbb{R}$ such that

$$p(x)^2 = a(x) + A(x,x)$$

for all $x \in X$.

Hence, if in addition p is even, we can infer that

$$a(x) + A(x,x) = p(x)^2 = p(-x)^2 = a(-x) + A(-x,-x) = -a(x) + A(x,x),$$

and thus $a(x) = 0$ for all $x \in X$. Therefore, we have $p(x)^2 = A(x, x)$ for all $x \in X$. Hence, if in addition p is nonnegative, we can infer that $p(x) = \sqrt{A(x, x)}$ for all $x \in X$.

In [12], we tried to prove the following generalization of Theorem 18.7 by induction. However, later Jens Schwaiger has suggested some more convenient proofs.

Theorem 18.8 *If P is a semi-inner product on X, then for any $n \in \mathbb{N}$, with $n > 1$, and $x = \sum_{i=1}^{n} x_i$, with $x_i \in X$, we have*

(1) $p(x)^2 = \sum_{i=1}^{n} P_1(x, x_i),$

(2) $p(x)^2 = \sum_{i=1}^{n} p(x_i)^2 + \sum_{1 \le i < j \le n} 2 P_1(x_i, x_j),$

(3) $p(x)^2 = \sum_{1 \le i < j \le n} p(x_i + x_j)^2 - (n-2) \sum_{i=1}^{n} p(x_i)^2.$

Remark 18.17 Meantime, we observed that the general solution and the generalized stability of the more general functional equation

$$f\left(\sum_{i=1}^{n} x_i\right) + (n-2) \sum_{i=1}^{n} f(x_i) = \sum_{1 \le i < j \le n} f(x_i + x_j),$$

where f is a function of one vector space X to another Y, were already established by Nakmahachalasint [88].

The $n = 3$ particular case of this fundamental equation was formerly intensively investigated by Kannappan [62], Jung [61], Rassias [99], Fechner [32], and later by Ng and Zhao [90]. Moreover, Popoviciu [98], Trif [128, 129], Smajdor [113], Brzdęk [16], Nakmahachalasint [89], and Najati and Rassias [87] considered some similar equations.

18.5   Parapreseminorms

Now, because of Theorems 18.3 and 18.7, we may also naturally introduce

Definition 18.3 Let X be an additively written group. Then, a function p of X to $\mathbb{R}$ will be called a *parapreseminorm* on X if for any $x, y, z \in X$ we have

(a) $0 \le p(x),$
(b) $p(-x) \le p(x),$
(c) $p(y + x) \le p(x + y),$
(d) $p(x+y+z)^2 \le p(x+y)^2 + p(x+z)^2 + p(y+z)^2 - p(x)^2 - p(y)^2 - p(z)^2.$

Remark 18.18 The above parapreseminorm p will called a *paraseminorm* if

(e) $p(nx) = np(x)$ for all $n \in \mathbb{N}$ and $x \in X$.

Moreover, a paraseminorm (parapreseminorm) p on X is called a *paranorm (paraprenorm)* if

(f) $p(x) = 0$ implies $x = 0$ for all $x \in X$.

This definition differs from that of Wilansky [136, p. 15]. However, it is in accordance with a former definition of the present author [118].

Now, by Theorems 18.3 and 18.7, we can at once state the following

Theorem 18.9 *If P is a semi-inner product on X and*

$$p(x) = \sqrt{P(x,x)}$$

for all $x \in X$, then p is a paraseminorm on X.

Remark 18.19 By Remark 18.10, in addition to the above theorem, we can also state that p is paranorm on X if and only if P is an inner product on X.

Moreover, by using Definition 18.3, we can also easily prove the following

Theorem 18.10 *If p is a parapreseminorm on X, then for any $x, y \in X$ we have*

(1) $p(0) = 0$,
(2) $p(x) = p(-x)$,
(3) $2p(x) \le p(2x)$,
(4) $p(x + y) = p(y + x)$,
(5) $2p(x)^2 + 2p^2(y) \le p(x+y)^2 + p(x-y)^2$.

Proof Assertions (2) and (4) can be immediately derived from properties (b) and (c) in Definition 18.3, respectively, by writing $-x$ in place of x in (b) and changing the roles x and y in (c).

To derive (1), by taking $z = 0$ in (d), we can note that

$$p(x+y)^2 \le p(x+y)^2 + p(x)^2 + p(y)^2 - p(x)^2 - p(y)^2 - p(0)^2$$
$$= p(x+y)^2 - p(0)^2.$$

Therefore, $p(0)^2 \le 0$, and thus $p(0)^2 = 0$. Therefore, (1) is also true.

Now, from (d), by taking $z = -y$ and using (1) and (2), we can also see that

$$p(x)^2 \le p(x+y)^2 + p(x-y)^2 + p(0)^2 - p(x)^2 - p(y)^2 - p(-y)^2$$
$$= p(x+y)^2 + p(x-y)^2 - p(x)^2 - 2p(y)^2.$$

Therefore, (5) is also true.

Finally, from (5), by taking $x = y$ and using (1), we can see that

$$2\,p\,(x)^2 + 2\,p\,(x)^2 \le p\,(2x)^2 + p\,(0)^2 = p\,(2x)^2,$$

and thus $\left(2\,p\,(x)\right)^2 \le p\,(2x)^2$. Hence, by (a), it is clear that (3) is also true. □

Remark 18.20 If in particular p is a paraseminorm on X, then by properties (1) and (2) and Remark 18.18 we have $p\,(0x) = p\,(0) = 0 = 0\,p\,(x)$ for all $x \in X$ and

$$p\left((-n)\,x\right) = p\,(-nx) = p\,(nx) = n\,p\,(x)$$

for all $n \in \mathbb{N}$ and $x \in X$. Therefore, $p\,(kx) = |k|\,p\,(x)$ also holds for all $k \in \mathbb{Z}$ and $x \in X$.

Remark 18.21 From property (5), we can at once see that if p is a parapreseminorm on X, then p^2 is a superquadratic function on X, and thus $-p^2$ is a subquadratic function on X.

Thus, some properties of parapreseminorms can be immediately derived from those of the subquadratic functions investigated by Kominek, Troczka-Pawelec, Gilányi and Kézi [49, 50, 65–67, 130].

By using property (5), we can also easily show that some important norms defined even on $\mathbb{R}^2$ fail to be paranorms.

Example 18.3 For any $x = (x_1, x_2) \in \mathbb{R}^2$, define

$$p_1(x) = |x_1| + |x_2|, \quad p_2(x) = \sqrt{x_1^2 + x_2^2} \quad \text{and} \quad p_\infty(x) = \max\left\{|x_1|, |x_2|\right\}.$$

Then, p_2 is a paranorm, but p_1 and p_∞ are not paranorms on $\mathbb{R}^2$.

Since p_2 can be derived from the usual inner product on $\mathbb{R}^2$, by Theorem 18.9 and Remark 18.19, it is clear that p_2 is a paranorm on $\mathbb{R}^2$.

Moreover, by taking

$$a = (1, 0), \qquad b = (0, 1), \qquad c = (1, 1), \qquad d = (-1, 1),$$

we can see that

$$2\,p_1(c)^2 + 2\,p_1(d)^2 = 2^4 \qquad \text{but} \qquad p_1(c + d)^2 + p_1(c - d)^2 = 2^3$$

and

$$2\,p_\infty(a)^2 + 2\,p_\infty(b)^2 = 2^2 \qquad \text{but} \qquad p_\infty(a + b)^2 + p_\infty(a - d)^2 = 2.$$

Therefore, by the assertion (5) of Theorem 18.10, the norms p_1 and p_∞ cannot be paranorms on $\mathbb{R}^2$.

18.6 Some Further Theorems on Parapreseminorms

Now, by using Theorem 18.10 and an argument given in the second part of [49, Remark 3.2] of Gilányi and Troczka-Pawelec, we can also prove the following

Theorem 18.11 *If p is a subadditive parapreseminorm on X, then for any $x, y \in X$ we have*

(1) $p(x+y)^2 - p(x-y)^2 \le 4p(x)p(y)$,
(2) $0 \le p(x+y)^2 + p(x-y)^2 - 2p(x)^2 - 2p^2(y) \le 4p(x)p(y)$.

Proof If $x, y \in X$, then by using the subadditivity of p, we can see that

$$p(x+y)^2 \le \big(p(x) + p(y)\big)^2 = p(x)^2 + p(y)^2 + 2p(x)p(y).$$

Hence, by using (5) in Theorem 18.10, we can infer that

$$2p(x+y)^2 \le 2p(x)^2 + 2p(y)^2 + 4p(x)p(y)$$
$$\le p(x+y)^2 + p(x-y)^2 + 4p(x)p(y).$$

Therefore, (1) also holds.

On the other hand, from (5) in Theorem 18.10, it is clear that the first part of (2) is true. Moreover, form the fact that

$$p(x+y)^2 \le p(x)^2 + p(y)^2 + 2p(x)p(y)$$

for all $x, y \in X$, by using (2) in Theorem 18.10, we can infer that

$$p(x-y)^2 \le p(x)^2 + p(-y)^2 + 2p(x)p(-y) = p(x)^2 + p(y)^2 + 2p(x)p(y)$$

for all $x, y \in X$. Therefore, we can also state that

$$p(x+y)^2 + p(x-y)^2 \le 2p(x)^2 + 2p(y)^2 + 4p(x)p(y)$$

for all $x, y \in X$. Thus, the second part of (2) is also true.   □

Moreover, by using Theorem 18.10 and an argument made with the help of Gy. Maksa and Z. Boros in the proof of [122, Theorem 8.1], we can also prove

Theorem 18.12 *If p is a parapreseminorm on X such that*
(1) $p(2x) \le 2p(x)$,   *(2)* $p(x+y+z) \le p(x+z+y)$,
for all $x, y, z \in X$, then the corresponding equalities also hold, and moreover we have

$$p(x+y)^2 + p(x-y)^2 = 2p(x)^2 + 2p(x)^2$$

for all $x, y \in X$.

Proof To check the latter equality, for any $x \in X$, define

$$f(x) = p(x)^2.$$

Then, by Theorem 18.10, for any $x, y \in X$ we have

$$2 f(x) + 2 f(y) \le f(x + y) + f(x - y).$$

Hence, it is clear that, for any $u, v \in X$, we have

$$2 f(u + v) + 2 f(u - v) \le f(u + v + u - v) + f\big(u + v - (u - v)\big).$$

Moreover, by using assumptions (2), (1) and the definition of f, we can see that

$$f(u + (v + u) - v) \le f(u - v + v + u) = f(2u) \le 4 f(u)$$

and

$$f\big(u + v - (u - v)\big) = f(u + 2v - u) \le f(u - u + 2v) = f(2v) \le 4 f(v).$$

Therefore,

$$2 f(u + v) + 2 f(u - v) \le 4 f(u) + 4 f(v),$$

and thus $f(u + v) + f(u - v) \le 2 f(u) + 2 f(v)$ also holds. This shows that we actually have

$$f(x + y) + f(x - y) = 2 f(x) + 2 f(y)$$

for all $x, y \in X$. Therefore, by the definition of f, the required equality is also true.

□

Remark 18.22 Note that if p is a nonnegative function of the group X such that

$$p(x + y)^2 + p(x - y)^2 \le 2 p(x)^2 + 2 p(y)^2$$

for all $x, y \in X$, then by putting $x = y$ we can already get

$$p(2x)^2 \le p(2x)^2 + p(0)^2 \le 4 p(x)^2 = \big(2 p(x)\big)^2,$$

and thus $p(2x) \le 2 p(x)$ for all $x \in X$. Therefore, condition (1) follows from the assertion of the theorem.

The necessity of the Kannappan condition (2) could certainly be demonstrated only with a much more difficult argument. In this respect, it is noteworthy that, by Stetkaer [114, Example 13.20 and Lemma B.4] there exists an integer-valued quadratic function f of a free group, with three generators, such that the Kannappan condition does not hold for f.

Now, by using Theorem 18.12 and an improvement of an argument of Ger [43], we can also prove the following addition to Theorem 18.11.

Corollary 18.1 *If p is as in Theorem 18.12, then for any $x, y \in X$, the following assertions are equivalent:*

(1) $p(x + y) \leq p(x) + p(y)$,   *(2) $p(x + y)^2 - p(x - y)^2 \leq 4 p(x) p(y)$.*

Proof By the nonnegativity of p and the assertion of Theorem 18.12, it is clear that the following inequalities are equivalent:

$$p(x + y) \leq p(x) + p(y),$$

$$p(x + y)^2 \leq \big(p(x) + p(y)\big)^2,$$

$$p(x + y)^2 \leq p(x)^2 + p(y)^2 + 2 p(x) p(y),$$

$$2 p(x + y)^2 \leq 2 p(x)^2 + 2 p(y)^2 + 4 p(x) p(y),$$

$$2 p(x + y)^2 \leq p(x + y)^2 + p(x - y)^2 + 4 p(x) p(y),$$

$$p(x + y)^2 - p(x - y)^2 \leq 4 p(x) p(y).$$

$\square$

Moreover, by using Theorem 18.12 and an argument of Ger [43], we can also prove the following

Theorem 18.13 *If p is as in Theorem 18.12, then p is already a seminorm on X.*

Proof By Theorem 18.12 and Remark 18.16, there exists a real-valued semi-inner product P on X such that $p(x)^2 = P(x, x)$ for all $x \in X$. Thus, by Remark 18.12, the required assertion is also true. $\square$

18.7 Characterizations of Additive and Jensen Functions

Now, we are ready to prove the following generalization of Theorem 18.1 of Maksa and Volkmann which is actually a hyperstability result. (See [17].)

Theorem 18.14 *If f is a function of one group X to another Y, q is a paraprenorm on Y, then the following assertions are equivalent:*

(1) $f(x) + f(y) = f(x + y)$ for all $x, y \in X$,
(2) $q\big(f(x) + f(y)\big) = q\big(f(x + y)\big)$ for all $x, y \in X$,
(3) $q\big(f(x) + f(y)\big) \leq q\big(f(x + y)\big)$ for all $x, y \in X$.

Proof Actually, only the implication (3) $\Longrightarrow$ (1) has to be proved. For this, note that if (3) holds, then in particular, by (3) in Theorem 18.10, we have

$$2q\left(f\left(0\right)\right) \leq q\left(2f\left(0\right)\right) \leq q\left(f\left(0\right)\right),$$

and thus $q\left(f\left(0\right)\right) \leq 0$. Therefore, $q\left(f\left(0\right)\right) = 0$, and thus $f\left(0\right) = 0$.

Now, by using (3), we can also see that

$$q\left(f\left(x\right) + f\left(-x\right)\right) \leq q\left(f\left(0\right)\right) = q\left(0\right) = 0,$$

and thus $f\left(x\right) + f\left(-x\right) = 0$ for all $x \in X$. Therefore, f is odd.

Hence, by using (2) in Theorem 18.10, we can see that

$$q\left(f\left(-x\right)\right) = q\left(-f\left(x\right)\right) = q\left(f\left(x\right)\right),$$

and thus

$$q\left(f\left(-y - x\right)\right) = q\left(f\left(-\left(x + y\right)\right)\right) = q\left(f\left(x + y\right)\right)$$

for all $x, y \in X$.

Now, by using the above facts, (d) in Definition 18.3, (4) in Theorem 18.10 and assertion (3), we can also see that

$$q\left(f\left(x + y\right) - f\left(y\right) - f\left(x\right)\right)^2 = q\left(f\left(x + y\right) + f\left(-y\right) + f\left(-x\right)\right)^2$$
$$\leq q\left(f\left(x+y\right) + f\left(-y\right)\right)^2 + q\left(f\left(x+y\right) + f\left(-x\right)\right)^2 + q\left(f\left(-y\right) + f\left(-x\right)\right)^2$$
$$- q\left(f\left(x + y\right)\right)^2 - q\left(f\left(-y\right)\right)^2 - q\left(f\left(-x\right)\right)^2 =$$
$$q\left(f\left(x + y\right) + f\left(-y\right)\right)^2 + q\left(f\left(-x\right) + f\left(x + y\right)\right)^2 + q\left(f\left(-y\right) + f\left(-x\right)\right)^2$$
$$- q\left(f\left(x + y\right)\right)^2 - q\left(f\left(y\right)\right)^2 - q\left(f\left(x\right)\right)^2 \leq$$
$$q\left(f\left(x\right)\right)^2 + q\left(f\left(y\right)\right)^2 + q\left(f\left(x + y\right)\right)^2$$
$$- q\left(f\left(x + y\right)\right)^2 - q\left(f\left(y\right)\right)^2 - q\left(f\left(x\right)\right)^2 = 0$$

for all $x, y \in X$. Therefore, we necessarily have

$$q\left(f\left(x + y\right) - f\left(y\right) - f\left(x\right)\right) = 0, \quad \text{and hence} \quad f\left(x + y\right) - f\left(y\right) - f\left(x\right) = 0$$

for all $x, y \in X$. Thus, (1) also holds. $\qquad\square$

From this theorem, following the ideas of Kwon et al. [75], we can derive

Corollary 18.2 *If f is a 2-homogeneous function of one uniquely 2-divisible group X to another Y and q is a 2-subhomogeneous paraprenorm on Y, then the following assertions are equivalent:*

(1) $2^{-1}\big(f(x) + f(y) \big) = f\big(2^{-1}(x+y) \big)$ *for all* $x, y \in X$,
(2) $q\big(2^{-1}(f(x) + f(y)) \big) = q\big(f(2^{-1}(x+y)) \big)$ *for all* $x, y \in X$,
(3) $q\big(2^{-1}(f(x) + f(y)) \big) \leq q\big(f(2^{-1}(x+y)) \big)$ *for all* $x, y \in X$.

Proof Actually now, by Theorem 18.10, q is also 2-homogeneous. Therefore, if (3) holds, then we have

$$q\big(f(x) + f(y) \big) = q\left(2\left(2^{-1}(f(x) + f(y)) \right) \right) = 2q\left(2^{-1}(f(x) + f(y)) \right)$$
$$\leq 2q\left(f\left(2^{-1}(x+y) \right) \right) = q\left(2f\left(2^{-1}(x+y) \right) \right)$$
$$= q\left(f\left(2(2^{-1}(x+y)) \right) \right) = q\big(f(x+y) \big)$$

for all $x, y \in X$. Hence, by using Theorem 18.14, we can infer that

$$f(x) + f(y) = f(x+y),$$

and thus

$$2^{-1}\big(f(x) + f(y) \big) = 2^{-1} f(x+y) = 2^{-1} f\left(2\left(2^{-1}(x+y) \right) \right)$$
$$= 2^{-1} 2 f\left(2^{-1}(x+y) \right) = f\left(2^{-1}(x+y) \right)$$

for all $x, y \in X$. Therefore, (1) also holds. $\square$

Remark 18.23 Note that if q is a parapreseminorm on a uniquely 2-divisible group Y, then by Theorem 18.10, for any $y \in Y$, we also have

$$q\left(2^{-1}y \right) = 2^{-1}\left(2q(2^{-1}y) \right) \leq 2^{-1} q\left(2(2^{-1}y) \right) = 2^{-1} q(y).$$

Therefore, in this case q is not only 2-superhomogeneous, but also 2^{-1}-subhomogeneous.

Remark 18.24 In this respect, it is also worth mentioning that if q is 2-subhomogeneous parapreseminorm on a uniquely 2-divisible commutative group Y, then by using two arguments of Rosenbaum [104], utilized also in [18], it can be shown that q is subadditive if and only if q is 2^{-1}-convex in the sense that

$$q\left(2^{-1}y + 2^{-1}z \right) \leq 2^{-1} q(y) + 2^{-1} q(z)$$

for all $y, z \in Y$.

Remark 18.25 Note that if in particular the groups X and Y considered in Corollary 18.2 are commutative, then the Jensen equation, considered in (1), can also be written in the more instructive form

$$2^{-1} f(x) + 2^{-1} f(y) = f\left(2^{-1} x + 2^{-1} y\right).$$

Namely, if for instance $x, y \in X$, then by the above assumptions

$$2\left(2^{-1} x + 2^{-1} y\right) = 2\left(2^{-1} x\right) + 2\left(2^{-1} y\right) = x + y,$$

and thus $2^{-1}(x + y) = 2^{-1} x + 2^{-1} y$ also holds. (See also Ger and Koclęga [45].)

Remark 18.26 In this respect, it is also worth noticing that by considering the modified Jensen equations

$$f(x + y) + f(x - y) = 2 f(x) \qquad \text{and} \qquad 2 f(x + y) = f(2x) + f(2y)$$

instead of the original one, one can certainly prove some counterparts of Corollary 18.2 under possibly weaker assumptions.

18.8 Further Characterizations of Additive and Jensen Functions

By using Theorem 18.14, we can also easily prove the following

Theorem 18.15 *If f is a function of one group X to another Y, Q is an inner product on Y and*

$$q(y) = \sqrt{Q(y, y)}$$

for all $y \in Y$, then the following assertions are equivalent:

(1) $f(x) + f(y) = f(x + y)$ *for all* $x, y \in X$,

(2) $2 Q_1\big(f(x), f(y)\big) = q\big(f(x + y)\big)^2 - q\big(f(x)\big)^2 - q\big(f(y)\big)^2$ *for all* $x, y \in X$,

(3) $2 Q_1\big(f(x), f(y)\big) \le q\big(f(x + y)\big)^2 - q\big(f(x)\big)^2 - q\big(f(y)\big)^2$ *for all* $x, y \in X$.

Proof If (1) holds, then by using (7) in Theorem 18.3 we can see that

$$2 Q_1\big(f(x), f(y)\big) = q\big(f(x) + f(y)\big)^2 - q\big(f(x)\big)^2 - q\big(f(y)\big)^2$$

$$= q\big(f(x + y)\big)^2 - q\big(f(x)\big)^2 - q\big(f(y)\big)^2$$

for all $x, y \in X$. Therefore, (2) also holds.

Now, since (2) trivially implies (3), it remains only to prove that (3) also implies (1). For this, note that if (3) holds, then by (7) in Theorem 18.3 we have

$$q\left(f(x)+f(y)\right)^2 = q\left(f(x)\right)^2 + q\left(f(y)\right)^2 + 2\,Q_1\left(f(x),\ f(y)\right)$$
$$\leq q\left(f(x)\right)^2 + q\left(f(y)\right)^2 + q\left(f(x+y)\right)^2 - q\left(f(x)\right)^2 - q\left(f(y)\right)^2$$
$$= q\left(f(x+y)\right)^2$$

for all $x,\,y \in X$. Therefore, by the nonnegativity of q, we also have

$$q\left(f(x)+f(y)\right) \leq q\left(f(x+y)\right)$$

for all $x,\,y \in X$. Hence, by Remark 18.19 and Theorem 18.14, we can see that (1) also holds.   □

From this theorem, analogously to Corollary 18.2, we can derive

Corollary 18.3 *If f is a 2-homogeneous function of one uniquely 2-divisible group X to another Y, Q is an inner product on Y and*

$$q(y) = \sqrt{Q(y,y)}$$

for all $y \in Y$, then the following assertions are equivalent:

(1) $2^{-1}\left(f(x)+f(y)\right) = f\left(2^{-1}(x+y)\right)$ *for all $x,\,y \in X$,*
(2) $2\,Q_1\left(2^{-1}f(x),\ 2^{-1}f(y)\right) = q\left(f\left(2^{-1}(x+y)\right)\right)^2 - q\left(2^{-1}f(x)\right)^2$
    $-q\left(2^{-1}f(y)\right)^2$ *for all $x,\,y \in X$,*
(3) $2\,Q_1\left(2^{-1}f(x),\ 2^{-1}f(y)\right) \leq q\left(f\left(2^{-1}(x+y)\right)\right)^2 - q\left(2^{-1}f(x)\right)^2$
    $-q\left(2^{-1}f(y)\right)^2$ *for all $x,\,y \in X$,*

Proof If (1) holds, then by using Theorem 18.3 and Remarks 18.5 and 18.23 we can see that

$$2\,Q_1\left(2^{-1}f(x),\ 2^{-1}f(y)\right)$$
$$= q\left(2^{-1}f(x)+2^{-1}f(y)\right)^2 - q\left(2^{-1}f(x)\right)^2 - q\left(2^{-1}f(y)\right)^2$$
$$= q\left(2^{-1}\left(f(x)+f(y)\right)\right)^2 - q\left(2^{-1}f(x)\right)^2 - q\left(2^{-1}f(y)\right)^2$$
$$= q\left(f\left(2^{-1}(x+y)\right)\right)^2 - q\left(2^{-1}f(x)\right)^2 - q\left(2^{-1}f(y)\right)^2$$

for all $x,\,y \in X$. Therefore, (2) also holds.

While, if (3) holds, then by using Theorems 18.2 and 18.3 we can see that

$$
\begin{aligned}
2\,Q_1\big(f(x),\ f(y)\big) &= 2\,Q_1\left(2\left(2^{-1}f(x)\right),\ 2\left(2^{-1}f(x)\right)\right) \\
&= 2^2\,2\,Q_1\left(2^{-1}f(x),\ 2^{-1}f(y)\right) \\
&\le 2^2\left(q\left(f\left(2^{-1}(x+y)\right)\right)^2 - q\left(2^{-1}f(x)\right)^2 - q\left(2^{-1}f(y)\right)^2\right) \\
&= \left(2q\left(f\left(2^{-1}(x+y)\right)\right)\right)^2 - \left(2q\left(2^{-1}f(x)\right)\right)^2 - \left(2q\left(2^{-1}f(y)\right)\right)^2 \\
&= q\left(2f\left(2^{-1}(x+y)\right)\right)^2 - q\left(f(x)\right)^2 - q\left(f(y)\right)^2 \\
&= q\left(f(x+y)\right)^2 - q\left(f(x)\right)^2 - q\left(f(y)\right)^2
\end{aligned}
$$

for all $x, y \in X$. $\square$

Remark 18.27 Note that the proofs of Theorems 18.14 and 18.15 do not require any particular trick. Therefore, they are more simple than the one given by Maksa and Volkmann [82] which was actually based on assertion (3) of Theorem 18.7.

In a former version of this paper [121], we have used assertion (4) of Theorem 18.7 to prove the implication (3) $\Longrightarrow$ (1) of Theorem 18.14. However, assertion (5) of Theorem 18.7 seems to be a more convenient tool than (3) and (4).

Remark 18.28 In this respect, it is also worth mentioning that particular cases of assertion (3) of Theorem 18.8 can certainly be also applied to some other important functional inequalities.

Such ones may be the quadratic and Drygas inequalities studied by Gillányi [47], Rätz [100], Fechner [31] and Manar and Elqorach [83]. And, the ones derivable from the modified quadratic and Cauchy equations considered by Kominek [64] and Stetkaer [114, p. 2.15].

Remark 18.29 However, note that if for instance f is an even, subadditive function of group X to $\mathbb{R}$, then f is nonnegative, and thus

$$
|f(x+y)| = f(x+y) \le f(x) + f(y) = |f(x) + f(y)|
$$

for all $x, y \in \mathbb{R}$.

Moreover, if f is an odd isometry of an arbitrary preseminormed group X to a 2-cancellable one Y, then $f(0) = 0$, and thus

$$
\begin{aligned}
\|f(x+y)\| &= \|f(x+y) - f(0)\| = \|x+y-0\| \\
&= \|x-(-y)\| = \|f(x) - f(-y)\| = \|f(x) + f(y)\|
\end{aligned}
$$

for all $x, y \in X$.

And, according to the easier part of a famous characterization theorem of Ger [42], the composition of an odd isometry and an additive function is also a solution of the corresponding equality.

Therefore, to prove certain counterparts of Theorems 18.14 and 18.15 some additional requirements will be needed. For some ideas in this respect, see [74, Theorem 2] of Kurepa which should also be proved with the help of assertion (5) of Theorem 18.7.

Remark 18.30 To justify the appropriateness of our present treatment, we can also note that an application of Theorem 18.14, to the proof of the left-invariance of some lower left-invariant generalized metrics, will be given in our forthcoming paper *Semimetrics and preseminorms on groups.*

Remark 18.31 Meantime, we have observed that, before us [14], Multarzyński [86] also considered semi-inner products on certain groupoids.

Moreover, before Boros [11] and Ger [43], Kurepa [71, 72] also proved Schwarz inequalities on groups.

The results of Kurepa strongly suggest that instead semi-inner products it may be more convenient to start with some quadratic functions.

Acknowledgements The work of the author has been supported by the Hungarian Scientific Research Fund (OTKA) Grant K-111651.

Moreover, the author is greatly indebted to Zoltán Boros, Gyula Maksa, Attila Gilányi and Jens Schwaiger for some inspiring conversations.

References

1. Aczél, J., Dhombres, J.: Functional Equations in Several Variables. Cambridge University Press, Cambridge (1989)
2. Alsina, C., Sikorska, J., Tomás, M. S.: Norm Derivatives and Characterizations of Inner Product Spaces. World Scientific, New Yersey (2010)
3. Amir, D: Characterizations of Inner Product Spaces. Birkhäuser, Besel (1986)
4. Antoine, J.-P., Grossmann, A.: Partial inner product spaces. I. General properties. J. Funct. Anal. **23**, 369–378 (1976)
5. Baker J.A.: On quadratic functionals continuous along rays. Glasnik Mat. **23**, 215–229 (1968)
6. Baron, K.: On additive involutions and Hamel bases. Aquationes Math. **87**, 159–163 (2014)
7. Baron, K.: Orthogonally additive bijections are additive. Aequationes Math. **89**, 297–299 (2015)
8. Batko, B, Tabor, J.: Stability of an alternative Cauchy equation on a restricted domain. Aequationes Math. **57**, 221–232 (1999)
9. Batko, B., Tabor, J.: Stability of the generalized alternative Cauchy equation. Abh. Math. Sem. Univ. Hamburg **69**, 67–73 (1999)
10. Bognár, J.: Indefinite Inner Product Spaces. Springer, Berlin (1974)
11. Boros, Z.: Schwarz inequality over groups. In: Talk Held at the Conference on Inequalities and Applications, Hajdúszoboszló, Hungary (2016)
12. Boros, Z., Száz, Á.: Semi-inner products and their induced seminorms and semimetrics on groups, Technical Report, 2016/6, 11 pp. Institute of Mathematics, University of Debrecen, Debrecen (2016)

13. Boros, Z., Száz, Á.: A weak Schwarz inequality for semi-inner products on groupoids, Rostock. Math. Kolloq. **71**, 28–40 (2016)

14. Boros, Z, Száz, Á.: Generalized Schwarz inequalities for generalized semi-inner products on groupoids can be derived from an equality. Novi Sad J. Math. **47**, 177–188 (2017)

15. Boros, Z., Száz, Á.: Infimum problems derived from the proofs of some generalized Schwarz inequalities, Teaching Math. Comput. Sci. **17**, 41–57 (2019)

16. Brzdęk, J.: A note on stability of the Popoviciu functional equation on restricted domain. Demonstration Math. **43**, 635–641 (2010)

17. Brzdęk, J., Ciepliński, K: Hyperstability and superstability. Abst. Appl. Anal. **2013**, 13 (2013)

18. Burai, P., Száz, Á.: Relationships between homogeneity, subadditivity and convexity properties. Univ. Beograd. Publ. Elektrotehn. Fak. Ser. Mat. **16**, 77–87 (2005)

19. Chmieliński, J.: Normed spaces equivalent to inner product spaces and stability of functional equations. Aequationes Math. **87**, 147–157 (2014)

20. Chmieliński, J.: A note on a characterization of metrics generated by norms. Rocky Mt. J. Math. **45**, 1801–1805 (2015)

21. Chmieliński, J.: On functional equations related to additive mappings and isometries. Aequationes Math. **89**, 97–105 (2015)

22. Chung, S.-C., Lee, S.-B., Park, W.-G.: On the stability of an additive functional inequality. Int. J. Math. Anal. **6**, 2647–2651 (2012)

23. Dhombres, J.G.: Some Aspects of Functional Equations. Chulalongkorn University, Bangkok (1979)

24. Dong, Y.: Generalized stabilities of two functional equations. Aequationes Math. **86**, 269–277 (2013)

25. Dong, Y., Chen, L.: On generalized Hyers-Ulam stability of additive mappings on restricted domains of Banach spaces. Aequationes Math. **90**, 871–878 (2016)

26. Dong, Y., Zheng, B.: On hyperstability of additive mappings onto Banach spaces. Bull. Aust. Math. Soc. **91**, 278–285 (2015)

27. Dragomir, S.S.: Semi-Inner Products and Applications. Nova Science Publishers, Hauppauge (2004)

28. Drygas, H.: Quasi-inner products and their applications. In: Gupta, A.K. (Ed.) Advances in Multivariate Statistical Analysis. Theory and Decision Library (Series B: Mathematical and Statistical Methods), vol. 5, pp. 13–30. Reidel, Dordrecht (1987)

29. Ebanks, B.R., Kannappan Pl., Sahoo, P.K.: A common generalization of functional equations characterizing normed and quasi-inner-product spaces. Canad. Math. Bull. **35**, 321–327 (1992)

30. Elqorachi, E., Manar, Y., Rassias, Th.M.: Hyers-Ulam stability of the quadratic functional equation. In: Rassias, Th.M., Brdek, J. (Eds.) Functional Equations in Mathematical Analysis. Springer Optimization and Its Applications, vol. 52, pp. 97–105 (2012)

31. Fechner, W.: Stability of a functional inequality associated with the Jordan-von Neumann functional equation. Aequationes Math. **71**, 149–161 (2006)

32. Fechner, W.: On the Hyers-Ulam stability of functional equations connected with additive and quadratic mappings. J. Math. Anal. Appl. **322**, 774–786 (2006)

33. Fechner, W.: Hlawka's functional inequality. Aequationes Math. **87**, 71–87 (2014)

34. Fechner, W.: Hlawka's functional inequality on topological groups. Banach J. Math. Anal. **11**, 130–142 (2017)

35. Fischer, P., Muszély, Gy.: Some generalizations of Cauchy's functional equations (Hungarian). Mat. Lapok **16**, 67–75 (1965)

36. Fischer, P., Muszély, Gy.: On some new generalizations of the functional equation of Cauchy. Can. Math. Bull. **10**, 197–205 (1967)

37. Fochi, M.: An alternative functional equation on restrictef domain. Aequationes Math. **70**, 201–212 (2005)

38. Fréchet, M.: Sur lá definition axiomatique d'une classe d'espaces vectoriels distanciés applicables vectoriellement sur l'espace de Hilbert. Ann. Math. **36**, 705–718 (1935)

39. Găvruta, P.: On the Hyers-Ulam-Rassias stability of the quadratic mappings. Nonlinear Funct. Anal. Appl. **9**, 415–428 (2004)
40. Ger, R.: On a characterization of strictly convex spaces. Atti Acad. Sci. Torino Cl. Sci. Fis. Mat. Natur. **127**, 131–138 (1993)
41. Ger, R.: A Pexider-type equation in normed linear spaces. Sitzungsber. Abt. II **206**, 291–303 (1997)
42. Ger, R.: Fischer–Muszély additivity on Abelian groups. Comment Math. Tomus Specialis in Honorem Juliani Musielak, pp. 83–96 (2004)
43. Ger, R.: On a problem of Navid Safaei. In: Talk Held at the Conference on Inequalities and Applications, Hajdúszoboszló, Hungary (2016)
44. Ger, R.: Fischer-Muszély additivity – a half century story. In: Brzdek, J., Ciepliński, K., Rassias, Th.M. (eds.) Developments in Functional Equations and Related Topics. Springer Optimization and Its Applications, pp. 69–102. Springer, Basel (2017)
45. Ger, R., Koclega, B.: Isometries and a generalized Cauchy equation. Aequationes Math. **60**, 72–79 (2000)
46. Ger, R., Koclega, B.: An interplay between Jensen's and Pexider's functional equations on semigroups. Ann. Univ. Sci. Budapest Sect Comp. **35**, 107–124 (2011)
47. Gilányi, A.: Eine zur Parallelogrammgleichung äquivalente Ungleichung. Aequationes Math. **62**, 303–309 (2001)
48. Gilányi, A.: On a problem by K. Nikodem. Math. Ineq. Appl. **5**, 707–710 (2002)
49. Gilányi, A., Troczka-Pawelec, K.: Regularity of weakly subquadratic functions. J. Math. Anal. Appl. **382**, 814–821 (2011)
50. Gilányi, A., Kézi, Cs., Troczka-Pawelec, K.: On two different concepts of subquadraticity. Inequalities and Applications 2010. International Series of Numerical Mathematics, vol. 161, pp. 209–215. Birkhäuser, Basel (2012)
51. Giles, J.R.: Classes of semi-inner-product spaces. Trans. Am. Math. Soc. **129**, 436–446 (1967)
52. Glavosits, T., Száz, Á.: Divisible and cancellable subsets of groupoids. Ann. Math. Inform. **43**, 67–91 (2014)
53. Golab, S., Światak, H.: Note on inner products in vector spaces. Aequationes Math. **8**, 74–75 (1972)
54. Haruki, H.: On the functional equations $| f (x+iy)| = | f (x)+ f (i y)|$ and $| f (x+iy)| = | f (x) - f (i y)|$ and on Ivory's theorem. Canad. Math. Bull. **9**, 473–480 (1966)
55. Haruki, H.: On the equivalence of Hille's and Robinson's functional equations. Ann. Polon. Math. **28**, 261–264 (1973)
56. Hosszú, M.: On an alternative functional equation (Hungarian). Mat. Lapok **14**, 98–102 (1963)
57. Hosszú, M.: A remark on the square norm. Aequationes Math. **2**, 190–193 (1969)
58. Istrǎtescu, V.I.: Inner Product Structures, Theory and Applications. Reidel, Dordrecht (1987)
59. Joichi, J.T.: Normed linear spaces equivalent to inner product spaces. Proc. Am. Math. Soc. **17**, 423–426 (1966)
60. Jordan, P., von Neumann, J.: On inner products in linear metric spaces. Ann. Math. **36**, 719–723 (1935)
61. Jung, S.-M.: On the Hyers-Ulam stability of the functional equations that have the quadratic property. J. Math. Anal. Appl. **222**, 126–137 (1998)
62. Kannappan, Pl.: Quadratic functional equation and inner product spaces. Results Math. **27**, 368–372 (1995)
63. Kannappan, Pl.: Functional Equations and Inequalities with Applications. Springer, Dordrecht (2009)
64. Kominek, Z.: Stability of a quadratic functional equation on semigroups. Publ. Math. Debrecen **75**, 173–178 (2009)
65. Kominek, Z.: On a Dygras inequality. Tatra Mt. Math. Publ. **52**, 65–70 (2012)
66. Kominek, Z., Troczka, K.: Some remarks on subquadratic functions. Demonstr. Math. **39**, 751–758 (2006)

67. Kominek, Z., Troczka, K.: Continuity of real valued subquaratic functions. Comment. Math. **51**, 71–75 (2011)
68. Kuczma, M.: On some alternative functional equations. Aequationes Math. **17**, 182–198 (1978)
69. Kuczma, M.: An Introduction to the Theory of Functional Equations and Inequalities. Państwowe Wydawnictwo Naukowe, Warszawa (1985)
70. Kurepa, S.: The Cauchy functional equation and scalar product in vector spaces. Glasnik Mat. **19**, 23–35 (1964)
71. Kurepa, S.: Quadratic and sesquilinear functionals. Glasnik Mat. **20**, 79–91 (1965)
72. Kurepa, S.: On bimorphisms and quadratic forms on groups. Aequationes Math. **9**, 30–45 (1973)
73. Kurepa, S.: On the definition of a quadratic form. Publ. Ints. Math (Beograd) **42**, 35–41 (1987)
74. Kurepa, S.: On P. Volkmann's paper. Glasnik Mat. **22**, 371–374 (1987)
75. Kwon, Y.H., Lee, H.M., Sim, J.S., Yand, J., Park, C.: Jordan-von Neumann type functional inequalities. J. Chungcheong Mah. Soc. **20**, 269–277 (2007)
76. Lee, Y.-H., Jung, S.-M.: A general uniqueness theorem concerning the stability of additive and quadratic equations. J. Funct. Spaces **2015**, 8 (2015). Art. ID 643969
77. Lee, J.R., Park, C., Shin, D.Y.: On the stability of generalized additive functional inequalities in Banach spaces. J. Ineq. Appl. **2008**, 13 (2008). Art. ID 210626
78. Lumer, G.: Semi-inner-product spaces. Trans. Am. Math. Soc. **100**, 29–43 (1961)
79. Makai, I.: Über invertierbare Lösungen der additive Cauchy-Functionalgleichung. Publ. Math. Debrecen **16**, 239–243 (1969)
80. Maksa, Gy.: A remark of symmetric biadditive functions having nonnegative diagonalization. Glasnik Mat. **15**, 279–282 (1980)
81. Maksa, Gy.: Remark on the talk of P. Volkmann. In: Proceedings of the Twenty-third International Symposium on Functional Equations. Gargano, Italy (1985)
82. Maksa, Gy., Volkmann, P.: Characterizations of group homomorphisms having values in an inner product space. Publ. Math. Debrecen **56**, 197–200 (2000)
83. Manar, Y., Elqorachi, E.: On functional inequalities associated with Drygas functional equation. Tbilisi Math. J. **7**, 73–78 (2014)
84. Marinescu, D.S., Monea, M., Opincariu, M., Stroe, M.: Some equivalent characterizations of inner product spaces and their consequences. Filomat **29**, 1587–1599 (2015)
85. Mitrinović, D.S., Pečarić, J.E., Fink, A.M.: Classical and New Inequalities in Analysis. Kluwer, Dordrecht (1993)
86. Multarziński, P.: Semi-inner product structures for groupoids (2013). arXiv:1301.0764v1 [math.GR]
87. Najati, A., Rassias, Th.M.: Stability of mixed functional equation in several variables on Banach modules. Nonlinear Anal. **72**, 1755–1767 (2010)
88. Nakmahachalasint, P.: On the generalized Ulam-Gavruta-Rassias stability of mixed-type linear and Euler-Lagrange-Rassias functional equations. Int. J. Math. Math. Sci. **2007**, 10 (2007). At. ID 63239
89. Nakmahachalasint, P.: Hyers-Ulam-Rassias and Ulam-Gavruta-Rassias stabilities of an additive functional equation in several variables. Int. J. Math. Math. Sci. **2007**, 6 (2007). At. ID 13437
90. Ng, C.T., Zhao, H.Y.: Kernel of the second order Cauchy difference on groups. Aequationes Math. **86**, 155–170 (2013)
91. Nikodem, K.: 7. Problem. Aequationes Math. **61**, 301 (2001)
92. Oikhberg, T., Rosenthal, H.: A metric characterization of normed linear spaces. Rocky Mont. J. Math. **37**, 597–608 (2007)
93. Oubbi, L.: Ulam-Hyers-Rassias stability problem for several kinds of mappings. Afr. Mat. **24**, 525–542 (2013)
94. Park, Ch., Cho, Y.S., Han, M.-H. Functional inequalities associated with Jordan–von Neumann type additive functional equations. J. Ineq. Appl. **2007**, 13 (2007). Art. ID 41820

95. Park, C., Lee, J.R., Rassias, Th.M.: Functional inequalities in Banach spaces and fuzzy Banach spaces. In: Rassias, Th.M., Pardalos, P.M. (eds.) Essays in Mathematics and Its Applications, pp. 263–310. Springer, Cham (2016)

96. Parnami, J.C., Vasudeva, H.L.: On Jensen's functional equation. Aequationes Math. **43**, 211–218 (1992)

97. Piejko, K.: Note on Robinson's Functional equation. Demonstratio Math. **32**, 713–715 (1999)

98. Popoviciu, T.: Sur certaines inéqualités qui caractérisentes fonctions convexes. An. Stiint. Univ. Al. I. Cuza Iasi **11**, 155–164 (1965)

99. Rassias, J.M.: On the Ulam stability of mixed type mappings on restricted domains. J. Math. Anal. Appl. **276**, 747–762 (2002)

100. Rätz, J.: On inequalities associated with the Jordan-von Neumann functional equation. Aequationes Math. **66**, 191–200 (2003)

101. Robinson, R.M.: A curious trigonometric identity. Am. Math. Monthly **64**, 83–85 (1957)

102. Roh, J., Chang, I.-S.: Functional inequalities associated with additive mappings. Abst. Appl. Anal. **2008**, 11 (2008). Art. ID 136592

103. Röhmer, J.: Ein Charakterisierung quadratischer Formen durch eine Functionalgleichung. Aequationes Math. **15**, 163–168 (1977)

104. Rosenbaum, R.A.: Sub-additive functions. Duke Math. J. **17**, 227–247 (1950)

105. Schöpf, P.: Solutions of $\| f(\xi + \eta) \| = \| f(\xi) + f(\eta) \|$. Math. Pannon. **8**, 117–127 (1997)

106. Šemrl, P.: A characterization of normed spaces. J. Math. Anal. Appl. **343**, 1047–1051 (2008)

107. Šemrl, P.: A characterization of normed spaces among metric spaces. Rocky Mt. J. Math. **41**, 293–298 (2011)

108. Sikorska, J.: Stability of the preservation of the equality of distance. J. Math. Anal. Appl. **311**, 209–217 (2005)

109. Sinopoulos, P.: Functional equations on semigroups. Aequationes Math. **59**, 255–261 (2000)

110. Skof, F.: On the functional equation $\| f(x+y) - f(x) \| = \| f(y) \|$. Atti Acad. Sci. Torino Cl. Sci. Fis. Mat. Natur. **127**, 229–237 (1993)

111. Skof, F.: On two conditional forms of the equation $\| f(x + y) \| = \| f(x) + f(y) \|$. Aequationes Math. **45**, 167–178 (1993)

112. Skof, F.: On the stability of functional equations on a restricted domain and a related topic. In: Tabor, J., Rassias, Th.M. (eds.) Stability of Mappings of Hyers-Ulam Type, pp. 141–151. Hadronic Press, Palm Harbor (1994)

113. Smajdor, W.: On a Jensen type functional equation. J. Appl. Anal. **13**, 19–31 (2007)

114. Stetkaer, H.: Functional Equations on Groups. World Scientific, New Jersey (2013)

115. Świątak, H.: On the functional equation $f(x+y)^2 = [f(x) + f(y)]^2$. Publ. Techn. Univ. Miskolc **30**, 307–308 (1970)

116. Świątak, H., Hosszú, M.: Remarks on the functional equation $e(x, y) f(x+y) = f(x) + f(y)$. Publ. Techn. Univ. Miskolc **30**, 323–325 (1970)

117. Szabó, Gy.: A conditional Cauchy equation on normed spaces. Publ. Math. Debrecen **42**, 265–271 (1993)

118. Száz, Á.: Preseminormed spaces. Publ. Math. Debrecen **30**, 217–224 (1983)

119. Száz, Á.: An instructive treatment of convergence, closure and orthogonality in semi-inner product spaces, Technical Report, 2006/2, 29 pp. Institute of Mathematics, University of Debrecen, Debrecen (2006)

120. Száz, Á.: Applications of fat and dense sets in the theory of additive functions, Technical Report, 2007/3, 29 pp. Institute of Mathematics, University of Debrecen, Debrecen (2007)

121. Száz, Á.: A generalization of a theorem of Maksa and Volkmann on additive functions, Technical Report, 2016/5, 6 pp. Institute of Mathematics, University of Debrecen (2016)

122. Száz, Á.: Remarks and Problems at the Conference on Inequalities and Applications, Hajdúszoboszló, Hungary, 2016, Technical Report, 2016/9, 34 pp. Institute of Mathematics, University of Debrecen, Debrecen (2016)

123. Száz, Á.: A natural Galois connection between generalized norms and metrics. Acta Univ. Sapientiae Math. **9**, 360–373 (2017)

124. Száz, Á.: Generalizations of an asymptotic stability theorem of Bahyrycz, Páles and Piszczek on Cauchy differences to generalized cocycles. Stud. Univ. Babes-Bolyai Math. **63**, 109–124 (2018)
125. Tabor, J.: Stability of the Fisher–Muszély functional equation. Publ. Math. Debrecen **62**, 205–211 (2003)
126. Tabor, J., Tabor, J.: 19. Remark (Solution of the 7. Problem posed by K. Nikodem.). Aequationes Math. **61**, 307–309 (2001)
127. Toborg, I., Volkmann, P.: On stability of the Cauchy functional equation in groupoids. Ann. Math. Sil. **31**, 155–164 (2017)
128. Trif, T.: Hyers-Ulam-Rassias stability of a Jensen type functional equation. J. Math. Anal. Appl. **250**, 579–588 (2000)
129. Trif, T.: Popoviciu's and related functional equations: a survey. In: Rassias, Th.M., Andrica, D. (eds.) Inequalities and Applications, pp. 273–286. Cluj-University Press, Cluj-Napoca (2008)
130. Troczka-Pawelec, K.: Some inequalities connected with a quadratic functional equation. Pr. Nauk. Akad. Jana Dlugosza Czest. Mat. **13**, 73–79 (2008)
131. Vincze, E.: On solutions of alternative functional equations (Humgarian). Mat. Lapok **15**, 179–195 (1964)
132. Vincze, E.: Beitrag zur Theorie der Cauchyschen Functionalgleichungen. Arch. Math. **15**, 132–135 (1964)
133. Vincze, E.: Über eine Verallgemeinerung der Cauchysche functionalgleichung. Funcialaj Ekvacioj **6**, 55–62 (1964)
134. Volkmann, P.: Pour une fonction réelle f l'inéquation $|f(x) + f(y)| \leq |f(x + y)|$ et l'équation de Cauchy sont équivalentes. In: Proceedings of the Twenty-third International Symposium on Functional Equations, Gargano, Italy (1985)
135. Vrbová, P.: Quadratic functionals and bilinear forms. Časopis Pěst. Mat. **98**, 159–161 (1973)
136. Wilanski, A.: Modern Methods in Topological Vector Spaces, McGraw-Hill, New York (1978)
137. Youssef, M., Elhoucien, E.: On functional inequalities associated with Drygas functional equation, 5 pp (2014). arXiv: 1405.7942

Chapter 19
Invariant Means in Stability Theory

László Székelyhidi

Abstract This is a survey paper about the use of invariant means in the theory of Ulam type stability of functional equations. We give a summary about invariant means and we present some typical recent applications concerning stability.

Keywords Hypergroup · Stability

Mathematics Subject Classification (2010) Primary 39B82; Secondary 43A07, 39B52

19.1 Amenable Groups and G-Sets

The concept of amenable group was introduced by J. von Neumann in his paper [39]. Roughly speaking, a group is amenable, if there exists a finitely additive probability measure defined on all subsets of the group which is invariant under the action of the group on itself. The exact definition follows: given a group G we say that G is *amenable* if there exists a function $\mu : \mathscr{P}(G) \to [0, 1]$ on the power set $\mathscr{P}(G)$ of G such that

1. $\mu(A \cup B) = \mu(A) + \mu(B)$ whenever $A, B \subseteq G$ and $A \cap B = \emptyset$;
2. $\mu(G) = 1$;
3. $\mu(gA) - \mu(A)$ holds for each $A \subseteq G$ and g in G, where gA is the left translate of A by the element g.

In fact, this is the concept of *left amenability* of G as we assume that μ is invariant with respect to the left translations. Clearly, we have the respective concept referring to right translations. Originally, von Neumann introduced this concept for G-sets.

L. Székelyhidi (✉)
Institute of Mathematics, University of Debrecen, Debrecen, Hungary

© Springer Nature Switzerland AG 2019
J. Brzdęk et al. (eds.), *Ulam Type Stability*,
https://doi.org/10.1007/978-3-030-28972-0_19

Given a group G with identity e by a *G-set* we mean a nonempty set X such that G *acts on* X: there is a function $\varphi : G \times X \to X$ such that

1. Identity: $\varphi(e, x) = x$ holds for each x in X;
2. Composition: $\varphi(g, \varphi(h, x)) = \varphi(gh, x)$ holds for each g, h in G and x in X.

Such an action is called a *left action*, and instead of $\varphi(g, x)$ we write $g \cdot x$, or simply gx. The set $Gx = \{gx : g \in G\}$ is the *orbit* of the element x in X. Similarly, we can define the *right action* of G on X, denoted by xg. Accordingly, we call X a *left G-set*, or a *right G-set*. Left and right actions can be converted into each other in a natural way. Indeed, if $\varphi_r : X \times G \to X$ is a right action of G on X, then we define $\varphi_l : G \times X \to G$ by $\varphi_l(g, x) = \varphi_r(x, g^{-1})$ to get a left action of G on X. Hence it is enough to consider e.g. left actions only. Then the original formulation of von Neumann reads as follows: given a group G, a left G-set X, and a nonempty subset $S \subseteq X$ we say that *the triple* (X, S, G) *is amenable*, or simply the *G-set X is amenable* if there exists a function $\mu : \mathscr{P}(X) \to [0, 1]$ such that

1. $\mu(A \cup B) = \mu(A) + \mu(B)$ whenever $A, B \subseteq X$ and $A \cap B = \emptyset$;
2. $\mu(S) = 1$;
3. $\mu(gA) = \mu(A)$ holds for each $A \subseteq X$ and g in G.

Such a function μ is called a *G-invariant measure for* (X, S, G), or simply an *invariant measure*. The set of all G-invariant measures on X will be denoted by $\mathfrak{M}_G(X)$. If $G = \{e\}$ is the trivial group, then we write simply $\mathfrak{M}(X)$ for $\mathfrak{M}_G(X)$, and its elements are called *measures* on G. We underline that, in fact, these measures are only finitely additive, hence they are not measures in the sense of Lebesgue theory. The space $\mathfrak{M}(X)$ is given the usual topology: (μ_i) converges to μ in $\mathfrak{M}(X)$ if and only if for each $\varepsilon > 0$ and for every finite collection $A_1, A_2, \ldots, A_k$ in $\mathscr{P}(X)$ we have an i_0 such that $|\mu_i(A_j) - \mu(A_j)| < \varepsilon$ for $j = 1, 2, \ldots, k$ and $i \geq i_0$. Given a map $f : X \to Y$ we have a natural map $f_* : \mathfrak{M}(X) \to \mathfrak{M}(Y)$ defined by

$$f_*(\mu)(B) = \mu\big(f^{-1}(B)\big) \quad \text{for each } B \subseteq Y.$$

Clearly, if G acts on X, then it also acts on $\mathfrak{M}(X)$: for a left action we have

$$g\mu(A) = \mu(gA) \quad \text{whenever } A \subseteq X.$$

Obviously, a G-set X is amenable if and only if $\mathfrak{M}_G(X) \neq \emptyset$.

Amenability can be considered as a finiteness condition: finite G-sets are amenable with the G-invariant measure

$$\mu(A) = \frac{|A|}{|S|},$$

where $|A|$ denotes the number of elements in A. More generally, the following simple statement holds.

Theorem 19.1 *Every G-set having a finite orbit is amenable.*

Proof Let x be in X such that Gx is finite; then the function μ defined on $\mathscr{P}(X)$ by

$$\mu(A) = \frac{|A \cap Gx|}{|S \cap Gx|}$$

is an invariant measure for (X, S, G).

We call the group G *amenable*, if the triple (G, G, G) is amenable.
The first well-known example for non-amenable group is the following one.

Theorem 19.2 *The free group F_k of rank $k \geq 2$ is not amenable.*

Proof We prove by contradiction. Assume that μ is a left invariant measure on F_k, and let $x \neq y$ denote two of the free generators of F_k. Let A denote in F_k the set of those words whose reduced form starts by a positive or negative power of x. Then $F_k = A \cup xA$, hence

$$1 = \mu(F_k) \leq \mu(A) + \mu(xA) = 2\mu(A),$$

consequently $\mu(A) \geq \frac{1}{2}$. On the other hand, the sets $y^{-1}A$, A, yA are pairwise disjoint, hence

$$1 = \mu(F_2) \geq \mu(y^{-1}A \cup A \cup yA) = \mu(y^{-1}A) + \mu(A) + \mu(yA) = 3\mu(A),$$

which implies $\frac{1}{2} \leq \mu(A) \leq \frac{1}{3}$, a contradiction.

We recall that given the G-set X, the H set Y and the surjective homomorphism $\Phi : G \to H$ the function $f : X \to Y$ is called Φ-*equivariant*, or simply *equivariant*, if

$$f(gx) = \Phi(g)f(x)$$

holds for each x in X and g in G. By the following theorem, equivariant mappings preserve amenability.

Theorem 19.3 *Let G, H be groups, $\Phi : G \to H$ a surjective homomorphism, X a G-set, Y an H-set, and $f : X \to Y$ a Φ-equivariant mapping. If X is amenable, then so is Y.*

Proof Let μ be a G-invariant measure on X; we show that $f_*(\mu)$ is an $H = \Phi(G)$-invariant measure on Y. Indeed, if $B \subseteq Y$, then

$$f^{-1}\big(\Phi(g)B\big) = gf^{-1}(B)$$

holds for each g in G. To prove this equality first let x be in $gf^{-1}(B)$, then $x = ga$ with some a in X such that $f(a)$ is in B. It follows $f(x) = f(ga) = \Phi(g)f(a)$, which is in $\Phi(g)B$, hence x is in $f^{-1}\big(\Phi(g)B\big)$.

For the reverse inclusion let x be in $f^{-1}\big(\Phi(g)B\big)$, then $f(x)$ is in $\Phi(g)B$, which implies that $f(g^{-1}x) = \Phi(g^{-1})f(x)$ is in B. Then $g^{-1}x$ is in $f^{-1}(B)$ and x is in $gf^{-1}(B)$ which completes the proof.

Now let $B \subseteq Y$ and h an arbitrary element of H. By the surjectivity of Φ, we have that $h = \Phi(g)$ for some g in G. Further,

$$f_*(\mu)(hB) = \mu\big(f^{-1}(hB)\big) = \mu\big(f^{-1}(\Phi(g)B)\big) = \mu\big(gf^{-1}(B)\big)$$
$$= \mu\big(f^{-1}(B)\big) = f_*(\mu)(B)$$

which was to be proved.

Corollary 19.1 *The group G is amenable if and only if every G-set is amenable.*

Proof If X is a G-set, then we take x in X, and define the map $f : G \to X$ as $f(g) = gx$, then f is equivariant with the homomorphism $\Phi = id$, and we apply the previous theorem.

19.2 Paradoxical Decompositions

We shall use $\sqcup$ for disjoint union. We say that the G-set X has a *paradoxical decomposition* if there are partitions

$$X = Y_1 \sqcup Y_2 \sqcup \cdots \sqcup Y_m = Z_1 \sqcup Z_2 \sqcup \cdots \sqcup Z_n \qquad (19.1)$$

and elements $g_1, g_2, \ldots, g_m, h_1.h_2, \ldots, h_n$ in G such that

$$X = g_1 Y_1 \sqcup g_2 Y_2 \sqcup \cdots \sqcup g_m Y_m \sqcup h_1 Z_1 \sqcup h_2 Z_2 \sqcup \cdots \sqcup h_n Z_n.$$

For instance, consider the free group F_2 with the two free generators a, b and let

$$Y_1 = \{\text{reduced words starting with } a\}, \quad Y_2 = F_2 \backslash Y_1$$
$$Z_1 = \{\text{reduced words starting with } b\} \cup \{1, b^{-1}, b^{-2}, \ldots\}, \quad Z_2 = F_2 \backslash Z_1.$$

Then

$$F_2 = Y_1 \sqcup Y_2 = Z_1 \sqcup Z_2 = Y_1 \sqcup a^{-1} Y_2 \sqcup Z_1 \sqcup b^{-1} Z_2$$

is a paradoxical decomposition of F_2.

Theorem 19.4 *The G-set X is amenable if and only if it has no paradoxical decomposition.*

Proof The proof can be found in [7] (see also [8]). In fact, the necessity of the condition is obvious: suppose that we have the paradoxical decomposition in the definition, and μ is a G-invariant measure on X. Then we have

$$\mu(X) = \sum_{i=1}^{m} \mu(Y_i) = \sum_{j=1}^{n} \mu(Z_j) = \sum_{i=1}^{m} \mu(Y_i) + \sum_{j=1}^{n} \mu(Z_j).$$

It follows $\sum_{i=1}^{m} \mu(Y_i) = \sum_{j=1}^{n} \mu(Z_j) = 0$, hence $\mu(X) = 0$, a contradiction.

Paradoxical decompositions are related to the Hausdorff–Banach–Tarski paradox. Consider the triple (X, S, G) where $X = S = S^{n-1}$, the unit sphere in $\mathbb{R}^n$ and $G = SO(n)$, the *special orthogonal group*. Hausdorff showed in [22] that an invariant measure fails to exist for $n = 3$. In fact, Hausdorff's example shows that S^2 can be decomposed into four disjoint sets A, B, C, D such that

1. D is denumerable, and it has measure zero with respect to any finite normalized invariant measure on S^2.
2. There is a $2\pi/3$ rotation in $SO(3)$ whose iterates carry A, B, C onto one another.
3. There is a π rotation in $SO(3)$ which carries each of A, B, C onto the union of the other two.

These investigations were generalized by Banach and Tarski in [6] and this phenomena is referred to as the *Hausdorff–Banach–Tarski Paradox*. The Tarski Alternative Theorem states that a group is either amenable or admits a paradoxical decomposition (see [8, 37, 38]).

Given the paradoxical decomposition (19.1) of the G-set X we define its *complexity* as $c = m + n$, and the number $T(X) = \inf c$, where the infimum is taken for all paradoxical decompositions of X, is called the *Tarski number* of X. If X has no paradoxical decomposition then we set $T(X) = +\infty$. By the Tarski Alternative Theorem, the group G is amenable if and only if $T(G) = +\infty$.

For more about paradoxical decompositions see [9].

19.3   Invariant Means

We recall that for any nonempty set X we denote by $\mathfrak{M}(X)$ the set of all finitely additive probability measures on X, that is, the set of all those functions $\mu : \mathscr{P}(X) \to [0, 1]$ satisfying the following two conditions:

1. $\mu(A \cup B) = \mu(A) + \mu(B)$ whenever $A, B \subseteq X$ and $A \cap B = \emptyset$;
2. $\mu(X) = 1$.

We introduced a topology on $\mathfrak{M}(X)$, which is clearly identical with the Tychonoff topology on product spaces. Consequently, we have the following result:

Theorem 19.5 $\mathfrak{M}(X)$ *is a compact Hausdorff space.*

Proof The set $\mathfrak{M}(X)$ is a subset of $[0, 1]^{\mathscr{P}(X)}$, and its topology coincides with the topology inherited from $[0, 1]^{\mathscr{P}(X)}$ as a subspace. As $[0, 1]^{\mathscr{P}(X)}$ is compact, it is enough to show that $\mathfrak{M}(X)$ is a closed subset. The two conditions

$$\mu(A \cup B) - \mu(A) - \mu(B) = 0, \quad \text{and} \quad \mu(X) - 1 = 0$$

—defining a measure—are zero sets of continuous functions, hence they are closed, and so is their intersection. For the same reason, as a subspace of a Hausdorff space, $\mathfrak{M}(X)$ is Hausdorff, too.

A simple example for measure is the *point mass* δ_x for each x in X defined by

$$\delta_x(A) = \begin{cases} 1 & \text{if } x \in A \\ 0 & \text{if } x \notin A, \end{cases}$$

whenever $A \subseteq X$. The mapping $\delta : X \to \mathfrak{M}(X)$ defined by $\delta(x) = \delta_x$ is injective and $\delta(X)$ is a discrete subset of $\mathfrak{M}(X)$. Consequently, $\delta(X)$ is not closed. Clearly, $\mathfrak{M}(X)$ is a convex set.

For each set X we denote by $l^\infty(X)$ the set of all bounded real valued functions on X, which is a Banach space when equipped with the pointwise linear operations and the sup norm $\|.\|_\infty$. The dual of $l^\infty(X)$ plays an important role, more exactly, the subspace $\mathscr{M}(X)$ of $l^\infty(X)$ consisting of all nonnegative, normalized linear functionals on $l^\infty(X)$:

$$\mathscr{M}(X) = \{m : m \in l^\infty(X)^*, m(f) \geq 0 \text{ if } f \geq 0, m(1) = 1\}.$$

The elements of $\mathscr{M}(X)$ are called *means*. Every mean m satisfies

$$\inf_{x \in X} f(x) \leq m(f) \leq \sup_{x \in X} f(x)$$

for each bounded function f. In other words, means are positive linear functionals on $l^\infty(X)$ satisfying the normalizing condition $m(1) = 1$. Clearly, $\mathscr{M}(X)$ is convex, and it is a weak*-closed subspace of the unit ball of $l^\infty(X)^*$, which is compact in the weak*-topology, hence $\mathscr{M}(X)$ is compact, too. The space $\mathscr{M}(X)$ is closely related to $\mathfrak{M}(X)$, as it is shown by the following theorem.

Theorem 19.6 *For each mean m on X and $A \subseteq X$ we let*

$$\mu_m(A) = m(\chi_A),$$

where χ_A is the characteristic function of the set A. Then the mapping $m \mapsto \mu_m$ is a homeomorphism between $\mathscr{M}(X)$ and $\mathfrak{M}(X)$.

Proof Let $\mathscr{S}(X)$ denote the space of *simple functions*, that is, the functions in $l^\infty(X)$ having finite range. It is easy to see that $\mathscr{S}(X)$ is a dense subspace in $l^\infty(X)$. If m in $l^\infty(X)^*$ satisfies $m(\chi_A) = 0$ for each $A \subseteq X$, then m vanishes on $\mathscr{S}(X)$, by linearity, hence $m = 0$, which proves injectivity.

For surjectivity, let μ be in $\mathfrak{M}(X)$; then we define for each f in $\mathscr{S}(X)$:

$$m_\mu(f) = \sum_{y \in f(X)} y\mu\left(f^{-1}(\{y\})\right).$$

Obviously, m_μ is linear on $\mathscr{S}(X)$. On the other hand, it is easy to see that m_μ is uniformly continuous on the dense subspace $\mathscr{S}(X)$, hence it extends uniquely and continuously to a linear function on $l^\infty(X)$: we denote the unique extension by m_μ, too. In other words, m_μ is in $l^\infty(X)^*$. It follows

$$m_\mu(\chi_A) = 0 \cdot \mu\left(\chi_A^{-1}(\{0\})\right) + 1 \cdot \mu\left(\chi_A^{-1}(\{1\})\right) = \mu(A),$$

which proves the surjectivity of the mapping $m \mapsto \mu_m$ between $\mathscr{M}(X)$ and $\mathfrak{M}(X)$, as $\mu = \mu_{m_\mu}$. Also, this mapping is weak*-continuous, hence, by bijectivity and compactness, its inverse is continuous, too.

The means in the convex hull of $\delta(X)$ are called *discrete means*—they form a weak*-dense subset $\mathscr{M}_d(X)$ in $\mathscr{M}(X)$, as it is shown by the following theorem.

Theorem 19.7 *The set of discrete means $\mathscr{M}_d(X)$ is weak*-dense in $\mathscr{M}(X)$.*

Proof We show that every mean m is the weak*-limit of some net of discrete means. Assuming the contrary, by the Hahn–Banach Theorem, there exists an $\varepsilon > 0$ and a function f in $l^\infty(X)$ such that

$$m(f) \geq \varepsilon + m'(f)$$

holds for each discrete mean m'. On the other hand, every point mass is a discrete mean, hence we have

$$m(f) > \sup_{\mathscr{M}_d(X)} m'(f) \geq \sup_{x \in X} f(x),$$

which contradicts to the property of the mean m.

If X is a G-set then G acts on $l^\infty(X)$ and also on $l^\infty(X)^*$ in a natural way: for each f in $l^\infty(X)$ we let $g \cdot f(x) = f(g \cdot x)$ and for each m in $l^\infty(X)^*$ we define $g \cdot m(f) = m(g \cdot f)$ for each f in $l^\infty(X)$. The mean m on X is called *G-invariant*, if $g \cdot m = m$ holds for each g in G. Of course, this is a *left G-invariant mean*, and *right G-invariant means* are defined analogously. The following theorem connects amenability with the existence of invariant means.

Theorem 19.8 *Let X be a G-set. Then X is amenable if and only if there exists a G-invariant mean on X.*

Proof Let the G-set X be amenable and let μ be a G-invariant measure on X. Then, by Theorem 19.6, there exists a mean m in $\mathcal{M}(X)$ such that $\mu = \mu_m$. Given f in $l^\infty(X)$ let (s_n) be a sequence of simple functions on X such that

$$\lim_{n\to\infty} \| f - s_n \|_\infty = 0,$$

then

$$\lim_{n\to\infty} |m(f) - m(s_n)| = 0. \tag{19.2}$$

Clearly, for each g in G we have

$$\lim_{n\to\infty} \| g \cdot f - g \cdot s_n \|_\infty = 0,$$

hence also

$$\lim_{n\to\infty} |m(g \cdot f) - m(g \cdot s_n)| = 0. \tag{19.3}$$

On the other hand, using the notation of Theorem 19.6, we have

$$m(g \cdot s_n) = m_\mu(g \cdot s_n) = \sum_{y \in g \cdot s_n(X)} y\mu\big((g \cdot s_n)^{-1}(\{y\})\big) =$$

$$\sum_{y \in s_n(X)} y\mu\big(g \cdot (s_n)^{-1}(\{y\})\big) = \sum_{y \in s_n(X)} y\mu\big(s_n^{-1}(\{y\})\big) = m_\mu(s_n) = m(s_n),$$

that is, by (19.2) and (19.3), we have $m(g \cdot f) = m(f)$, which proves that m is a G-invariant mean on X.

Conversely, if m is a G-invariant mean on X, then it is obvious, that the measure μ_m, defined in Theorem 19.6, is a G-invariant measure on X, hence X is amenable.

The following criterium for the existence of invariant means is useful (see [12, 16]).

Theorem 19.9 (Dixmier) *Given the group G and the G-set X there is an invariant mean on $l^\infty(X)$ if and only if for every functions $f_1, f_2, \ldots, f_n$ in $l^\infty(X)$ and every elements $g_1, g_2, \ldots, g_n$ in G we have*

$$\inf_{x \in X} \sum_{k=1}^{n} (f_k(x) - f_k(g_k \cdot x)) \le 0.$$

Proof Let Y denote the subspace in $l^\infty(X)$ generated by all functions of the form

$$h = \sum_{k=1}^{n} (f_k - g_k \cdot f_k)$$

with $f_1, f_2, \ldots, f_n$ in $l^\infty(X)$ and $g_1, g_2, \ldots, g_n$ in G. Then Y is annihilated by any invariant mean m, hence for h in Y we have

$$\inf_{x \in X} h(x) \le m(h) = 0.$$

For the converse assume that the given condition is satisfied, and we consider the set K of those real valued bounded functions Φ on X for which $\inf \Phi > 0$. This is a convex open set in $l^\infty(X)$, which is disjoint from Y, hence, by the Hahn–Banach Theorem there is an m in $l^\infty(X)^*$ such that $m(Y) = 0$ and $m(K) > 0$. By rescaling we have $m(1) = 1$ which is then an invariant mean on X.

19.4   Elementary Groups

In this section we study algebraic properties of amenable groups. First we recall that the group G is *left amenable* if there is a left invariant mean on G, or, what is the same, there is a finitely additive normalized left invariant measure (simply: measure) defined on all subsets of G. Analogously, we call G *right amenable*, if there is a right invariant mean on G, or, what is the same, there is a finitely additive normalized right invariant measure defined on all subsets of G. Finally, we call G *amenable*, if there exists a two-sided invariant mean on G, or, what is the same, there is a finitely additive normalized two-sided invariant measure defined on all subsets of G. By virtue of the following theorem these concepts coincide.

Theorem 19.10 *Let G be a group. Then the following statements are equivalent:*

1. *G is amenable;*
2. *G is left amenable;*
3. *G is right amenable.*

Proof The implication 1.⇒2. is obvious. If m is a left invariant mean on G, then we define

$$\tilde{m}(f) = m(\check{f})$$

for each f in $l^\infty(G)$, where

$$\check{f}(g) = f(g^{-1})$$

for each g in G. Then we have

$$\tilde{m}(f \cdot g) = m\big((f \cdot g)^{\vee}\big) = m\big(g \cdot \check{f}\big) = m(\check{f}) = \tilde{m}(f),$$

hence $\tilde{m}$ is a right invariant mean on G and 2. $\Rightarrow$ 3. is proved.

Finally, let m be a right invariant mean on G, and for each f in $l^{\infty}(G)$ and x in G we write

$$F_f(x) = m(x \cdot f).$$

Then, clearly, F_f is bounded, hence we can define

$$\tilde{m}(f) = m(F_f).$$

We prove that $\tilde{m}$ is two-sided invariant. First we note that

$$F_{g \cdot f} = F_f \cdot g$$

holds for each g in G. Indeed, for x in G we have

$$F_{g \cdot f}(x) = m\big(x \cdot (g \cdot f)\big) = m\big((xg) \cdot f\big) = F_f(xg) = F_f \cdot g(x).$$

Using this we obtain

$$\tilde{m}(g \cdot f) = m(F_{g \cdot f}) = m(F_f \cdot g) = m(F_f) = \tilde{m}(f),$$

by the right invariance of m. On the other hand, we have

$$F_{f \cdot g} = F_f.$$

Indeed,

$$F_{f \cdot g}(x) = m\big(x \cdot (f \cdot g)\big) = m\big((x \cdot f) \cdot g\big) = m(x \cdot f) = F_f(x)$$

for each x in G, by the right invariance of m. Hence

$$\tilde{m}(f \cdot g) = m(F_{f \cdot g}) = m(F_f) = \tilde{m}(f),$$

which proves the two-sided invariance of $\tilde{m}$ and the implication 3. $\Rightarrow$ 1.

Corollary 19.2 *Every finite group is amenable.*

Proof This is obvious: for instance, it follows from Theorem 19.1.

Theorem 19.11 *The additive group $\mathbb{Z}$ of integers is amenable.*

Proof Let $S : l^\infty(\mathbb{Z}) \to l^\infty(\mathbb{Z})$ denote the *shift operator* defined by

$$(Sx)_n = x_{n+1}$$

whenever $x = (\ldots, x_{-1}, x_0, x_1, \ldots)$ is in $l^\infty(\mathbb{Z})$. Let e denote the constant 1 sequence on $\mathbb{Z}$ and we define the closed subspace

$$Y = \{Sx - x : x \in l^\infty(\mathbb{Z})\}.$$

Then dist $(Y, e) \geq 1$; otherwise we had an x in $l^\infty(\mathbb{Z})$ with

$$x_{n+1} - x_n \geq \varepsilon$$

for some $\varepsilon > 0$, hence x cannot be bounded. We define for each x in the subspace $\mathbb{R}e + Y$

$$m(x) = m(\lambda e + y) = \lambda$$

for each real λ; in fact, this formula defines uniquely m, and it is a linear functional on $\mathbb{R}e + Y$ with norm 1 and $m(1) = 1$, vanishing on Y. By the Hahn–Banach Theorem, m extends to $l^\infty(\mathbb{Z})$ to a linear functional with normed 1 and $m(1) = 1$— the shift invariance of m follows from the construction.

It turns out that the class of amenable groups, which will be denoted by AG, is closed under some elementary algebraic operations.

Theorem 19.12 *Every subgroup of an amenable group is amenable.*

Proof Let μ be a left invariant measure on G, and let R denote a complete set of representatives of the right cosets of the subgroup H. Then we define

$$\tilde{\mu}(A) = \mu\left(\bigcup_{r \in R} Ar\right) \text{ for } A \subseteq H.$$

Obviously, $\tilde{\mu}$ is normalized as

$$\tilde{\mu}(H) = \mu\left(\bigcup_{r \in R} Hr\right) = \mu(G) = 1.$$

If A and B are disjoint subsets of H, then the sets $\bigcup_{r \in R} Ar$ and $\bigcup_{r \in R} Br$ are disjoint, too, hence we have

$$\tilde{\mu}(A \cup B) = \mu\left(\bigcup_{r \in R}(A \cup B)r\right) = \mu\left(\bigcup_{r \in R}(Ar \cup Br)\right) =$$

$$\mu\left(\left(\bigcup_{r \in R} Ar\right) \cup \left(\bigcup_{r \in R} Br\right)\right) = \mu\left(\bigcup_{r \in R} Ar\right) + \mu\left(\bigcup_{r \in R} Br\right) = \tilde{\mu}(A) + \tilde{\mu}(B),$$

thus $\tilde{\mu}$ is a measure. Finally, for each h in H and $A \subseteq H$ we derive

$$\tilde{\mu}(hA) = \mu\left(\bigcup_{r \in R}(hA)r\right) = \mu\left(h\bigcup_{r \in R} Ar\right) = \mu\left(\bigcup_{r \in R} Ar\right) = \tilde{\mu}(A),$$

that is, $\tilde{\mu}$ is left invariant.

By Theorem 19.2 we get the following corollary:

Corollary 19.3 *If a group has a free subgroup on two generators, then the group is non-amenable.*

If we denote by NF the family of groups having no subgroup on two generators, then we have the inclusion $AG \subseteq NF$. In fact, in [16] the conjecture has been formulated that we have equality in this inclusion. Even in the eighties, the only known examples for non-amenable groups had a subgroup on two free generators. For instance, $SL(2, \mathbb{Z})$ is non-amenable. Indeed, it is possible to show (see e.g. [8, Lemma 2.3.2]) that the matrices

$$a = \begin{pmatrix} 1 & 2 \\ 0 & 1 \end{pmatrix}$$

and

$$b = \begin{pmatrix} 1 & 0 \\ 2 & 1 \end{pmatrix}$$

generate a free subgroup in $SL(2, \mathbb{Z})$. Accordingly, no matrix group is amenable which includes $SL(2, \mathbb{Z})$ as a subgroup.

Theorem 19.13 *Every factor group of an amenable group is amenable.*

Proof Let μ be an invariant measure on the group G, and let $\Phi : G \to G/N$ denote the canonical homomorphism of G onto the factor group with respect to the normal subgroup N. Then we define

$$\tilde{\mu}(A) = \mu\left(\Phi^{-1}(A)\right) \text{ if } A \subseteq G/N.$$

We have

$$\tilde{\mu}(G/N) = \mu\left(\Phi^{-1}(G/N)\right) = \mu(G) = 1,$$

so that $\tilde{\mu}$ is normalized. If A, B are disjoint subsets in G/N, then $\Phi^{-1}(A)$ and $\Phi^{-1}(B)$ are disjoint, hence we obtain

$$\tilde{\mu}(A \cup B) = \mu\left(\Phi^{-1}(A \cup B)\right) = \mu\left(\Phi^{-1}(A) \cup \Phi^{-1}(B)\right) =$$

$$\mu\left(\Phi^{-1}(A)\right) + \mu\left(\Phi^{-1}(B)\right) = \tilde{\mu}(A) + \tilde{\mu}(B).$$

Finally, if g is in G and $A \subseteq G/N$, then

$$\tilde{\mu}(gNA) = \mu\big(\Phi^{-1}(gNA)\big) = \mu\big(g\Phi^{-1}(NA)\big) = \mu\big(g\Phi^{-1}(A)\big) = \mu\big(\Phi^{-1}(A)\big) = \tilde{\mu}(A),$$

hence $\tilde{\mu}$ is a left invariant measure on G/N.

Corollary 19.4 *Every homomorphic image of an amenable group is amenable.*

Proof By the Fundamental Theorem on homomorphisms between groups, every homomorphic image of a group is isomorphic to some factor group.

Recall that the group G is said to be the *extension of the group N by the group Q*, if there is a short exact sequence

$$1 \to N \to G \to Q \to 1.$$

In more details this means that N is a normal subgroup of G and the factor group G/N is isomorphic to Q. The following theorem says that if Q and N are amenable, then so is G.

Theorem 19.14 *Every extension of an amenable group by an amenable group is amenable.*

Proof Let m, M be left invariant means on N and G/N, respectively. If f is in $l^\infty(G)$ and g, h are lying in the same coset of N, then $Ng = Nh$, hence $g = nh$ for some n in N. It follows $g \cdot f(x) = nh \cdot f(x)$ holds for each x in N. This implies that if we restrict the functions $g \cdot f$ and $nh \cdot f$ to N, then

$$m(g \cdot f) = m\big(n \cdot (h \cdot f)\big) = m(h \cdot f).$$

So, we can define the function $F_f : G/N \to \mathbb{R}$ by

$$F_f(gN) = m(g \cdot f),$$

as the right side depends only on gN. Finally, we define

$$\tilde{m}(f) = M(F_f).$$

Clearly, $\tilde{m}$ is a normalized positive linear functional on $l^\infty(G)$. We have to prove the invariance, only. Let g be in G, then for each f in $l^\infty(G)$ we have

$$F_{g \cdot f}(h \cdot N) = m\big(h \cdot (g \cdot f)\big) = m(gh \cdot f) = F_f(ghN) = g \cdot F_f(hN),$$

hence

$$\tilde{m}(g \cdot f) = M(F_{g \cdot f}) = M(g \cdot F_f) = M(F_f) = \tilde{m}(f),$$

consequently, $\tilde{m}$ is a left invariant mean on G.

Corollary 19.5 *The direct sum of finitely many amenable groups is amenable.*

Proof Indeed, if G is the direct sum of the amenable groups H and N, then G is the extension of N by H.

We recall that the set I is called *directed* if it is partially ordered by $\leq$ and for each i, j in I there is a k in I such that $i \leq k$ and $j \leq k$. A family of sets is called *directed* if it is a directed set equipped with the inclusion as partial order. The union of a directed family of sets is called *direct union*.

Theorem 19.15 *The direct union of amenable groups is amenable.*

Proof Let m_α be a left invariant mean on the group H_α and let G be the direct union of the directed family (H_α). Then for f in $l^\infty(G)$

$$\tilde{m}_\alpha(f) = m_\alpha(f\big|_{H_\alpha})$$

defines a mean on G which is invariant under translations by elements of H_α. Let $\mathscr{M}_\alpha$ denote the set of all means on G which are invariant under translations by elements of H_α; then $\mathscr{M}_\alpha$ is weak*-compact in $l^\infty(G)^*$, and the family $(\mathscr{M}_\alpha)$ has the finite intersection property: indeed, we have

$$\bigcap_{i=1}^{N} \mathscr{M}_{\alpha_i} \supseteq \mathscr{M}_\beta \neq \emptyset,$$

whenever

$$\bigcup_{i=1}^{N} H_{\alpha_i} \subseteq H_\beta.$$

The existence of such a β is guaranteed by the directed property of the family (H_α) for any choice of the finitely many indices $\alpha_1, \alpha_2, \ldots, \alpha_N$. By compactness, the intersection $\bigcap \mathscr{M}_\alpha$ is nonempty, and each element of this set is a left invariant mean on G.

Theorem 19.16 *Every commutative group is amenable.*

Proof First we remark that every group has maximal amenable subgroups. Indeed, this follows from Zorn's Lemma and Theorem 19.15. Let G be commutative, and let M be a maximal amenable subgroup. If x is an element of G not in M, then the cyclic group generated by x is amenable: if x is of finite order, then it generates a finite group, and if it is of infinite order, then it generates a group isomorphic to $\mathbb{Z}$, which is amenable, by Theorem 19.11. We conclude that the subgroup of G generated by M and x is the extension of M by an amenable group, which is amenable, too, by Theorem 19.14. This is a contradiction, which proves that $M = G$, and G is amenable.

Corollary 19.6 *Every solvable group is amenable.*

Proof Solvable groups can be obtained from commutative groups via finitely many extensions by commutative groups, hence they are amenable, by Theorems 19.16 and 19.14.

The group G is called *polycyclic*, if there is a finite sequence of subgroups

$$\{e\} = H_0 \lhd H_1 \lhd \cdots \lhd H_n = G$$

such that H_i is normal in H_{i+1}, and H_{i+1}/H_i is cyclic for $i = 0, 1, \ldots, n - 1$. It is not difficult to show that every finitely generated nilpotent group is polycyclic. Clearly, every polycyclic groups is solvable. Hence we have

Corollary 19.7 *Every polycyclic group is amenable.*

Let $\mathscr{P}$ be a family of groups. We say that a group G is *locally* $\mathscr{P}$, if every finitely generated subgroup of G belongs to $\mathscr{P}$.

Theorem 19.17 *Every locally amenable group is amenable.*

Proof Every group is the direct union of its finitely generated subgroups, hence if every finitely generated subgroup is amenable, then, by Theorem 19.15, the group itself is amenable.

Corollary 19.8 *Every locally finite group is amenable.*

Proof Every locally finite group is locally amenable, hence the previous theorem applies.

Corollary 19.9 *The direct sum of any family of amenable groups is amenable.*

Proof Every finitely generated subgroup of the direct sum is included in the direct sum of some finite subfamily, which is amenable, by Theorem 19.5. It follows that the direct sum of any family of amenable groups is locally amenable, hence Theorem 19.17 applies.

We have seen that starting with finite and commutative groups we can create amenable groups when applying any of the following constructions:

1. taking subgroups;
2. taking homomorphic images;
3. taking amenable extensions;
4. taking direct unions.

The elements of the smallest family of groups including finite groups and commutative groups, and is closed under these four operations are called *elementary amenable groups*, and the family is denoted by EG. Summarizing the above results we have proved the inclusion $EG \subseteq AG \subseteq NF$. Von Neumann asked in [39] whether $AG = NF$ and Day in [11] noted that it is not known if $EG = AG$, or even $EG = NF$. In [10] the author shows that the groups in EG can be constructed from finite and commutative groups using merely steps 3. and 4. above. He also shows that every torsion group is EG is locally finite (see [10, Theorem 2.3]). Combining this with the existence of non-locally finite torsion groups (see [15, 27]) it is proved that $EG \neq NF$, hence $EG \neq AG$, or $AG \neq NF$, or both.

19.5   Growth of Groups

To discuss asymptotic growth properties of groups we use the word metric. Suppose that a group G and a symmetric generating set S is given, that is, for s in S we have s^{-1} is in S, too. A *word* over S is a finite ordered sequence $w = s_1 s_2 \ldots s_n$ whose terms are elements of S. The number n is called the *length* of the word w. An *evaluation* of this word will be called the element $\overline{w} = s_1 \cdot s_2 \cdots \cdot s_n$ of G. The *empty word* is the empty sequence $w = \emptyset$, its length is 0, and its evaluation is the identity element of G: $\overline{\emptyset} = e$.

Given an element g in G the *word norm* $|g|$ of g with respect to S is the shortest length of a word whose evaluation is g. Given two elements g, h in G the *word distance* $d_S(g, h)$ with respect to S is the word norm of $g^{-1}h$. It is easy to verify that the word metric satisfies the axioms of the metric; the symmetry is the consequence of the fact that S is supposed to be a symmetric set.

A simple example is presented by the free group F_2 on two generators a, b. A symmetric generating set is $S = \{a, b, a^{-1}, b^{-1}\}$. A word is called *reduced* if a and a^{-1}, or b and b^{-1} do not occur next to each other. Every element in F_2 can be written in a unique way in reduced form, and this is the shortest word representing the element. For instance, $b^{-1}a$ is the reduced form of this element, hence $|b^{-1}a| = 2$, and also $d_S(a, b) = 2$. The left action of G on itself is defined by $l : g \mapsto lg$ for each l, g in G. This mapping is an isometry with respect to the word metric as $d_S(lg, lh) = |(lh)^{-1} \cdot (lg)| = |h^{-1}g| = d_S(g, h)$ for each l, g, h in G.

Let G be finitely generated, S a finite symmetric generating set, and let $B_n(G, S)$ denote the set of those elements which can be represented as a word of length at most n:

$$B_n(G, S) = \{g : g = s_1 \cdot s_2 \cdots \cdot s_k \text{ with } s_1, s_2, \ldots, s_k \in S, k \leq n\},$$

further let $\nu_{G,S}(n)$ denote the number of elements of $B_n(G, S)$. It can be shown that if S_1 is another finite symmetric generating set, then $\nu_{G,S}$ and ν_{G,S_1} are *equivalent* in the sense that

$$\nu_{G,S}(n/C) \leq \nu_{G,S_1}(n) \leq \nu_{G,S}(Cn)$$

holds for each n with some positive number $C > 0$.

We say that G has *exponential growth*, if $\nu_{G,S}(n) \geq B^n$ holds for some $B > 1$; otherwise it has *subexponential growth*. It has *polynomial growth*, if $\nu_{G,S}(n) \leq p(n)$ holds for some polynomial p, and it has *intermediate growth* if its growth is neither exponential, nor polynomial. It is easy to see that these properties are independent of the choice of the generating set. It follows that every finitely generated group has at least polynomial growth. Also, every finitely generated Abelian group has polynomial growth (see [8, Corollary 6.6.12]).

As an example we consider $G = \mathbb{Z}$, the group of integers with the symmetric generating set $S = \{-1, 1\}$. Then we have

$$B_n(G, S) = \{-n, -(n-1), \ldots, -1, 0, 1, \ldots, n-1, n\}$$

hence $v_{G,S}(n) = 2n + 1$. It follows that $\mathbb{Z}$ has polynomial growth.

Groups with polynomial growth are related to nilpotent groups. A group G is said to be nilpotent if it has an upper central series which terminates with G. An *upper central series* is an ascending sequence

$$\{e\} = Z_0 \lhd Z_1 \lhd \cdots \lhd Z_n \lhd \ldots$$

such that the successive subgroups Z_n are defined as

$$Z_{n+1} = \{z \in G : [z, g] \in Z_n \text{ for each } g \in G\} \ (n = 0, 1, \ldots).$$

Here $[z, g] = z^{-1}g^{-1}zg$ is the *commutator* of z and g. Hence Z_1 is the *center* of G: the set of all elements of G which commute with every element of G. Consequently, G is nilpotent if and only if for some $n \geq$ we have

$$\{e\} = Z_0 \lhd Z_1 \lhd \cdots \lhd Z_n = G.$$

The length of an upper central sequence describes "non-commutativity" of the group; if—and only if—the length is ≤ 1, then the group is commutative. The following theorem describes some relation between nilpotency and growth (see [8, Theorem 6.8.1]).

Theorem 19.18 *Every finitely generated nilpotent group has polynomial growth.*

In [26] John Milnor posed a problem about the existence of a finitely generated group of intermediate growth. An important contribution in the subject is provided by Gromov's celebrated theorem (see [21], also [7, Theorem 4.1]).

Theorem 19.19 (Gromov) *A finitely generated group has polynomial growth if and only if it is virtually nilpotent, that is, it has a nilpotent subgroup with finite index.*

We have the following corollary (see [7, Corollary 4.2]):

Corollary 19.10 *Every finitely generated group with polynomial growth is amenable.*

In fact, the following more general theorem holds true (see [7]):

Theorem 19.20 *Every finitely generated group with subexponential growth is amenable.*

We denote the smallest family of groups containing finite groups and commutative groups having subexponential growth and closed under the four elementary operations by SG. We call the groups in SG *subexponentially amenable groups*, and we have

$$EG \subseteq SG \subseteq AG \subseteq NF$$

and $EG \neq NF$. We also have the following theorem (see [8, Corollary 6.6.5]).

Theorem 19.21 *Every finitely generated group which contains a subgroup isomorphic to the free group F_2 has exponential growth.*

In his 1980 paper (see [17]) R. I. Grigorchuk constructed a group which was an infinite, finitely generated torsion group, but not locally finite. This example answered Burnside's problem in the negative: not every torsion group is locally finite. In the next section we exhibit further important properties of the Grigorchuk group related to amenability.

19.6 The Grigorchuk Group and the Burnside Groups

Let Sym A denote the set of all bijections of the set A. The *Grigorchuk group* G is a subgroup of Sym $\cup_{n\in\mathbb{N}}\{0, 1\}^n$ generated by the bijections a, b, c, d defined recursively as follows:

$$a(x_0 x_1 \ldots) = (1 - x_0)x_1 \ldots$$

$$b(x_0 x_1 \ldots) = \begin{cases} x_0 a(x_1 \ldots), & \text{if } x_0 = 0 \\ x_0 c(x_1 \ldots), & \text{if } x_0 = 1 \end{cases}$$

$$c(x_0 x_1 \ldots) = \begin{cases} x_0 a(x_1 \ldots), & \text{if } x_0 = 0 \\ x_0 d(x_1 \ldots), & \text{if } x_0 = 1 \end{cases}$$

$$d(x_0 x_1 \ldots) = \begin{cases} x_0 x_1 \ldots, & \text{if } x_0 = 0 \\ x_0 b(x_1 \ldots), & \text{if } x_0 = 1. \end{cases}$$

The elements of G can be realized as words on the alphabet $\{0, 1\}$. Every word w can be written in a unique way as $w = s_1 s_2 \ldots s_n$ with s_i in $\{0, 1\}$, and $n = l(w)$, the length of w.

Obviously, $a^2 = b^2 = c^2 = d^2 = id$, hence $S = \{a, b, c, d\}$ is a symmetric generating set of G. Also we have the relations $bc = cb = d$, $dc = cd = b$, $db = bd = c$, which can be proved by the induction on the length of the words. We give a simple example. Let $w = 10110$. Then we have

$$a(w) = a(10110) = 00110$$

$$b(w) = b(10110) = 1c(0110) = 10a(110) = 10010$$

$$c(w) = c(10110) = 1d(0110) = 10110$$

$$d(w) = d(10110) = 1b(0110) = 10a(110) = 10010.$$

The following theorem summarizes the most important properties of the Grigorchuk group.

Theorem 19.22 *The Grigorchuk group has the following properties:*

1. *The Grigorchuk group is an infinite finitely generated torsion group (see [8]).*
2. *The Grigorchuk group has intermediate growth (see [18, 19]).*
3. *The Grigorchuk group has no free subgroup on two generators.*

Hence the Grigorchuk group answers Milnor's question about the existence of groups with intermediate growth. Also it provides a solution to the problem formulated in [11] as it is presented in the following theorem:

Theorem 19.23 *The Grigorchuk group is amenable (see [8, Corollary 6.11.3]) but not elementary amenable (see [20]).*

It follows that we can update our chain of inclusions:

$$EG \subsetneq SG \subseteq AG \subseteq NF.$$

The *free Burnside group* $B(m, n)$ is the factor group of the free group F_m on m generators with respect to the normal subgroup generated by all nth powers. The order of every element of $B(m, n)$ divides n, hence $B(m, n)$ is a torsion group. It follows that $B(m, n)$ cannot have a free subgroup on two generators, that is, $B(m, n)$ belongs to NF. On the other hand, S. I. Adyan proved in [1] that the free Burnside group is non-amenable for $m \geq 2$ if $n \geq 665$ is odd. In other words, we can modify our chain of inclusions to

$$EG \subsetneq SG \subseteq AG \subsetneq NF.$$

In fact, in the paper [28] A. Ju. Olšanskiĭ proved that there exists a non-amenable group such that every proper subgroup of it is cyclic. It follows that such a group cannot have a free subgroup on two generators, that is, it belongs to NF but not to AG.

Finally, to fill the gap about the inclusion $SG \subseteq AG$ we note that there is a general construction of amenable groups (see [7, Section 7.2] which produces groups of "bounded tree automorphisms" with exponential growth (see [7, Proposition 7.14], also [7, Example 7.15]). Finally, we conclude

$$EG \subsetneq SG \subsetneq AG \subsetneq NF.$$

19.7   Ulam Stability

The concept of "Ulam stability" arises form the problem of Stanislaw Ulam he presented in 1940 at the Mathematics Club of the University of Wisconsin. The problem is formulated by Ulam in The Scottish Book (see [25] on p. 11. as follows: "As an example of this 'epsilon stability,' consider the simple functional equation: $f(x + y) = f(x) + f(y)$, i.e., the equation defining the automorphism of the group of real numbers under addition. The 'epsilonic' analogue of this equation is $|g(x + y) - g(x) - g(y)| < \epsilon$. The question is then: Is the solution g necessarily near some solution f of the strictly linear equation? As D. H. Hyers and I showed, the answer is yes. In fact, $|g - f| < \epsilon$, with the same epsilon as above." In [14] G.-L. Forti uses the following definition (with somewhat different notation): Let G be a group (or a semigroup) and B a Banach space. We say that the couple (G, B) has the property of the *stability of homomorphisms* if for each $\varepsilon > 0$ there exists a $\delta > 0$ such that for every function $f : G \to B$, satisfying

$$\|f(xy) - f(x) - f(y)\| \le \varepsilon$$

whenever x, y is in G, there exist a homomorphism $h : G \to B$ such that

$$\|f(x) - h(x)\| \le \delta \tag{19.4}$$

for all x in G. In this case we say that the functional equation

$$f(xy) = f(x) + f(y)$$

is, or simply the homomorphisms are *Ulam-stable* for the couple (G, B). Of course, the functional equation and the pair G, B can be replaced by other equations, resp. structures, but this problem was the starting point of the flourishing stability theory. As Ulam mentioned in his remark above, Hyers became the first contributor to this theory by his basic theorem as follows (see [23]):

Theorem 19.24 (Hyers) *Let E and E' be Banach spaces and let $f : E \to E'$ be such that*

$$\|f(x + y) - f(x) - f(y)\| \le \varepsilon$$

for x, y in E. Then the limit $l(x) = \lim_{n \to \infty} 2^{-n} f(2^n x)$ exists for each x in E, l is an additive function, and the inequality $\|f(x) - l(x)\| \le \varepsilon$ is true for all x in E. Moreover l is the only additive function satisfying this inequality.

From the original proof of Hyers it is clear that Banach space E can be replaced by any Abelian group. In fact, the existence of the limit $l(x)$ also follows on any group, if $f(2^n x)$ is replaced by $f(x^{2^n})$—although the function l is not necessarily a homomorphism, in general. Further, we observe in the theorem of Hyers that δ can be taken as ε; in fact, in [14, Theorem 2] the following simple result is proved:

Theorem 19.25 *If G is any group and B is a Banach space, further homomorphisms are Ulam-stable for the couple (G, B), then the smallest δ satisfying (19.4) is ε.*

On the other hand, the following theorem shows that the Banach space E' can be replaced by the space of real, or complex numbers.

Theorem 19.26 *Let $\mathbb{K}$ denote $\mathbb{R}$ or $\mathbb{C}$. If the homomorphisms are Ulam-stable for the pair $(G, \mathbb{K})$, then the homomorphisms are Ulam-stable for the pair (G, B) with any $\mathbb{K}$-Banach space B.*

Proof Let $f : G \to B$ be a function satisfying

$$\|f(xy) - f(x) - f(y)\| \le \varepsilon$$

for each x, y in G; then, for every linear functional L on B we have

$$\|(L \circ f)(xy) - (L \circ f)(x) - (L \circ f)(y)\| \le \varepsilon \|L\|$$

for each x, y in G. By assumption, there exists a homomorphism $h_L : G \to \mathbb{K}$ such that, by Theorem 19.25

$$|(L \circ f)(x) - h_L(x)| \le \varepsilon \|L\| \tag{19.5}$$

holds for each x in G. We define

$$h(x) = \lim_{n \to \infty} 2^{-n} f(x^{2^n}),$$

which limit exists, by the above remarks. By the continuity of L, we have

$$h_L(x) = \lim_{n \to \infty} 2^{-n} (L \circ f)(x^{2^n}) = L(h(x))$$

for each L in the dual of B, and for every x in G. It follows

$$L(h(xy)) = h_L(xy) = h_L(x) + h_L(y) = L(h(x)) + L(h(y)) = L(h(x) + h(y)),$$

hence $h(xy) = h(x) + h(y)$ holds for each x, y in G, as the dual of B is a separating family for B. By Eq. (19.5), we have

$$|L(f(x)) - L(h(x))| \leq \varepsilon \|L\|$$

for each x in G. If there exists an x_0 such that

$$\|f(x_0) - h(x_0)\| > \varepsilon,$$

then, by the Hahn–Banach Theorem, we take a linear functional L_0 with $\|L_0\| = 1$ in the dual of B such that

$$L_0(f(x_0) - h(x_0)) = \|f(x_0) - h(x_0)\|.$$

Then

$$\varepsilon < \|f(x_0) - h(x_0)\| = |L_0(f(x_0) - h(x_0))| \leq \varepsilon,$$

a contradiction. It follows $\|f(x) - h(x)\| \leq \varepsilon$ which was to be proved.

In the light of these results we shall consider the stability problem only for pairs $(G, \mathbb{K})$ where G is any group.

The stability property of homomorphisms can also be formulated in the following, slightly different way: given a group G and a function $f : G \to \mathbb{K}$ we define the *Cauchy-difference* $\mathscr{C} f$ of f as

$$\mathscr{C} f(x, y) = f(xy) - f(x) - f(y)$$

for each x, y in G. We say that the pair $(G, \mathbb{K})$ has the HS-property, if for every function $f : G \to \mathbb{K}$ with bounded Cauchy-difference there exists a homomorphism $h : G \to \mathbb{K}$ such that $f - h$ is bounded. The relation between the HS-property and the Ulam-stability of homomorphisms for the pair $(G, \mathbb{K})$ is clarified in the following theorem.

Theorem 19.27 *Let G be a group. The pair $(G, \mathbb{K})$ has the HS-property if and only if the homomorphisms are Ulam-stable for the pair $(G, \mathbb{K})$.*

Proof The sufficiency is obvious. Conversely, suppose that the pair $(G, \mathbb{K})$ has the HS-property and let $f : G \to \mathbb{K}$ be a function with

$$|f(xy) - f(x) - f(y)| \leq \varepsilon$$

for each x, y in G. By the HS-property, there is a homomorphism $h : G \to \mathbb{K}$ such that $f - h$ is bounded; say $|f(x) - h(x)| \leq K$ for each x in G. By Theorem 19.25, we can take $K = \varepsilon$, and our statement is proved.

By the above results, we introduce the class HS of groups G for which the pair $(G, \mathbb{R})$ has the HS-property: clearly, finite and commutative groups are included in

H S. In the subsequent paragraphs we shall study the location of *H S* in the chain
of families

$$EG \subsetneq SG \subsetneq AG \subsetneq NF.$$

Of course, similar results can be obtained if $\mathbb{R}$ is replaced by $\mathbb{C}$.

19.8 Invariant Means and Ulam-Stability

First we present a proof of Hyers' theorem and this proof can be considered as a
first step toward the application of invariant means in stability theory.

Theorem 19.28 *Let G be an Abelian group and* $f : G \to \mathbb{R}$ *such that*

$$|f(x + y) - f(x) - f(y)| \le \varepsilon$$

for each x, y in G. Then there exists a unique homomorphism $h : G \to \mathbb{R}$ *such that*
$|f(x) - h(x)| \le \varepsilon$ *holds for each x in G.*

Proof We note that the uniqueness is obvious: if there were two homomorphisms
with the given property, then their difference—which is a homomorphism, too—
were bounded, and it is impossible unless it is zero.

We prove that the sequence $\{2^{-n} f(2^n x)\}$ is bounded for each x in G. More
exactly, we show that

$$|2^{-n} f(2^n x) - f(x)| \le (1 - 2^{-n})\varepsilon.$$

We have $|f(2x) - 2f(x)| \le \varepsilon$, by the assumption on f, with $x = y$. This is our
statement for $n = 1$. Supposing that we have proved the statement for n, we put $2^n x$
for x in the previous inequality to obtain

$$|f(2 \cdot 2^n x) - 2f(2^n x)| \le \varepsilon.$$

By assumption, we have

$$|2f(2^n x) - 2^{n+1} f(x)| \le (2^{n+1} - 2)\varepsilon,$$

hence, by adding these inequalities

$$|f(2^{n+1} x) - 2^{n+1} f(x)| \le (2^{n+1} - 1)\varepsilon,$$

which is our statement. In particular,

$$|2^{-n} f(2^n x) - f(x)| \le \varepsilon.$$

Let LIM_n be a Banach-limit on $\mathbb{N}$, and we define

$$h(x) = LIM_n\left(2^{-n} f(2^n x)\right),$$

then $h : G \to \mathbb{R}$ is well-defined. On the other hand, by the properties of LIM_n, we have

$$|f(x) - h(x)| = |LIM_n\left(f(x) - 2^{-n} f(2^n x)\right)| \leq LIM_n\left(|f(x) - 2^{-n} f(2^n x)|\right) \leq \varepsilon.$$

Now we show that h is a homomorphism. For x, y in G we have

$$h(x + y) - h(x) - h(y) = LIM_n\left(2^{-n}[f\left(2^n(x + y)\right) - f\left(2^n x\right) - f\left(2^n y\right)]\right) \leq$$

$$\varepsilon \cdot LIM_n(2^{-n}) = 0,$$

as $LIM_n(c_n) = \lim_{n \to \infty} c_n$, if $\{c_n\}$ is convergent.

Of course, the Banach-limit LIM_n can be replaced by any invariant mean on $\mathbb{Z}$, the additive group of integers.

This proof suggests that similar argument can be used on amenable groups. In fact, the first application of invariant means in stability theory was the following result (see [33]) presented at the 22nd International Symposium on Functional Equations:

Theorem 19.29 *Every amenable group is in* HS.

Proof Let $f : G \to \mathbb{R}$ be a function satisfying

$$|f(xy) - f(x) - f(y)| \leq \varepsilon$$

for each x, y in the amenable group G. Let M be any right invariant mean on G and we define

$$h(y) = M_x[f(xy) - f(x)]$$

for each y in G. Here M_x means that M is applied on the function $x \mapsto f(xy) - f(x)$, which is obviously bounded on G. We have

$$|f(y) - h(y)| = |M_x[f(y) - f(x) - f(xy)]| \leq M_x[|f(y) - f(x) - f(xy)|] \leq \varepsilon,$$

by the properties of M. On the other hand, we have

$$h(yz) - h(y) - h(z) = M_x[f(xyz) - f(x) - f(xy) + f(x) - f(xz) + f(x)] =$$

$$M_x[f(xyz) - f(xy) - f(xz) + f(x)] = M_x\left[[f(xyz) - f(xy)] - [f(xz) - f(x)]\right] = 0,$$

as the function $x \mapsto f(xyz) - f(xy)$ is the right translate of $x \mapsto f(xz) - f(x)$ by y, and M is right invariant. Hence h is a homomorphism. Further, we have

$$|f(y) - h(y)| = |M_x[f(y) - f(x) + f(xy)]| \leq M_x[|f(y) - f(x) + f(xy)|] \leq \varepsilon,$$

and the proof is complete.

By this result, we have the inclusion $AG \subseteq HS$. The natural question arises: are there any groups not in HS? The first example for such a group was presented by G. L. Forti at the same 22nd ISFE:

Theorem 19.30 *The free group F_2 on two generators is not in HS.*

Proof The proof here is taken from [14]. Let a, b the two generators of F_2 and suppose that every element of F_2 is written in reduced form: it does not contain $aa^{-1}, a^{-1}a, bb^{-1}, b^{-1}b$, and it is written without exponents different from -1 and 1. For x in F_2 we define $f(x)$ as

$$f(x) = r(x) - s(x),$$

where $r(x)$ is the number of words of the form ab contained in x, and $s(x)$ is the number of words of the form $b^{-1}a^{-1}$ contained in x. Then $f : F_2 \to \mathbb{R}$ is well-defined and unbounded. Further, we have

$$f(xy) - f(x) - f(y) \in \{-1, 0, 1\},$$

hence $|f(xy) - f(x) - f(y)| \leq 1$. Assume that $h : F_2 \to \mathbb{R}$ is a homomorphism. Then h is completely determined by the values $h(a)$ and $h(b)$. On the other hand, clearly f vanishes on the two subgroups generated by a, and b, respectively. Hence, if $f - h$ is bounded on these two subgroups, then h vanishes on these two subgroups, as well. In particular, $h(a) = h(b) = 0$, hence $h = 0$, which contradicts the boundedness of $f - h$.

The appearance of the free group on two generators arises the natural question: is there any close connection between amenability and Ulam-stability? Another problem is whether we have $HS \subseteq NF$? In fact, from the above result it follows only that the equation of homomorphism is not Ulam-stable on the free group on two generators, not on groups having a subgroup isomorphic to F_2. This leads to the question: does HS have the property that every subgroup of a group belonging to HS belongs to HS, too? In [14] the related question is formulated as an open problem in the following way: given a group G, a subgroup $H \subseteq G$ and a function $f : H \to \mathbb{R}$ with $\mathscr{C}f$ is bounded does f have an extension $\tilde{f} : G \to \mathbb{R}$ such that $\mathscr{C}\tilde{f}$ is bounded?

19.9   Elementary Operations and Stability

In this section we study the behavior of the family HS under the elementary operations listed in Sect. 19.4. In the previous section we observed that the following problem is still open: is it true, that if the group G is in HS, then every subgroup $H \subseteq G$ is in HS? Nevertheless, we have:

Theorem 19.31 *Let G be a group in HS. Then every homomorphic image of G is in HS.*

Proof Let $\Phi : G \to H$ be a surjective homomorphism with G in HS, and let $f : H \to \mathbb{C}$ be a function satisfying

$$|f(uv) - f(u) - f(v)| \leq \varepsilon$$

for each u, v in H. Let x, y be in G, then

$$|f; \big(\Phi(x)\Phi(y)\big) - f\big(\Phi(x)\big) - f\big(\Phi(y)\big)| \leq \varepsilon,$$

hence

$$|(f \circ \Phi)(xy) - (f \circ \Phi)(x) - (f \circ \Phi)(y)| \leq \varepsilon,$$

that is, by the property of G, there exists a homomorphism $a : G \to H$ such that

$$|(f \circ \Phi)(x) - a(x)| \leq \varepsilon$$

holds for each x in G. It follows for each natural number n

$$|f\big(\Phi(x)^n\big) - na(x)| = |f\big(\Phi(x^n)\big) - na(x)| \leq \varepsilon,$$

and

$$\left|\frac{1}{n} f\big(\Phi(x)^n\big) - a(x)\right| \leq \frac{1}{n}\varepsilon \to 0,$$

hence

$$a(x) = \lim_{n \to \infty} \frac{1}{n} f\big(\Phi(x)^n\big)$$

for each x in G. It follows that $\Phi(x) = \Phi(y)$ implies $a(x) = a(y)$, hence we can define the function A as

$$A\big(\Phi(x)\big) = a(x)$$

for each x in G. Clearly, $A : H \to \mathbb{C}$ is a homomorphism, and we have for each
$u = \Phi(x)$

$$|f(u) - A(u)| = |(f \circ \Phi)(x) - A(\Phi(x))| = |(f \circ \Phi)(x) - a(x)| \le \varepsilon,$$

and the proof is complete.

Corollary 19.11 *Let G be a group in HS. Then every factor group of G is in HS.*

Proof By the Homomorphism Theorem, every factor group of G is isomorphic to
a homomorphic image of G: in fact, the factor group with respect to the normal
subgroup of N is isomorphic to the homomorphic image of G under the natural
homomorphism.

Theorem 19.32 *The direct sum of any two groups in HS is in HS.*

Proof Let $G = H \oplus K$ with H, K in HS and let $f : H \oplus K \to \mathbb{C}$ be a function
satisfying

$$|f((x, u)(y, v)) - f(x, u) - f(y, v)| \le \varepsilon$$

for each x, y in H and u, v in K. Recall that the operation in $H \oplus K$ is defined as

$$(x, u)(y, v) = (xy, uv),$$

hence we have

$$|f(xy, uv) - f(x, u) - f(y, v)| \le \varepsilon \qquad (19.6)$$

for each x, y in H and u, v in K. Putting $y = e_H, u = e_K$, the identity elements in
H and K, respectively, we have

$$|f(x, v) - f(x, e_K) - f(e_H, v)| \le \varepsilon$$

for each x in H and v in K. On the other hand, putting $u = v = e_K$ in (19.6), we
have

$$|f(xy, e_K) - f(x, e_K) - f(y, e_k)| \le \varepsilon, \qquad (19.7)$$

hence by the assumption on H, there is a homomorphism $h_H : H \to \mathbb{C}$ such that

$$|f(x, e_K) - h_H(x)| \le \varepsilon$$

for each x in H. Similarly, there exists a homomorphism $h_K : K \to \mathbb{C}$ such that

$$|f(e_H, y) - h_K(y)| \le \varepsilon$$

for each y in K. Hence, we have

$$|f(x, v) - h_H(x) - h_K(v)| \leq \qquad (19.8)$$

$$|f(x, v) - f(x, e_K) - f(e_H, v)| + |f(x, e_K) - h_H(x)| + |f(e_H, v) - h_K(v)| \leq 3\varepsilon$$

for each (x, v) in $H \oplus K$. On the other hand, with the notation $h(x, u) = h_H(x) + h_K(u)$ we obtain

$$h\big((x, u)(y, v)\big) = h(xy, uv) = h_H(xy) + h_K(uv) =$$

$$h_H(x) + h_H(y) + h_K(u) + h_K(v) = h(x, u) + h(y, v),$$

that is, $h : H \oplus K \to \mathbb{C}$ is a homomorphism. Finally, from (19.8) we infer

$$|f(x, v) - h(x, v)| \leq 3\varepsilon$$

for each (x, v) in $H \oplus K$, which proves the theorem.

Theorem 19.33 *Let G be the direct union of a family of groups in HS. Then G is in HS.*

Proof Let G be the union of the directed family $\{H_\alpha\}$ of subgroups in HS and let $f : G \to \mathbb{C}$ be a function satisfying

$$|f(xy) - f(x) - f(y)| \leq \varepsilon$$

for each x, y in G. If f_α denotes the restriction of f to H_α, then clearly

$$|f_\alpha(xy) - f_\alpha(x) - f_\alpha(y)| \leq \varepsilon$$

holds for each x, y in H_α, hence, by assumption, there are homomorphisms $h_\alpha : H_\alpha \to \mathbb{C}$ such that

$$|f_\alpha(x) - h_\alpha(x)| \leq \varepsilon$$

holds for x in H_α. We show that the family $\{h_\alpha\}$ is directed with respect to extension: if for some α, β, γ we have $H_\alpha \cup H_\beta \subseteq H_\gamma$, then $h_\gamma(x) = h_\alpha(x)$ for each x in H_α, and $h_\gamma(x) = h_\beta(x)$ for each x in H_β. Indeed, by the properties of the homomorphisms h_α, we have

$$|f_\alpha(x) - h_\alpha(x)| \leq \varepsilon$$

for each x in H_α, and

$$|f_\gamma(x) - h_\gamma(x)| \leq \varepsilon$$

for each x in H_γ. As $H_\alpha \subseteq H_\gamma$, and $f_\gamma(x) = f_\alpha(x) = f(x)$ for x in H_α, by the uniqueness statement in Hyers' Theorem, we infer $h_\gamma(x) = h_\alpha(x)$ for x in H_α. Hence the direct limit $h : G \to \mathbb{C}$ of the family of functions $\{h_\alpha\}$ exists, it is clearly a homomorphism of G, and it obviously satisfies

$$|f(x) - h(x)| \leq \varepsilon$$

for each x in G.

We can summarize these results by establishing that the family HS enjoys similar properties like EG, but we are unable to prove that it is closed under forming subgroups and group extensions with groups in HS. We note that the construction of Forti in Theorem 19.30 also works on the free Burnside group $B(m, n)$ providing an example for a group which is not in AG (for $m \geq 2$ and $n \geq 665$ odd), not in HS, and it is in NF: it has no free subgroup on two generators. This means that violation of the Ulam-stability property does not mean that the group must contain free subgroup of rank 2. We note, that $B(m, n)$ is a torsion group, hence every homomorphism of it is zero. In [14], Forti studied the connection between AG and HS, and he proved the following result:

Theorem 19.34 *The group G is amenable if and only if for every positive integer n and for every functions $f_1, f_2, \ldots, f_n : G \to \mathbb{R}$ with bounded Cauchy-differences there exist homomorphisms $h_1, h_2, \ldots, h_n : G \to \mathbb{R}$ such that $f_i - h_i$ is bounded for each $i = 1, 2, \ldots, n$, further for every $x_1, x_2, \ldots, x_n$ in G we have*

$$\inf_y \sum_{i=1}^n \mathscr{C} f_i(x_i, y) \leq \sum_{i=1}^n h_i(x_i) - f_i(x_i) \leq \sup_y \sum_{i=1}^n \mathscr{C} f_i(x_i, y). \qquad (19.9)$$

Proof First assume that G is amenable and M is a left invariant mean on G. If $f_1, f_2, \ldots, f_n : G \to \mathbb{R}$ are functions with bounded Cauchy-differences, then we let

$$h_i(x) = M_y[f_i(xy) - f_i(y)]$$

for $i = 1, 2, \ldots, n$ and x in G. By Theorem 19.29, $h_i : G \to \mathbb{R}$ is a homomorphism, and $f_i - h_i$ is bounded for $i = 1, 2, \ldots, n$. Then, for each $x_1, x_2, \ldots, x_n$ in G we derive

$$\inf_y \sum_{i=1}^n [f_i(x_i y) - f_i(y)] \leq M_y \left[\sum_{i=1}^n [f_i(x_i y) - f_i(y)] \right] \leq \sup_y \sum_{i=1}^n [f_i(x_i y) - f_i(y)],$$

or

$$\inf_y \sum_{i=1}^n \mathscr{C} f_i(x_i, y) + \sum_{i=1}^n f_i(x_i) \leq \sum_{i=1}^n h_i(x_i) \leq \sup_y \sum_{i=1}^n \mathscr{C} f_i(x_i, y) + \sum_{i=1}^n f_i(x_i),$$

by the properties of the mean. This latter inequality is exactly (19.9).

For the converse we show that the condition in this theorem implies the condition for amenability in Theorem 19.9. Indeed, let $f_1, f_2, \ldots, f_n : G \to \mathbb{R}$ be bounded functions and let $x_1, x_2, \ldots, x_n$ be arbitrary elements in G. Clearly, the Cauchy-differences of the f_i's are bounded, and the corresponding homomorphisms h_i are zero. Assume that the condition of Dixmier is not satisfied, that is

$$\inf_{x \in G} \sum_{i=1}^{n} (f_i(x) - f_i(x_i \cdot x)) = \varepsilon > 0.$$

In other words, we have

$$\sum_{i=1}^{n} (-f_i(x_i) - \mathscr{C} f_i(x_i \cdot x)) \geq \varepsilon, \quad \text{or} \quad \sum_{i=1}^{n} \mathscr{C} f_i(x_i \cdot x) \leq - \sum_{i=1}^{n} f_i(x_i) - \varepsilon,$$

hence, by (19.9) and $h_i \equiv 0$

$$- \sum_{i=1}^{n} f_i(x_i) \leq - \sum_{i=1}^{n} f_i(x_i) - \varepsilon,$$

a contradiction.

Seemingly, the condition in this theorem is stronger than the HS-property. Nevertheless, we were unable to find a group having the HS-property but not amenable. If we denote the class of groups having the property in the previous theorem by MS, then

$$AG = MS \subseteq HS.$$

Whether the inclusion is proper is an open question.

19.10   Stability of Linear Functional Equations

Invariant means can be used to prove stability theorems for a wide class of functional equations. In this section we consider linear functional equations. The setting is the following: let G be an Abelian group, n a natural number, and let $\varphi_i, \psi_i : G \to G$ be homomorphisms such that the range of φ_i is contained in the range of ψ_i: $\operatorname{Ran} \varphi_i \subseteq \operatorname{Ran} \psi_i$ for $i = 1, 2, \ldots, n$. We consider the functional equation

$$f_0(x) + \sum_{i=1}^{n+1} f_i(\varphi_i(x) + \psi_i(y)) = 0, \tag{19.10}$$

where $f_i : G \to \mathbb{C}$ are the unknown functions and the equation holds for each x, y in G. For $n = 1$, $\varphi_1 = 0$, $\varphi_2 = \psi_1 = \psi_2 = id$ and $f_0 = f_1 = -f_2$ we have the equation of homomorphisms:

$$f_0(x) + f_0(y) = f_0(x + y),$$

and with the choice $f_i = (-1)^{n+1-i} \binom{n+1}{i} f_0$, $\varphi_i = id$, $\psi_i(x) = ix$ for $i = 1, 2, \ldots, n + 1$ we have the Fréchet functional equation:

$$\Delta_y^{n+1} f(x) = \sum_{i=0}^{n+1} (-1)^{n+1-i} \binom{n+1}{i} f(x + iy) = 0. \tag{19.11}$$

We recall that solutions $f : G \to \mathbb{C}$ of the latter functional equation are called *generalized polynomials* of degree at most n. The difference operators are defined in the usual way:

$$\Delta_y f(x) = f(x + y) - f(x)$$

and the iterates

$$\Delta_{y_1, y_2, \ldots, y_{n+1}} f(x) = \Delta_{y_{n+1}} [\Delta_{y_1, y_2, \ldots, y_n} f](x)$$

for each function $f : G \to \mathbb{C}$ and elements $x, y, y_1, y_2, \ldots, y_{n+1}$ in G. In particular, for $y = y_1 = y_2 = \cdots = y_{n+1}$ we write

$$\Delta_y^{n+1} f(x) = \Delta_{y_1, y_2, \ldots, y_{n+1}} f(x).$$

We use the notation $\Delta_y^0 f(x) = f(x)$ for each x, y in G.

Linear functional equations have a huge literature. For references the reader should consult with handbooks on functional equations. Here we focus on the Ulam-stability of these functional equations. First we show that the Fréchet functional equation (19.11) possesses the Ulam-stability property in the following way (see [31]):

Theorem 19.35 *Let G be an Abelian group, n a natural number, and $f : G \to \mathbb{C}$ a function such that the function*

$$(x, y) \mapsto \Delta_y^{n+1} f(x)$$

is bounded. Then there exists a generalized polynomial $P : G \to \mathbb{C}$ of degree at most n such that $f - P$ is bounded.

Proof First we note that, by the results in [13], the expression $\Delta_{y_1, y_2, \ldots, y_{n+1}} f(x)$ is a linear combination of terms $\Delta_y^{n+1} f(z)$, where the number of terms depends on n only, and the y, z's depend on $x, y_1, y_2, \ldots, y_{n+1}$ linearly. It follows, that the function

$$(x, y_1, y_2, \ldots, y_{n+1}) \mapsto \Delta_{y_1, y_2, \ldots, y_{n+1}} f(x)$$

is bounded on G^{n+2}.

Without loss of generality we may assume that $f(0) = 0$. Let M be an invariant mean on G. Then, for each $x, y_1, y_2, \ldots, y_{n+1}$ in G we can calculate as follows:

$$M_x[\Delta_{y_1, y_2, \ldots, y_{n+1}} f(x)] = M_x[\Delta_{y_1, y_2, \ldots, y_n} f(x+y_{n+1})] - M_x[\Delta_{y_1, y_2, \ldots, y_n} f(x)] = 0,$$

by the invariance of M.

By induction, we can prove for $k = 1, 2, \ldots$ that

$$M_{y_{n+2}} M_{y_{n+3}} \cdots M_{y_{n+k+1}} [\Delta_{y_1, y_2, \ldots, y_{n+k+1}} f](x) = (-1)^k \Delta_{y_1, y_2, \ldots, y_{n+1}} f(x).$$

We define for x in G:

$$f_0(x) = (-1)^n M_{y_1} M_{y_2} \cdots M_{y_n} [\Delta_{y_1, y_2, \ldots, y_n, x} f](0),$$

then $f_0 : G \to \mathbb{C}$ is a bounded function. On the other hand, we have for each $u_1, u_2, \ldots, u_{n+1}$ in G:

$$[\Delta_{u_1, u_2, \ldots, u_{n+1}} (f - f_0)](x) =$$

$$\Delta_{u_1, u_2, \ldots, u_{n+1}} f(x) + (-1)^{n+1} \Delta_{u_1, u_2, \ldots, u_{n+1}} \left[M_{y_1} M_{y_2} \cdots M_{y_n} [\Delta_{y_1, y_2, \ldots, y_n, x} f](0) \right] =$$

$$\Delta_{u_1, u_2, \ldots, u_{n+1}} f(x) +$$

$$(-1)^{n+1} \Delta_{u_1, u_2, \ldots, u_{n+1}} \left[M_{y_1} M_{y_2} \cdots M_{y_n} [\Delta_{y_1, y_2, \ldots, y_n} f(x) - \Delta_{y_1, y_2, \ldots, y_n} f(0)] \right] =$$

$$\Delta_{u_1, u_2, \ldots, u_{n+1}} f(x) + (-1)^{n+1} M_{y_1} M_{y_2} \cdots M_{y_n} [\Delta_{u_1, u_2, \ldots, u_{n+1}} \Delta_{y_1, y_2, \ldots, y_n} f](x) =$$

$$\Delta_{u_1, u_2, \ldots, u_{n+1}} f(x) + (-1)^{n+1} (-1)^n [\Delta_{u_1, u_2, \ldots, u_{n+1}} f](x) = 0$$

hence $P = f - f_0$ is a generalized polynomial of degree at most n.

Corollary 19.12 *Let G be an Abelian group, n a natural number, and $f : G \to \mathbb{C}$ a function such that the function*

$$(x, y_1, y_2, \ldots, y_{n+1}) \mapsto \Delta_{y_1, y_2, \ldots, y_{n+1}} f(x)$$

is bounded. Then there exists a generalized polynomial $P : G \to \mathbb{C}$ of degree at most n such that $f - P$ is bounded.

We shall denote by LS the family of those Abelian groups G for which the following property holds: for each natural number n, for each homomorphisms $\varphi_i, \psi_i : G \to G$ $(i = 1, 2, \ldots, n + 1)$ with the above properties and for each $\varepsilon > 0$ there exists a $\delta > 0$ such that if $f_0 : G \to \mathbb{C}$ is a function such that

$$|f_0(x) + \sum_{i=1}^{n+1} f_i(\varphi_i(x) + \psi_i(y))| \le \varepsilon, \tag{19.12}$$

then there exists a generalized polynomial $P_0 : G \to \mathbb{C}$ such that $|f_0(x) - P_0(x)| \le \delta$ for each x in G. Roughly speaking, LS-property for G means that linear functional equations have a stability property which is analogous to Ulam-stability of homomorphisms. First we show that We underline that here we consider commutative groups, only. Now we prove that for commutative groups, HS-property and LS-property are equivalent.

Theorem 19.36 *Every Abelian group is in LS.*

Proof We assume that G is an Abelian group in HS, and let n, φ_i, ψ_i and ε be given with the properties as above. We introduce the function

$$F_{n+1}(x, y) = f_0(x) + \sum_{i=1}^{n+1} f_i(\varphi_i(x) + \psi_i(y))$$

for each x, y in G; then $F_{n+1} : G \times G \to \mathbb{C}$ is bounded: $|F_{n+1}(x, y)| \le \varepsilon$. Let z_{n+1} be arbitrary in G; by assumption, $\varphi_{n+1}(z_{n+1})$ is in the range of ψ_{n+1}, hence there is a w_{n+1} in G such that $\psi_{n+1}(w_{n+1}) = \varphi_{n+1}(z_{n+1})$. Putting $x + z_{n+1}$ for x and $y - w_{n+1}$ for y in F_{n+1} we have

$$F_{n+1}(x + z_{n+1}, y - w_{n+1}) =$$

$$f_0(x + z_{n+1}) + \sum_{i=1}^{n} f_i(\varphi_i(x) + \varphi_i(z_{n+1}) + \psi_i(y) - \psi_i(w_{n+1})) + f_{n+1}(\varphi_{n+1}(x) + \psi_{n+1}(y)),$$

hence

$$F_{n+1}(x + z_{n+1}, y - w_{n+1}) - F_{n+1}(x, y) =$$

$$\Delta_{z_{n+1}} f_0(x) + \sum_{i=1}^{n} \Delta_{\varphi_i(z_{n+1}) - \psi_i(w_{n+1})} f_i(\varphi_i(x) + \psi_i(y))$$

for each x, y in G. If we write $F_n(x, y) = F_{n+1}(x + z_{n+1}, y - w_{n+1}) - F_{n+1}(x, y)$ for each x, y in G (while z_{n+1} is arbitrary in G, but fixed), then $F_n : G \times G \to \mathbb{C}$ is a bounded function; in fact, $|F_n(x, y)| \le 2\varepsilon$. Also, we write

$$g_i(x) = \Delta_{\varphi_i(z_{n+1}) - \psi_i(w_{n+1})} f_i(x)$$

for $i = 1, 2, \ldots, n$. Then we conclude that the functions $\Delta_{z_{n+1}} f_0, g_1, g_2, \ldots, g_n$ satisfy the condition

$$F_n(x, y) = \Delta_{z_{n+1}} f_0 + \sum_{i=1}^{n} g_i(\varphi_i(x) + \psi_i(y))$$

holds for x, y in G, and F_n is a bounded function. Continuing this process, after $n+1$ steps we arrive at the following conclusion: for every choice of $x, z_1, z_2, \ldots, z_{n+1}$ in G we have

$$|\Delta_{z_1, z_2, \ldots, z_{n+1}} f_0(x)| \leq 2^{n+1} \varepsilon.$$

By the previous theorem, our statement follows.

The previous theorem can be used to describe all solutions of inhomogeneous linear functional equations with bounded right hand side. More exactly, we have (see [30]):

Corollary 19.13 *Let G be an Abelian group, n a natural number, $\varphi_i, \phi_i : G \to G$ homomorphisms with $\mathrm{Ran}\, \varphi_i \subseteq \psi_i$, and let c_i be complex numbers $(i = 1, 2 \ldots, n + 1)$. If $F : G \times G \to \mathbb{R}$ is a bounded function, then every solution $f : G \to \mathbb{R}$ of the functional equation*

$$f(x) + \sum_{i=1}^{n+1} c_i f(\varphi_i(x) + \psi_i(y)) = F(x, y) \tag{19.13}$$

has the form $f = f_0 + P$, where $f_0 : G \to \mathbb{R}$ is a bounded solution of (19.13), and $P : G \to \mathbb{R}$ is a generalized polynomial of degree at most n satisfying the homogeneous equation

$$P(x) + \sum_{i=1}^{n+1} c_i P(\varphi_i(x) + \psi_i(y)) = 0 \tag{19.14}$$

corresponding to (19.13).

Proof Clearly, every function $f = f_0 + P$ of the given form is a solution of the functional equation (19.13).

For the converse, we know, by Theorem 19.36, that every function $f : G \to \mathbb{R}$ satisfying (19.13) has the form $f = f_0 + P$ with $f_0 : G \to \mathbb{R}$ is bounded, and $P : G \to \mathbb{R}$ is a polynomial of degree at most n. Substituting into (19.13) gives

$$P(x) + \sum_{i=1}^{n+1} c_i P(\varphi_i(x) + \psi_i(y)) = F(x, y) - [f_0(x) + \sum_{i=1}^{n+1} c_i f_0(\varphi_i(x) + \psi_i(y))]$$

$$\tag{19.15}$$

for each x, y in G. The left side is a generalized polynomial in both x and y, and the right side is bounded, hence both sides are constant:

$$P(x) + \sum_{i=1}^{n+1} c_i P\big(\varphi_i(x) + \psi_i(y)\big) = K \tag{19.16}$$

with some real number K. Now we have two cases.

If $\sum_{i=1}^{n+1} c_i = -1$, then we write $P = Q + P(0)$, where $Q : G \to \mathbb{R}$ is a generalized polynomial of degree at most n with $Q(0) = 0$. Substitution into (19.13) gives

$$Q(x) + \sum_{i=1}^{n+1} c_i Q\big(\varphi_i(x) + \psi_i(y)\big) = K,$$

and putting $x = y = 0$ we obtain $K = 0$. It follows, by (19.16), that P is a solution of (19.14). In this case, by (19.15) and (19.16), we infer that f_0 is a bounded solution of the functional equation (19.13), and our statement is proved.

If $c = \sum_{i=1}^{n+2} c_i \neq -1$, then we define for x in G:

$$Q(x) = P(x) - K(1+c)^{-1}, \quad \text{and} \quad g_0(x) = f_0(x) + K(1+c)^{-1}.$$

By (19.16), we get:

$$Q(x) + \sum_{i=1}^{n+1} c_i Q\big(\varphi_i(x) + \psi_i(y)\big) =$$

$$P(x) - K(1+c)^{-1} + \sum_{i=1}^{n+1} c_i P\big(\varphi_i(x) + \psi_i(y)\big) - \sum_{i=1}^{n+1} c_i K(1+c)^{-1} = K - K = 0,$$

hence Q is a solution of (19.14). Clearly, Q is a generalized polynomial of degree at most n and $f = g_0 + Q$, where g_0 is bounded. Finally, we conclude

$$g_0(x) + \sum_{i=1}^{n+1} c_i g_0\big(\varphi_i(x) + \psi_i(y)\big) = f_0(x) + \sum_{i=1}^{n+1} c_i f_0\big(\varphi_i(x) + \psi_i(y)\big) + K =$$

$$F(x, y) - [P(x) + \sum_{i=1}^{n+1} c_i P\big(\varphi_i(x) + \psi_i(y)\big)] + K = F(x, y),$$

that is, g_0 is a bounded solution of the functional equation (19.13). The theorem is proved.

19.11   Functional Equations Involving Nonlinearity

If a functional equation is nonlinear in the above sense, then, in some cases, other types of stability appear. Prominent examples are provided by the multiplicative Cauchy equation, the trigonometric functional equations, etc. One of those new phenomena is the so-called *superstability*. The multiplicative Cauchy equation and its pexiderized versions posses this superstability property—or some variant of it—which means that if the multiplicative Cauchy difference

$$(x, y) \mapsto f(xy) - f(x)f(y)$$

is bounded, then f is either bounded, or it is an exact solution of the equation (see e.g. [32]). It is clear that we cannot expect that an exact solution plus a bounded function will not—in general—have the property, that its multiplicative Cauchy difference is bounded. Another interesting example is presented by the addition formulas of some trigonometric functions. For instance, assume that a pair of functions f, g has the property that the function

$$(x, y) \mapsto f(xy) - f(x)g(y) - f(y)g(x)$$

is bounded. What can be said about the pair (f, g) in connection with the sine functional equation? It turns out that, in the presence of some additive terms, amenability plays a role in the Ulam-stability, but it is not the case for "pure" multiplicative equations.

In the subsequent sections we exhibit the stability properties of the sine and the cosine functional equations on amenable groups, and we also present an Ulam-stability result for some functional equations related to spherical functions.

19.12   The Stability of the Sine and the Cosine Equations

First we study the sine functional equation. The following theorem—also for complex valued functions—was proved in [34, Theorem 2.3] below. We recall that a function $m : G \to \mathbb{R}$ is called *exponential*, resp. *additive*, if

$$m(xy) = m(x)m(y), \quad \text{resp.} \quad a(xy) = a(x) + a(y)$$

holds for each x, y in G.

Theorem 19.37 *Let G be an amenable group and let $f, gG :\to \mathbb{R}$ be functions. The function*

$$(x, y) \mapsto f(xy) - f(x)g(y) - f(y)g(x) \tag{19.17}$$

is bounded if and only if one of the following possibilities holds:

1. $f = 0$ *and g is arbitrary;*
2. f, g *are bounded;*
3. $f = am + b, \quad g = m,$ *where m is a bounded exponential, a is additive, and* $b : G \to \mathbb{R}$ *is a bounded function;*
4. $f = \lambda(m - b), \quad g = \frac{1}{2}(m + b),$ *where m is an exponential,* $b : G \to \mathbb{R}$ *is a bounded function, and* λ *is a complex number;*
5. $f(xy) = f(x)g(y) + f(y)g(x)$ *holds for each x, y in G.*

Proof It is clear that function in (19.17) is bounded in the cases 1. and 2. In the third case we obtain, by elementary calculation

$$f(xy) - f(x)g(y) - f(y)g(x) = b(xy) - b(x)m(y) - b(y)m(x),$$

which is bounded. Similarly, in case 4. we have

$$f(xy) - f(x)g(y) - f(y)g(x) = \lambda\big(b(x)b(y) - b(xy)\big),$$

which is bounded, as well. The case 5. is obvious: in this case the function in (19.17) is identically zero.

Now suppose that f is unbounded and we use [34, Lemma 2.2] which says, that, if $\mathscr{F}$ denotes the space of all bounded real functions on G and f is unbounded, then either 4., or 5. holds, or $g = m$ is a bounded exponential. We show that in this case we have possibility 3. We may assume $m \neq 0$, otherwise f is bounded, then $m(x)^{-1} = m(x^{-1})$, consequently

$$(x, y) \mapsto \frac{f(xy) - f(x)g(y) - f(y)g(x)}{m(xy)} = f(xy)m\big((xy)^{-1}\big) - f(x)m(x^{-1}) - f(y)m(y^{-1})$$

is bounded, hence, by Hyers' Theorem,

$$f(x)m(x)^{-1} = f(x)m(x^{-1}) = a(x) + b_0(x)$$

with some additive function a and bounded function $b_0 : G \to \mathbb{R}$. It follows $f(x) = a(x)m(x) + b_0(x)m(x) = a(x)m(x) + b(x)$, and our statement is proved, as b is bounded.

The cosine equation was considered also in [34, Theorem 3.3] and the corresponding result is as follows:

Theorem 19.38 *Let G be an amenable group and let* $f, g :\to \mathbb{R}$ *be functions. The function*

$$(x, y) \mapsto f(xy) - f(x)f(y) + g(x)g(y) \tag{19.18}$$

is bounded if and only if one of the following possibilities holds:

1. $f = 0$ *and g is arbitrary;*
2. f *is an exponential, g is bounded;*

3. $f = (1+a)m + b$, $g = am + b$, or $f = am + b$, $g = (1-a)m - b$, where m is a bounded exponential, a is additive, and $b : G \to \mathbb{R}$ is a bounded function;
4. $f = \frac{\lambda^2}{\lambda^2-1}m - \frac{1}{\lambda^2-1}b$, $g = \frac{\lambda}{\lambda^2-1}m + \frac{\lambda}{\lambda^2-1}b$, where m is an exponential, $b :$ $G \to \mathbb{R}$ is a bounded function, and λ is a complex number with $\lambda^2 \neq 1$;
5. $f(xy) = f(x)f(y) - g(x)g(y)$ holds for each x, y in G.

19.13 A Functional Equation Related to Spherical Functions

In this section we consider the functional equation

$$\int_K f(xky) \, dm_K(k) = g(x)h(y) + p(y). \tag{19.19}$$

Here we suppose that G is a group, K is a compact subgroup in G with normalized Haar measure m_K, and the unknown functions $f, g, h, p : G \to \mathbb{C}$ are continuous. This functional equation was investigated in [36], where the general solution was described and its stability was proved under different conditions. Clearly, if $K = \{e\}$ is the trivial subgroup, then the equation reduces to

$$f(xy) = g(x)h(y) + p(y),$$

which is a special Levi–Civitá equation, including the pexiderized additive and multiplicative Cauchy functional equations. If K is a finite group, then the left side reduces to a linear combination:

$$\sum_{j=1}^n f(xk_j y) = g(x)h(y) + p(x).$$

If in (19.19) p vanishes and $f = g = h$, then the nonzero K-invariant solutions of

$$\int_K f(xky) \, dm_K(k) = f(x)f(y) \tag{19.20}$$

are called *generalized K-spherical functions*. Here we recall the following result:

Theorem 19.39 *Suppose that G is an amenable group, K is finite and let f, g, h, p be continuous functions with f and h unbounded. Then the function*

$$(x, y) \mapsto \int_K f(xky) dm_K(k) - g(x)h(y) - p(y)$$

is bounded if and only if we have

$$f(x) = h(e)[\varphi(x) + \psi(x)] + b_1(x)$$

$$g(x) = \varphi(x) + \psi(x)$$

$$h(x) = h(e)\omega(x)$$

$$p(x) = h(e)\varphi(x) + b_2(x)$$

where $\omega : G \to \mathbb{C}$ is a generalized K-spherical function, $b_1, b_2 : G \to \mathbb{C}$ are bounded functions, $h(e)$ is a nonzero complex number, $\varphi : G \to \mathbb{C}$ is a function satisfying

$$\int_K \varphi(xky)dm_K(k) = \varphi(x)\omega(y) + \varphi(y) \tag{19.21}$$

and $\psi : G \to \mathbb{C}$ is a function satisfying

$$\int_K \psi(xky)dm_K(k) = \psi(x)\omega(y) \tag{19.22}$$

for each x, y in G.

Proof As f is unbounded, hence g is unbounded, too, and the function

$$x \mapsto \int_K f(xky)dm_K(k) - g(x)h(y)$$

is bounded for every fixed y in G. By Theorem [36, Theorem 5], it follows that $h = c\omega$, where $c = h(e) \neq 0$, and ω is a generalized K-spherical function on G. Replacing h by $h/h(e)$ we may suppose that $h(e) = 1$. Putting $y = e$ in the condition we have that $f - g$ is bounded. Let M be a right invariant mean on G and we define

$$\varphi(y) = M_x[\int_K g(xky)dm_K(k) - g(x)\omega(y)]$$

for each y in G. Then, since ω is a generalized K-spherical function, we have

$$\int_K \varphi(ylz)dm_K(l) - \varphi(y)\omega(z) - \varphi(z) =$$

$$\int_K M_x[\int_K g(xkylz)dm_K(k) - g(x)\omega(y(z))]dm_K(l) -$$

$$\omega(z)M_x[\int_K g(xky)dm_K(k) - g(x)\omega(y)] - M_x[\int_K g(xkz)dm_K(k) - g(x)\omega(z)] =$$

$$\int_K M_x[\int_K g(xkylz)dm_K(l) - g(xky)\omega(z)]dm_K(k)-$$

$$\int_K M_x[\int_K g(xlz)dm_K(l) - g(x)\omega(z)]dm_K(k) = 0,$$

by Fubini's Theorem about interchanging the order of integration, and by the right invariance of the mean M. Now we obtain

$$\varphi(y) - p(y) = M_x[\int_K g(xky)dm_K(k) - g(x)\omega(y) - p(y)] =$$

$$M_x[\int_K f(xky)dm_K(k) - g(x)\omega(y) - p(y)] + M_x[\int_K (g(xky) - f(xky)dm_K(k)]$$

and here both terms are bounded. It follows that $p - \varphi$ is bounded.

As $f - g$ is bounded we have

$$(x, y) \mapsto \int_K f(xky)dm_K(k) - \int_K g(xky)dm_K(k)$$

is bounded, hence we have that the function

$$(x, y) \mapsto \int_K g(xky)dm_K(k) - g(x)\omega(y) - \varphi(y)$$

is bounded, too. We let

$$\left|\int_K g(xky)dm_K(k) - g(x)\omega(y) - \varphi(y)\right| \le L$$

for each x, y in G with some constant L. It follows

$$\left|\int_K \int_K g(xlykz)dm_K(k)dm_K(l) - \omega(z)\int_K g(xly)dm_K(l) - \varphi(z)\right| \le L$$

and

$$\left|\int_K \int_K g(xlykz)dm_K(l)dm_K(k)-\right.$$

$$\left.-g(x)\int_K \omega(ykz)dm_K(k) - \int_K \varphi(ykz)dm_K(k)\right| \le L.$$

From these two inequalities, by (19.21) and the property of ω, we infer

$$\left| \omega(z) \left(\int_K g(xly)dm_K(l) - g(x)\omega(y) - \varphi(y) \right) \right| \le 2L$$

for each x, y, z in G. As $\omega = h$ is unbounded it follows that we have

$$\int_K g(xly)dm_K(l) = g(x)\omega(y) + \varphi(y)$$

for each x, y in G. Hence and from (19.21), we have

$$\int_K (g(xly) - \varphi(xly))dm_K(l) = (g(x) - \varphi(x))\omega(y),$$

that is, $g = \varphi + \psi$, where $\psi : G \to \mathbb{C}$ satisfies (19.22) for each x, y in G. The theorem is proved.

19.14 Generalizations

As we said above several results in the preceding paragraphs can be extended to more general situations. The theory of amenability and invariant means is well-developed even on semigroups (see [11, 16]), and groupoids (see [2, 3]). The concept of amenability and invariant means on hypergroups has also been invented and applied. In [24] the author proves that every commutative hypergroup is amenable. In [29] a systematic study of amenability of hypergroups was initiated. The first applications of invariant means to functional equations on hypergroups can be found in [35].

In [5] the author verifies the existence of left and right invariant means acting on certain function spaces over an amenable semigroup that are essentially larger than the spaces of bounded functions (see also [4]).

Amenability on groups can be used to prove stability of some linear functional equations even if the group is non-commutative. In [40] the author proves the following stability result for the square norm equation using exactly the same idea of [33] (see Theorem 19.29 above).

Theorem 19.40 *Let G be an amenable group. If $f : G \to \mathbb{C}$ satisfies*

$$|f(xy) - f(xy^{-1}) - 2f(x) - 2f(y)| \le \delta$$

for each x, y in G, then there exists a quadratic function $q : G \to \mathbb{C}$, that is

$$q(xy) + q(xy^{-1}) = 2q(x) + 2q(y),$$

such that $|f - q| \le \frac{5}{2}\delta$.

The main step in the proof is to show that the function

$$y \mapsto M_x[f(yx) + f(xy^{-1}) - 2f(x)]$$

is quadratic.

Acknowledgement Research was supported by OTKA Grant No. K111651.

References

1. Adyan, S.I.: Random walks on free periodic groups. Izv. Akad. Nauk SSSR Ser. Mat. **46**(6), 1139–1149, 1343 (1982)
2. Anantharaman, C., Renault, J.: Amenable Groupoids. L'Enseignement Mathématique, Geneva (2000)
3. Anantharaman, C., Renault, J.: Amenable groupoids. In: Groupoids in Analysis, Geometry, and Physics. Contemporary Mathematics, vol. 282, pp. 35–46. Amerian Mathematical Society, Providence (2001)
4. Badora, R.: Invariant means, set ideals and separation theorems. J. Inequal. Pure Appl. Math. **6**(1), Article 18, 9pp (2005)
5. Badora, R.: On generalized invariant means and separation theorems. J. Inequal. Pure Appl. Math. **7**(1), Article 12, 8pp (2006)
6. Banach, S., Tarski, A.: Sur la décomposition des ensembles de points en parties respectivement congruentes. Fundam. Math. **6**(1), 244–277 (1924)
7. Bartholdi, L.: Amenability of groups and G-sets. In: Sequences, Groups, and Number Theory, pp. 433–544. Birkhäuser/Springer, Cham (2018)
8. Ceccherini-Silberstein, T., Coornaert, M.: Cellular Automata and Groups. Springer Monographs in Mathematics. Springer, Berlin (2010)
9. Ceccherini-Silberstein, T., Grigorchuk, R.I., de la Harpe, P.: Amenability and paradoxical decompositions for pseudogroups and discrete metric spaces. Tr. Mat. Inst. Steklova **224**, 68–111 (1999)
10. Chou, C.: Elementary amenable groups. Ill. J. Math. **24**(3), 396–407 (1980)
11. Day, M.M.: Amenable semigroups. Ill. J. Math. **1**, 509–544 (1957)
12. Dixmier, J.: Les moyennes invariantes dans les semi-groupes et leurs applications. Acta Sci. Math. Szeged **12**, 213–227 (1950)
13. Djokovič, D.Ž.: A representation theorem for $(X_1 - 1)(X_2 - 1) \cdots (X_n - 1)$ and its applications. Ann. Polon. Math. **22**, 189–198 (1969/1970)
14. Forti, G.-L.: The stability of homomorphisms and amenability, with applications to functional equations. Abh. Math. Semin. Univ. Hambg. **57**, 215–226 (1987)
15. Golod, E.S.: On nil-algebras and finitely approximable p-groups. Izv. Akad. Nauk SSSR Ser. Mat. **28**, 273–276 (1964)
16. Greenleaf, F.P.: Invariant Means on Topological Groups and Their Applications. Van Nostrand Mathematical Studies, No. 16. Van Nostrand Reinhold Co., New York (1969)
17. Grigorchuk, R.I.: On Burnside's problem on periodic groups. Funktsional. Anal. i Prilozhen. **14**(1), 53–54 (1980)
18. Grigorchuk, R.I.: On the Milnor problem of group growth. Dokl. Akad. Nauk SSSR **271**(1), 30–33 (1983)
19. Grigorchuk, R.I.: Degrees of growth of finitely generated groups and the theory of invariant means. Izv. Akad. Nauk SSSR Ser. Mat. **48**(5), 939–985 (1984)

20. Grigorchuk, R.I.: An example of a finitely presented amenable group that does not belong to the class EG. Sbornik Math. **189**(1), 79–100 (1998)
21. Gromov, M.: Groups of polynomial growth and expanding maps. Inst. Hautes Études Sci. Publ. Math. **53**, 53–73 (1981)
22. Hausdorff, F.: Grundzüge der Mengenlehre. Chelsea Publishing Company, New York (1949)
23. Hyers, D.H.: On the stability of the linear functional equation. Proc. Natl. Acad. Sci. U. S. A. **27**, 222–224 (1941)
24. Lau, A.: Analysis on a class of Banach algebras with applications to harmonic analysis on locally compact groups and semigroups. Fundam. Math. **118**(3), 161–175 (1983)
25. Mauldin, R.D. (ed.): The Scottish Book. Birkhäuser, Boston (1981)
26. Milnor, J.: Problems and solutions: advanced problem 5603. Am. Math. Mon. **75**(6), 685–686 (1968)
27. Novikov, P.S., Adjan, S.I.: Infinite periodic groups I, II, III. Izv. Akad. Nauk SSSR Ser. Mat. **32**:212–244, 251–524, 709–731 (1968)
28. Olšanskiĭ, A.J.: On the question of the existence of an invariant mean on a group. Usp. Mat. Nauk **35**(4), 199–200, 214 (1980)
29. Skantharajah, M.: Amenable hypergroups. Ill. J. Math. **36**(1), 15–46 (1992)
30. Székelyhidi, L.: The stability of linear functional equations. C. R. Math. Rep. Acad. Sci. Can. **3**(2), 63–67 (1981)
31. Székelyhidi, L.: Note on a stability theorem. Can. Math. Bull. **25**(4), 500–501 (1982)
32. Székelyhidi, L.: On a theorem of Baker, Lawrence and Zorzitto. Proc. Am. Math. Soc. **84**(1), 95–96 (1982)
33. Székelyhidi, L.: Remark 17. In: Report of Meeting: The Twenty-Second International Symposium on Functional Equations, 16–22 December 1984. Oberwolfach, Germany. Aequationes Mathematicae, vol. 29(1), pp. 62–111 (1985)
34. Székelyhidi, L.: The stability of the sine and cosine functional equations. Proc. Am. Math. Soc. **110**(1), 109–115 (1990)
35. Székelyhidi, L.: Functional Equations on Hypergroups. World Scientific Publishing Co. Pte. Ltd., Hackensack (2013)
36. Székelyhidi, L.: Superstability of functional equations related to spherical functions. Open Math. **15**, 427–432 (2017)
37. Tarski, A.: Sur les fonctions additives dans les classes abstraites et leur application au problème de la mesure. CR. Soc. Sc. Varsovie **22**, 114–117 (1929)
38. Tarski, A.: Algebraische Fassung des Massproblems. Fundam. Math. **31**, 207–223 (1938)
39. von Neumann, J.: Zur allgemeinen Theorie des Masses. Fundam. Math. **13**, 73–116 (1929)
40. Yang, D.: The stability of the quadratic functional equation on amenable groups. J. Math. Anal. Appl. **291**(2), 666–672 (2004)

Chapter 20
On Geometry of Banach Function Modules: Selected Topics

Paweł Wójcik

Abstract The aim of the paper is to present results concerning the geometry of Banach function modules. In particular, we characterize the k-smooth points in Banach function modules and we compute the norm derivatives in Banach function modules. Using the notion of the norm derivatives, we apply our results to characterize orthogonality in the sense of Birkhoff in $\mathscr{C}(K; X)$, and to give a new characterization of smooth points in $\mathscr{C}(K)$. Moreover, the stability of the orthogonality equation in normed spaces is considered.

Keywords Function module · Extreme point · k-Smooth point · Norm derivatives · Stability · Orthogonality equation

Mathematics Subject Classification (2010) Primary 46B20, 39B82; Secondary 46E15, 39B52, 46C50

20.1 Preliminaries

Let X be a real Banach space. The closed unit ball of X is denoted by $B(X)$. The unit sphere of X is denoted by $S(X)$. A subset $A \subset X$ is *absorbing* if for each x in X there is an $\varepsilon > 0$ such that $tx \in A$ for $0 < t < \varepsilon$. Fix $x \in X \setminus \{0\}$. We consider the set $J(x)$ defined as follows

$$J(x) := \{x^* \in X^* : \|x^*\| = 1, \ x^*(x) = \|x\|\}.$$

The set $J(x)$ is convex and closed, and $J(x) \subset S(X^*)$. By the Hahn-Banach theorem we get $J(x) \neq \emptyset$ for all $x \in X \setminus \{0\}$.

P. Wójcik (✉)
Department of Mathematics, Pedagogical University of Cracow, Kraków, Poland
e-mail: pawel.wojcik@up.krakow.pl

© Springer Nature Switzerland AG 2019
J. Brzdęk et al. (eds.), *Ulam Type Stability*,
https://doi.org/10.1007/978-3-030-28972-0_20

20.1.1 Function Modules

If K is a compact Hausdorff space and X a real normed linear space, we let $\mathscr{C}(K; X)$ denote the space of all continuous functions f from K to X with

$$\|f\|_\infty := \sup\{\|f(t)\|_X : t \in K\}.$$

In particular, $\mathscr{C}(K) := \mathscr{C}(K; \mathbb{R})$. In this paper we consider a class of spaces, called *Banach function modules*, which are more general than the spaces $\mathscr{C}(K; X)$. The space $\bigoplus_{t\in K} Y_t$ denotes the functions $y \colon K \to \prod_{t\in K} Y_t$ for which

$$\|y\|_\infty := \sup\{\|y(t)\|_{Y_t} : t \in K\} < \infty,$$

and to shorten the notation we will write $\|y\| = \sup\{\|y(t)\| : t \in K\}$.

A *Banach function module*, or *function module* is a triple $(K, (Y_t)_{t\in K}, Y)$, where K is a nonempty compact Hausdorff space, $(Y_t)_{t\in K}$ a family of Banach spaces, and Y a closed subspace of $\bigoplus_{t\in K} Y_t$ such that the following conditions are satisfied:

(FM1) if $h \in \mathscr{C}(K)$ and $y \in Y$, then $hy \in Y$; here $(hy)(t) := h(t) \cdot y(t)$;
(FM2) $t \to \|y(t)\|$ is an upper semicontinuous function for every $y \in Y$;
(FM3) $Y_t = \{y(t) : y \in Y\}$ for every $t \in K$;
(FM4) $\mathrm{cl}\,\{t \in K : Y_t \neq \{0\}\} = K$.

So, we will say that Y is a function module. The following notation will be frequently used:

$$Y = (K, (Y_t)_{t\in K}, Y) \text{ and } Y^* = (K, (Y_t)_{t\in K}, Y)^*.$$

A natural example of a function module is a space $\mathscr{C}(K; X)$ where K is compact and we take $Y_t = X$ for each $t \in K$. Another example of a function module is a space $Y_1 \oplus_\infty Y_2 \oplus_\infty Y_3$, where Y_1, Y_2, Y_3 are Banach spaces and $K = \{1, 2, 3\}$. A property of function modules that will be of importance to us is that for any $t \in K$, there is some element F of the module which attains its norm on t. Even a bit more can be said.

Theorem 20.1 *If* $(K, (Y_t)_{t\in K}, Y)$ *is a function module and* $t \in K$, $u \in Y_t$ *are given, there exists* $F \in Y$ *such that*

$$F(t) = u \text{ and} \|F\| = \|F(t)\| = \|u\|.$$

Furthermore, if U *is a given neighborhood of* t, F *may be chosen so that* $F(r) = 0$ *for* $r \in K \setminus U$.

20.1.2 Extreme Points

The useful tool in this paper is a theorem due to Behrends [2] which characterizes the extremal points of the closed unit ball in a space $(K, (Y_t)_{t \in K}, Y)^*$ (or briefly Y^*) in terms of extremal points of the closed unit ball in Y_t. By $\text{Ext}(L)$ we will denote the set of all extremal points of a given set L. For $t \in K$, let $\psi_t : Y \to Y_t$ denote the evaluation function defined by $\psi_t(y) := y(t)$. We are ready to describe the extreme points of the closed unit ball of the dual of the function module Y.

Theorem 20.2 ([2, p. 80]) *If $(K, (Y_t)_{t \in K}, Y)$ is a function module, then γ is an extreme point of $B(Y^*)$ if and only if*

$$\text{there are } t \in K \text{ and } y^* \in \text{Ext}(B(Y_t^*)) \text{ such that } \gamma = y^* \circ \psi_t.$$

20.2 Multismoothness in Banach Function Modules

Let X be a real normed linear space. The point $x \in S(X)$ is a *smooth point* if $J(x)$ consists exactly of one point. In this paper, motivated by the results published by Lin and Rao [9] and Saleh Hamarsheh [11], we study the notion of *k-smooth points*. In [9, 11] they generalize the notion of smoothness by calling a unit vector x in a Banach space X a *k-smooth point*, or a *multismooth point of order k* if $\dim \text{span} J(x) = k$. For a natural number k, the set of k-smooth points in X is denoted by $\mathcal{N}_{sm}^k(X)$. It is easy to see that the point x is smooth point if and only if x is a 1-smooth point. Note that $J(x)$ is a weak*-compact convex set and hence it is easy to see that $x \in \mathcal{N}_{sm}^k(X)$ if and only if $\dim \text{span} \text{Ext} J(x) = k$ because

$$\dim \text{span} J(x) = \dim \text{span} \text{Ext} J(x). \tag{20.1}$$

Lin and Rao characterized in [9] multismoothness in l^∞-direct sums and proved the following theorem.

Theorem 20.3 ([9]) *Let $\{X_i : i \in I\}$ be an infinite family of nonzero Banach spaces. Let $X = \bigoplus_{i \in I} X_i$ (i.e., X is a l^∞ direct sum) and let $x = (x_i)$ be a unit vector in X. Let*

$I_1 := \{i \in I : \|x_i\| < 1\}$, *let*
$I_2 := \{i \in I : \|x_i\| = 1, \text{ and } x_i \in \mathcal{N}_{sm}^{m_i}(X_i)\}$, *and let*
$I_3 := I \setminus (I_1 \cup I_2)$.
Then the following statements are equivalent:

(a) $x \in \mathcal{N}_{sm}^k(X)$,
(b) $I_3 = \emptyset$, I_2 *is finite*, $\sup\{\|x_i\| : i \in I_1\} < 1$, $k = \sum_{i \in I_2} m_i$.

Similar investigations have been carried out by Saleh Hamarsheh in l^1-direct sums (see [11, Theorem 7, p. 3]). The paper [10] also presents an interesting result. Namely, Saleh Hamarsheh [10] proved that if $f \in L^1(\mu, X)$ is not a smooth point, (i.e., $f \notin \mathcal{N}_{sm}^1(L^1(\mu, X))$), then f is not a multismooth point of any finite order, which means that $f \notin \mathcal{N}_{sm}^k(L^1(\mu, X))$ for all $k \in \{2, 3, 4, \ldots\}$.

In a recent paper [14] the author has proved the following characterization of multismoothness in $\mathcal{K}(H_1, H_2)$, i.e., the space of all compact operators between two Hilbert spaces.

Theorem 20.4 ([14]) *Let H_1, H_2 be real (resp. complex) Hilbert spaces. Suppose that $A \in \mathcal{K}(H_1, H_2)$, $\|A\| = 1$. Then the following statements are equivalent:*

(a) $A \in \mathcal{N}_{sm}^k(\mathcal{K}(H_1, H_2))$,
(b) $k = \binom{n+1}{2}$, *(resp. $k = n^2$)*
    where $n = \dim \operatorname{span}\{x \in S(H_1) : \|Ax\| = \|A\|\}$.

Theorem 20.3 motivates this section. Now we return to the geometry of Banach function modules to prove the main result of this section. Let $(K, (Y_t)_{t \in K}, Y)$ be a function module. Given $y \in Y$, we denote

$$\mathcal{M}(y) := \{t \in K : \|y(t)\| = \|y\|\}.$$

By (FM2) and by compactness of K we have $\mathcal{M}(y) \neq \emptyset$. There is a useful observation that can be made here. Because of (FM1) (and Theorem 20.1) it can be shown that if $t_1, \ldots, t_n$ are distinct points in K and $z_k^* \in Y_{t_k}^*$, then

$$z_1^* \circ \psi_{t_1}, \ldots, z_n^* \circ \psi_{t_n} \text{ are linearly independent.} \tag{20.2}$$

The main result in this section is the following:

Theorem 20.5 *Let $(K, (Y_t)_{t \in K}, Y)$ be a function module. Let $k \in \mathbb{N}$. Suppose that $y \in Y$ and $\|y\| = 1$. Then the following statements are equivalent:*

(a) $y \in \mathcal{N}_{sm}^k(Y)$,

(b) $\mathcal{M}(y) = \{t_1, \ldots, t_p\}$, $y(t_j) \in \mathcal{N}_{sm}^{m_j}(Y_{t_j})$, $k = \sum_{j=1}^{p} m_j$.

Before launching into the proof, a few words motivating the proof are appropriate. Theorem 20.3 motivates this section. It is worth mentioning that Theorems 20.3 and 20.5 are not equivalent (i.e., both theorems are independent). Indeed, in general case, $(K, (Y_t)_{t \in K}, Y) \subsetneq \bigoplus_{t \in K} Y_t$. In fact, they are not equal unless $K = \{1, \ldots, n\}$. Moreover, if $y \in Y$, i.e., $y \in (K, (Y_t)_{t \in K}, Y)$, then the following implication holds true:

$$y \in \mathcal{N}_{sm}^k\left(\bigoplus_{t \in K} Y_t\right) \quad \Rightarrow \quad y \in \mathcal{N}_{sm}^j((K, (Y_t)_{t \in K}, Y)) \text{ and } j \leqslant k. \tag{20.3}$$

The reverse implication (more precisely, $y \in \mathcal{N}_{sm}^k(Y) \Rightarrow y \in \mathcal{N}_{sm}^k(\bigoplus Y_t))$ is, generally, not true unless $K = \{1, \ldots, n\}$, which means $\bigoplus Y_t = Y$.

To summarize, this discussion shows that Theorem 20.3 and the "property"

$$y \in \mathcal{N}_{sm}^k(Y) \nRightarrow y \in \mathcal{N}_{sm}^k(\bigoplus Y_t)$$

motivate Theorem 20.5. The method of proof presented here is different from that of [9, 11].

Proof To prove (a)$\Leftrightarrow$(b) suppose that $y \in S(Y)$. If we consider

$$J(y) = \{\varphi \in Y^* : \|\varphi\| = 1, \ \varphi(y) = 1\},$$

it is a straightforward computation to show that $J(y)$ is an extremal subset of $B(Y^*)$. Therefore

$$\mathrm{Ext}J(y) \subset \mathrm{Ext}B(Y^*).$$

It follows from this and Theorem 20.2 that

$$\mathrm{Ext}J(y) \subset \{z^* \circ \psi_t : t \in K, z^* \in \mathrm{Ext}B(Y_t^*)\}. \tag{20.4}$$

Now we will show that

$$\mathrm{Ext}J(y) \subset \{z^* \circ \psi_t \in S(Y^*) : t \in \mathcal{M}(y), \ z^* \in \mathrm{Ext}J(y(t))\}. \tag{20.5}$$

Let $\varphi \in \mathrm{Ext}J(y)$. By (20.4) for some $t \in K$ and $z^* \in \mathrm{Ext}B(Y_t^*)$, we get

$$\varphi = z^* \circ \psi_t.$$

Thus $1 = \varphi(y) = z^*(y(t))$ and so $\|y(t)\| = 1$. It follows that

$$t \in \mathcal{M}(y) \text{ and } z^* \in J(y(t)).$$

Since

$$z^* \in J(y(t)) \subset B(Y_t^*) \text{ and } z^* \in \mathrm{Ext}B(Y_t^*),$$

we have $z^* \in \mathrm{Ext}J(y(t))$, so

$$\varphi \in \{z^* \circ \psi_t \in S(Y^*) : t \in \mathcal{M}(y), \ z^* \in \mathrm{Ext}J(y(t))\},$$

i.e., (20.5) holds.

A moment's reflection shows that

$$\{z^* \circ \psi_t \in S(Y^*) : t \in \mathcal{M}(y), \ z^* \in \mathrm{Ext} J(y(t))\} \subset J(y). \tag{20.6}$$

Now we show $k = \sum\limits_{t \in \mathcal{M}(y)} \dim \mathrm{span} J(y(t))$. We get that

$$\sum_{t \in \mathcal{M}(y)} \dim \mathrm{span} J(y(t)) \overset{(20.1)}{=} \sum_{t \in \mathcal{M}(y)} \dim \mathrm{span} \mathrm{Ext} J(y(t)) \overset{(20.2)}{=}$$

$$\overset{(20.2)}{=} \dim \mathrm{span} \bigcup_{t \in \mathcal{M}(y)} \{z^* \circ \psi_t \in S(Y^*) : z^* \in \mathrm{Ext} J(y(t))\} =$$

$$= \dim \mathrm{span}\{z^* \circ \psi_t \in S(Y^*) : t \in \mathcal{M}(y), \ z^* \in \mathrm{Ext} J(y(t))\} \overset{(20.5)}{\geqslant}$$

$$\overset{(20.5)}{\geqslant} \dim \mathrm{span} \mathrm{Ext} J(y) \overset{(20.1)}{=} \dim \mathrm{span} J(y) = k.$$

On the other hand we get

$$k = \dim \mathrm{span} J(y) \overset{(20.6)}{\geqslant}$$

$$\overset{(20.6)}{\geqslant} \dim \mathrm{span}\{z^* \circ \psi_t \in S(Y^*) : t \in \mathcal{M}(y), \ z^* \in \mathrm{Ext} J(y(t))\} =$$

$$= \dim \mathrm{span} \bigcup_{t \in \mathcal{M}(y)} \{z^* \circ \psi_t \in S(Y^*) : z^* \in \mathrm{Ext} J(y(t))\} \overset{(20.2)}{=}$$

$$\overset{(20.2)}{=} \sum_{t \in \mathcal{M}(y)} \dim \mathrm{span} \mathrm{Ext} J(y(t)) \overset{(20.1)}{=} \sum_{t \in \mathcal{M}(y)} \dim \mathrm{span} J(y(t)).$$

It follows from the above inequalities that $k = \sum\limits_{t \in \mathcal{M}(y)} \dim \mathrm{span} J(y(t))$. In this case,

$$k = \sum_{t \in \mathcal{M}(y)} \dim \mathrm{span} J(y(t)) = \sum_{j=1}^{p} m_j$$

for some numbers $p, m_1, \ldots, m_p$ such that $\mathcal{M}(y) = \{t_1, \ldots, t_p\}$, where

$$m_1 = \dim \mathrm{span} J(y(t_1)), \ldots, m_p = \dim \mathrm{span} J(y(t_p)),$$

i.e., $y(t_1) \in \mathcal{N}_{sm}^{m_1}(Y_{t_1}), \ldots, y(t_p) \in \mathcal{N}_{sm}^{m_p}(Y_{t_p})$. The theorem is proved.

20.3 Norm Derivatives in Banach Function Module

In a real normed linear space $(X, \|\cdot\|)$, the Gateaux derivatives of the norm are given for fixed x and y in X by the two expressions $\lim\limits_{\lambda \to 0^\pm} \frac{\|x + \lambda y\| - \|x\|}{\lambda}$. Note that instead of considering the above norm derivatives, it is more convenient to introduce the functionals

$$\rho'_\pm(x, y) := \lim_{\lambda \to 0^\pm} \frac{\|x + \lambda y\|^2 - \|x\|^2}{2\lambda} = \|x\| \cdot \lim_{\lambda \to 0^\pm} \frac{\|x + \lambda y\| - \|x\|}{\lambda}, \qquad x, y \in X,$$

because when the norm comes from an inner product $\langle \cdot | \cdot \rangle \colon X \times X \to \mathbb{R}$, we obtain $\rho'_+(x, y) = \langle x | y \rangle = \rho'_-(x, y)$, i.e., functionals ρ'_+, ρ'_- are perfect generalizations of inner products. Convexity of the norm yields that the above definition is meaningful. The mappings ρ'_+ and ρ'_- are called *the norm derivatives* and their following properties, which will be useful in the present note, can be found, e.g., in [1, 4]:

(ND1) $\forall_{x, y \in X} \, \forall_{\alpha \in \mathbb{R}} \quad \rho'_\pm(x, \alpha x + y) = \alpha \|x\|^2 + \rho'_\pm(x, y);$
(ND2) $\forall_{x, y \in X} \, \forall_{\alpha \geqslant 0} \quad \rho'_\pm(\alpha x, y) = \alpha \rho'_\pm(x, y) = \rho'_\pm(x, \alpha y);$
(ND3) $\forall_{x, y \in X} \, \forall_{\alpha < 0} \quad \rho'_\pm(\alpha x, y) = \alpha \rho'_\mp(x, y) = \rho'_\pm(x, \alpha y);$
(ND4) $\forall_{x \in X} \quad \rho'_\pm(x, x) = \|x\|^2;$
(ND5) $\forall_{x, y \in X} \quad |\rho'_\pm(x, y)| \leqslant \|x\| \cdot \|y\|;$
(ND6) $\forall_{x, y \in X} \quad \rho'_-(x, y) \leqslant \rho'_+(x, y).$

Moreover, mappings ρ'_+, ρ'_- are continuous with respect to the second variable, but not necessarily with respect to the first one. If the norm is generated by an inner product $\langle \cdot | \cdot \rangle$, we consider the standard orthogonality relation: $x \perp y :\Leftrightarrow \langle x | y \rangle = 0$. In general case, we may consider the definition introduced by Birkhoff [3]:

$$x \perp_B y :\Leftrightarrow \forall_{\lambda \in \mathbb{R}} \, \|x\| \leqslant \|x + \lambda y\|.$$

Now, we define ρ_+-*orthogonality* and ρ_--*orthogonality*:

$$x \perp_{\rho_+} y :\Leftrightarrow \rho'_+(x, y) = 0, \qquad\qquad x \perp_{\rho_-} y :\Leftrightarrow \rho'_-(x, y) = 0.$$

In a real normed space X, we have for arbitrary $x, y \in X$ (cf. [1, 4]):

$$x \perp_B y \quad \Leftrightarrow \quad \rho'_-(x, y) \leqslant 0 \leqslant \rho'_+(x, y). \qquad (20.7)$$

It yields

$$\perp_{\rho_-}, \perp_{\rho_+} \subset \perp_B \qquad (20.8)$$

and, if X is smooth, then also $\perp_{\rho_-} = \perp_{\rho_+} = \perp_B$. Let us recall the following result containing a representation of the norm derivatives $\rho'_\pm$ in terms of supporting functionals.

Theorem 20.6 ([4, Theorem 15, p. 36]) *Let X be a real normed space. Then one has the representation:*

$$\rho'_+(x, y) = \|x\| \cdot \sup \{x^*(y) : x^* \in J(x)\} \text{ and}$$
$$\rho'_-(x, y) = \|x\| \cdot \inf \{x^*(y) : x^* \in J(x)\} \quad for \ all \ x, y \in X. \tag{20.9}$$

20.3.1 Examples

The following examples motivate this section.

Example 20.1 Let c_0 be the space of all sequences convergent to 0 with supremum norm. Then for $x = (x_n)$, $y = (y_n)$ we have (see [6]):

$$\rho'_+(x, y) = \max \{x_n y_n : |x_n| = \|x\|\} \quad and$$
$$\rho'_-(x, y) = \min \{x_n y_n : |x_n| = \|x\|\} . \tag{20.10}$$

Example 20.2 Let $\mathscr{C}[0, 1]$ be the space of all continuous real-valued functions equipped with the norm $\|x\| = \sup\{|x(u)| : u \in [0, 1]\}$. Then for x, y in $\mathscr{C}[0, 1]$ we have (see [5]):

$$\rho'_+(x, y) = \max \{x(u)y(u) : |x(u)| = \|x\|\} ,$$
$$\rho'_-(x, y) = \min \{x(u)y(u) : |x(u)| = \|x\|\} . \tag{20.11}$$

In this section we will generalize these examples (see our main result, i.e., Theorem 20.8). The method of proof presented here is different from that of [5, 6]. We obtain the earlier results (i.e., (20.10) and (20.11)) as particular cases.

20.3.2 Main Results

A well-known theorem of Singer [12] will be useful in this section.

Theorem 20.7 ([12, p. 170]) *Let X be a real normed linear space, E an n-dimensional subspace of X, $x \in X \setminus E$. The following statements are equivalent:*

(a) $x \perp_B E$;

(b) *There exist $\varphi_1, \ldots, \varphi_{n+1} \in \text{Ext}(B(X^*))$ and $\lambda_1, \ldots, \lambda_{n+1} \geqslant 0$ with*
$$\sum_{j=1}^{n+1} \lambda_j = 1, \text{ such that } \quad \forall_{y \in E} \sum_{j=1}^{n+1} \lambda_j \varphi_j(y) = 0, \text{ and } \forall_j \ \varphi_j(x) = \|x\|.$$

Let Y be a function module and let $x \in Y$. Recall that we denote $\mathscr{M}(x) := \{t \in K : \|x(t)\| = \|x\|\}$. By (FM2) and by compactness of K we have $\mathscr{M}(x) \neq \emptyset$.

Lemma 20.1 *Let* $(K, (Y_t)_{t \in K}, Y)$ *be a function module. Suppose that* $y, z \in Y$, $y \neq 0 \neq z$. *Then the following conditions are equivalent:*

(a) $y \perp_B z$,
(b) *There exist* $\lambda \in [0, 1]$, $t_1, t_2 \in \mathcal{M}(y)$ *and* $y_1^* \in J(y(t_1))$, $y_2^* \in J(y_2(t_2))$ *such that*
$\lambda y_1^*(z(t_1)) + (1 - \lambda) y_2^*(z(t_2)) = 0$.

Proof Suppose that $y \perp_B z$. Clearly dim span$\{z\} = 1$. Applying Theorem 20.7 we obtain

$$\lambda \varphi_1(z) + (1 - \lambda) \varphi_2(z) = 0 \text{ and } \varphi_1(y) = \|y\|, \ \varphi_2(y) = \|y\| \tag{20.12}$$

for some $\lambda \in [0, 1]$ and for some $\varphi_1, \varphi_2 \in \text{Ext}(B(Y^*))$.
By Theorem 20.2, we have

$$\varphi_1 = y_1^* \circ \psi_{t_1}, \quad \varphi_2 = y_2^* \circ \psi_{t_2}$$

for some $t_1, t_2 \in K$, $y_1^* \in \text{Ext} B(Y_{t_1}^*)$ and $y_2^* \in \text{Ext} B(Y_{t_2}^*)$. Now the condition (20.12) becomes

$$\lambda y_1^*(z(t_1)) + (1 - \lambda) y_2^*(z(t_2)) = 0$$

and

$$y_1^*(y(t_1)) = \|y\|, \quad y_2^*(y(t_2)) = \|y\|.$$

Thus we obtain $\|y(t_1)\| = \|y\|$ and $\|y(t_2)\| = \|y\|$. It follows that

$$y_1^* \in J(y(t_1)) \text{ and } y_2^* \in J(y(t_2)).$$

The proof of this implication is complete.
We prove (b)$\Rightarrow$(a). Fix an arbitrary $\beta \in \mathbb{R}$. From (b) we get

$$\|y\| = \lambda \|y\| + (1 - \lambda) \|y\| \overset{(b)}{=} \lambda y_1^*(y(t_1)) + (1 - \lambda) y_2^*(y(t_2)) =$$

$$= \lambda y_1^*(y(t_1)) + (1 - \lambda) y_2^*(y(t_2)) + \beta \cdot 0 \overset{(b)}{=}$$

$$\overset{(b)}{=} \lambda y_1^*(y(t_1)) + (1 - \lambda) y_2^*(y(t_2)) +$$

$$+ \beta \cdot \left(\lambda y_1^*(z(t_1)) + (1 - \lambda) y_2^*(z(t_2))\right) =$$

$$= \lambda y_1^* (y(t_1) + \beta \cdot z(t_1)) + (1 - \lambda) y_2^* (y(t_2) + \beta \cdot z(t_2)) \leqslant$$

$$\leqslant \lambda \|y + \beta z\| + (1 - \lambda) \lambda \|y + \beta z\| = \|y + \beta z\|.$$

That means $y \perp_B z$. The proof of Lemma 20.1 is complete.

Now we prove the main result of this section. The same symbol, ρ'_+, will be used to denote the norm derivatives on Y_t and on Y.

Theorem 20.8 *Let $(K, (Y_t)_{t\in K}, Y)$ be a function module. Suppose that $x, y \in Y$ and $x \neq 0 \neq y$. Then*

$$\rho'_+(x, y) = \sup\left\{\rho'_+(x(t), y(t)) : t \in \mathscr{M}(x)\right\},$$
$$\rho'_-(x, y) = \inf\left\{\rho'_-(x(t), y(t)) : t \in \mathscr{M}(x)\right\}. \tag{20.13}$$

Proof Without loss of generality, we may assume that $\|x\|_\infty = 1$ (see (ND2)). Since the proofs are similar we present only $\rho'_+(x, y) = \sup\{\ldots\}$. Fix $\lambda \in (0, 1)$. Fix $t \in \mathscr{M}(x)$ to obtain

$$\frac{\|x(t)+\lambda y(t)\|^2-\|x(t)\|^2}{2\lambda} = \frac{\|x(t)+\lambda y(t)\|^2-\|x\|^2}{2\lambda} \leqslant \frac{\|x+\lambda y\|^2-\|x\|^2}{2\lambda}.$$
$$\tag{20.14}$$

Since λ was arbitrarily chosen from the interval $(0, 1)$, letting $\lambda \to 0^+$ in (20.14) we obtain

$$\rho'_+(x(t), y(t)) \leqslant \rho'_+(x, y).$$

Since t was arbitrarily chosen from the set $\mathscr{M}(x)$, we get

$$\sup\left\{\rho'_+(x(t), y(t)) : t \in \mathscr{M}(x)\right\} \leqslant \rho'_+(x, y).$$

It follows from the above inequality that

$$\rho'_+(x, y) \geqslant \sup\left\{\rho'_+(x(t), y(t)) : t \in \mathscr{M}(x)\right\} \overset{(20.9)}{=}$$
$$= \sup\left\{\|x(t)\| \cdot \sup\{x^*(y(t)) : x^* \in J(x(t))\} : t \in \mathscr{M}(x)\right\} =$$
$$= \sup\left\{\|x\| \cdot \sup\{x^*(y(t)) : x^* \in J(x(t))\} : t \in \mathscr{M}(x)\right\} =$$
$$= \sup\left\{\sup\{x^*(y(t)) : x^* \in J(x(t))\} : t \in \mathscr{M}(x)\right\} =: c. \tag{20.15}$$

To end the proof, it remains to prove $\rho'_+(x, y) \leqslant c$. By (20.15) we have

$$\forall_{t\in\mathscr{M}(x)} \forall_{x^*\in J(x(t))} \; x^*(y(t)) \leqslant c. \tag{20.16}$$

From (ND1) we have for $\alpha := \dfrac{-\rho'_+(x,y)}{\|x\|^2}$

$$\rho'_+(x, \alpha x + y) = 0, \text{ i.e., } x \perp_{\rho_+} \alpha x + y.$$

By (20.8) we get

$$x \perp_B \alpha x + y.$$

Using Lemma 20.1 we obtain

$$\lambda x_1^*(\alpha x(t_1)+y(t_1))+(1-\lambda)x_2^*(\alpha x(t_2)+y(t_2))=0 \qquad (20.17)$$

for some $t_1, t_2 \in \mathcal{M}(x)$, $x_1^* \in J(x(t_1))$, $x_2^* \in J(x(t_2))$ and for some $\lambda \in [0, 1]$. It follows from (20.17) that

$$\begin{aligned}
0 &= \lambda x_1^*(\alpha x(t_1)+y(t_1))+(1-\lambda)x_2^*(\alpha x(t_2)+y(t_2)) = \\
&= \alpha\lambda x_1^*(x(t_1))+\lambda x_1^*(y(t_1)) + \alpha(1-\lambda)x_2^*(x(t_2))+(1-\lambda)x_2^*(y(t_2)) = \\
&= \alpha\lambda\|x(t_1)\|+\lambda x_1^*(y(t_1)) + \alpha(1-\lambda)\|x(t_2)\|+(1-\lambda)x_2^*(y(t_1)) \overset{(20.16)}{\leqslant} \\
&\leqslant \alpha\lambda\|x\|+\lambda c+\alpha(1-\lambda)\|x\|+(1-\lambda)c = \\
&= \alpha\|x\|+c = \frac{-\rho'_+(x, y)}{\|x\|^2}\|x\|+c = \frac{-\rho'_+(x, y)}{1^2}\cdot 1+c = \\
&= -\rho'_+(x, y)+c,
\end{aligned}$$

and so $\rho'_+(x, y) \leqslant c$. The proof is complete.

As an illustration of the applications of our theorems we explore here the Birkhoff orthogonality in function modules. It is a straightforward verification to show that the relation $\perp_B$ is homogeneous, i.e., $x\perp_B y$ implies $\alpha x\perp_B\beta y$ (for arbitrary $\alpha, \beta \in \mathbb{R}$). We will extend this result.

Theorem 20.9 *Let $(K, (Y_t)_{t\in K}, Y)$ be a function module. Suppose that $x, y \in Y$ and $x \neq 0 \neq y$. Let $f, h \in \mathscr{C}(K)$. Suppose that*

$$\mathcal{M}(x) \subset \mathcal{M}(f) \quad and \quad f(t)\cdot h(t) \geqslant 0 \qquad (20.18)$$

for $t \in \mathcal{M}(x)$. Then

$$x\perp_B y \quad \Rightarrow \quad f\cdot x\perp_B h\cdot y.$$

Proof Assume that $x\perp_B y$. Condition $\mathcal{M}(x) \subset \mathcal{M}(f)$ yields that

$$\mathcal{M}(x) \subset \mathcal{M}(f\cdot x). \qquad (20.19)$$

Combining (20.7) and (20.13), we immediately get

$$\inf\left\{\rho'_-(x(t), y(t)) : t \in \mathcal{M}(x)\right\} \leqslant 0 \leqslant \sup\left\{\rho'_+(x(t), y(t)) : t \in \mathcal{M}(x)\right\}.$$

Next, from (ND2), (ND3) and (20.18), it follows that

$$\begin{aligned}
\inf\left\{\rho'_-(f(t)x(t), h(t)y(t)) : t \in \mathcal{M}(x)\right\} \leqslant 0 \leqslant \\
\leqslant \sup\left\{\rho'_+(f(t)x(t), h(t)y(t)) : t \in \mathcal{M}(x)\right\},
\end{aligned} \qquad (20.20)$$

and consequently,

$$\rho'_-(f \cdot x, h \cdot y) \overset{(20.13)}{=} \inf \left\{ \rho'_-(f(t) \cdot x(t), h(t) \cdot y(t)) : t \in \mathcal{M}(f \cdot x) \right\} \overset{(20.19)}{\leqslant}$$

$$\leqslant \inf \left\{ \rho'_-(f(t) \cdot x(t), h(t) \cdot y(t)) : t \in \mathcal{M}(x) \right\} \overset{(20.20)}{\leqslant} 0 \overset{(20.20)}{\leqslant}$$

$$\leqslant \sup \left\{ \rho'_+(f(t) \cdot x(t), h(t) \cdot y(t)) : t \in \mathcal{M}(x) \right\} \overset{(20.19)}{\leqslant}$$

$$\leqslant \sup \left\{ \rho'_+(f(t) \cdot x(t), h(t) \cdot y(t)) : t \in \mathcal{M}(f \cdot x) \right\} \overset{(20.13)}{=}$$

$$\overset{(20.13)}{=} \rho'_+(f \cdot x, h \cdot y),$$

whence using (20.7) again we deduce $f \cdot x \perp_B h \cdot y$.

Let X be a real normed linear space. We define $\{x\}^{\perp_B}$ as being Birkhoff's orthogonal set of x, i.e., $\{x\}^{\perp_B} := \{y \in X : x \perp_B y\}$. It is known (e.g. [4, Theorem 48, p. 127]) that

$$x \text{ is smooth if and only if } \{x\}^{\perp_B} \text{ is a linear subspace of } X. \tag{20.21}$$

Indeed, The next theorem clarifies completely the relation between condition $x \in \mathcal{N}^1_{sm}(X)$ and Birkhoff's orthogonal sets. Namely, the condition (20.21) can be strengthen as follows.

Theorem 20.10 *Let $(K, (Y_t)_{t \in K}, Y)$ be a function module. If $x \in Y$, then*

$$x \text{ is smooth } \Leftrightarrow \{x\}^{\perp_B} \text{ is a function module (precisely, submodule).}$$

Proof Suppose that the vector x is smooth, i.e., $x \in \mathcal{N}^1_{sm}(Y)$. According to (20.21), $\{x\}^{\perp_B}$ is a linear subspace of Y. It remains to show (FM1). It follows from Theorem 20.5 that card $\mathcal{M}(x) = 1$ (for example, let $\mathcal{M}(x) = \{t_1\}$) and $x(t_1) \in \mathcal{N}^1_{sm}(Y_{t_1})$. Since the point $x(t_1)$ is smooth in Y_{t_1}, we have

$$\rho'_+(x(t_1), \cdot) = \rho'_-(x(t_1), \cdot)$$

and

$$\rho'_+(x(t_1), \alpha z(t_1)) = \alpha \rho'_+(x(t_1), z(t_1)) \tag{20.22}$$

for $\alpha \in \mathbb{R}, z \in Y$. It follows that

$$x \perp_B z \Leftrightarrow \rho_+(x, z) = 0.$$

Observe first that by Theorem 20.8, we get

$$\rho'_-(x, z) = \rho'_-(x(t_1), z(t_1)) = \rho'_+(x(t_1), z(t_1)) = \rho'_+(x, z)$$

for all $z \in Y$. Therefore, it follows that

$$x \perp_B z \quad \Leftrightarrow \quad \rho_+(x, z) = 0 \quad \Leftrightarrow \quad \rho_+(x(t_1), z(t_1)) = 0. \tag{20.23}$$

Fix $f \in \mathscr{C}(K)$. Thus we obtain the following implications:

$$y \in \{x\}^{\perp_B} \Rightarrow x \perp_B y \overset{(20.23)}{\Rightarrow} \rho'_+(x(t_1), y(t_1)) = 0 \Rightarrow$$

$$\Rightarrow f(t_1) \cdot \rho'_+(x(t_1), y(t_1)) = 0 \overset{(20.22)}{\Rightarrow}$$

$$\Rightarrow \rho'_+(x(t_1), f(t_1) \cdot y(t_1)) = 0 \overset{(20.23)}{\Rightarrow}$$

$$\Rightarrow x \perp_B f \cdot y \Rightarrow f \cdot y \in \{x\}^{\perp_B},$$

so (FM1) holds. The converse implication is trivial; see (20.21).

Using the concept of the norm derivatives, we apply our results to characterize orthogonality in the sense of Birkhoff in $\mathscr{C}(K; X)$. The mappings ρ'_-, ρ'_+ are continuous with respect to the second variable. If a normed space X is smooth, then the mappings ρ'_-, ρ'_+ are also continuous with respect to the first variable and $\rho'_- = \rho'_+$.

Theorem 20.11 *Suppose that K is a compact Hausdorff space. Let X be a smooth normed linear. Assume that $f, g \in \mathscr{C}(K; X)$. If $\mathscr{M}(f)$ is connected then*

$$f \perp_B g \quad \Leftrightarrow \quad \exists_{t_o \in \mathscr{M}(f)} f(t_o) \perp_B g(t_o).$$

Proof We start with proving "$\Leftarrow$". Suppose $\exists_{t_o \in \mathscr{M}(f)} f(t_o) \perp_B g(t_o)$. Fix an arbitrary $\lambda \in \mathbb{R}$. It follows that

$$\|f\| = \|f(t_o)\| \leqslant \|f(t_o) + \lambda g(t_o)\| \leqslant \|f + \lambda g\|.$$

That means $f \perp_B g$.

We prove "$\Rightarrow$". Assume $f \perp_B g$. Combining (20.7) and (20.13), we get

$$\inf \{\rho'_-(f(t), g(t)) : t \in \mathscr{M}(f)\} \leqslant 0 \leqslant$$
$$\leqslant \sup \{\rho'_+(f(t), g(t)) : t \in \mathscr{M}(f)\}. \tag{20.24}$$

The smoothness of X implies that

$$\rho'_-(f(t), g(t)) = \rho'_+(f(t), g(t)).$$

It is helpful to recall that $\mathscr{M}(f)$ is compact. Define

$$\varphi \colon \mathscr{M}(f) \to \mathbb{R} \text{ by } \varphi(\cdot) := \rho'_+(f(\cdot), g(\cdot)).$$

Now the condition (20.24) becomes

$$\min\{\varphi(t) : t \in \mathscr{M}(f)\} \leqslant 0 \leqslant \max\{\varphi(t) : t \in \mathscr{M}(f)\}.$$

Since X is smooth, φ is continuous; indeed, see [4, Theorem 1, p. 4] and [4, Corollary 5, p. 38]. Moreover, $\varphi(t_o)$ is connected. Using the Darboux property we get for some $t_o \in \mathscr{M}(f)$ that $\varphi(t_o) = 0$. Thus for the element t_o we have $\rho'_+(f(t_0), g(t_o)) = 0$. Hence $f(t_o) \perp_B g(t_o)$. The proof of Theorem 20.11 is complete.

Theorem 20.12 *Let* $f \in \mathscr{C}(K)$. *The following conditions are equivalent:*

(a) $f \in \mathscr{N}^1_{sm}(\mathscr{C}(K))$,
(b) $\{f\}^{\perp_B}$ *is an ideal in the algebra* $\mathscr{C}(K)$,
(c) $\{f\}^{\perp_B}$ *is a maximal ideal in the algebra* $\mathscr{C}(K)$.

Proof The implication (c)$\Rightarrow$(b) has a trivial verification. It is easy to see that $\mathscr{C}(K)$ is a function module and submodules of $\mathscr{C}(K)$ are ideals. So, applying Theorem 20.10 we obtain the implication (b)$\Rightarrow$(a). In order to prove (a)$\Rightarrow$(c), assume that $f \in \mathscr{N}^1_{sm}(\mathscr{C}(K))$. Then $\rho'_+(f, \cdot)$ is linear. Notice that $\rho'(f, \cdot) \in \mathscr{C}(K)^*$. Since

$$\{f\}^{\perp_B} = \ker \rho'_+(f, \cdot),$$

we get

$$\operatorname{co\,dim}\{f\}^{\perp_B} = 1.$$

Hence $\{f\}^{\perp_B}$ is a maximal ideal in $\mathscr{C}(K)$.

20.4 Stability of Orthogonality Equation in Normed Spaces

The aim of this section is to present results concerning the norm derivatives and its preservation (both accurate and approximate) by mapping. Recently, the author prove the following theorem.

Theorem 20.13 ([15, Theorem 4]) *Let* X *and* Y *be real Banach spaces. Assume that* Y *is separable. Let* $f \colon X \to Y$ *be a mapping satisfying the functional equation*

$$\forall_{x,y \in X} \quad \rho'_+(f(x), f(y)) = \rho'_+(x, y). \tag{20.25}$$

If f *is surjective, then* f *is linear.*

Now, suppose that a function $f : X \to Y$ is, in some sense, an approximate solution of (20.25). The natural problems are: to describe such a class of mapping mappings and to answer the stability question. More precisely, the question arises: how much f differs from an exact solution of (20.25). This is a classical problem in the theory of stability (see, e.g.: [7, 8, 13]). For the orthogonality equation (20.25), we would like to consider the following type of stability.

Theorem 20.14 *Let X and Y be real smooth Banach spaces. Assume that Y is separable. Let $f : X \to Y$ be a mapping satisfying the functional inequality*

$$\forall_{x,y \in X} \quad |\rho'_+(f(x), f(y)) - \rho'_+(x, y)| \leq \varepsilon. \tag{20.26}$$

If f is surjective, then

$$\forall_{x,y \in X} \quad \|f(x+y) - f(x) - f(y)\| \leq \sqrt{3\varepsilon}. \tag{20.27}$$

Proof Since X, Y are smooth, $\rho'_+(a, \cdot)$ and $\rho'_+(b, \cdot)$ are linear for all $a \in X, b \in Y$. For arbitrary $x, y, z \in X$ we have

$$
\begin{aligned}
|\rho'_+(f(a), f(x+y) - f(x) - f(y))| &= |\rho'_+(f(a), f(x+y)) - \rho'_+(f(a), f(x)) \\
&\quad - \rho'_+(f(a), f(y)) \\
&\quad - \rho'_+(a, x+y) + \rho'_+(a, x) + \rho'_+(a, y)| \leq \\
&\leq |\rho'_+(f(a), f(x+y)) - \rho'_+(a, x+y)| \\
&\quad + |\rho'_+(f(a), f(x)) - \rho'_+(a, x)| \\
&\quad + |\rho'_+(f(a), f(y)) - \rho'_+(a, y)| \leq 3\varepsilon.
\end{aligned}
\tag{20.28}
$$

Let $x, y \in X$. Since f is surjective, there is $a \in X$ such that $f(a) = f(x+y) - f(x) - f(y)$. It follows from (20.28) that

$$
\begin{aligned}
\|f(x+y) - f(x) - f(y)\|^2 &= |\rho'_+(f(x+y) - f(x) - f(y), f(x+y) - f(x) - f(y))| \\
&= |\rho'_+(f(a), f(x+y)) - f(x) - f(y))| \leq 3\varepsilon.
\end{aligned}
$$

The proof is complete.

Finally, we present a result concerning the stability of the orthogonality equation in normed spaces. Applying the assumption of surjectivity of a mapping f we can derive some stability result for Eq. (20.25).

Theorem 20.15 *Let X, Y be Banach smooth spaces and $f : X \to Y$ satisfy (20.26) (with some $\varepsilon \geq 0$). If f is surjective, then there is a mapping $f_\varepsilon : X \to Y$ such that f_ε satisfies (20.25) and $\|f(x) - f_\varepsilon(x)\| \leq \sqrt{3\varepsilon}$.*

Proof By Theorem 20.14, f is $\sqrt{3\varepsilon}$-additive, i.e. f satisfies (20.27). The well known Hyers' Theorem (see [7]) implies that the mapping

$$\forall_{x \in X} \quad f_\varepsilon(x) := \lim_{n \to +\infty} \frac{1}{2^n} f(2^n x)$$

is well-defined and additive. Moreover $\| f(x) - f_\varepsilon(x)\| \leq \sqrt{3\varepsilon}$.

Since X, Y is smooth, the both mappings ρ'_+ and ρ'_+ (in X and in Y, respectively) are continuous with respect to the first and second variable. Putting in (20.26), the vectors $2^n u$, $2^n w$ in place of x, y respectively, and dividing by 4^n and using (ND2) we obtain

$$\forall_{u,w \in X} \quad \left| \rho'_+ \left(\frac{1}{2^n} f(2^n u), \frac{1}{2^n} f(2^n w) \right) - \rho'_+ \left(\frac{1}{2^n} 2^n u, \frac{1}{2^n} 2^n w \right) \right| \leq \frac{\varepsilon}{4^n},$$

whence, letting $n \to \infty$, we obtain $\rho'_+(f_\varepsilon(u), f_\varepsilon(w)) = \rho'_+(u, w)$ for all $u, w \in X$. The proof is complete.

References

1. Alsina, C., Sikorska, J., Santos Tomás, M.: Norm Derivatives and Characterizations of Inner Product Spaces. World Scientific, Hackensack (2009)
2. Behrends, E.: M Structure and Banach-Stone Theorem, Lecture Notes in Mathematics, vol. 736. Springer, Berlin (1979)
3. Birkhoff, G.: Orthogonality in linear metric spaces. Duke Math. J. **1**, 169–172 (1935)
4. Dragomir, S.S.: Semi-Inner Products and Applications. Nova Science Publishers, Inc., Hauppauge (2004)
5. Dragomir, S.S., Koliha, J.J.: The mappings $\Psi^p_{x,y}$ in normed linear spaces and its applications in the theory of inequalities. Math. Inequal. Appl. **2**, 367–381 (1999)
6. Dragomir, S.S., Koliha, J.J.: The mappings $\delta_{x,y}$ in normed linear spaces and refinements of the Cauchy-Schwarz inequality. Nonlinear Anal. **41**, 205–220 (2000)
7. Hyers, D.H.: On the stability of the linear functional equation. Proc. Natl. Acad. Sci. U. S. A. **27**, 222–224 (1941)
8. Hyers, D.H. , Isac, G., Rassias, T.M., Stability of Functional Equations in Several Variables. Progress in Nonlinear Differential Equations and Their Applications, vol. 34. Birkhäuser, Boston (1998)
9. Lin, B.L., Rao, T.S.S.R.K.: Multismoothness in Banach spaces. Int. J. Math. Math. Sci. **2007**, Article ID 52382, 12 pp. (2007)
10. Saleh Hamarsheh, A.: Multismoothness in $L^1(\mu, X)$. Int. Math. Forum **9**(33), 1621–1624 (2014)
11. Saleh Hamarsheh, A.: k-smooth points in some Banach spaces. Int. J. Math. Math. Sci. **2015**, Article ID 394282, 4 pp. (2015)
12. Singer, I.: Best Approximation in Normed Linear Spaces by Elements of Linear Subspaces. Springer, Berlin (1970)
13. Ulam, S.M.: Problems in Modern Mathematics. Chapter VI, Some Questions in Analysis: §1, Stability, Science Editions. Wiley, New York (1964)
14. Wójcik, P.: k-smoothness: an answer to an open problem. Math. Scand. **123**(1), 85–90 (2018)
15. Wójcik, P.: On an orthogonality equation in normed spaces. Funct. Anal. Appl. **52**, 224–227 (2018)

Chapter 21
On Exact and Approximate Orthogonalities Based on Norm Derivatives

Ali Zamani and Mahdi Dehghani

Abstract We survey mainly recent results on the orthogonality relations in normed linear spaces related to norm derivatives. We will focus on fundamental properties of norm derivatives orthogonality, differences and connections between these orthogonality types, and geometric results and problems closely related to them.

Keywords Norm derivative · Orthogonality · Approximate orthogonality · Orthogonality preserving mappings · Approximate orthogonality preserving property · Stability

Mathematics Subject Classification (2010) Primary 46B20; Secondary 46C50, 47B49

21.1 Introduction

One of the most well-known concept in study of the geometry of normed linear spaces is the notion of orthogonality. This concept and its connection with several geometric properties of normed linear spaces, like strict convexity (rotundity) and smoothness has been studied extensively. It is known that in an inner product space $(H, \langle \cdot, \cdot \rangle)$ there is one orthogonality relation derived from inner product. In fact, the vectors $x, y \in H$ are orthogonal (written as $x \perp y$) if and only if $\langle x, y \rangle = 0$.

The situation is completely different in general normed linear spaces. However, there is not a unique way to define the notion of orthogonality in general normed

A. Zamani (✉)
Department of Mathematics, Farhangian University, Tehran, Iran

M. Dehghani
Department of Pure Mathematics, Faculty of Mathematical Sciences, University of Kashan, Kashan, Iran
e-mail: m.dehghani@kashanu.ac.ir

© Springer Nature Switzerland AG 2019
J. Brzdęk et al. (eds.), *Ulam Type Stability*,
https://doi.org/10.1007/978-3-030-28972-0_21

linear spaces. Since 1934 many mathematicians have introduced different generalized orthogonality in normed linear spaces for which, all of them are generalizations of orthogonality in an inner product space.

In 1934, Roberts [67] introduced the first orthogonality in real normed linear spaces. Let $(X, \| \cdot \|)$ be a normed linear space over $\mathbb{K} \in \{\mathbb{R}, \mathbb{C}\}$, whose dimension is at least 2. A vector $x \in X$ is said to be orthogonal in the sense of Roberts to a vector $y \in X$, denoted by $x \perp_R y$ if

$$\|x - ty\| = \|x + ty\| \qquad (t \in \mathbb{K}).$$

Later, in 1935 Birkhoff [8] introduced one of the most important orthogonality type. This notion of orthogonality was developed by James in [38, 39]. (Actually, this notion was much earlier considered by Carathéodory, see [2].) A vector $x \in X$ is said to be orthogonal to a vector $y \in X$ in the sense of Birkhoff–James, written as $x \perp_B y$, if

$$\|x + ty\| \geq \|x\| \qquad (t \in \mathbb{K}).$$

The geometrical interpretation is that the line passing through x in the direction of y supports (at the point x) the ball centred at 0 and with radius $\|x\|$. Note that Roberts orthogonality implies Birkhoff–James orthogonality. In [39] James elaborated how the notions like smoothness, rotundity, etc., of a normed linear space can be studied using Birkhoff–James orthogonality.

Also, James showed an example of a normed plane in which at least one of any two vectors, which are Roberts orthogonal to each other, must be the origin cf. [38]. Due to this situation, James introduced in 1945 isosceles orthogonality and Pythagorean orthogonality [38]. A vector $x \in X$ is said to be isosceles orthogonal to a vector $y \in X$ denoted by $x \perp_I y$ if

$$\|x + y\| = \|x - y\|.$$

Furthermore, a vector $x \in X$ is said to be Pythagorean orthogonal to a vector $y \in X$ denoted by $x \perp_P y$ if

$$\|x + y\|^2 = \|x\|^2 + \|y\|^2.$$

For normed linear spaces, isosceles and Pythagorean orthogonality are not equivalent. They are also not equivalent to Roberts orthogonality. Of course, in an inner product space we have

$$\perp_B = \perp_I = \perp_P = \perp_R = \perp.$$

However, properties like symmetry, homogeneity, additivity, etc., of the orthogonality in inner product spaces do not always carry over to generalized orthogonalities. For example, it is known that Birkhoff–James orthogonality is homogeneous and not symmetric, while isosceles orthogonality and Pythagorean orthogonality are symmetric but not homogeneous, which shows (besides further properties) that these types of orthogonalities are different. We refer the reader to [2, 4, 38–40] and the references therein for basic properties of these type of orthogonalities. Also, a classification of different types of orthogonality in normed linear spaces, their main properties, and the relations between them can be found in e.g., survey paper [68] (see also [24, 25, 37]).

Recall that a normed linear space X is called smooth if each point of the unit sphere $\mathbb{S}_X$ has a unique supporting hyperplane to the closed unit ball $\mathbb{B}_X$, or equivalently, if to each nonzero $x \in X$ there exits a unique $x^* \in X^*$ satisfying $\|x^*\| = 1$ and $x^*(x) = \|x\|$ (see e.g., [3, 31]). Here, X^* denotes as usual the (topological) dual of X. In the case of real normed linear space $(X, \|\cdot\|)$, it has been proved that X is smooth if the $\|\cdot\|$ has the Gateaux derivative in X, i.e.,

$$G_{\pm}(x, y) := \lim_{t \to 0^{\pm}} \frac{\|x + ty\| - \|x\|}{t}$$

exists for each $x, y \in X$; see e.g., [48].

One of the prominent reasons for importance of Birkhoff–James orthogonality is its application to characterize smooth normed linear spaces. Considering existence properties of Birkhoff–James orthogonality, we recall here the following result from [39].

Lemma 21.1 ([39, Corollary 2.2 and Theorem 4.1]) *Let X be a normed linear space and let $x, y \in X$ with $x \neq 0$. Then there exists $t \in \mathbb{K}$ such that $x \perp_B (tx + y)$. In particular, t is unique if and only if X is smooth.*

The concept of semi-inner product space was introduced by Lumer [53] and then the main properties of it were discovered in [34, 55, 64]. It has been proved in [53] that in any normed linear space $(X, \|\cdot\|)$ there exists a mapping $[\cdot|\cdot] : X \times X \to \mathbb{K}$ satisfying the properties:

(i) $[\alpha x + y | z] = \alpha[x|z] + [y|z]$ for all $x, y, z \in X$ and all $\alpha \in \mathbb{K}$;
(ii) $[x|\beta y] = \bar{\beta}[x|y]$ for all $x, y \in X$ and all $\beta \in \mathbb{K}$;
(iii) $[x|x] = \|x\|^2$ for all $x \in X$;
(iv) $|[x|y]| \leq \|x\|\|y\|$ for all $x, y \in X$.

Such a mapping is called a semi-inner product in X. It is known, however, that in a normed linear space there exists exactly one semi-inner product if and only if the space is smooth. More characterizations of smooth normed linear spaces by the notion of semi-inner products could be found in [31].

For vectors $x, y \in X$, the semi-inner product orthogonality is defined as follows:

$$x \perp_s y \quad \text{if and only if} \quad [y|x] = 0.$$

We remark that for any semi-inner product that generates the norm, we have $\perp_s \subset \perp_B$. Nevertheless, the reverse implication is generally not true; see e.g, [31].

For more information about semi-inner product spaces and its relation with Birkhoff–James orthogonality the reader is refereed to [31] and the references therein.

In 1986 norm derivatives were defined by Amir [4] in a real normed linear space $(X, \| \cdot \|)$ as follows:

$$\rho_\pm(x, y) := \lim_{t \to 0^\pm} \frac{\|x + ty\|^2 - \|x\|^2}{2t} = \|x\| \lim_{t \to 0^\pm} \frac{\|x + ty\| - \|x\|}{t}.$$

These functionals extend inner products and many geometrical properties of inner product spaces could be formulated in normed linear spaces by means of norm derivatives. The problem of finding necessary and sufficient conditions for a normed linear space to be an inner product one has been investigated by many mathematicians. There are many different ways to characterize inner product spaces among normed linear spaces. In 1935, Jordan and von Neumann [42] proved that the norm on a linear space X is induced by an inner product if and only if it satisfies the parallelogram law. Another way to obtain characterizations of inner product spaces is to force the orthogonality relation on a normed linear space to fulfill some properties of the natural orthogonality of inner product spaces. Day [26] and James [40] obtained some new characterizations of inner product spaces by means of isosceles and Birkhoff–James orthogonality. For instance, they proved that a normed linear space X, whose dimension is at least three, is an inner product space if and only if Birkhoff–James orthogonality is symmetric in X. Also, it has been proved in [38] that isosceles orthogonality is homogeneous in a normed linear space if and only if this space is an inner product space. In particular, Tapia [70, 71] characterized inner product spaces in terms of norm derivatives. More precisely, he proved that a normed linear space X is an inner product space if and only if $G_+(\cdot, \cdot)$ is linear in the first variable if and only if $G_+(\cdot, \cdot)$ is symmetric.

Norm derivatives play an important role in describing the geometric properties of normed linear spaces. The basic geometric properties such as strict convexity and smoothness of normed linear spaces have been characterized by many mathematicians using the notion of norm derivatives. As the most famous descriptions for smooth real normed linear spaces based on norm derivatives, we point out here the following result from [3].

Lemma 21.2 ([3, Remark 2.1.1]) *Let X be a real normed linear space. Then X is smooth if and only if $\rho_-(x, y) = \rho_+(x, y)$ for all $x, y \in X$.*

Orthogonality relations which are taken from norm derivatives provide a good framework for developing studies of the geometric structure of normed linear spaces. During the last years many papers concerning various aspects of orthogonalities related to norm derivatives have appeared. In this paper we want to give some overview on these results as well as to collect a number of items from the literature dealing with the subject. Our paper can also be taken as an update of existing surveys and monographs; see [16, 22, 68].

21.2 Exact and Approximate Norm Derivatives Orthogonalities

In this section we assume that the considered normed linear spaces are real and their dimensions are not less than 2.

21.2.1 ρ-Orthogonality

Let $(X, \| \cdot \|)$ be a normed linear space and let $x, y \in X$. The orthogonality relations associated to the functionals ρ_- and ρ_+ are defined by

$$x \perp_{\rho_-} y \quad \text{if and only if} \quad \rho_-(x, y) = 0;$$

$$x \perp_{\rho_+} y \quad \text{if and only if} \quad \rho_+(x, y) = 0.$$

In 1987, Miličić [56] introduced a new orthogonality relation as follows

$$x \perp_\rho y \quad \text{if and only if} \quad \rho(x, y) = 0,$$

where the functional $\rho(\cdot, \cdot) := \langle \cdot, \cdot \rangle_g : X \times X \to \mathbb{R}$ was defined by

$$\rho(x, y) = \langle y, x \rangle_g = \frac{\rho_-(x, y) + \rho_+(x, y)}{2}.$$

Among the just defined three orthogonality relations only $\perp_\rho$ is homogeneous (i.e., for all $x, y \in X$ and all $\alpha, \beta \in \mathbb{R}$, if $x \perp_\rho y$, then $\alpha x \perp_\rho \beta y$) and none of them is symmetric. It has been proved in [3] that the relations $\perp_{\rho_\pm}$ and $\perp_\rho$ in a normed linear space X are symmetric if and only if X is an inner product space. First, we remind several properties of these functions, which are used to obtain different characterizations of inner product spaces and smooth normed linear spaces.

Theorem 21.1 *Let $(X, \| \cdot \|)$ be a normed linear space, and let $x, y \in X$. Then*

(i) $\rho_\pm(x, x) = \|x\|^2$ *and* $\rho_-(x, y) \le \rho_+(x, y)$.

(ii) For all $t \in \mathbb{R}$, $\rho_\pm(tx, y) = \rho_\pm(x, ty) = \begin{cases} t\rho_\pm(x, y) & t \ge 0 \\ t\rho_\mp(x, y) & t \le 0. \end{cases}$

(iii) $|\rho_\pm(x, y)| \le \|x\| \|y\|$.

(iv) For all $t \in \mathbb{R}$, $\rho_\pm(x, tx + y) = t\|x\|^2 + \rho_\pm(x, y)$.

In [3], Alsina et al. provided a complete description of these orthogonality relations. The relation of Birkhoff–James orthogonality, $\rho_\pm$-orthogonality and ρ-orthogonality has been obtained in [3] as follows:

Theorem 21.2 ([3, Propositions 2.2.2-3]) *Let X be a normed linear space. Then* $\perp_{\rho_\pm} \subset \perp_B$ *and* $\perp_\rho \subset \perp_B$.

In particular, the equalities $\perp_B = \perp_{\rho_-}$, $\perp_B = \perp_{\rho_+}$ and $\perp_B = \perp_\rho$ in X are equivalent to the smoothness of X.

Now, we recall that norm derivatives characterize Birkhoff–James orthogonality in the following sense.

Theorem 21.3 ([4, 39]) *Let* $(X, \| \cdot \|)$ *be a normed linear space,* $x, y \in X$ *and let* $\alpha \in \mathbb{R}$. *Then the following conditions are equivalent:*

(i) $x \perp_B (y - \alpha x)$.
(ii) $\rho_-(x, y) \le \alpha \|x\|^2 \le \rho_+(x, y)$.

In particular, $x \perp_B y$ *if and only if* $\rho_-(x, y) \le 0 \le \rho_+(x, y)$.

There is a deep connection of smooth normed linear spaces and the orthogonality relations related norm derivatives. Chmieliński and Wójcik in [21] clarified that the relations $\perp_{\rho_\pm}$ and $\perp_\rho$ are generally incomparable. More precisely, they proved that these orthogonality relations are comparable in a normed linear space X if and only if X is smooth.

Theorem 21.4 ([21, Theorem 1]) *Let* X *be a normed linear space. Then the following conditions are equivalent:*

(i) $\perp_{\rho_+} \subset \perp_{\rho_-}$. *(ii)* $\perp_{\rho_-} \subset \perp_{\rho_+}$. *(iii)* $\perp_{\rho_+} = \perp_{\rho_-}$.

(iv) $\perp_{\rho_+} \subset \perp_\rho$. *(v)* $\perp_\rho \subset \perp_{\rho_+}$. *(vi)* $\perp_{\rho_+} = \perp_\rho$.

(vii) $\perp_{\rho_-} \subset \perp_\rho$. *(viii)* $\perp_\rho \subset \perp_{\rho_-}$. *(ix)* $\perp_{\rho_-} = \perp_\rho$. *(x)* X *is smooth.*

Finally, we remark that the connection between the relations $\perp_\rho$ and $\perp_s$ were given in [21].

Theorem 21.5 ([21, Theorem 2]) *Let* X *be a normed linear space and let* $[\cdot|\cdot]$ *be a given semi-inner product in* X. *Then the following conditions are equivalent:*

(i) $\perp_\rho \subset \perp_s$. *(ii)* $\perp_s \subset \perp_\rho$. *(iii)* $\perp_\rho = \perp_s$. *(iv)* $\rho(\cdot, \cdot) = [\cdot|\cdot]$.

21.2.2 ρ_*-*Orthogonality*

Another type of an orthogonality relation connected to norm derivatives that was introduced in [11] is ρ_*-orthogonality. In this section we will review elementary properties of ρ_*-orthogonality. Also, some characterizations of smooth normed linear spaces in terms of ρ_*-orthogonality which has been obtained in [60] are reviewed.

Definition 21.1 ([11]) Let X be a normed linear space. Then a vector $x \in X$ is called ρ_*-orthogonal to a vector $y \in X$, denoted by $x \perp_{\rho_*} y$ if

$$\rho_*(x, y) := \rho_-(x, y)\rho_+(x, y) = 0.$$

First, we represent some elementary properties of the functional ρ_*.

Proposition 21.1 ([60, Proposition 2.1]) *Let $(X, \| \cdot \|)$ be a normed linear space. Then*

(i) $\rho_*(tx, y) = \rho_*(x, ty) = t^2\rho_*(x, y)$ *for all $x, y \in X$ and all $t \in \mathbb{R}$.*
(ii) $|\rho_*(x, y)| \leq \|x\|^2\|y\|^2$ *for all $x, y \in X$.*
(iii) *For all nonzero vectors $x, y \in X$, if $x \perp_{\rho_*} y$, then x and y are linearly independent.*
(iv) $\rho_*(x, tx + y) = t^2\|x\|^4 + 2t\|x\|^2\rho(x, y) + \rho_*(x, y)$ *for all $x, y \in X$ and all $t \in \mathbb{R}$.*

It is clear that $\perp_{\rho_-} \cup \perp_{\rho_+} = \perp_{\rho_*} \subset \perp_B$ and so the equality $\perp_B = \perp_{\rho_*}$ implies the smoothness of the norm. Also, it is noticed in [60] that the relations $\perp_\rho$ and $\perp_{\rho_*}$ are incomparable. In fact, according to the following theorem, these orthogonality relations in a normed linear space X are comparable if and only if X is smooth.

Theorem 21.6 ([60, Theorem 3.1]) *Let X be a normed linear space. Then the following conditions are equivalent:*

(i) $\perp_B \subset \perp_{\rho_*}$.   *(ii)* $\perp_B = \perp_{\rho_*}$.   *(iii)* $\perp_\rho \subset \perp_{\rho_*}$.

(iv) $\perp_{\rho_*} \subset \perp_\rho$.   *(v)* $\perp_{\rho_*} = \perp_\rho$.   *(vi)* $\perp_{\rho_*} \subset \perp_{\rho_+}$.

(vii) $\perp_{\rho_*} \subset \perp_{\rho_-}$.   *(viii)* $\perp_{\rho_*} = \perp_{\rho_-}$.   *(ix)* X *is smooth.*

Moreover, the connection between semi-inner product orthogonality and ρ_*-orthogonality has been established in the following theorem.

Theorem 21.7 ([60, Proposition 2.2]) *Let X be a normed linear space and let $[\cdot|\cdot]$ be a given semi-inner product in X. Then the following conditions are equivalent:*

(i) $\perp_{\rho_*} = \perp_s$.
(ii) $\perp_{\rho_*} \subset \perp_s$.
(iii) $\rho_*(x, y) = [y|x]^2$ *for all $x, y \in X$.*

Let us now suppose that $\perp$ is a binary relation on a real vector space X satisfying

(O1) Totality of $\perp$ for zero: $x \perp 0$ and $0 \perp x$ for all $x \in X$;
(O2) Independence: if $x, y \in X \setminus \{0\}$ and $x \perp y$, then x and y are linearly independent;
(O3) Homogeneity: if $x, y \in X$ and $x \perp y$, then $\alpha x \perp \beta y$ for all $\alpha, \beta \in \mathbb{R}$;
(O4) The Thalesian property: let P be a two-dimensional subspace of X. If $x \in P$ and $\mu \geq 0$, then there exists $y \in P$ such that $x \perp y$ and $x + y \perp \mu x - y$.

The pair $(X, \perp)$ is called an orthogonality space in the sense of Rätz [66]. Inner product spaces and normed linear spaces with Birkhoff–James orthogonality are typical examples of orthogonality spaces. Also, it has been proved in [3] that ρ-orthogonality is an orthogonality space. Using Proposition 21.1, it easy to check that the conditions (O1)–(O3) are true for ρ_*-orthogonality and the following theorem ensures that ρ_*-orthogonality has the Talesian property. Therefore a normed linear space with ρ_*-orthogonality is an orthogonality space in the sense of Rätz.

Theorem 21.8 ([60, Theorem 4.2]) *For any two-dimensional subspace P of a normed linear space X and for every $x \in P$, $\mu \geq 0$, there exists a vector $y \in P$ such that*

$$x \perp_{\rho_*} y \quad \text{and} \quad x + y \perp_{\rho_*} \mu x - y.$$

Let X be a normed linear space and let $(G, +)$ be an Abelian group. Let us recall that a mapping $A : X \longrightarrow G$ is called additive if $A(x + y) = A(x) + A(y)$ for all $x, y \in X$, a mapping $B : X \times X \longrightarrow G$ is called biadditive if it is additive in both variables and a mapping $Q : X \longrightarrow G$ is called quadratic if $Q(x+y)+Q(x-y) = 2Q(x) + 2Q(y)$ for all $x, y \in X$. As an immediate consequence of Theorem 21.8 and [3, Theorem 2.8.1], we deduce the following assertion.

Corollary 21.1 *Let X be a normed linear space and let $(G, +)$ be an Abelian group. A mapping $f : X \longrightarrow G$ satisfies the condition*

$$x \perp_{\rho_*} y \Longrightarrow f(x + y) = f(x) + f(y) \qquad (x, y \in X)$$

if and only if there exist an additive mapping $A : X \longrightarrow G$ and a biadditive and symmetric mapping $B : X \times X \longrightarrow G$ such that

$$f(x) = A(x) + B(x, x) \qquad (x \in X)$$

and

$$x \perp_{\rho_*} y \Longrightarrow B(x, y) = 0 \qquad (x, y \in X).$$

Finally, as a consequence of Theorem 21.8 and [58, Theorem 3], we have the following result.

Corollary 21.2 *Let X be a normed linear space and let $(G, +)$ be an Abelian group. Suppose that Y is a real Banach space. If $f : X \longrightarrow G$ is a mapping fulfilling*

$$x \perp_{\rho_*} y \Longrightarrow \| f(x + y) - f(x) - f(y) \| \leq \varepsilon \qquad (x, y \in X)$$

for some $\varepsilon > 0$, then there exist exactly an additive mapping $A : X \longrightarrow Y$ and exactly a quadratic mapping $Q : X \longrightarrow Y$ such that

$$\| f(x) - f(0) - A(x) - Q(x) \| \leq \frac{68}{3} \varepsilon \qquad (x \in X).$$

21.2.3   Some Generalized Norm Derivatives Orthogonality

In [88] an orthogonality relation as an extension of $\rho_\pm$ and ρ-orthogonality that is called ρ_λ-orthogonality has been introduced. We start this section by reviewing some main results which obtained about this orthogonality relation in [88].

Let X be a normed linear space and let $\lambda \in [0, 1]$. Then a vector $x \in X$ is said to be ρ_λ-orthogonal to a vector $y \in X$ denoted by $x \perp_{\rho_\lambda} y$ if

$$\rho_\lambda(x, y) := \lambda \rho_-(x, y) + (1 - \lambda)\rho_+(x, y) = 0.$$

It is evident that ρ_0 and ρ_1-orthogonality coincide with ρ_+ and ρ_--orthogonality, respectively. Also, $\rho_{\frac{1}{2}}$-orthogonality is equivalent to ρ-orthogonality. As an extension of Theorem 21.2, it has been proved that ρ_λ-orthogonality always implies Birkhoff–James orthogonality.

Proposition 21.2 ([88, Theorem 2.5]) *Let X be a normed linear space and let $\lambda \in [0, 1]$. Then $\perp_{\rho_\lambda} \subset \perp_B$.*

It is noticed in [88, Example 2.8] that for nonsmooth normed linear spaces, the orthogonalities $\perp_{\rho_\lambda}$ and $\perp_B$ may not coincide. However, analogously to Theorem 21.2, the equality $\perp_{\rho_\lambda} = \perp_B$ in a normed linear space X implies the smoothness of X.

Theorem 21.9 ([88, Theorem 2.7]) *Let X be a normed linear space and let $\lambda \in [0, 1]$. Then the following statements are equivalent:*

(i) $\perp_B \subset \perp_{\rho_\lambda}$. (ii) $\perp_B = \perp_{\rho_\lambda}$. (iii) *X is smooth.*

Moreover, the relations $\perp_{\rho_\pm}$, $\perp_\rho$ and $\perp_{\rho_\lambda}$ are generally incomparable; cf. [88, Example 2.10]. The following theorems give some characterizations of smooth normed linear spaces in terms of ρ_λ-orthogonality.

Theorem 21.10 ([88, Theorem 2.12]) *Let X be a normed linear space and let $\lambda \in (0, 1]$. Then the following conditions are equivalent.*

(i) $\perp_{\rho_\lambda} \subset \perp_{\rho_+}$. (ii) $\perp_{\rho_+} \subset \perp_{\rho_\lambda}$. (iii) $\perp_{\rho_\lambda} = \perp_{\rho_+}$. (iv) *X is smooth.*

Theorem 21.11 ([88, Theorem 2.13]) *Let X be a normed linear space and let $\lambda \in [0, 1)$. Then the following conditions are equivalent:*

$(i)\ \perp_{\rho_\lambda} \subset \perp_{\rho_-}$. $(ii)\ \perp_{\rho_-} \subset \perp_{\rho_\lambda}$. $(iii)\ \perp_{\rho_\lambda} = \perp_{\rho_-}$. $(iv)\ X$ *is smooth.*

Theorem 21.12 ([88, Theorem 2.11]) *Let X be a normed linear space and let $\lambda \in [0, 1]$ such that $\lambda \neq \frac{1}{2}$. Then the following conditions are equivalent:*

$(i)\ \perp_{\rho} \subset \perp_{\rho_\lambda}$. $(ii)\ \perp_{\rho_\lambda} \subset \perp_{\rho}$. $(iii)\ \perp_{\rho_\lambda} = \perp_{\rho}$. $(iv)\ X$ *is smooth.*

More generally, a new orthogonality relation based on norm derivatives which is a generalization of the above orthogonalities has been introduced and studied in [28]. We will continue this section to review this orthogonality and its relation with other types of orthogonality relations which have already introduced.

Definition 21.2 ([28]) Let X be a normed linear space, and let $\lambda \in [0, 1]$, $\upsilon = \frac{1}{2k-1}$ with $k \in \mathbb{N}$. For $x, y \in X$, consider the functional $\rho_\lambda^\upsilon : X \times X \to \mathbb{R}$ which is defined by

$$\rho_\lambda^\upsilon(x, y) := \lambda \rho_-^\upsilon(x, y)\rho_+^{1-\upsilon}(x, y) + (1 - \lambda)\rho_+^\upsilon(x, y)\rho_-^{1-\upsilon}(x, y).$$

A vector $x \in X$ is called ρ_λ^υ-orthogonal to a vector $y \in X$, denoted by $x \perp_{\rho_\lambda^\upsilon} y$, if $\rho_\lambda^\upsilon(x, y) = 0$.

It is obvious that for a real inner product space, ρ_λ^υ-orthogonality coincides with the standard orthogonality given by the inner product. Therefore ρ_λ^υ-orthogonality can be considered as a generalization of orthogonality of inner product spaces in real normed linear spaces. We have

$$\rho_0^\upsilon(x, y) = \rho_+^\upsilon(x, y)\rho_-^{1-\upsilon}(x, y) \text{ and } \rho_1^\upsilon(x, y) = \rho_-^\upsilon(x, y)\rho_+^{1-\upsilon}(x, y)$$

for all $\upsilon = \frac{1}{2k-1}$ $(k \in \mathbb{N})$. Hence it is easy to see that $\perp_{\rho_0^\upsilon} = \perp_{\rho_*}$ and $\perp_{\rho_1^\upsilon} = \perp_{\rho_*}$ for all $\upsilon = \frac{1}{2k+1}$ $(k \in \mathbb{N})$. On the other hand, we have $\rho_\lambda^1(x, y) = \rho_\lambda(x, y)$ and therefore ρ_λ^1-orthogonality coincides with ρ_λ-orthogonality for all $\lambda \in [0, 1]$. We point out here the elementary properties of the functional ρ_λ^υ.

Proposition 21.3 ([28, Theorem 2.1]) *Let $(X, \| \cdot \|)$ be a normed linear space and let $x, y \in X$. Then*

(i) $\rho_\lambda^\upsilon(x, x) = \|x\|^2$.

(ii) $\rho_\lambda^\upsilon(tx, y) = \rho_\lambda^\upsilon(x, ty) = \begin{cases} t\rho_\lambda^\upsilon(x, y) & t \geq 0 \\ t\rho_{1-\lambda}^\upsilon(x, y) & t \leq 0. \end{cases}$

(iii) $|\rho_\lambda^\upsilon(x, y)| \leq \|x\|\|y\|$.

(iv) *Let t be a real number such that $\rho_*(x, tx + y) \neq 0$. If*

$$K := K(x, y, t) = \frac{\rho_-(x, tx + y)}{\rho_+(x, tx + y)},$$

then

$$\rho_\lambda^v(x, tx + y) = t\|x\|^2(\lambda K^v + (1 - \lambda)K^{-v}) + \lambda K^v \rho_+(x, y)$$
$$+ (1 - \lambda)K^{-v}\rho_-(x, y).$$

It is clear that $\perp_{\rho_\pm} \subset \perp_{\rho_\lambda^v}$ and so $\perp_{\rho_*} \subset \perp_{\rho_\lambda^v}$. However, some illustrative example have been prepared in [28] which show that the relations $\perp_\rho$, $\perp_{\rho_\lambda}$ and $\perp_{\rho_\lambda^v}$ are generally incomparable. This fact lead us to the following descriptions of smooth normed linear spaces.

Theorem 21.13 ([28, Theorem 2.14]) *Let X be a normed linear space and let $\lambda \in [0, 1] \setminus \{\frac{1}{2}\}$ and $v = \frac{1}{2k-1}$ ($k \in \mathbb{N}$). Then the following conditions are equivalent:*

(i) $\perp_\rho \subset \perp_{\rho_\lambda^v}$.   (ii) $\perp_{\rho_\lambda^v} \subset \perp_\rho$.   (iii) $\perp_{\rho_\lambda^v} = \perp_\rho$.   (iv) X is smooth.

It is worth noting that the situation is different for the case $\lambda = \frac{1}{2}$ and in this case, we have $\perp_\rho \subset \perp_{\rho_{\frac{1}{2}}^v}$. Indeed, for each $x, y \in X$, if $x \perp_\rho y$, then $\rho_-(x, y) = -\rho_+(x, y)$. Hence

$$\rho_{\frac{1}{2}}^v(x, y) = \frac{1}{2}\left[(-1)^v \rho_+^v(x, y)\rho_+^{1-v}(x, y) + (-1)^{1-v}\rho_+^v(x, y)\rho_+^{1-v}(x, y)\right]$$
$$= \frac{1}{2}[-\rho_+(x, y) + \rho_+(x, y)] = 0.$$

Theorem 21.14 ([28, Theorem 2.16]) *Let X be a normed linear space and let $\lambda \in (0, 1)$ and $v = \frac{1}{2k+1}$ ($k \in \mathbb{N}$). Then the following conditions are equivalent:*

(i) $\perp_{\rho_\lambda^v} \subset \perp_{\rho_\lambda}$.   (ii) $\perp_{\rho_\lambda} \subset \perp_{\rho_\lambda^v}$ ($\lambda \neq \frac{1}{2}$).   (iii) $\perp_{\rho_\lambda} = \perp_{\rho_\lambda^v}$.   (iv) X is smooth.

Theorem 21.15 ([28, Theorem 2.17]) *Let X be a normed linear space and let $\lambda \in [0, 1]$ and $v = \frac{1}{2k-1}$ ($k \in \mathbb{N}$). Then the following conditions are equivalent:*

(i) $\perp_{\rho_\lambda^v} \subset \perp_{\rho_-}$ (except for $\perp_{\rho_1^1} = \perp_{\rho_-}$).
(ii) $\perp_{\rho_\lambda^v} \subset \perp_{\rho_+}$ (except for $\perp_{\rho_0^1} = \perp_{\rho_+}$).
(iii) X is smooth.

The following result is an analogue of Theorems 21.2 and Proposition 21.2 which describes the relation between Birkhoff–James orthogonality and ρ_λ^v-orthogonality.

Proposition 21.4 ([28, Proposition 2.9]) *Let X be a normed linear space and let $\lambda \in [0, 1]$ and $v = \frac{1}{2k-1}$ ($k \in \mathbb{N}$). Then $\perp_{\rho_\lambda^v} \subset \perp_B$.*

Also, as stated in [28], for non-smooth normed linear spaces, Birkhoff–James orthogonality and ρ_λ^v-orthogonality may not coincide. Now, as an analogue of Theorems 21.2 and 21.9, we prove that the equality $\perp_{\rho_\lambda^v} = \perp_B$ in normed linear spaces yields the smoothness of the norm. In fact, all the results which mentioned in Theorem 21.2 and Proposition 21.2 are given from the next theorem for the particular modes of λ and v.

Theorem 21.16 *Let X be a normed linear space, $\lambda \in [0, 1]$ and let $v = \frac{1}{2k-1}$ $(k \in \mathbb{N})$. Then the following conditions are equivalent:*

(i) X is smooth.

(ii) $\perp_B \subset \perp_{\rho_\lambda^v}$.

Proof The implication (i)$\Rightarrow$(ii) is clear. Now, we prove the implication (ii)$\Rightarrow$(i). Suppose that $\lambda \in [0, 1]$ such that $\lambda \neq \frac{1}{2}$ and (ii) holds. It follows from (ii) and Theorem 21.2 that $\perp_\rho \subset \perp_B \subset \perp_{\rho_\lambda^v}$ and so Theorem 21.13 concludes that X is smooth.

Now, assume that $\lambda = \frac{1}{2}$. If $x, y \in X$ and $x \neq 0$, then we obtain from Lemma 21.1 that there is $t \in \mathbb{R}$ such that $x \perp_B (tx + y)$ and so (ii) implies that there is $t \in \mathbb{R}$ such that $\rho_\lambda^v(x, tx + y) = 0$. If $\rho_*(x, tx + y) \neq 0$, then it follows from Proposition 21.3 (iv) that

$$K^v \rho_+(x, tx + y) + K^{-v} \rho_-(x, tx + y) = 0.$$

So, we have $K^{2v-1} = -1$. Accordingly, $K = \frac{\rho_-(x, tx+y)}{\rho_+(x, tx+y)} = -1$ and so $t = \frac{-\rho(x,y)}{\|x\|^2}$. Consequently, Birkhoff–James orthogonality is right-unique. Therefore X is smooth, by Lemma 21.1.

Also, if $\rho_*(x, tx + y) = 0$, then $\rho_-(x, tx + y)\rho_+(x, tx + y) = 0$. Therefore, we obtain $t = \frac{-\rho_\pm(x,y)}{\|x\|^2}$. Hence Birkhoff–James orthogonality is right-unique, and so X is smooth.

21.2.4 The λ-Angularly Property of Norms

The concept of angle and the question how to measure angles are interesting from the geometrical view points; see e.g. [7, 61, 62] and the references therein. In this section, we study an angle function based on ρ_λ. Let us begin with some observations. In a real inner product space $(H, \langle \cdot, \cdot \rangle)$, the angle $\theta(x, y)$ between two non-zero elements x, y is defined by

$$\theta(x, y) = \arccos\left(\frac{\langle x, y \rangle}{\|x\| \|y\|}\right).$$

Now, let $(X, \| \cdot \|)$ be a real normed linear space, and let $\lambda \in [0, 1]$. For all non-zero elements $x, y \in X$ we have $-1 \leq \frac{\rho_\lambda(x,y)}{\|x\|\|y\|} \leq 1$. Hence we can define the notion of λ-angle between the non-zero elements x and y.

Definition 21.3 The number

$$\theta_\lambda(x, y) := \arccos \left(\frac{\rho_\lambda(x, y)}{\|x\|\|y\|} \right).$$

is called the λ-angle between the element x and the element y in a normed linear space.

We will refrain from referring to the λ-angle between x and y, since the λ-angle from x to y may not coincide with the λ-angle from y to x. Notice that $\theta_\lambda(x, y)$ does not depend on the lengths of x and y. Also, if the norm in X arises from an inner product, it is easy to see that λ-angles agree with angles defined by the inner product.

Definition 21.4 Two norms, $\| \cdot \|_1$ and $\| \cdot \|_2$, on X have the λ-angularly property if there exists a constant C such that for all non-zero elements $x, y \in X$,

$$\tan \left(\frac{\theta_{\lambda,2}(x, y)}{2} \right) \leq C \tan \left(\frac{\theta_{\lambda,1}(x, y)}{2} \right).$$

Here $\theta_{\lambda,1}(x, y)$ and $\theta_{\lambda,2}(x, y)$ are the λ-angles from x to y relative to $\| \cdot \|_1$ and $\| \cdot \|_2$, respectively. Also, $\tan(\frac{\pi}{2})$ is taken to be $+\infty$.

Our definition is motivated by the Wielandt and generalized Wielandt inequalities, which can be applied in matrix analysis and multivariate analysis, where angles between elements correspond to statistical correlation; see e.g. [74].

Remark 21.1 Suppose the norms $\| \cdot \|_1$ and $\| \cdot \|_2$ have the λ-angularly property on X. Then the norms $\| \cdot \|_2$ and $\| \cdot \|_1$ have the $(1-\lambda)$-angularly property on X. Indeed, for every non-zero $x, y \in X$ we have

$$\tan \left(\frac{\theta_{1-\lambda,1}(x, y)}{2} \right) = -\tan \left(\frac{\theta_{\lambda,1}(x, -y)}{2} \right)$$

$$\leq -\frac{1}{C} \tan \left(\frac{\theta_{\lambda,2}(x, -y)}{2} \right) = \frac{1}{C} \tan \left(\frac{\theta_{1-\lambda,2}(x, y)}{2} \right).$$

In the following theorem we show that λ-angularly property of norms share a geometric property.

Recall that a normed linear space $(X, \| \cdot \|)$ is strictly convex (rotund) if and only if $x \neq y$ and $\|x\| = \|y\| = 1$ together imply that $\|tx + (1 - t)y\| < 1$ for all $0 < t < 1$. To get the next result we use some ideas of [44].

Theorem 21.17 *Suppose the norms* $\| \cdot \|_1$ *and* $\| \cdot \|_2$ *have the* λ-*angularly property on* X. *Then the following statements are equivalent:*

(i) $(X, \| \cdot \|_1)$ *is strictly convex.*
(ii) $(X, \| \cdot \|_2)$ *is strictly convex.*

Proof (i)$\Rightarrow$(ii) Since a normed linear space is strictly convex if every boundary point of the unit ball is an extreme point (see [31]), hence it is enough to show that if $\frac{x}{\|x\|_1}$ is an extreme point of the $\| \cdot \|_1$-unit ball, then $\frac{x}{\|x\|_2}$ is an extreme point of the $\| \cdot \|_2$-unit ball. Suppose $\frac{x}{\|x\|_2}$ is not an extreme point of the $\| \cdot \|_2$-unit ball. Then there are points y and z in X such that $\frac{x}{\|x\|_2} = \frac{y+z}{2}$ and the closed line segment from y to z is contained in the $\| \cdot \|_2$-unit ball. If $s \in [0, 1]$ then the points $(1 - s)y + sz$ and $sy + (1 - s)z$ are on the line segment and hence in the $\| \cdot \|_2$-unit ball. Thus,

$$2 = \|y + z\|_2 = \|(1 - s)y + sz + sy + (1 - s)z\|_2$$
$$\leq \|(1 - s)y + sz\|_2 + \|sy + (1 - s)z\|_2 \leq 1 + 1 = 2.$$

It follows that $\|(1 - s)y + sz\|_2 = \|sy + (1 - s)z\|_2 = 1$. In particular, we observe that $\|y\|_2 = \|z\|_2 = 1$. Hence

$$\rho_{\lambda,2}(y, z) = \lambda \rho_{-,2}(y, z) + (1 - \lambda)\rho_{+,2}(y, z)$$

$$= \lambda \|y\|_2 \lim_{t \to 0^-} \frac{\|y + tz\|_2 - \|y\|_2}{t} + (1-\lambda)\|y\|_2 \lim_{t \to 0^+} \frac{\|y + tz\|_2 - \|y\|_2}{t}$$

$$= \lambda \lim_{s \to 0^-} \frac{\left\|y + \frac{s}{1-s}z\right\|_2 - 1}{\frac{s}{1-s}} + (1 - \lambda) \lim_{s \to 0^+} \frac{\left\|y + \frac{s}{1-s}z\right\|_2 - 1}{\frac{s}{1-s}}$$

$$= \lambda \lim_{s \to 0^-} \frac{\|(1 - s)y + sz\|_2 - (1 - s)}{s}$$

$$+ (1 - \lambda) \lim_{s \to 0^+} \frac{\|(1 - s)y + sz\|_2 - (1 - s)}{s}$$

$$= \lambda + (1 - \lambda) = 1.$$

It follows that $\rho_{\lambda,2}(y, z) = 1$, $\cos\left(\theta_{\lambda,2}(y, z)\right) = 1$, and $\tan\left(\frac{\theta_{\lambda,2}(x,y)}{2}\right) = 0$. By the λ-angularly property, $\tan\left(\frac{\theta_{\lambda,1}(x,y)}{2}\right) = 0$ as well. This implies $\cos\left(\theta_{\lambda,1}(y, z)\right) = 1$ and hence $\rho_{\lambda,1}(y, z) = \|y\|_1\|z\|_1$. From [88, Theorem 2.2] we obtain

$$\|y\|_1\|z\|_1 = \rho_{\lambda,1}(y, z) \leq (\|y + z\|_1 - \|y\|_1)\|y\|_1 \leq \|z\|_1\|y\|_1,$$

and hence $(\|y + z\|_1 - \|y\|_1)\|y\|_1 = \|z\|_1\|y\|_1$, i.e., $\|y + z\|_1 = \|y\|_1 + \|z\|_1$. On the other hands, we have

$$\frac{x}{\|x\|_1} = \frac{\frac{y+z}{2}\|x\|_2}{\left\|\frac{y+z}{2}\|x\|_2\right\|_1} = \frac{y+z}{\|y+z\|_1} = \frac{\|y\|_1}{\|y\|_1 + \|z\|_1}\frac{y}{\|y\|_1} + \frac{\|z\|_1}{\|y\|_1 + \|z\|_1}\frac{z}{\|z\|_1},$$

which is a convex combination of the points $\frac{y}{\|y\|_1}$ and $\frac{z}{\|z\|_1}$. Thus, $\frac{x}{\|x\|_1}$ is an interior point of the line segment from $\frac{y}{\|y\|_1}$ to $\frac{z}{\|z\|_1}$. Since the endpoints of this segment lie in the $\|\cdot\|_1$-unit ball, so the convexity shows that the entire line segment lies in the $\|\cdot\|_1$-unit ball. Thus $\frac{x}{\|x\|_1}$ is not an extreme point of the $\|\cdot\|_1$-unit ball, which is a contradiction.

By using a similar argument we get (ii)$\Rightarrow$(i). $\qquad\blacksquare$

The next theorem may be viewed as a stability result for the λ-angularly property of norms.

Theorem 21.18 *Suppose the norms $\|\cdot\|_1$ and $\|\cdot\|_2$ have the λ-angularly property on X and let $\|\cdot\|_3 = \|\cdot\|_1 + \|\cdot\|_2$. Then the following statements hold.*

(i) The norms $\|\cdot\|_3$ and $\|\cdot\|_1$ have the λ-angularly property.
(ii) The norms $\|\cdot\|_3$ and $\|\cdot\|_2$ have the $(1-\lambda)$-angularly property.

Proof

(i) Let $x, y \in X \setminus \{0\}$. Let $\rho_{\lambda,i}(x, y)$ and $\theta_{\lambda,i}(x, y)$ be the functional ρ_λ and the λ-angle from x to y with respect to the norm $\|\cdot\|_i$, for $i = 1, 2, 3$. We have

$$\rho_{\lambda,3}(x, y) = \lambda\rho_{-,3}(x, y) + (1-\lambda)\rho_{+,3}(x, y)$$

$$= \lambda\|x\|_3 \lim_{t\to 0^-}\frac{\|x+ty\|_3 - \|x\|_3}{t} + (1-\lambda)\|x\|_3 \lim_{t\to 0^+}\frac{\|x+ty\|_3 - \|x\|_3}{t}$$

$$= \lambda\|x\|_3 \lim_{t\to 0^-}\frac{\|x+ty\|_1 + \|x+ty\|_2 - \|x\|_1 - \|x\|_2}{t}$$

$$\qquad + (1-\lambda)\|x\|_3 \lim_{t\to 0^+}\frac{\|x+ty\|_1 + \|x+ty\|_2 - \|x\|_1 - \|x\|_2}{t}$$

$$= \lambda\|x\|_3\frac{\rho_{-,1}(x, y)}{\|x\|_1} + \lambda\|x\|_3\frac{\rho_{-,2}(x, y)}{\|x\|_2}$$

$$\qquad + (1-\lambda)\|x\|_3\frac{\rho_{+,1}(x, y)}{\|x\|_1} + (1-\lambda)\|x\|_3\frac{\rho_{+,2}(x, y)}{\|x\|_2}$$

$$= \frac{\|x\|_3}{\|x\|_1}\Big(\lambda\rho_{-,1}(x, y) + (1-\lambda)\rho_{+,1}(x, y)\Big)$$

$$\qquad + \frac{\|x\|_3}{\|x\|_2}\Big(\lambda\rho_{-,2}(x, y) + (1-\lambda)\rho_{+,2}(x, y)\Big)$$

$$= \frac{\|x\|_3}{\|x\|_1}\rho_{\lambda,1}(x, y) + \frac{\|x\|_3}{\|x\|_2}\rho_{\lambda,2}(x, y).$$

Therefore

$$\rho_{\lambda,3}(x, y) = \frac{\|x\|_3}{\|x\|_1}\rho_{\lambda,1}(x, y) + \frac{\|x\|_3}{\|x\|_2}\rho_{\lambda,2}(x, y),$$

whence

$$\cos\theta_{\lambda,3}(x, y) = \frac{\rho_{\lambda,3}(x, y)}{\|x\|_3 \|y\|_3}$$

$$= \frac{\rho_{\lambda,1}(x, y)}{\|x\|_1 \|y\|_3} + \frac{\rho_{\lambda,2}(x, y)}{\|x\|_2 \|y\|_3}$$

$$= \frac{\|y\|_1}{\|y\|_3}\cos\theta_{\lambda,1}(x, y) + \frac{\|y\|_2}{\|y\|_3}\cos\theta_{\lambda,2}(x, y).$$

Thus

$$\cos\theta_{\lambda,3}(x, y) = \frac{\|y\|_1}{\|y\|_3}\cos\theta_{\lambda,1}(x, y) + \frac{\|y\|_2}{\|y\|_3}\cos\theta_{\lambda,2}(x, y). \qquad (21.1)$$

Now, by (21.1) and the fact that $\frac{1+r}{1+t} \le 1 + \frac{r}{t}$ for all $r, t > 0$, we have

$$\tan\left(\frac{\theta_{\lambda,3}(x, y)}{2}\right) = \sqrt{\frac{1 - \cos\theta_{\lambda,3}(x, y)}{1 + \cos\theta_{\lambda,3}(x, y)}}$$

$$\le \sqrt{1 + \frac{\tan\left(\frac{\theta_{\lambda,2}^2(x,y)}{2}\right)}{\tan\left(\frac{\theta_{\lambda,1}^2(x,y)}{2}\right)}}\,\tan\left(\frac{\theta_{\lambda,1}(x, y)}{2}\right)$$

$$\le \sqrt{1 + C^2}\,\tan\left(\frac{\theta_{\lambda,1}(x, y)}{2}\right).$$

Hence

$$\tan\left(\frac{\theta_{\lambda,3}(x, y)}{2}\right) \le \sqrt{1 + C^2}\,\tan\left(\frac{\theta_{\lambda,1}(x, y)}{2}\right).$$

So, the norms $\| \cdot \|_3$ and $\| \cdot \|_1$ have the λ-angularly property.

(ii) Since the norms $\| \cdot \|_1$ and $\| \cdot \|_2$ have the λ-angularly property, Remark 21.1 shows that the norms $\| \cdot \|_2$ and $\| \cdot \|_1$ have the $(1 - \lambda)$-angularly property. Thus from (i) we conclude that the norms $\| \cdot \|_3$ and $\| \cdot \|_2$ have the $(1 - \lambda)$-angularly property.

21.2.5 *Approximate Norm Derivatives Orthogonalities*

In an inner product space $(H, \langle \cdot, \cdot \rangle)$ an approximate orthogonality (ε-orthogonality) of vectors $x, y \in H$ was naturally defined in [13, 30] by

$$x \perp^\varepsilon y \quad \text{if and only if} \quad |\langle x, y \rangle| \leq \varepsilon \|x\| \|y\|.$$

For $\varepsilon \geq 1$, it is clear that every pair of vectors are ε-orthogonal, so the interesting case is when $\varepsilon \in [0, 1)$.

Now, let $(X, \| \cdot \|)$ be a normed linear space and let $x, y \in X$. Analogously, for a given semi-inner product $[\cdot|\cdot]$ on X the approximate semi-orthogonality relation was defined in [15, 30] by

$$x \perp_s^\varepsilon y \quad \text{if and only if} \quad |[y|x]| \leq \varepsilon \|x\| \|y\|.$$

The first notion of approximate Birkhoff–James orthogonality has been proposed by Dragomir [29] as follows:

$$x \perp_D^\varepsilon y \quad \text{if and only if} \quad \|x + ty\| \geq (1 - \varepsilon)\|x\| \qquad (t \in \mathbb{K}).$$

Chmieliński [12] also introduced another notion of approximate Birkhoff–James orthogonality, defined in the following way:

$$x \perp_B^\varepsilon y \quad \text{if and only if} \quad \|x + ty\|^2 \geq \|x\|^2 - 2\varepsilon \|x\| \|ty\| \qquad (t \in \mathbb{K}).$$

We would like to remark that in a normed linear space, both types of approximate Birkhoff–James orthogonality are homogeneous. For more information about these types of approximate orthogonality and their properties the reader is referred to [12, 29].

Inspired by approximate Birkhoff–James orthogonality, for a normed linear space, others notions of approximate orthogonality were considered. One of them is the approximate Roberts orthogonality. In fact, the authors in [86] introduced two versions of approximate Roberts orthogonality as follows:

$$x \perp_R^\varepsilon y \Leftrightarrow \left| \|x + ty\|^2 - \|x - ty\|^2 \right| \leq 4\varepsilon \|x\| \|ty\| \qquad (t \in \mathbb{R})$$

and

$$x^\varepsilon \perp_R y \Leftrightarrow \left| \|x + ty\| - \|x - ty\| \right| \leq \varepsilon (\|x + ty\| + \|x - ty\|) \qquad (t \in \mathbb{R}).$$

It can be remarked that these two orthogonality relations are related to analogous definitions for isosceles orthogonality introduced in [20] (see also [85]). Another one is the approximate Pythagorean orthogonality which has been investigated in [77]:

$$x \perp_P^\varepsilon y \Leftrightarrow \left| \|x + y\|^2 - \|x\|^2 - \|y\|^2 \right| \le 2\varepsilon \|x\| \|y\|.$$

Also, we remember two generalized types of approximate isosceles orthogonality, namely approximate cI-orthogonality, in normed linear spaces were considered in [87]. For a fixed $c \ne 0$, the first one is

$$x^\varepsilon \perp_{cI} y \Leftrightarrow \left| \|x + cy\|^2 - \|x - cy\|^2 \right| \le 4\varepsilon \|x\| \|cy\|,$$

and the second one is

$$x \perp_{cI}^\varepsilon y \Leftrightarrow \left| \|x + cy\| - \|x - cy\| \right| \le \varepsilon (\|x + cy\| + \|x - cy\|).$$

In a similar way Chmieliński and Wójcik [22] introduced the notions of an approximate $\rho_\pm$ and ρ-orthogonality as follows:

$$x \perp_{\rho_\pm}^\varepsilon y \quad \text{if and only if} \quad |\rho_\pm(x, y)| \le \varepsilon \|x\| \|y\|,$$

$$x \perp_\rho^\varepsilon y \quad \text{if and only if} \quad |\rho(x, y)| \le \varepsilon \|x\| \|y\|.$$

Similarly, the approximate ρ_*-orthogonality has been defined and studied in [27]:

$$x \perp_{\rho_*}^\varepsilon y \quad \text{if and only if} \quad |\rho_*(x, y)| \le \varepsilon^2 \|x\|^2 \|y\|^2.$$

Obviously, if the norm in X comes from an inner product, then

$$\perp^\varepsilon = \perp_s^\varepsilon = \perp_B^\varepsilon = \perp_R^\varepsilon = \perp_P^\varepsilon = {}^\varepsilon \perp_{cI} = \perp_{\rho_\pm}^\varepsilon = \perp_\rho^\varepsilon = \perp_{\rho_*}^\varepsilon.$$

Also, it is clear that for $\varepsilon = 0$ all the above approximate orthogonalities coincide with the related exact orthogonalities.

Chmieliński and Wójcik generalized Theorem 21.3 for approximate Birkhoff–James orthogonality in [22] as follows:

Theorem 21.19 ([22, Thorem 3.1]) *Let $(X, \| \cdot \|)$ be a normed linear space and let $\varepsilon \in [0, 1)$. Then, for arbitrary $x, y \in X$ and $\alpha \in \mathbb{R}$ the following condition are equivalent:*

(i) $x \perp_B^\varepsilon (y - \alpha x)$.
(ii) $\rho_-(x, y) - \varepsilon \|x\| \|y - \alpha x\| \le \alpha \|x\|^2 \le \rho_+(x, y) + \varepsilon \|x\| \|y - \alpha x\|$.

In particular, $x \perp_B^\varepsilon y$ if and only if

$$\rho_-(x, y) - \varepsilon \|x\| \|y\| \le 0 \le \rho_+(x, y) + \varepsilon \|x\| \|y\|.$$

They also identified the relationship between $\perp_s^\varepsilon$, $\perp_{\rho_\pm}^\varepsilon$, $\perp_\rho^\varepsilon$ and $\perp_B^\varepsilon$ in the following theorem.

Theorem 21.20 ([21, 22]) *Let $\varepsilon \in [0, 1)$. For an arbitrary normed linear space X and $\diamond \in \{s, \rho_-, \rho_+, \rho\}$ we have $\perp_\diamond^\varepsilon \subset \perp_B^\varepsilon$.*

Of course, for non-smooth normed linear spaces, the approximate orthogonalities $\perp_{\rho_\pm}^\varepsilon$ and $\perp_\rho^\varepsilon$ are incomparable. The following generalization of Theorem 21.4 has been proved in [22].

Theorem 21.21 ([22, Theorem 3.3]) *Let X be a normed linear space and let $\varepsilon \in [0, 1)$. Then the following conditions are equivalent:*

(i) $\perp_{\rho_+}^\varepsilon \subset \perp_{\rho_-}^\varepsilon$.   (ii) $\perp_{\rho_-}^\varepsilon \subset \perp_{\rho_+}^\varepsilon$.   (iii) $\perp_{\rho_+}^\varepsilon = \perp_{\rho_-}^\varepsilon$.

(iv) $\perp_{\rho_+}^\varepsilon \subset \perp_\rho^\varepsilon$.   (v) $\perp_\rho^\varepsilon \subset \perp_{\rho_+}^\varepsilon$.   (vi) $\perp_{\rho_+}^\varepsilon = \perp_\rho^\varepsilon$.

(vii) $\perp_{\rho_-}^\varepsilon \subset \perp_\rho^\varepsilon$.   (viii) $\perp_\rho^\varepsilon \subset \perp_{\rho_-}^\varepsilon$.   (ix) $\perp_{\rho_-}^\varepsilon = \perp_\rho^\varepsilon$.   (x) X is smooth.

Some illustrated examples were provided in [21, 22] which show that equalities in Theorem 21.20 need not to hold in non-smooth normed linear spaces. Actually, using this fact and Theorem 21.21 it has been proved in [22] that the smoothness of a normed linear space X resulted also from $\perp_{\rho_\pm}^\varepsilon = \perp_B^\varepsilon$ and $\perp_\rho^\varepsilon = \perp_B^\varepsilon$ for some $\varepsilon \in [0, 1)$. In fact, the following theorem is a generalization of Theorem 21.2.

Theorem 21.22 ([22, Theorem 3.4]) *Let X be a normed linear space and let $\varepsilon \in [0, 1)$. If $\perp_{\rho_\pm}^\varepsilon = \perp_B^\varepsilon$ or $\perp_\rho^\varepsilon = \perp_B^\varepsilon$, then X is smooth.*

Moreover, an approximate version of Theorem 21.5 has been prepared as follows:

Theorem 21.23 ([22, Theorem 3.5]) *Let X be a normed linear space and let $[\cdot|\cdot]$ be a fixed semi-inner product in X. For $\varepsilon \in [0, 1)$ the following conditions are equivalent:*

(i) $\perp_\rho^\varepsilon \subset \perp_s^\varepsilon$.   (ii) $\perp_s^\varepsilon \subset \perp_\rho^\varepsilon$.   (iii) $\perp_\rho^\varepsilon = \perp_s^\varepsilon$.   (iv) $\langle \cdot, \cdot \rangle_g = [\cdot|\cdot]$.

Another characterization of smooth normed linear spaces using comparison of approximate $\rho_\pm$-orthogonality and approximate semi-inner product has been presented in [78].

Theorem 21.24 *Let X be a normed linear space and let $[\cdot|\cdot]$ be a fixed semi-inner product in X. For $\varepsilon \in [0, 1)$ the following conditions are equivalent.*

(i) $\perp_{\rho_+}^\varepsilon \subset \perp_s^\varepsilon$.   (ii) $\perp_s^\varepsilon \subset \perp_{\rho_+}^\varepsilon$.   (iii) $\perp_{\rho_+}^\varepsilon = \perp_s^\varepsilon$.

(iv) $\perp_{\rho_-}^\varepsilon \subset \perp_s^\varepsilon$.   (v) $\perp_s^\varepsilon \subset \perp_{\rho_-}^\varepsilon$.   (vi) $\perp_{\rho_-}^\varepsilon = \perp_s^\varepsilon$.

(vii) X is smooth.

In [69] Stypuła and Wójcik by introducing the constants

$$\mathscr{E}^{\rho}(X) := \inf \left\{ \varepsilon \in [0, 1] : \quad \perp_{\rho_+} \subset \perp_{\rho_-}^{\varepsilon} \right\}$$

and

$$\mathscr{R}(X) := \sup \left\{ \|x - y\| : \quad \mathrm{conv}\{x, y\} \subset \mathbb{S}_X \right\}$$

provided some different characterizations of rotundity and smoothness of dual spaces. We have, of course, $0 \le \mathscr{E}^{\rho}(X) \le 1$ and $0 \le \mathscr{R}(X) \le 2$. Observe that,

$$\mathscr{E}^{\rho}(X) = 0 \quad \text{if and only if} \quad X \text{ is smooth}$$

and

$$\mathscr{R}(X) = 0 \quad \text{if and only if} \quad X \text{ is rotund.}$$

A well-known theorem states that if X^* is rotund, then X is smooth. The following theorem states this well-known result in terms of constants $\mathscr{E}^{\rho}(X)$ and $\mathscr{R}(X)$.

Theorem 21.25 ([69, Corollary 2.6]) *Let X be a real normed linear space. Then*

$$\mathscr{E}^{\rho}(X) \le \mathscr{R}(X^*).$$

Moreover, if X is a reflexive Banach space, then

$$\mathscr{E}^{\rho}(X) \le \mathscr{R}(X^*) \le 2\mathscr{E}^{\rho}(X).$$

Hence,

(i) if X is reflexive, X^* is rotund if and only if X is smooth;
(ii) if X is reflexive, X^* is smooth if and only if X is rotund.

Now, let us review the results obtained related to approximate ρ_*-orthogonality from [27]. It is easy to check that the approximate ρ_*-orthogonality is homogenous. Also, if $x \perp_{\rho_+}^{\varepsilon} y$ and $x \perp_{\rho_-}^{\varepsilon} y$, then $x \perp_{\rho_*}^{\varepsilon} y$. Indeed, by the arithmetic-geometric means inequality, we get

$$|\rho_*(x, y)| = |\rho_-(x, y)\rho_+(x, y)| \le \left(\frac{|\rho_-(x, y)| + |\rho_+(x, y)|}{2} \right)^2 \le \varepsilon^2 \|x\|^2 \|y\|^2.$$

We notice that the relations $\perp_{\rho_\pm}^{\varepsilon}$, $\perp_{\rho}^{\varepsilon}$ and $\perp_{\rho_*}^{\varepsilon}$ are generally incomparable, see [27, Example 2.1]. Also, the relation between $\perp_{\rho_*}^{\varepsilon}$ and $\perp_B^{\varepsilon}$ has been identified as follows:

Theorem 21.26 ([27, Theorem 2.3]) *Let X be a normed linear space and let $\varepsilon \in [0, 1)$. Then $\perp_{\rho_*}^{\varepsilon} \subset \perp_B^{\varepsilon}$.*

It is noticed in [27, Example 2.4] that for nonsmooth normed linear spaces, the orthogonalities $\perp_{\rho_*}^{\varepsilon}$ and $\perp_B^{\varepsilon}$ may not coincide.

Theorem 21.27 ([27, Remark 2.5]) *Let X be a normed linear space and let $\varepsilon \in [0, 1)$. If $\perp_B^{\varepsilon} \subset \perp_{\rho_*}^{\varepsilon}$, then X is smooth.*

To finish this section we consider analogously, the notion of approximate ρ_{λ}^{v}-orthogonality which is studied in [1]. In fact, naturally, for $\varepsilon \in [0, 1)$, $\lambda \in [0, 1]$ and $v = \frac{1}{2k-1}$ ($k \in \mathbb{N}$), we say that a vector $x \in X$ is approximate ρ_{λ}^{v}-orthogonal to a vector $y \in X$, in short $x \perp_{\rho_{\lambda}^{v}}^{\varepsilon} y$, if

$$|\rho_{\lambda}^{v}(x, y)| \leq \varepsilon \|x\| \|y\|.$$

In particular, for $v = 1$, we have $x \perp_{\rho_{\lambda}}^{\varepsilon} y$ if and only if $|\rho_{\lambda}(x, y)| \leq \varepsilon \|x\| \|y\|$. Note that the relations $x \perp_{\rho_0}^{\varepsilon} y$, $x \perp_{\rho_1}^{\varepsilon} y$ and $x \perp_{\rho_{\frac{1}{2}}}^{\varepsilon} y$ coincide with the relations $x \perp_{\rho_+}^{\varepsilon} y$, $x \perp_{\rho_-}^{\varepsilon} y$ and $x \perp_{\rho}^{\varepsilon} y$, respectively.

In [1] some illustrated examples have been presented to show that the relations $\perp_{\rho_+}^{\varepsilon}$, $\perp_{\rho}^{\varepsilon}$, $\perp_{\rho_*}^{\varepsilon}$, $\perp_{\rho_{\lambda}}^{\varepsilon}$ and $\perp_{\rho_{\lambda}^{v}}^{\varepsilon}$ are incomparable in general normed linear spaces.

The following result is a generalization of Theorem 21.20.

Theorem 21.28 ([1, Theorem 2.4]) *Let X be a normed linear space and let $\varepsilon \in [0, 1)$, $\lambda \in [0, 1]$ and $v = \frac{1}{2k-1}$ ($k \in \mathbb{N}$). Then $\perp_{\rho_{\lambda}^{v}}^{\varepsilon} \subset \perp_B^{\varepsilon}$.*

According to [1], there are non-smooth normed linear spaces such that $\perp_B^{\varepsilon} \not\subset \perp_{\rho_{\lambda}}^{\varepsilon}$. Analogously to Theorem 21.22, it has been proved in the following theorem that approximate Birkhoff–James orthogonality and approximate ρ_{λ}^{v}-orthogonality in a normed linear space X are equivalent if and only if X is smooth.

Theorem 21.29 ([1, Theorem 2.7]) *Let X be a normed linear space and let $\varepsilon \in [0, 1)$, $\lambda \in [0, 1]$ and $v = \frac{1}{2k-1}$ ($k \in \mathbb{N}$). If $\perp_B^{\varepsilon} \subset \perp_{\rho_{\lambda}^{v}}^{\varepsilon}$, then X is smooth.*

21.3 Orthogonality Preserving Property and Applications in the Geometry of Normed Linear Spaces

21.3.1 Linear Mappings Preserving Orthogonality

The problem of determining the structure of linear mappings between normed linear spaces, which leave certain properties invariant, has been considered in several papers. These are the so-called linear preserver problems, see [10, 52] and the references therein. The study on linear orthogonality preserving mappings can be considered as a part of the theory of linear preservers. The orthogonality preserving property have been intensively studied recently in connection with functional analysis and operator theory; cf. [13, 23, 47, 72, 81, 89, 91].

Let H and K be inner product spaces. A mapping $T : H \to K$ is called orthogonality preserving if

$$x \perp y \Rightarrow Tx \perp Ty \qquad (x, y \in X).$$

Such mappings can be very irregular, far from being continuous or linear (see [13, Example 2]). For that reason we restrict ourselves to linear mappings only. On the other hand, for linear orthogonality preserving mappings we have a simple characterization.

Theorem 21.30 ([13, Theorem 1]) *Let H and K be (real or complex) inner product spaces. For a nonzero linear mapping $T : H \to K$ the following conditions are equivalent (with some $\gamma > 0$):*

(i) T is a similarity (scalar multiple of a linear isometry), i.e., $\|Tx\| = \gamma\|x\|$ for all $x \in H$.

(ii) $\langle Tx, Ty \rangle = \gamma^2 \langle x, y \rangle$ for all $x, y \in H$.

(iii) T is orthogonality preserving.

Orthogonality preserving mappings have been widely studied in the setting of inner product C^*-modules, see [5, 6, 32, 35, 43, 49–51, 59, 89]. In particular, further generalizations of Theorem 21.30 can be found in [18, 33]. Similar investigations have been carried out in normed linear spaces for sesquilinear form (instead of inner products) in paper [79].

Let X and Y be normed linear spaces and let $T : X \to Y$ be a linear and continuous operator. The norm of T is defined as usual:

$$\|T\| = \sup \left\{ \|Tx\| : \|x\| = 1 \right\} = \inf \left\{ M > 0 : \|Tx\| \leq M\|x\|, \ x \in X \right\}.$$

Similarly, we define

$$[T] := \inf \left\{ \|Tx\| : \|x\| = 1 \right\} = \sup \left\{ m \geq 0 : \|Tx\| \geq m\|x\|, \ x \in X \right\}.$$

Now, let $\diamond, \heartsuit \in \{B, I, s, \rho_\pm, \rho, \rho_*, \rho_\lambda, \rho_\lambda^\nu\}$. We say that a mapping $T : X \to Y$ (exactly) preserves $(\diamond, \heartsuit)$-orthogonality if

$$x \perp_\diamond y \Rightarrow Tx \perp_\heartsuit Ty \qquad (x, y \in X).$$

In particular, we say that T is $\diamond$-orthogonality preserving if

$$x \perp_\diamond y \Rightarrow Tx \perp_\diamond Ty \qquad (x, y \in X).$$

Koehler and Rosenthal [45, Theorem 1] showed that a linear operator from a normed linear space into itself is an isometry if and only if it preserves some semi-inner product. Blanco and Turnšek [9, Remark 3.2] and Chmieliński [15, Theorem 2.5] extended it to different normed linear spaces. Namely, we have the following result.

Theorem 21.31 *Let X and Y be normed linear spaces. For a linear mapping T : $X \longrightarrow Y$ and some $\gamma > 0$ the following conditions are equivalent:*

(i) T is a similarity.
(ii) $[Tx, Ty]_Y = \gamma^2[x, y]_X$ for all $x, y \in X$.
(iii) T is s-orthogonality preserving.

The conditions (ii) *and* (iii) *should be understood that they are satisfied with respect to some semi-inner products $[\cdot, \cdot]_X$ and $[\cdot, \cdot]_Y$ in X and Y, respectively.*

It has been proved by Koldobsky [46] that a linear mapping $T : X \to X$ preserving B-orthogonality has to be a similarity. In [36, Theorem 1], Ionică using the connections between the Birkhoff–James orthogonality and norm derivatives gave an alternative proof of the above results in the case of different real normed linear spaces (see also [65]). The respective result for both real and complex cases was given by Blanco and Turnšek in [9, Theorem 3.1]. Very recently, Wójcik in [83] presented a somewhat simpler proof of this theorem.

Theorem 21.32 *Let X and Y be (real or complex) normed linear spaces. A linear mapping $T : X \to Y$ is B-orthogonality preserving if and only if it is a scalar multiple of a linear isometry.*

The following result gives a characterization of inner product spaces.

Theorem 21.33 ([15, Theorem 2.9]) *Let X be a normed linear space. Suppose that there exists an inner product space K and a linear mapping T from X into K or from K onto X such that T preserves B-orthogonality. Then X is an inner product space.*

Martini and Wu [54, Lemma 4] proved the following result.

Theorem 21.34 *Let X and Y be two normed linear spaces. If a linear mapping $T : X \longrightarrow Y$ preserves I-orthogonality, then it also preserves B-orthogonality.*

Combining Theorems 21.32 and 21.34 actually lead us to the following result.

Corollary 21.3 *Let X and Y be normed linear spaces, and let $T : X \longrightarrow Y$ be a nonzero linear mapping. Then the following conditions are equivalent:*

(i) T is I-orthogonality preserving.
(ii) T is a scalar multiple of a linear isometry.

Remark 21.2 Notice that Corollary 21.3 also has been proved in [16, Theorem 4.5].

The next theorem gives characterizations of inner product spaces by properties of linear operators related to B-orthogonality and I-orthogonality.

492 A. Zamani and M. Dehghani

Theorem 21.35 ([84, Theorems 9, 10]) *Let X and Y be two normed linear spaces. Each one of the following conditions implies that X and $T(X)$ are inner product spaces.*

(i) *There exists a nonzero linear mapping $T : X \longrightarrow Y$ which preserve (I, B)-orthogonality.*

(ii) *There exists a nonzero linear mapping $T : X \longrightarrow Y$ which preserve (B, I)-orthogonality.*

The orthogonality preserving mappings have been considered also in [63]. The paper [63] shows another way to consider the orthogonality preserving mappings. Some other results on B-orthogonality preserving mapping can be found in [19, 82, 84].

21.3.2 Mappings Which Exactly Preserve Norm Derivatives Orthogonality

The aim of this subsection is to present results concerning the linear mappings which preserve norm derivatives orthogonality. We survey on the results presented in [11, 21, 22, 28, 60, 75, 88], as well as give some new and more general ones. In 2010, Chmieliński and Wójcik [21] studied norm derivatives orthogonality preserving mappings. They proved that for arbitrary normed linear spaces X and Y, if a linear mapping $T : X \longrightarrow Y$ preserves ρ_--orthogonality or preserves ρ_+-orthogonality then it is a similarity. Later, Wójcik [75] showed that a linear mapping preserving ρ-orthogonality has to be a similarity. These results give

Theorem 21.36 *Let X and Y be normed linear spaces, and let $T : X \longrightarrow Y$ be a nonzero linear mapping. Then the following conditions are equivalent:*

(i) *T preserves ρ_+-orthogonality.*

(ii) *T preserves ρ_--orthogonality.*

(iii) *T preserves ρ-orthogonality.*

(iv) *$\|Tx\| = \|T\| \, \|x\|$ for all $x \in X$.*

(v) *$\rho_+(Tx, Ty) = \|T\|^2 \rho_+(x, y)$ for all $x, y \in X$.*

(vi) *$\rho_-(Tx, Ty) = \|T\|^2 \rho_-(x, y)$ for all $x, y \in X$.*

(vii) *$\rho(Tx, Ty) = \|T\|^2 \rho(x, y)$ for all $x, y \in X$.*

As for the ρ_*-orthogonality preserving mapping the following characterization has been given in [60] (see also [11]).

Theorem 21.37 *Let X, Y be normed linear spaces and let $T : X \longrightarrow Y$ be a nonzero linear mapping. Then the following conditions are equivalent:*

(i) *T preserves ρ_*-orthogonality.*

(ii) *T preserves (ρ_*, B)-orthogonality.*

(iii) *T preserves (B, ρ_*)-orthogonality.*

(iv) $\|Tx\| = \|T\|\,\|x\|$ *for all* $x \in X$.

(v) $\rho_*(Tx, Ty) = \|T\|^4 \rho_*(x, y)$ *for all* $x, y \in X$.

If $X = Y$, *then each one of these assertions is also equivalent to*

(vi) *there exists a semi-inner product* $[\cdot|\cdot] : X \times X \longrightarrow \mathbb{R}$ *satisfying*

$$[Tx, Ty]_X = \|T\|^2[x, y]_X \qquad (x, y \in X).$$

Recall that a normed linear space $(X, \|\cdot\|)$ satisfies the δ-parallelogram law for some $\delta \in [0, 1)$, if the double inequality

$$2(1 - \delta)\|z\|^2 \leq \|z + w\|^2 + \|z - w\|^2 - 2\|w\|^2 \leq 2(1 + \delta)\|z\|^2$$

holds for all $z, w \in X$; cf. [17]. Also a normed linear space $(X, \|\cdot\|)$ is equivalent to an inner product space if there exist an inner product in X and a norm $|||\cdot|||$ generated by this inner product such that

$$\frac{1}{k}\|x\| \leq |||x||| \leq k\|x\| \qquad (x \in X)$$

holds for some $k \geq 1$; see [41].

Corollary 21.4 ([60, Corollary 2.7]) *Any one of the following assertions implies that* X *is equivalent to an inner product space.*

(i) *There exist a normed linear space* Y *satisfying the* δ-parallelogram law for some $\delta \in [0, 1)$ *and a nonzero linear mapping* $T : X \longrightarrow Y$ *such that* T *preserves* ρ_*-orthogonality.

(ii) *There exist a normed linear space* Y *satisfying the* δ-parallelogram law for some $\delta \in [0, 1)$ *and a nonzero surjective linear mapping* $S : Y \longrightarrow X$ *such that* S *preserves* ρ_*-orthogonality.

We remark that the converse of Corollary 21.4 holds also true. Indeed, if X is equivalent to an inner product space, then we can choose $\delta = 0$, $Y = X$ and $T = id$, the identity operator on X. Recall that a normed linear space $(X, \|\cdot\|)$ is called uniformly smooth if X satisfies the property that for every $\varepsilon > 0$ there exists $\delta > 0$ such that if $x, y \in X$ with $\|x\| = 1$ and $\|y\| \leq \delta$, then $\|x+y\|+\|x-y\| \leq 2+\varepsilon\|y\|$; cf. [3].

The modulus of smoothness of X is the function ϱ_X defined for every $t > 0$ by the formula

$$\varrho_X(t) := \sup\left\{\frac{\|x + y\| + \|x - y\|}{2} - 1 : \|x\| = 1,\ \|y\| = t\right\}.$$

Furthermore, X is called uniformly convex if for every $0 < \varepsilon \leq 2$ there is some $\delta > 0$ such that for any two vectors with $\|x\| = \|y\| = 1$, the condition $\|x - y\| \geq \varepsilon$ implies that $\left\|\frac{x+y}{2}\right\| \leq 1 - \delta$.

The modulus of convexity of X is the function σ_X defined by

$$\sigma_X(\varepsilon) := \inf\left\{ 1 - \left\| \frac{x+y}{2} \right\| \ : \ \|x\| = \|y\| = 1, \|x - y\| \geq \varepsilon \right\}.$$

Let X, Y be normed linear spaces. If a linear mapping $T : X \longrightarrow Y$ preserves ρ_*-orthogonality, then from Theorem 21.37 we conclude that T must be a similarity. Thus, the spaces X and Y have to share some geometrical properties. In particular, the modulus of convexity σ_X and modulus of smoothness ϱ_X must be preserved, i.e., $\sigma_X = \sigma_{T(X)}$ and $\varrho_X = \varrho_{T(X)}$. As a consequence, we have the following result.

Corollary 21.5 *Let X be a normed linear space. Suppose that there exists a normed linear space Y which is a uniformly convex (uniformly smooth) space, a strictly convex space, or an inner product space and a nontrivial linear mapping T from X into Y (or from Y onto X) such that T preserves ρ_*-orthogonality. Then X is, respectively, a uniformly convex (uniformly smooth) space, a strictly convex space, an inner product space.*

Recently, the authors of the paper [88] considered the class of linear mappings preserving ρ_λ-orthogonality. They showed that each such a mapping must be a similarity. Namely, they proved the following result.

Theorem 21.38 ([88, Theorem 3.4]) *Let X and Y be normed linear spaces and $\lambda \in [0, 1]$. Let $T : X \longrightarrow Y$ be a nonzero linear mapping. Then the following conditions are equivalent:*

(i) T preserves ρ_λ-orthogonality.
(ii) $\|Tx\| = \|T\| \, \|x\|$ for all $x \in X$.
(iii) $\rho_\lambda(Tx, Ty) = \|T\|^2 \, \rho_\lambda(x, y)$ for all $x, y \in X$.

Let X be a normed linear space endowed with two norms $\| \cdot \|_1$ and $\| \cdot \|_2$, which generate respective functionals $\rho_{\diamond,1}$ and $\rho_{\diamond,2}$, where $\diamond \in \{\lambda, *\}$. Following [3, Definition 2.4.1], we say that functionals $\rho_{\diamond,1}$ and $\rho_{\diamond,2}$ are equivalent if there exist constants $0 < m \leq M$ such that

$$m|\rho_{\diamond,1}(x, y)| \leq |\rho_{\diamond,2}(x, y)| \leq M|\rho_{\diamond,1}(x, y)| \qquad (x, y \in X).$$

Taking $X = Y$ and $T = id$, one obtains, from Theorems 21.37 and 21.38, the following result.

Corollary 21.6 *Let X be a normed linear space endowed with two norms $\| \cdot \|_1$ and $\| \cdot \|_2$, which generate respective functionals $\rho_{\diamond,1}$ and $\rho_{\diamond,2}$ with $\diamond \in \{\lambda, *\}$ and $\lambda \in [0, 1]$. Then the following conditions are equivalent:*

(i) The functionals $\rho_{\diamond,1}$ and $\rho_{\diamond,2}$ are equivalent.
(ii) The spaces $(X, \| \cdot \|_1)$ and $(X, \| \cdot \|_2)$ are isometrically isomorphic.

Next, we formulate one of our main results.

Theorem 21.39 *Let X, Y be normed linear spaces, $\lambda \in [0, 1]$ and let $\upsilon = \frac{1}{2k-1}$ $(k \in \mathbb{N})$. If $T : X \longrightarrow Y$ be a nonzero linear mapping, then the following conditions are equivalent:*

(i) T preserves ρ_λ^υ-orthogonality.
(ii) T preserves $(\rho_\lambda^\upsilon, B)$-orthogonality.
(iii) $\|Tx\| = \|T\| \|x\|$ for all $x \in X$.
(iv) $\rho_\lambda^\upsilon(Tx, Ty) = \|T\|^2 \rho_\lambda^\upsilon(x, y)$ for all $x, y \in X$.

Proof (i)$\Rightarrow$(ii) Suppose that $x, y \in X$ and $x \perp_{\rho_\lambda^\upsilon} y$. Then $Tx \perp_{\rho_\lambda^\upsilon} Ty$, by (i). It follows from Proposition 21.4 that $Tx \perp_B Ty$. Thus T preserves $(\rho_\lambda^\upsilon, B)$-orthogonality.

(ii)$\Rightarrow$(iii) Suppose that (ii) holds and fix $x, y \in X \setminus \{0\}$. If x and y are linearly dependent, then $\frac{\|Tx\|}{\|x\|} = \frac{\|Ty\|}{\|y\|}$. Now, assume that x and y are linearly independent. For any $t \in \mathbb{R}$, it is easy to see that $\rho_\pm\left(x + ty, \frac{-\rho_\pm(x+ty, y)}{\|x+ty\|^2}(x + ty) + y\right) = 0$ and hence

$$\rho_\lambda^\upsilon\left(x + ty, \frac{-\rho_\pm(x + ty, y)}{\|x + ty\|^2}(x + ty) + y\right) = 0.$$

It follows from Proposition 21.4 that $Tx + tTy \perp_B \frac{-\rho_\pm(x+ty,y)}{\|x+ty\|^2}(Tx + tTy) + Ty$. By Theorem 21.3, we get

$$\rho_-\left(Tx + tTy, \frac{-\rho_\pm(x + ty, y)}{\|x + ty\|^2}(Tx + tTy) + Ty\right)$$
$$\leq 0$$
$$\leq \rho_+\left(Tx + tTy, \frac{-\rho_\pm(x + ty, y)}{\|x + ty\|^2}(Tx + tTy) + Ty\right).$$

This implies

$$\frac{-\rho_-(x + ty, y)}{\|x + ty\|^2}\|Tx + tTy\|^2 + \rho_-(Tx + tTy, Ty) \leq 0 \qquad (t \in \mathbb{R}) \qquad (21.2)$$

and

$$0 \leq \frac{-\rho_+(x + ty, y)}{\|x + ty\|^2}\|Tx + tTy\|^2 + \rho_+(Tx + tTy, Ty) \qquad (t \in \mathbb{R}). \qquad (21.3)$$

Let us define

$$\varphi_{x,y}(t) := \frac{\|Tx + tTy\|}{\|x + ty\|} \qquad (t \in \mathbb{R}).$$

Then simple computations show that

$$(\varphi_{x,y})'_\pm(t) = \frac{\rho_\pm(Tx+tTy, Ty)\|x+ty\| - \rho_\pm(x+ty, y)\|Tx+tTy\|}{\|x+ty\|^2}.$$

From (21.2) and (21.3) it follows that

$$0 \le (\varphi_{x,y})'_-(t) \quad \text{and} \quad (\varphi_{x,y})'_+(t) \le 0 \qquad (t \in \mathbb{R}).$$

Hence $\varphi_{x,y}$ is constant on $\mathbb{R}$. Therefore,

$$\frac{\|Tx\|}{\|x\|} = \varphi_{x,y}(0) = \lim_{t\to\infty} \varphi_{x,y}(t) = \frac{\|Ty\|}{\|y\|}.$$

Now, we fix a unit vector y_0 in X. For every nonzero vector $x \in X$, we conclude that $\frac{\|Tx\|}{\|x\|} = \|Ty_0\|$. Hence $\|Tx\| = \|Ty_0\|\|x\|$ for all $x \in X$. Therefore (iii) is valid.

The other implications are trivial.

Let us adopt the notion of Birkhoff orthogonal set of x from [3]:

$$[x]^B_{\|\cdot\|} = \{y \in X : x \perp_B y\}.$$

We now define the $\diamond$-orthogonal set of x as follows:

$$[x]^\diamond_{\|\cdot\|} = \{y \in X : x \perp_\diamond y\},$$

where $\diamond \in \{I, s, \rho_*, \rho_\lambda^\upsilon\}$.

Theorem 21.40 *Let X be a normed linear space endowed with two norms $\|\cdot\|_1$ and $\|\cdot\|_2$, and let $\lambda \in [0, 1]$ and $\upsilon = \frac{1}{2k-1}$ ($k \in \mathbb{N}$). For every $x \in X$, the following conditions are equivalent:*

(i) $[x]^B_{\|\cdot\|_1} = [x]^B_{\|\cdot\|_2}$.   *(ii)* $[x]^s_{\|\cdot\|_1} = [x]^s_{\|\cdot\|_2}$.

(iii) $[x]^I_{\|\cdot\|_1} = [x]^I_{\|\cdot\|_2}$.   *(iv)* $[x]^{\rho_*}_{\|\cdot\|_1} = [x]^{\rho_*}_{\|\cdot\|_2}$.

(v) $[x]^{\rho_\lambda^\upsilon}_{\|\cdot\|_1} = [x]^{\rho_\lambda^\upsilon}_{\|\cdot\|_2}$.

Proof (i)$\Rightarrow$(v) Suppose that (i) holds and define $T = id : (X, \|\cdot\|_1) \to (X, \|\cdot\|_2)$ to be the identity map. Then T is B-orthogonal preserving. It follows from Theorem 21.32 that there exists $M > 0$ such that $\|Tx\|_2 = \|x\|_2 = M\|x\|_1$, which implies that $[x]^{\rho_\lambda^\upsilon}_{\|\cdot\|_1} = [x]^{\rho_\lambda^\upsilon}_{\|\cdot\|_2}$ ($x \in X$).

(v)$\Rightarrow$(i) If (v) holds, then $T = id : (X, \|\cdot\|_1) \to (X, \|\cdot\|_2)$ is ρ_λ^υ-orthogonal preserving. It follows from Theorem 21.39 that there exists $M > 0$ such that $\|x\|_2 = \|Tx\|_2 = M\|x\|_1$, which ensures that $[x]^B_{\|\cdot\|_1} = [x]^B_{\|\cdot\|_2}$ ($x \in X$).

The other implications can be proved similarly.

In [75, Theorem 5.1] Wójcik proved that a real normed linear space X is smooth if and only if there exist a real normed linear space Y and a nonvanishing linear mapping $T : X \longrightarrow Y$, such that T preserves $(\diamondsuit, \heartsuit)$-orthogonality for some $\diamondsuit, \heartsuit \in \{\rho_-, \rho_+, \rho\}$ with $\diamondsuit \neq \heartsuit$.

In the sequel, from [60, Theorems 3.2-3] and [28, Theorems 2.20-22] we are going to provide some characterizations of smooth real normed linear spaces in terms of linear mappings that preserve ρ_* and ρ_λ^v-orthogonality to other types of orthogonality relations.

Theorem 21.41 *Let X be a real normed linear space and let $\lambda \in [0, 1]$ and $\upsilon = \frac{1}{2k-1}$ ($k \in \mathbb{N}$). Then the following conditions are equivalent:*

(i) *X is smooth.*

(ii) *There exist a normed linear space Y and a nonvanishing linear mapping $T :$ $X \longrightarrow Y$ such that T preserves (ρ_*, ρ_+)-orthogonality.*

(iii) *There exist a normed linear space Y and a nonvanishing linear mapping $T :$ $X \longrightarrow Y$ such that T preserves (ρ_*, ρ_-)-orthogonality.*

(iv) *There exist a normed linear space Y and a nonvanishing linear mapping $T :$ $X \longrightarrow Y$ such that T preserves (ρ_*, ρ)-orthogonality.*

(v) *There exist a normed linear space Y and a nonvanishing linear mapping $T :$ $X \longrightarrow Y$ such that T preserves (ρ, ρ_*)-orthogonality.*

(vi) *There exist a normed linear space Y and a nonvanishing linear mapping $T :$ $X \longrightarrow Y$ such that T preserves (ρ_λ^v, ρ_-)-orthogonality.*

(vii) *There exist a normed linear space Y and a nonvanishing linear mapping $T :$ $X \longrightarrow Y$ such that T preserves (ρ_λ^v, ρ_+)-orthogonality.*

(viii) *There exist a normed linear space Y and a nonvanishing linear mapping $T :$ $X \longrightarrow Y$ such that T preserves $(\rho_\lambda^v, \rho_\lambda)$-orthogonality.*

(ix) *There exist a normed linear space Y and a nonvanishing linear mapping $T :$ $X \longrightarrow Y$ such that T preserves (ρ_λ^v, ρ)-orthogonality.*

21.3.3 Approximate Orthogonality Preserving Mapping

Ulam [73] raised the general problem of when a mathematical object which satisfies a certain property approximately must be close, in some sense, to one that satisfies this property accurately. Approximately orthogonality preserving mappings in the framework of inner product spaces have been studied in this setting, see [13, 14, 23, 47, 72, 81, 90, 91].

Let H and K be two inner product spaces and let $\delta, \varepsilon \in [0, 1)$. A mapping $T : H \to K$ is called a (δ, ε)-orthogonality preserving if

$$x \perp^\delta y \Rightarrow Tx \perp^\varepsilon Ty \qquad (x, y \in H).$$

Often $\delta = 0$ has been considered. Therefore, we say that T is ε-orthogonality preserving if

$$x \perp y \Rightarrow Tx \perp^{\varepsilon} Ty \qquad (x, y \in H).$$

Obviously, if $\delta = \varepsilon = 0$, then T is orthogonality preserving. Hence, the natural question is whether a (δ, ε)-orthogonality preserving linear mapping T must be close to a linear orthogonality preserving mapping. The following result was proved in [13, Theorem 2] (see also [72, Remark 2.1]).

Theorem 21.42 ([13, Theorem 2]) *Let H and K be two Hilbert spaces, and let T : $H \to K$ be a nonzero linear ε-orthogonality preserving mapping with $\varepsilon \in [0, 1)$. Then T is injective, continuous and, with some $\gamma > 0$, T satisfies the functional inequality*

$$\left| \langle Tx, Ty \rangle - \gamma \langle x, y \rangle \right| \le \frac{4\varepsilon}{1 + \varepsilon} \min \left\{ \gamma \|x\| \|y\|, \|Tx\| \|Ty\| \right\} \qquad (x, y \in H).$$

Conversely, if $T : H \to K$ satisfies

$$\left| \langle Tx, Ty \rangle - \gamma \langle x, y \rangle \right| \le \varepsilon \min \left\{ \gamma \|x\| \|y\|, \|Tx\| \|Ty\| \right\} \qquad (x, y \in H)$$

with $\varepsilon \ge 0$ and with $\gamma > 0$, then T is a quasi-linear mapping and ε-orthogonality preserving.

Recently, Moslehian et al. [61] have been obtained the following result.

Theorem 21.43 ([61, Theorem 3.10]) *Let H and K be two real Hilbert spaces and $\dim H < \infty$. Let $T : H \to K$ be a linear mapping with $0 < [T]$. Then there exists γ such that T satisfies*

$$\left| \langle Tx, Ty \rangle - \gamma \langle x, y \rangle \right| \le \left(1 - \frac{[T]^2}{\|T\|^2} \right) \|T\|^2 \|x\| \|y\| \qquad (x, y \in H).$$

Moreover, $[T]^2 \le |\gamma| \le 2\|T\|^2 - [T]^2$.

In 2007, Turnšek proved the following

Theorem 21.44 ([72, Theorem 2.3]) *Let H and K be two Hilbert spaces, T : $H \to K$ be a nonzero ε-orthogonality preserving linear mapping, $\varepsilon \in [0, 1)$, and $T = U|T|$ be its polar decomposition. Then U is an isometry and*

$$\left\| T - \|T\| U \right\| \le \left(1 - \sqrt{\frac{1 - \varepsilon}{1 + \varepsilon}} \right) \|T\|.$$

Wójcik extended Theorem 21.44 as follows. (The same result is later obtained in [91] by using a different approach.)

Theorem 21.45 ([81, Theorem 5.4]) *Let H be Hilbert space, $T : H \to H$ be a nonzero linear ε-orthogonality preserving mapping and let $\varepsilon \in [0, 1)$. Then there exists linear mapping $S : H \to H$ preserving orthogonality such that*

$$\left\| T - S \right\| \leq \frac{1}{2}\left(1 - \sqrt{\frac{1 - \varepsilon}{1 + \varepsilon}}\right) \|T\|.$$

Moreover, $\|S\| = \frac{1}{2}(\|T\| + [T])$ and $\|T - S\| = \frac{1}{2}(\|T\| - [T])$.

Kong and Cao [47] considered the class of (δ, ε)-orthogonality preserving linear mappings. They proved the following result.

Theorem 21.46 *Let $\delta, \varepsilon \in [0, 1)$. Let H, K be Hilbert spaces and let $T : H \to K$ be a nonzero (δ, ε)-orthogonality preserving linear mapping. Then there exists $\lambda_0 \in \{z \in \mathbb{C} : \frac{\delta+1}{2} \leq |z| \leq \delta + 2\}$ such that*

$$\sqrt{\frac{|\lambda_0|^2 - \varepsilon|\lambda_0|^2}{(\delta + 1)^2 + \varepsilon(\delta + 1)^2}} \|T\|\|x\| \leq \|Tx\| \leq \|Tx\|\|x\| \qquad (x \in H).$$

An stronger version of the previous theorem proved by Wójcik in [81].

Theorem 21.47 ([81, Theorem 3.4]) *Let $\delta, \varepsilon \in [0, 1)$. Let H, K be Hilbert spaces and let $T : H \to K$ be a nonzero (δ, ε)-orthogonality preserving linear mapping. Then T is injective, continuous and $\delta \leq \varepsilon$. Moreover the following inequality is true:*

$$\sqrt{\frac{1 - \varepsilon}{1 + \varepsilon}}\sqrt{\frac{1 + \delta}{1 - \delta}} \|T\|\|x\| \leq \|Tx\| \leq \|Tx\|\|x\| \qquad (x \in H).$$

As for a stability problem, we would like to know, whether each (δ, ε)-orthogonality preserving linear mapping T can be approximated by a linear orthogonality preserving map S. Kong and Cao [47] proved the following result.

Theorem 21.48 *Let $\delta, \varepsilon \in [0, 1)$. Let H, K be Hilbert spaces and let $T : H \to K$ be linear mapping (δ, ε)-orthogonality preserving. Let $T = U|T|$ be its polar decomposition. Then U is an isometry and there exists $\lambda_0 \in \mathbb{C}$ such that*

$$\left\| T - \|T\|U \right\| \leq \left(1 - \sqrt{\frac{|\lambda_0|^2 - \varepsilon|\lambda_0|^2}{(\delta + 1)^2 + \varepsilon(\delta + 1)^2}}\right) \|T\|.$$

Wójcik extended Theorem 21.48 as follows.

Theorem 21.49 *Let $\delta, \varepsilon \in [0, 1)$. Let H be Hilbert space, and let $T : H \to K$ be linear mapping (δ, ε)-orthogonality preserving. Then there exists linear mapping $S : H \to H$ preserving orthogonality such that*

$$\left\| T - S \right\| \leq \frac{1}{2}\left(1 - \sqrt{\frac{1-\varepsilon}{1+\varepsilon}}\sqrt{\frac{1+\delta}{1-\delta}}\right)\|T\|.$$

Moreover, $\|S\| = \frac{1}{2}(\|T\| + [T])$ *and* $\|T - S\| = \frac{1}{2}(\|T\| - [T])$.

For Hilbert C^*-modules some analogous results can be found in [35, 59]. Recently, Chmieliński et al. [23] have been verified the approximate orthogonality preserving property for two linear mappings. Similar investigations have been carried out for pairs of mappings on inner product C^*-modules in [33].

Approximate orthogonality preserving property has been considered also in the setting of normed linear spaces with respect to various definitions of orthogonality in general normed linear spaces.

Let X, Y be two normed linear spaces, $\delta, \varepsilon \in [0, 1)$ and

$$\diamond, \heartsuit \in \{B, cI, R, P, \rho_{\pm}, \rho, \rho_*, \rho_{\lambda}, \rho_{\lambda}^{v}\}.$$

A mapping $T : X \rightarrow Y$ is called a (δ, ε)-$(\diamond, \heartsuit)$-orthogonality preserving if

$$x \perp_{\diamond}^{\delta} y \Rightarrow Tx \perp_{\heartsuit}^{\varepsilon} Ty, \qquad (x, y \in X).$$

In particular, we say that T is ε-$\diamond$-orthogonality preserving if

$$x \perp_{\diamond} y \Rightarrow Tx \perp_{\diamond}^{\varepsilon} Ty, \qquad (x, y \in X).$$

Mojškerc and Turnšek [57] considered the class of linear mappings approximately preserving the Birkhoff–James orthogonality. They proved the following result.

Theorem 21.50 ([57, Theorem 3.5 and Remark 3.1]) *Let* X, Y *be two normed linear spaces,* $\varepsilon \in [0, \frac{1}{16})$ *and let* $T : X \rightarrow Y$ *be a nonzero* ε-*B-orthogonality preserving linear mapping. Then* T *is injective, continuous and*

$$(1 - 16\varepsilon)\|T\|\|x\| \leq \|Tx\| \leq \|T\|\|x\| \qquad (x \in X).$$

Also, if X *and* Y *are real normed linear spaces, then the constant* $(1 - 16\varepsilon)$ *can be replaced by* $(1 - 8\varepsilon)$ *with* $\varepsilon \in [0, \frac{1}{8})$.

The following result was proved in [87]:

Theorem 21.51 ([87, Theorem 3.2]) *Let* X, Y *be two real normed linear spaces, and let* $0 < b \leq a$ *and* $\delta, \varepsilon \in [0, \frac{b}{a})$. *Let* $T : X \longrightarrow Y$ *be a nonzero linear* (δ, ε)-(aI, bI)-*orthogonality preserving mapping. Then* $\delta \leq \frac{a-b+\varepsilon(a+b)}{a+b-\varepsilon(a-b)}$ *and* T *is injective, continuous and satisfies*

$$\frac{(1+\delta)(b-\varepsilon a)}{(1-\delta)(a+\varepsilon b)}\gamma\|x\| \leq \|Tx\| \leq \frac{(1-\delta)(a+\varepsilon b)}{(1+\delta)(b-\varepsilon a)}\gamma\|x\|$$

for all $x \in X$ *and for all* $\gamma \in [[T], \|T\|]$.

Taking $a = b = 1$ and $\delta = 0$ we get from Theorem 21.51 the following result.

Theorem 21.52 ([20, Theorem 3.2]) *Let X and Y be two real normed linear spaces, and let $\varepsilon \in [0, 1)$. Let $T : X \longrightarrow Y$ be a nonzero linear ε-I-orthogonality preserving mapping. Then T is injective, continuous and satisfies*

$$\frac{1 - \varepsilon}{1 + \varepsilon} \|T\| \|x\| \leq \|Tx\| \leq \frac{1 + \varepsilon}{1 - \varepsilon} [T] \|x\| \qquad (x \in X). \tag{21.4}$$

In the next Theorem we formulate a result from Theorem 21.52.

Theorem 21.53 ([20, Theorem 3.6]) *Let X and Y be two real normed linear spaces, and let $\varepsilon \in [0, 1)$. For a nontrivial linear mapping $T : X \longrightarrow Y$ the following conditions are equivalent:*

(i) *T preserves ε-I-orthogonality.*
(ii) *$\frac{1-\varepsilon}{1+\varepsilon} \|T\| \|x\| \leq \|Tx\| \leq \frac{1+\varepsilon}{1-\varepsilon} [T] \|x\|$ for all $x \in X$.*
(iii) *$\frac{1-\varepsilon}{1+\varepsilon} \gamma \|x\| \leq \|Tx\| \leq \frac{1+\varepsilon}{1-\varepsilon} \gamma \|x\|$ for all $x \in X$ and for all $\gamma \in [[T], \|T\|]$.*
(iv) *$\|Tx\| \|y\| \leq \frac{1+\varepsilon}{1-\varepsilon} \|Ty\| \|x\|$ for all $x, y \in X$.*
(v) *$\|T\| \leq \frac{1+\varepsilon}{1-\varepsilon} [T]$.*

As consequences of Theorem 21.51, we have the following results.

Corollary 21.7 ([87, Corollary 3.3]) *Let X, Y be two real normed linear spaces, and let $0 < b \leq a$ and $\varepsilon, \delta \in [0, \frac{b}{a})$. Let $T : X \longrightarrow Y$ be a linear (δ, ε)-(aI, bI)-orthogonality preserving mapping with $0 \leq \frac{a-b+\varepsilon(a+b)}{a+b-\varepsilon(a-b)} < \delta$. Then $T = 0$.*

Corollary 21.8 ([87, Corollary 3.4]) *Let X, Y be two real normed linear spaces, and let $0 < b \leq a$ and $\varepsilon, \delta \in [0, \frac{b}{a})$. Let $T : X \longrightarrow Y$ be a nonzero linear (δ, ε)-(aI, bI)-orthogonality preserving mapping. If a linear mapping $S : X \to Y$ satisfies $\|S - T\| \leq \theta \|T\|$, then $\|S\| \leq \eta[S]$, where $\eta = \frac{(1-\delta)^2(a+\varepsilon b)^2 + \theta(1-\delta^2)(a+\varepsilon b)(b-\varepsilon a)}{(1+\delta)^2(b-\varepsilon a)^2 - \theta(1-\delta^2)(a+\varepsilon b)(b-\varepsilon a)}$.*

Wójcik in [77] was obtain the following result for the stability of the orthogonality preserving mappings for the finite-dimensional spaces.

Theorem 21.54 ([77]) *Let X and Y be finite-dimensional real normed linear spaces, and let $\diamondsuit \in \{D, I, R, P\}$. Then, for an arbitrary $\theta > 0$, there exists $\varepsilon > 0$ such that for any linear ε-$\diamondsuit$-orthogonality preserving mapping $T : X \longrightarrow Y$ there exists a linear $\diamondsuit$-orthogonality preserving mapping $S : X \longrightarrow Y$ such that*

$$\|T - S\| \leq \theta \min\{\|T\|, \|S\|\}.$$

Some other results for the stability of the orthogonality preserving property in normed linear spaces can be found in [16, 20, 57, 68, 76].

Approximate orthogonality preserving mappings have been also considered for norm derivatives orthogonality relations.

Remark 21.3 Let X, Y be normed linear spaces and let $T : X \to Y$ be linear mapping. Then, T approximately preserves ρ_+-orthogonality if and only if T approximately preserves ρ_--orthogonality. Indeed, suppose that T approximately preserves ρ_+-orthogonality and let $x \perp_{\rho_-} y$. Thus $-x \perp_{\rho_+} y$, hence $-Tx \perp^\varepsilon_{\rho_+} Ty$ and finally $Tx \perp^\varepsilon_{\rho_-} Ty$, i.e., T approximately preserves ρ_--orthogonality. The proof of the reverse is the same.

Chmieliński and Wójcik in [22, Theorem 5.1] proved that an approximate $\rho_\pm$-orthogonality preserving mapping is an approximate B-orthogonality preserving mapping. Next, Wójcik [80, Theorem 5.5] obtained a same result for approximate ρ-orthogonality preserving mappings. More precisely, he proved that the property that a linear mapping approximately preserves the Birkhoff–James orthogonality is equivalent to that it approximately preserves the ρ and $\rho_\pm$-orthogonality (the proof of which is by no means elementary). The same result was proved in [27, Theorem 2.7] for approximate ρ_*-orthogonality preserving mappings. Thus, from Theorem 21.50, we obtain the following characterization of linear mappings approximately preserving the orthogonality relations.

Theorem 21.55 *Let* X, Y *be two real normed linear space and let* $\varepsilon \in [0, \frac{1}{8})$. *If* $T : X \to Y$ *is a nonzero linear mapping, then the following conditions are equivalent:*

(i) T *is* ε-ρ_--orthogonality preserving.
(ii) T *is* ε-ρ_+-orthogonality preserving.
(iii) T *is* ε-ρ-orthogonality preserving.
(iv) T *is* ε-ρ_*-orthogonality preserving.
(v) T *is* ε-B-orthogonality preserving.

Moreover, each of the above conditions implies that T *is injective, continuous and*

$$(1 - 8\varepsilon)\|T\|\|x\| \le \|Tx\| \le \|T\|\|x\| \qquad (x \in X).$$

Note that, in particular for $\diamond, \heartsuit \in \{B, \rho_\pm, \rho, \rho_*\}$ with $\diamond \ne \heartsuit$, the property that a linear mapping approximately preserves the $\diamond$-orthogonality is equivalent to that it approximately preserves the $\heartsuit$-orthogonality. Although $\perp^\varepsilon_\diamond$ and $\perp^\varepsilon_\heartsuit$ need not be equivalent unless we assume the smoothness of the norm.

Taking $X = Y$ and $T = id$, one obtains, from Theorem 21.55, the following result.

Corollary 21.9 *Let* $\varepsilon \in [0, \frac{1}{8})$ *and* $\diamond \in \{B, \rho_\pm, \rho, \rho_*\}$. *Let* $\|\cdot\|_1$ *and* $\|\cdot\|_2$ *be two norms in a linear space* X. *By* $\perp_1$ *and* $\perp_2$ *we denote the* $\diamond$-*orthogonality with respect to one of the two norms. If* $\perp_1 \subset \perp^\varepsilon_2$, *then both norms are equivalent and, with some* $\gamma > 0$, *we have*

$$(1 - 8\varepsilon)\gamma\|x\|_1 \le \|x\|_2 \le \gamma\|x\|_1 \qquad (x \in X).$$

Recently, the class of linear mappings approximately preserving ρ_λ^v-orthogonality has been studied in [1]. Although $\perp_{\rho_\lambda^v}^\varepsilon$ need not be equivalent to $\perp_B^\varepsilon$, unless we assume the smoothness of the norm, it has been proved in [1] the following result.

Theorem 21.56 ([1, Theorem 3.4]) *Let X and Y be normed linear spaces and let $\varepsilon \in [0, 1)$, $\lambda \in [0, 1]$ and $v = \frac{1}{2k+1}$ ($k \in \mathbb{N}$). If $T : X \to Y$ is a nonzero linear mapping, then the following conditions are equivalent:*

(i) T is ε-ρ_λ^v-orthogonality preserving.
(ii) T is ε-B-orthogonality preserving.

Moreover, each of the above conditions implies that T is injective, continuous and

$$(1 - 8\varepsilon)\|T\|\|x\| \le \|Tx\| \le \|T\|\|x\| \qquad (x \in X).$$

As an special case, the authors in [1] proved a similar result for approximately ρ_λ-orthogonality preserving linear mappings.

Corollary 21.10 ([1, Corollary 3.6]) *Let X and Y be normed linear spaces and let $\varepsilon \in [0, 1)$ and $\lambda \in [0, 1]$. If $T : X \to Y$ is a nonzero linear mapping, then the following conditions are equivalent:*

(i) T is ε-ρ_λ-orthogonality preserving.
(ii) T is ε-B-orthogonality preserving.

Moreover, each of the above conditions implies that T is injective, continuous and

$$(1 - 8\varepsilon)\|T\|\|x\| \le \|Tx\| \le \|T\|\|x\| \qquad (x \in X).$$

In particular, if we take $\varepsilon = 0$ in the foregoing corollary, then we obtain that every ρ_λ-orthogonality preserving mapping is a similarity. We should notify that this result was already shown in [88, Theorem 3.4] with a different approach.

Acknowledgement The authors would like to thank the referee for her/his valuable suggestions and comments.

References

1. Abed, M.Y., Dehghani, M., Jahanipur, R.: Approximate ρ_λ^v-orthogonality and its preservation. Ann. Funct. Anal. https://projecteuclid.org/accepted/euclid.afa
2. Alonso, J., Martini, H., Wu, S.: On Birkhoff orthogonality and isosceles orthogonality in normed linear spaces. Aequationes Math. **83**, 153–189 (2012)
3. Alsina, C., Sikorska, J., Tomás, M.S.: Norm Derivatives and Characterizations of Inner Product Spaces. World Scientific, Hackensack (2010)
4. Amir, D.: Characterizations of Inner Products Spaces. Birkauser, Basel (1986)
5. Ansari-piri, E., Sanati, R.G., Kardel, M.: A characterization of orthogonality preserving operators. Bull. Iran. Math. Soc. **43**(7), 2495–2505 (2017)

6. Arambašić, L., Rajić, R.: Operators preserving the strong Birkhoff–James orthogonality on **B(H)**. Linear Algebra Appl. **471**, 394–404 (2015)
7. Balestro, V., Horváth, Á.G., Martini, H., Teixeira, R.: Angles in normed spaces. Aequationes Math. **91**(2), 201–236 (2017)
8. Birkhoff, G.: Orthogonality in linear metric spaces. Duke Math. J. **1**, 169–172 (1935)
9. Blanco, A., Turnšek, A.: On maps that preserve orthogonality in normed spaces. Proc. R. Soc. Edinb. Sect. A **136**(4), 709–716 (2006)
10. Brešar, M., Chebotar, M.A.: Linear preserver problems. In: Functional Identities, pp. 189–219. Birkhäuser, Basel (2007)
11. Chen, C., Lu, F.: Linear maps preserving orthogonality. Ann. Funct. Anal. **6**(4), 70–76 (2015)
12. Chmieliński, J.: On an ε-Birkhoff orthogonality. J. Inequal. Pure Appl. Math. **6**(3), Art. 79 (2005)
13. Chmieliński, J.: Linear mappings approximately preserving orthogonality. J. Math. Anal. Appl. **304**, 158–169 (2005)
14. Chmieliński, J.: Stability of the orthogonality preserving property in finite-dimensional inner product spaces. J. Math. Anal. Appl. **318**(2), 433–443 (2006)
15. Chmieliński, J.: Remarks on orthogonality preserving mappings in normed spaces and some stability problems. Banach J. Math. Anal. **1**(1), 117–124 (2007)
16. Chmieliński, J.: Orthogonality preserving property and its Ulam stability. In: Rassias, T.M., Brzdęk, J. (eds.) Functional Equations in Mathematical Analysis. Springer Optimization and Its Applications, vol. 52, pp. 33–58. Springer, Berlin (2011)
17. Chmieliński, J.: Normed spaces equivalent to inner product spaces and stability of functional equations. Aequationes Math. **87**, 147–157 (2014)
18. Chmieliński, J.: Orthogonality equation with two unknown functions. Aequationes Math. **90**, 11–23 (2016)
19. Chmieliński, J.: Operators reversing orthogonality in normed spaces. Adv. Oper. Theory **1**(1), 8–14 (2016)
20. Chmieliński, J., Wójcik, P.: Isosceles-orthogonality preserving property and its stability. Nonlinear Anal. **72**, 1445–1453 (2010)
21. Chmieliński, J., Wójcik, P.: On ρ-orthogonality. Aequationes Math. **80**, 45–55 (2010)
22. Chmieliński, J., Wójcik, P.: ρ-orthogonality and its preservation–revisited. Recent developments in functional equations and inequalities, 17–30, Banach Center Publications, vol. 99, Polish Academy of Sciences, Institute of Mathematic, Warsaw (2013)
23. Chmieliński, J., Łukasik, R., Wójcik, P.: On the stability of the orthogonality equation and the orthogonality-preserving property with two unknown functions. Banach J. Math. Anal. **10**(4), 828–847 (2016)
24. Dadipour, F., Sadeghi, F., Salemi, A.: Characterizations of inner product spaces involving homogeneity of isosceles orthogonality. Arch. Math. **104**, 431–439 (2015)
25. Dadipour, F., Sadeghi, F., Salemi, A.: An orthogonality in normed linear spaces based on angular distance inequality. Aequationes Math. **90**, 281–297 (2016)
26. Day, M.M.: Some characterizations of inner-product spaces. Trans. Am. Math. Soc. **62**, 320–337 (1947)
27. Dehghani, M., Zamani, A.: Linear mappings approximately preserving ρ∗-orthogonality. Indag. Math. **28**, 992–1001 (2017)
28. Dehghani, M., Abed, M., Jahanipur, R.: A generalized orthogonality relation via norm derivatives in real normed linear spaces. Aequationes Math. **93**(4), 651–667 (2019)
29. Dragomir, S.S.: On approximation of continuous linear functionals in normed linear spaces. An. Univ. Timisoara Ser. Stiint. Mat. **29**, 51–58 (1991)
30. Dragomir, S.S.: Continuous linear functionals and norm derivatives in real normed spaces. Univ. Beograd. Publ. Elektrotehn. Fak. Ser. Mat. **3**, 5–12 (1992)
31. Dragomir, S.S.: Semi-Inner Products and Applications. Nova Science Publishers, Inc., Hauppauge (2004)

32. Frank, M., Mishchenko, A.S., Pavlov, A.A.: Orthogonality-preserving, C^*-conformal and conformal module mappings on Hilbert C^*-modules. J. Funct. Anal. **260**, 327–339 (2011)
33. Frank, M., Moslehian, M.S., Zamani, A.: Orthogonality preserving property for pairs of operators on Hilbert C^*-modules (2017). arXiv:1711.04724, Nov 10
34. Giles, J.R.: Classes of semi-inner product spaces. Trans. Am. Math. Soc. **116**, 436–446 (1967)
35. Ilišević, D., Turnšek, A.: Approximately orthogonality preserving mappings on C^*-modules. J. Math. Anal. Appl. **341**, 298–308 (2008)
36. Ionică, I.: On linear operators preserving orthogonality. Ann. Ştiinţ. Univ. Al. I. Cuza Iaşi. Mat. (N.S.) **58**, 325–332 (2012)
37. Jahn, T.: Orthogonality in generalized Minkowski spaces. J. Convex Anal. **26**(1), 49–76 (2019)
38. James, R.C.: Orthogonality in normed linear spaces. Duke Math. J. **12**(2), 291–302 (1945)
39. James, R.C.: Orthogonality and linear functionals in normed linear spaces. Trans. Am. Math. Soc. **61**(2), 265–292 (1947)
40. James, R.C.: Inner product in normed linear spaces. Bull. Am. Math. Soc. **53**, 559–566 (1947)
41. Joichi, J.T.: Normed linear spaces equivalent to inner product spaces. Proc. Am. Math. Soc. **17**, 423–426 (1966)
42. Jordan, P., von Neumann, J.: On inner products in linear, metric spaces. Ann. Math. **36**(3), 719–723 (1935)
43. Karia, D.J., Parmar, Y.M.: Orthogonality preserving maps and pro-C^*-modules. J. Anal. **26**, 1–10 (2018)
44. Kikianty, E., Sinnamon, G.: Angular equivalence of normed spaces. J. Math. Anal. Appl. **454**(2), 942–960 (2017)
45. Koehler, D., Rosenthal, P.: On isometries of normed linear spaces. Stud. Math. **36**, 213–216 (1970)
46. Koldobsky, A.: Operators preserving orthogonality are isometries. Proc. R. Soc. Edinb. Sect. A. **123**(5), 835–837 (1993)
47. Kong, L., Cao, H.: Stability of orthogonality preserving mapping and the orthogonality equation. J. Shaanxi Normal Univ. Nat. Sci. Ed. **36**(5), 10–14 (2008)
48. Köthe, G.: Topological Vector Spaces I. Springer, Berlin (1969)
49. Leung, C.-W., Ng, C.-K., Wong, N.-C.: Linear orthogonality preservers of Hilbert bundles. J. Aust. Math. Soc. **89**(2), 245–254 (2010)
50. Leung, C.-W., Ng, C.-K., Wong, N.-C.: Linear orthogonality preservers of Hilbert C^*-modules over C^*-algebras with real rank zero. Proc. Am. Math. Soc. **140**(9), 3151–3160 (2012)
51. Leung, C.-W., Ng, C.-K., Wong, N.-C.: Linear orthogonality preservers of Hilbert C^*-modules. J. Oper. Theory **71**(2), 571–584 (2014)
52. Li, C.K., Pierce, S.: Linear preserver problems. Am. Math. Mon. **108**(7), 591–605 (2001)
53. Lumer, G.: Semi-inner product spaces. Trans. Am. Math. Soc. **100**, 29–43 (1961)
54. Martini, H., Wu, S.: On maps preserving isosceles orthogonality in normed linear spaces. Note Mat. **29**(1), 55–59 (2009)
55. Miličić, P.M.: Sur le semi-produit scalaire dans quelques espaces vectorial norŕnes. Mat. Vesn. **8**(55), 181–185 (1971)
56. Miličić, P.M.: Sur la G-orthogonalité dans les espéaceés normés. Mat. Vesn. **39**, 325–334 (1987)
57. Mojškerc, B., Turnšek, A.: Mappings approximately preserving orthogonality in normed spaces. Nonlinear Anal. **73**, 3821–3831 (2010)
58. Moslehian, M.S.: On the stability of the orthogonal Pexiderized Cauchy equation. J. Math. Anal. Appl. **318**, 211–223 (2006)
59. Moslehian, M.S., Zamani, A.: Mappings preserving approximate orthogonality in Hilbert C^*-modules. Math. Scand. **122**, 257–276 (2018)

60. Moslehian, M.S., Zamani, A., Dehghani, M.: Characterizations of smooth spaces by ρ_*-orthogonality. Houst. J. Math. **43**(4), 1187–1208 (2017)
61. Moslehian, M.S., Zamani, A., Wójcik, P.: Approximately angle preserving mappings. Bull. Aust. Math. Soc. **99**(3) 485–496 (2019)
62. Nabavi Sales, S.M.S.: On mappings which approximately preserve angles. Aequationes Math. **92**, 1079–1090 (2018)
63. Pambuccian, V.: A logical look at characterizations of geometric transformations under mild hypotheses. Indag. Math. **11**(3), 453–462 (2000)
64. Papini, P.L.: Uñasservatione sui prodotti semi-scalari negli spasi di Banach. Boll. Un. Mat. Ital. **6**, 684–689 (1969)
65. Precupanu, T.: Duality mapping and Birkhoff orthogonality. An. Stiint, Univ. Al. I. Cuza Iasi. Mat. (S.N.) **59**(1), 103–112 (2013)
66. Rätz, J.: On orthogonally additive mappings. Aequationes Math. **28**, 35–49 (1985)
67. Roberts, B.D.: On the geometry of abstract vector spaces. Tôhoku Math. J. **39**, 42–59 (1934)
68. Sikorska, J.: Orthogonalities and functional equations. Aequationes Math. **89**, 215–277 (2015)
69. Stypuła, T., Wójcik, P.: Characterizations of rotundity and smoothness by approximate orthogonalities. Ann. Math. Sil. **30**, 193–201 (2016)
70. Tapia, R.A.: A characterization of inner product spaces. Bull. Am. Math. Soc. **79**, 530–531 (1973)
71. Tapia, R.A.: A characterization of inner product spaces. Proc. Am. Math. Soc. **41**, 569–574 (1973)
72. Turnšek, A.: On mappings approximately preserving orthogonality. J. Math. Anal. Appl. **336**, 625–631 (2007)
73. Ulam, S.M.: A Collection of Mathematical Problems. Interscience Publishers, New York (1960)
74. Wang, S.-G., Ip, W.-C.: A matrix version of the Wielandt inequality and its applications to statistics. Linear Algebra Appl. **296**, 171–181 (1999)
75. Wójcik, P.: Linear mappings preserving ρ-orthogonality. J. Math. Anal. Appl. **386**, 171–176 (2012)
76. Wójcik, P.: Operators preserving and approximately preserving orthogonality and similar relations. Doctorial dissertation (in Polish), Katowice (2013)
77. Wójcik, P.: On mappings approximately transferring relations in finite-dimensional normed spaces. Linear Algebra Appl. **460**, 125–135 (2014)
78. Wójcik, P.: Characterizations of smooth spaces by approximate orthogonalities. Aequationes Math. **89**, 1189–1194 (2015)
79. Wójcik, P.: Operators preserving sesquilinear form. Linear Algebra Appl. **469**, 531–538 (2015)
80. Wójcik, P.: Linear mappings approximately preserving orthogonality in real normed spaces. Banach J. Math. Anal. **9**(2), 134–141 (2015)
81. Wójcik, P.: On certain basis connected with operator and its applications. J. Math. Anal. Appl. **423**(2), 1320–1329 (2015)
82. Wójcik, P.: Operators reversing orthogonality and characterization of inner product spaces. Khayyam J. Math. **3**(1), 23–25 (2017)
83. Wójcik, P.: Mappings preserving B-orthogonality. Indag. Math. **30**, 197–200 (2019)
84. Wu, S., He, C., Yang, G.: Orthogonalities, linear operators, and characterization of inner product spaces. Aequationes Math. **91**(5), 969–978 (2017)
85. Zamani, A.: Approximately bisectrix-orthogonality preserving mappings. Comment. Math. **54**(2), 167–176 (2014)
86. Zamani, A., Moslehian, M.S.: Approximate Roberts orthogonality. Aequationes Math. **89**, 529–541 (2015)
87. Zamani, A., Moslehian, M.S.: Approximate Roberts orthogonality sets and (δ, ε)-(a, b)-isosceles-orthogonality preserving mappings. Aequationes Math. **90**, 647–659 (2016)

88. Zamani, A., Moslehian, M.S.: An extension of orthogonality relations based on norm derivatives. Q. J. Math. **70**(2), 379–393 (2019)
89. Zamani, A., Moslehian, M.S., Frank, M.: Angle preserving mappings. Z. Anal. Anwend. **34**,
485–500 (2015)
90. Zhang, Y.: An identity for (δ, ε)-approximately orthogonality preserving mappings. Linear
Algebra Appl. **554**, 358–370 (2018)
91. Zhang, Y., Chen, Y., Hadwin, D., Kong, L.: AOP mappings and the distance to the scalar
multiples of isometries. J. Math. Anal. Appl. **431**(2), 1275–1284 (2015)

Index

A

Absolute stability, 247
Accounting robustness, DPI, 274
Additive functions, 397–403
Additive mapping, 200
Additive ρ-functional inequalities, 200
Algebraic semigroups, 34
Amenable groups
    algebraic properties, 417–424
    commutative group, 422, 423
    direct union, 422
    G-invariant measure, 410
    left action, 410
    left amenability of G, 409
    locally finite group, 423
    polycyclic group, 423
    right action, 410
    simply equivariant, 411
    solvable group, 423
Amenable semigroups, 170–175
Aoki-Rolewicz theorem, 101, 106
Approximate Birkhoff–James orthogonality, 485, 486
Approximate isosceles orthogonality, 486
Approximately orthogonality preserving (AOP), 61–62
Approximately orthogonality reversing (AOR), 61–63
Approximate norm derivatives orthogonalities, 485–489
Approximate orthogonality, 58, 66, 485, 486
Approximate orthogonality preserving mapping, 497–503

Approximate orthogonality preserving property, 500
Approximate symmetry, 63
Approximation of approximation, 247–248
Approximation of functions
    additive mappings, 155–157
    cubic mappings, 157–163
    quadratic mappings, 155–157
Asymptotic approximation, 90
Augmented matrix, 276, 283, 284, 309

B

Banach function modules, geometry of
    class of spaces, 454
    extreme points, 455
    multismoothness, 455–458
    norm derivatives, 459–466
Banach lattice
    cone-related functional, 3–5
    continuous function, 4, 5
    join homomorphism, 3
    maximum preserving functional equation, 3
    semi-homogeneity, 3
    triangle inequality, 8
Banach spaces, 97, 99, 232
    convexity and smoothness, 334
    finite-dimensional strictly convex and smooth, 335
    geometry, 331, 334
    symmetric linear operators, 333
Bi-additive s-functional inequality, 201–206
Bifurcation behaviour, 304–305

© Springer Nature Switzerland AG 2019
J. Brzdęk et al. (eds.), *Ulam Type Stability*,
https://doi.org/10.1007/978-3-030-28972-0

Printed in the United States
By Bookmasters